Lead-Free Solder

INTERCONNECT RELIABILITY

Edited by
Dongkai Shangguan

ASM International®
Materials Park, Ohio 44073-0002
www.asminternational.org

Copyright © 2005
by
ASM International®
All rights reserved

No part of this book may be reproduced, stored in a retrieval system, or transmitted, in any form or by any means, electronic, mechanical, photocopying, recording, or otherwise, without the written permission of the copyright owner.

First printing, August 2005

Great care is taken in the compilation and production of this book, but it should be made clear that NO WARRANTIES, EXPRESS OR IMPLIED, INCLUDING, WITHOUT LIMITATION, WARRANTIES OF MERCHANTABILITY OR FITNESS FOR A PARTICULAR PURPOSE, ARE GIVEN IN CONNECTION WITH THIS PUBLICATION. Although this information is believed to be accurate by ASM, ASM cannot guarantee that favorable results will be obtained from the use of this publication alone. This publication is intended for use by persons having technical skill, at their sole discretion and risk. Since the conditions of product or material use are outside of ASM's control, ASM assumes no liability or obligation in connection with any use of this information. No claim of any kind, whether as to products or information in this publication, and whether or not based on negligence, shall be greater in amount than the purchase price of this product or publication in respect of which damages are claimed. THE REMEDY HEREBY PROVIDED SHALL BE THE EXCLUSIVE AND SOLE REMEDY OF BUYER, AND IN NO EVENT SHALL EITHER PARTY BE LIABLE FOR SPECIAL, INDIRECT OR CONSEQUENTIAL DAMAGES WHETHER OR NOT CAUSED BY OR RESULTING FROM THE NEGLIGENCE OF SUCH PARTY. As with any material, evaluation of the material under end-use conditions prior to specification is essential. Therefore, specific testing under actual conditions is recommended.

Nothing contained in this book shall be construed as a grant of any right of manufacture, sale, use, or reproduction, in connection with any method, process, apparatus, product, composition, or system, whether or not covered by letters patent, copyright, or trademark, and nothing contained in this book shall be construed as a defense against any alleged infringement of letters patent, copyright, or trademark, or as a defense against liability for such infringement.

Comments, criticisms, and suggestions are invited, and should be forwarded to ASM International.

Prepared under the direction of the ASM International Technical Book Committee (2004–2005), Yip-Wah Chung, FASM, Chair.

ASM International staff who worked on this project include Scott Henry, Senior Manger of Product and Service Development; Bonnie Sanders, Manager of Production; Madrid Tramble, Senior Production Coordinator, and Kathryn Muldoon, Production Assistant

Library of Congress Cataloging-in-Publication Data
Lead-free solder interconnect reliability / edited by Dongkai Shangguan.
p. cm.
Includes bibliographical references and index.
ISBN: 0-87170-816-7
1. Microelectronic packaging—Reliability. 2. Solder and soldering. 3. Lead-free electronics manufacturing processes. I. Shangguan, Dongkai, 1963–

TK7870.15.L425 2005
621.381′046—cd22   2005050106
SAN: 204-7586

ASM International®
Materials Park, OH 44073-0002
www.asminternational.org

Printed in the United States of America

Contents

Foreword by Gary F. Shade and Thomas S. Passek vi
Foreword by Nicholas Brathwaite vii
Preface ix

Chapter 1 Lead-Free Soldering and Environmental Compliance: An Overview 1
D. Shangguan, Flextronics International
   Introduction 1
   Materials for Lead-Free Soldering 2
   Processes, Equipment, and Quality For Lead-Free Soldering 6
   Lead-Free Solder Interconnect Reliability 10
   Design For Lead-Free Soldering and Environmental Compliance 17
   Summary 21

Chapter 2 Microstructural Evolution and Interfacial Interactions in Lead-Free Solder Interconnects 29
Z. Mei, Cisco Systems
   Introduction 29
   Microstructural Evolution in Lead-Free Solders 29
   Interactions Between Substrates and Solders 33
   Molten Solder-Substrate Interactions 33
   Solid Solder-Substrate Reactions 48
   Interface Reliability 48

Chapter 3 Fatigue and Creep of Lead-Free Solder Alloys: Fundamental Properties 67
P.T. Vianco, Sandia National Laboratories
   Introduction 67
   Material Deformation 69
   Creep Deformation 87
   Summary 103

Chapter 4 Lead-Free Solder Joint Reliability Trends 107
J.-P. Clech, Electronics Packaging Solutions International Inc.
   Introduction 107
   Empirical Correlations of SAC Thermal Cycling Test Data 108
   Lead-Free to SnPb Comparison 109
   Critical Component Data 111
   Impact of Pb Contaminant or Sn-Pb Alloy on Lead-Free Reliability 113
   Discussion 116
   Conclusions 123

Chapter 5 Chemical Interactions and Reliability Testing for Lead-Free Solder Interconnects 129
L.J. Turbini, Centre for Microelectronics Assembly and Packaging, Canada

Introduction 129
Background on Solder Flux Chemistry 129
Electrochemical Migration 131
Surface Insulation Resistance 132
Corrosion Test Method 135
Conductive Anodic Filament Formation 137
Flux Residues and RF Signal Integrity 142
Summary 143

Chapter 6 Tin Whisker Growth on Lead-Free Solder Finishes 147
K.N. Tu, J.O. Suh, and A.T. Wu, UCLA

Introduction 147
Morphology of Sn Whisker on Lead-Free Solder Finishes 148
The Stress Generation (Driving Force) in Sn Whisker Growth 149
Room Temperature Reaction Between Sn and Cu to Form Cu_6Sn_5 151
The Stress Relaxation (Kinetic Process) in Sn Whisker Growth 152
Measurement of the Parameters Which Affect Sn Whisker Growth 153
Suppression of Sn Whisker Growth 161
Accelerated Testing of Sn Whisker Growth 161
Conclusions 162

Chapter 7 Accelerated Testing Methodology for Lead-Free Solder Interconnects 165
G. Grossmann, Swiss Federal Laboratories for Materials Testing and Research, Switzerland

Introduction 165
Metallurgical Background 165
Deformation of Tin-Based Solder Alloys 168
Acceleration 169
Designing a Test 176

Chapter 8 Thermomechanical Reliability Prediction of Lead-Free Solder Interconnects 181
K. Kacker and S.K. Sitaraman, Georgia Institute of Technology

Introduction 181
Constitutive Models for Lead-Free Solder Alloy 181
Geometric Modeling 186
Loading Conditions and Thermomechanical Stresses 188
Deformation Mechanisms 190
Thermomechanical Reliability in Packages 191
Model Validation 194
Conclusions 196

Chapter 9 Design for Reliability: Finite-Element Modeling of Lead-Free Solder Interconnects 199
W. Dauksher, Agilent Technologies

Introduction 199
Modeling 200
Modeled Geometry 201

Materials Discussion 202
Failure Criteria Discussion 206
Modeling Difficulties 208
Suggested Geometric Finite Element Model 208
An Example Using the 256 PBGA 209
An Examination of Failure Theories with Literature Data 209
Design for Reliability 214
Future Research Needs 220
Concluding Remarks 222
....222

Chapter 10 Characterization and Failure Analyses of Lead-Free Solder Defects 227
R. Ghaffarian, Jet Propulsion Laboratory, California Institute of Technology

Introduction 227
Tin-Lead and Lead-Free Alloys 227
Characterization and Analytical Techniques 228
Void Levels for Lead-Free and Mixed Assembly with Various Surface Finishes 239
Microstructure Before and After Drop Test 241
X-ray Inspection Systems 241
Conclusions 245

Chapter 11 Reliability of Interconnects with Conductive Adhesives 249
J. Liu and Z. Mo, Chalmers University of Technology, Sweden

Introduction to Conductive Adhesive Technology 249
Reliability Concerns of ICA Interconnects 251
Reliability of ACA Interconnects 255
Theoretical Studies and Numeric Simulations 263
Conclusions 273

Chapter 12 Lead-Free Solder Interconnect Reliability Outlook 277
D. Shangguan, Flextronics International

Solder Alloy Characteristics and Interfacial Interactions 277
Tin Whisker Growth 277
PWB Reliability 278
Solder Constitutive Equation and Thermal Fatigue Reliability Prediction 278
Dynamic Mechanical Loading Conditions 278
Accelerated Testing Profile and Acceleration Factor 278
Complex Loading Conditions and Total Reliability Optimization 279
Reliability Degradation Assessment for Component Reutilization and Repurposing 279

Appendix 1 Selected acronyms and abbreviations related to surface mount and lead-free solder interconnect technology 283

Subject Index 285

Foreword

By Gary F. Shade and Thomas S. Passek

The worldwide drive to remove lead from electronic assemblies adds challenges to an already complex and difficult technology. Well beyond simple structures, solder interconnects provide at least as many challenges as the components they connect. Constant new demands of size, speed, power, and new materials continually direct the developer onto the frontiers of science to provide reliable interconnects.

Despite these challenges for materials, mechanical and thermal practitioners, most universities have largely overlooked solder interconnects as a prime area for teaching and research. To fill this void, Dongkai Shangguan has assembled a team of leading experts representing international government research facilities, universities, and industrial companies. Together they provide a wealth of up-to-date knowledge and data on the reliability of lead-free solder interconnects.

Solder interconnects have been a constant challenge in the past and remain that way today. Industry is putting tremendous resources into action to provide reliable lead-free interconnects, and this will undoubtedly continue for the foreseeable future as packages, boards, and materials continue to evolve to meet the needs for improved reliability, performance, and cost.

The Electronic Device Failure Analysis Society (EDFAS), an affiliate society of ASM International, is pleased to see this fine work made available for practitioners in reliability and failure analysis of lead-free interconnects. EDFAS and ASM, along with the editors, the authors, and the reviewers have collaborated to produce a book that meets high technical standards. To all who contributed toward the completion of this task, we extend our sincere thanks.

Gary F. Shade
President
Electronic Device Failure Analysis Society

Thomas S. Passek
Executive Director
Electronic Device Failure Analysis Society

Foreword

By Nicholas Brathwaite

It's only recently that inhabitants of planet Earth have begun attempting to conscientiously, proactively, and responsibly manage the environment in which we live. A truly global, environmentally conscious culture has emerged across all tiers of society.

The fact is, we have little choice.

Take the electronics industry, for example, upon which our society increasingly depends for safety, security, comfort, and convenience. As a result of our ever-growing appetite for electronic products, we now generate millions of tons of electronic waste annually.

In addition to challenging the capacity of landfills, this trend continues to deplete our resources and increase environmental pollution. The Restriction of the Use of Certain Hazardous Substances in Electrical and Electronic Equipment (RoHS) and Waste Electrical and Electronic Equipment (WEEE) directives, introduced by the European Union, are intended to protect human health and the environment.

By demanding reuse, recycling, and separate collection, the primary purpose of the WEEE directive is reduction in the amount of electrical and electronic equipment entering landfills. The RoHS directive is intended to eliminate or reduce harmful substances at the source, ensuring that these hazardous substances are not a threat to human life.

Similar legislation is emerging in other parts of the world (most notably, China). While there are still differing opinions regarding the merits of these directives, their global impact on the electronics industry is undeniably clear. It is probably no exaggeration to say that these are leading to revolutionary changes in the electronics industry.

Of the six substances restricted by RoHS, lead is arguable the one that has the greatest impact on the electronics packaging and assembly industry. Approximately 100 million pounds of solder (mostly with tin-lead alloys) are consumed annually to generate 10 trillion solder joints, which is the primary interconnect between the integrated circuit and the printed circuit board. The implementation of lead-free solder assembly processes, therefore, is a critical element in the global effort towards RoHS compliance.

Significant volumes of research and development work on manufacturing issues associated with lead-free assembly have been conducted and published in the past decade by the industry, national laboratories, consortia, and academia worldwide. Reliability studies of lead-free solder interconnects, however, are still emerging.

As we begin the process of putting more and more lead-free electronics products in the marketplace, it is vital that we have an in-depth and comprehensive understanding of lead-free solder interconnect reliability to help the industry safeguard the reliability of the electronics systems and products.

As an industry, we are facing increasingly demanding customers and mounting competitive pressures. As such, industry-wide collaboration is key to a successful journey towards environmental compliance. The industry is fortunate to have a few dedicated experts who can lead and guide us through these challenges, and we are proud to count Dongkai Shangguan among these experts.

While our company has benefited tremendously from Dr. Shangguan's world-leading expertise in packaging and assembly technology, reliablity, and environmentally conscious design and manufacturing, it is gratifying to see that he has now assembled a global team of renowned experts, across industry and academia, to address the issue of lead-free solder interconnect reliability in great depth and breadth.

On this journey towards environmental compliance and leadership, expert advice in the form of this book is not only welcomed but is vitally necessary.

Nicholas Brathwaite
Chief Technology Officer
Flextronics

Preface

As I sit down to put the finishing touches on this book, the year 2004 is drawing to a close. Through the window of my study, I can see green grass, blue sky, and bright sunshine, here at the heart of the Silicon Valley. This brought my memory back to 1991, when at about the same time of the year, I sat in my study in a Detroit suburb staring at the snow outside the window, trying to put together a plan for a lead-free solder project at Ford Electronics.

The electronics industry has come a long way since then. Over the past 13 years, there has been a great deal of development in the lead-free arena. The passage of the European Union "Directive on the Restriction of the Use of Certain Hazardous Substances in Electrical and Electronic Equipment" (RoHS) legislation has made the drive towards worldwide adoption of lead-free solder unstoppable for packaging, board assembly, and manufacturing of electronics products. Although there is still debate about the merits of RoHS legislation, the transition to lead-free soldering is already underway worldwide.

The global progress towards lead-free soldering includes solder alloy selection and evaluation, process development, and infrastructural development. A significant volume of R&D work has been published by industry, national labs, consortia, and academia worldwide in these areas. Nonetheless, the reliability of lead-free solder interconnects is still an emerging science. The existence of a large volume of reliability data on tin-lead solders under different use conditions has been a key prerequisite for the successful use of tin-lead solder in a multitude of applications. Correspondingly, similar data on lead-free solders are critically needed by the industry, before across-the-board conversion can take place.

The issue of reliability, however, is complicated by the wide variety of application environments and requirements, which give rise to different stress conditions (thermomechanical, dynamical, electrochemical, electrical, etc.). The physics of failure, which is directly related to reliability, is a critical topic still under intense investigation for lead-free solders. The topic is further complicated by the fact that the relative reliability comparison between the tin-lead solder and the lead-free solder alloys varies with the loading conditions. These complications create great difficulties for the development of appropriate accelerated testing profiles and for reliability prediction, both of which are critical elements of reliable product design. Due to the complexity of the subject, a great deal of confusion still exists in the industry, which hampers lead-free solder implementation and may even lead to catastrophe if reliability is not properly managed for critical products.

Researchers worldwide have undertaken serious efforts to study lead-free solder reliability under different loading conditions. While the results of these studies have been reported rather sporadically at various technical conferences, a book is critically needed which is dedicated to this topic of paramount importance to the electronics industry. Consequently, the objective of this book is to disseminate the most up-to-date knowledge and data on the reliability of lead-free solder interconnects, under various application conditions. The book attempts to cover the complex topic of lead-free solder interconnect reliability in a comprehensive manner, and both fundamental research and practical considerations are addressed. While the book will have archival reference value for academic researchers and educators alike, it is primarily targeted towards practitioners in the electronics business, who need to understand the reliability of solder interconnects, for product design, testing, and assurance.

Chapter 1 offers an overview of lead-free soldering (materials, processes and reliability), as well as a perspective of overall environmental compliance for electronics products. The intricate interplay among the various use conditions, physics of failure, failure modes, and testing, analysis and prediction methodology, is outlined for the interconnect system (which includes the components, the substrate, and the solder). This is followed by chapters with detailed discussions on the fundamental microstructural evolution and creep and fatigue properties of the lead-free solder alloys, and the overall comparison of reliability between tin-lead and lead-free solders. Chemical interactions and reliability, and tin whisker growth, are then presented in considerable depth to address both the fundamental understanding and the application-oriented issues. Accelerated testing methodology, reliability prediction and design for reliability, as well as characterization and failure analyses, are reviewed primarily from the practical considerations. While a number of different lead-free solder alloys exist, the data presented in this book are focused primarily on Sn-Ag-Cu and Sn-Cu alloys, as these are believed to be the primary lead-free solder alloys to be used by the worldwide electronics industry. Many of the methodologies and analyses presented in the book, however, are applicable to other lead-free solder alloy systems as well. The reliability of conductive adhesives is included toward the end of the book as another important alternative to tin-lead solders. The very last chapter of the book attempts to outline the most critical issues on lead-free solder interconnect reliability for future research.

It is no exaggeration to say that we have only taken the very first step of a long journey towards reliable lead-free solder interconnect systems for electronics products worldwide. It is hoped that this book can provide help for our industry and academic colleagues worldwide in this endeavor. No doubt, many questions still remain, and much more work needs to be done and will be done. It is my personal wish that this book can play some role in stimulating further exploration in the field of lead-free reliability.

ACKNOWLEDGMENTS

Since I submitted the book proposal to ASM International 16 months ago, I have had the privilege to work with many friends and colleagues on this venture. I would like to thank the chapter authors for their contributions and cooperation over the past year and for their confidence in me. I am grateful to Keith Newman for his help with reviewing the manuscripts and for his many valuable and constructive comments.

My past and current colleagues at Ford, Visteon, and Flextronics deserve special thanks; it has been a joy and privilege to work with them. I am grateful for the opportunity to learn from them and get to know them on a personal level.

Numerous industry and academic colleagues worldwide deserve my sincere thanks, as they have been the source of stimulation and inspiration for my work in this area. Specifically, I want to mention my colleagues at the EMS Forum, and the many leaders at IPC, Soldertec, HDPUG, and other consortia as well as my friends from Shanghai University. I have also learned a great deal from the many people I met at various conferences; their comments, questions and feedback served as a constant reminder that what we know today is far from what we need to know.

Thanks are also due to Scott Henry and the editorial and production staff at ASM International for their cooperation and support.

This project would not have been possible without the support of my family. With a job that takes me to distant countries away from home for about half of the time, and long work hours wherever I may be, the "extra" time I needed for this project has necessitated sacrifice on their part. As I am writing this paragraph, my younger son just completed his pinewood derby car, but my travel schedule for the next couple of weeks will not allow me to be there when the race takes place. I am counting on his older brother to videotape the event for me. My family's love constantly motivates me to keep marching forward on this truth-seeking journey that may in some small way benefit our society. Their support and understanding gave me the time to complete this important project, before I move on to the next one.

Dongkai Shangguan, Ph.D., MBA
San Jose, CA
December 2004

CHAPTER 1

Lead-Free Soldering and Environmental Compliance: An Overview

Dongkai Shangguan, Flextronics International

Introduction

Solder interconnects perform three major functions: electrical, mechanical, and thermal. They provide the electrical connection path from the silicon chip to the circuitry on the substrate within a package (first level interconnection), between the different packages (as in packaging stacking, see Ref 1), and between the package and the copper (Cu) traces on the printed wiring board, or PWB (second level interconnection). At the same time, they also serve as the mechanical support for the various elements which are interconnected. As the power of the chips rises, heat dissipation is becoming an increasingly critical issue that can affect the performance of the electronic system; solder interconnects, along with other thermal management tools, also serve the function of heat dissipation. The successful functioning of electronic products depends on the *reliable* interconnections provided by these tiny and numerous solder joints, over the life of the product, under vastly different use conditions.

From the very beginning of the electronics industry (not to mention ancient solder applications dating back 5,000 years), solder joints have been made primarily of alloys of tin (Sn) and lead (Pb). In particular, the eutectic tin-lead alloy (63%Sn and 37%Pb by weight, eutectic temperature 183 °C, or 361 °F) has been used almost exclusively in electronics, due to its unique characteristics (cost, availability, ease of use, and electrical/thermal/mechanical/chemical characteristics). Although our understanding of the tin-lead solder is not yet complete, a large body of information does exist as a result of research over the past 50 years.

The ever-increasing amount of electronic waste (e-waste), most of which end up in landfills, has become a serious worldwide concern. The harmful effect of lead to humans is well known (Ref 2). There are still differing opinions, however, on the significance of the environmental impact of lead from electronics. The merits of lead elimination from electronics are beyond the scope of this book; nonetheless, a lifecycle assessment (LCA) is often recommended to provide a holistic view (Ref 3). The recent passage of the European Union (EU) "Directive on the Restriction of the Use of Certain Hazardous Substances (RoHS) in Electrical and Electronic Equipment (EEE)" (Ref 4) has made the drive towards worldwide adoption of lead-free soldering unstoppable for electronics packaging, board assembly, and manufacturing of electronics products, along with the elimination of mercury (Hg), cadmium (Cd), hexavalent chromium (Cr^{+6}), polybrominated biphenyls (FBB), and polybrominated diphenyl ethers (PBDE).

The global landscape for lead-free soldering is undergoing significant transformation. It is moving from the laboratory to the manufacturing floor, from R&D to factory implementation, from low volume to high volume, and from consumer products to most product categories. As the worldwide electronics industry implements lead-free solder in PWB (also known as printed circuit board or PCB) assembly, compatibility issues are critical to a smooth transition from the current Sn-Pb soldering to lead-free soldering in volume manufacturing for PWB assembly (Ref 5–11). A systematic and holistic approach, as illustrated in Fig. 1, encompassing design, materials, processes, quality and reliability, equipment, and operations and business considerations, is critical to successful industry-wide implementation of lead-free solders. The reliability of the interconnection system using the lead-free solder is central to the transition to lead-free.

Solder interconnect reliability is the probability of the solder interconnects to perform the intended functions (electrical, mechanical and thermal), for a prescribed product life, under applicable use conditions (such as temperature, humidity, cyclic temperature variations, voltage, current density, static and dynamic mechanical loading, and corrosion), without failures. Failures may manifest themselves in different modes, such as electrochemical or mechanical, and may occur at various system interconnection locations (components, substrates, and/or solder joints).

Materials for Lead-Free Soldering

Lead-Free Solder Alloys

Over the last decade, the industry has studied a wide range of alloys to replace the tin-lead alloy. The alloy selection has been based on the following considerations (Ref 12–15): toxicity, physical properties (melting temperature, surface tension and wettability, thermal and electrical conductivity), mechanical properties, microstructural characteristics, electrochemical properties (corrosion, oxidation and dross formation, and compatibility with no-clean fluxes), manufacturability, cost, and availability. Yet another important consideration for selecting the lead-free solder alloy for commercial use is whether or not the alloy may be covered by any patents. Lead-free alloy selection, as well as associated patent issues, have been described in detail in the literature in Ref 16–20.

There are many solder alloys that do not contain lead, and various lead-free solder alloys will continue to be used in the industry. Currently, the industry (Ref 17) is converging on the ternary eutectic Sn-Ag-Cu (SAC) alloy (eutectic temperature approximately 217 °C, or 422 °F) for reflow, and Sn-Ag-Cu or Sn-0.7Cu alloy (eutectic temperature 227 °C, or 440 °F) for wave soldering. All alloy concentrations are by weight percentage, unless otherwise specified. The microstructural characteristics of the lead-free solder alloys will be discussed in great detail in a later chapter.

It is generally believed that the different variations of the Sn-Ag-Cu alloy, with silver content from 3.0% to 4.0%, are all acceptable compositions. Recent studies by IPC, solder suppliers, and electronic manufacturing services (EMS) companies (Ref 21), in an effort to standardize the alloy composition, have concluded that there is no significant difference in the process performance and thermomechanical reliability among these different alloy compositions.

The tin-copper alloy has been found to be inferior to SAC in terms of wettability, dross formation, and reliability under typical loading conditions; however, its much lower cost as compared with SAC makes it an attractive alternative alloy for wave soldering, especially for cost sensitive products. Even though most manufacturers may prefer to use the same alloy for all of the solder interconnects on the entire board (including reflow and wave soldering), there are products in volume production today which use SAC for reflow and tin-copper for wave soldering on the same board. As such, methods for inspection, rework, and accelerated testing must be compatible with both alloys. Several variants of the tin-copper alloy have also been intro-

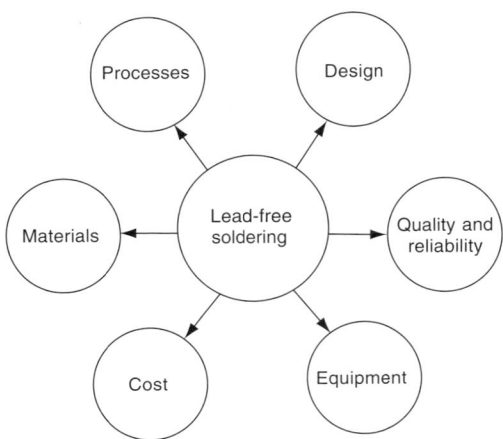

Fig. 1 A holistic approach to lead-free soldering

duced, including silver (Ag), nickel (Ni), and other alloying additions (Ref 22). The steep slope of the tin-copper binary phase diagram (approximately 30 °C, or 55 °F, change in the liquidus temperature for every percent change in composition, roughly 20 times the slope for tin-lead) suggests that the chemical composition of the tin-copper solder pot will need to be closely controlled.

During the initial transition period, some products may be assembled with lead-free solder using lead-containing component termination. For wave soldering, the level of lead in the solder pot, due to dissolution of tin-lead-plated component finishes, will need to be monitored. The long-term reliability impact of lead in lead-free solder is a subject for further study and will be discussed in later chapters. Preliminary studies have indicated that the impact varies with the amount of lead in the solder joint, due to the formation of segregated phases (e.g., coarse lead grains) in the last-to-solidify interdendritic tin grain boundaries, where cracks may initiate and propagate under cyclic loading (Ref 23).

The potential compatibility concerns of bismuth (Bi) containing solder alloys and tin-lead component terminations (or tin-lead solder alloys and bismuth containing component terminations), due to the formation of the low melting temperature (98 °C, or 208 °F) Sn-Pb-Bi ternary eutectic phase, have often been discussed in the literature; however, the exact concentration of bismuth necessary to cause an actual reliability degradation (based on kinetics considerations) is not yet clear, and component terminations with tin alloys with low bismuth (1–3%) have been used with the tin-lead solder in actual products. As will be discussed later, bismuth is often used as an alloy addition to tin to reduce the risk of tin whisker growth.

Other Lead-Free Alternatives

Other lead-free solder alloys which have been developed and evaluated by the industry include Sn-Ag, Sn-Ag-Bi, Sn-Bi-In, and other ternary and quaternary alloys (Ref 24–33). The use of alloys containing indium (In) may be limited due to the scarcity of indium and its high cost. The eutectic Sn-Ag alloy (melting temperature 221 °C, or 429 °F) was qualified for automotive applications in the 1990s on ceramic thick film substrates and PWBs (Ref 3, 12–15, 34–45). The microstructural characteristics of the Sn-Ag-Cu system under different reflow conditions and solid state thermal aging conditions, as well as reliability testing and crack initiation and propagation, were studied both experimentally and numerically (Ref 12–15, 35, 37–38, 40, 42, 44), along with the physical properties (Ref 39), mechanical properties (Ref 28), and the microstructural changes during creep of the Sn-Ag alloy (Ref 41, 43).

For high temperature applications, the eutectic Au-20Sn alloy (melting temperature 278 °C, or 532 °F) has been used for die attach, especially for optoelectronics packaging. Tin-antimony (melting temperature 235 °C, or 455 °F) is another example of a high temperature lead-free solder alloy. For low temperature applications, alternatives include Sn-Zn, Sn-Zn-Bi, Sn-Bi-In, and Sn-Bi-Ag (Ref 46–48).

The discussions in the following sections of this chapter, as well as the other chapters of this book, will be focused primarily on the SAC and tin-copper lead-free solders, as these are currently the leading lead-free solder alloy choices for the worldwide electronics industry. There are still ongoing discussions about the merits of these and other alloy choices; nonetheless, a significant amount of industry resources have been and will continue to be devoted to building up the infrastructure around the characteristics of these alloys.

Conductive adhesives are another alternative to tin-lead solder interconnects without lead. The relatively high cost of conductive adhesives, at the present time however, limits their primary use to niche applications. The reliability of conductive adhesive interconnects will be discussed in great detail in Chapter 11.

Flux

Early attempts to simply mix no-clean flux (developed for tin-lead) with lead-free solder alloys yielded miserable results. No-clean flux required reformulation for lead-free solder alloys in order to accommodate their unique characteristics. The chemical reaction between the flux and the solder alloy in the paste affects the rheological characteristics of the solder paste (which is critical for printing performance). The differences in density between the lead-free and tin-lead solder alloys mean that the metal loading of the solder paste needs to be different.

The higher soldering temperature needed for lead-free solders will require greater stability of the flux at higher temperatures. The performance of flux residues after reflow, in terms of in-circuit test (ICT) probe ability and electromigration, is also an important consideration. Similarly, no-clean and volatile organic compound (VOC) free fluxes need to be formulated specifically for lead-free wave soldering. Water-soluble fluxes for lead-free solder paste and wave soldering applications will also be needed for certain applications.

In wireless communication products, the effect of no-clean fluxes on the radio frequency (RF) signal integrity of a product is a matter of concern. As RF frequencies increase, every part of a circuit, such as the conductor traces, the solder mask, and the flux residue, all add to the overall circuit design and tolerances. The RF impact of lead-free solder and flux residues (Ref 49) needs further study.

Printed Wiring Boards

There are three primary issues related to PWB for lead-free soldering: laminate materials for reliability (to be discussed in a later section), halogen content, and surface finishes for solderability.

Halogen Content. Halogen is used as flame retardant in the PWB. The European Union RoHS legislation calls for the elimination of certain halogen, specifically PBB and PBDE. However, the halogen typically used in the PWB is Tetrabromobisphenol A (TBBPA), which is not covered under RoHS. At the present time, halogen-free PWB is generally more costly than regular PWB materials, and the RF performance of halogen-free PWB is under further evaluation.

Surface Finishes. The search for alternatives to hot air solder leveling (HASL) has been on-going for several years, primarily because of the inherent inconsistency in the quality of the HASL finish. For example, the thickness (and therefore, solderability) of HASL is difficult to control. In areas with a very thin layer of HASL, consumption of Sn by the formation of tin-copper intermetallics will render the areas non-wettable. The HASL finish is typically non-flat (with a dome shape), making it difficult to deposit a consistent amount of solder paste during solder paste printing and difficult to place fine pitch (<25 mil) devices. The HASL process itself is not as clean and easy to control as some plating processes. The current move towards lead-free solder has provided the additional impetus towards alternative surface finishes.

Lead-free HASL, using lead-free alloys (such as tin-copper) in place of tin-lead, is commercially available; however, whisker growth is a concern for some users. Organic solderability preservative (OSP), a finish which provides a flat pad surface, has also been in use for many years. Both OSP and HASL are relatively low cost in comparison with other finishes. Board storage and handling, as well as solderability degradation through multiple soldering processes, are some of the drawbacks for OSP (Ref 13). Electroless nickel and immersion gold (ENIG, or Ni-Au) provides good solderability and contact/switch interfaces for most applications, as well as bondability with aluminum wires; however, tight plating process control is necessary in order to prevent the occurrence of catastrophic "black pad" failures.

Briefly, the black pad phenomenon is due to the oxidation of the nickel layer during the immersion gold (Au) process, which, by nature, is a galvanic corrosion process through an electrochemical reaction. The susceptibility of the nickel layer to oxidation depends on the microstructure of the nickel layer, such as grain size, and crystalline versus amorphous structure, which in turn is influenced by the metal turn over, or MTO, and the chemistry, such as the phosphorus (P) content of the nickel plating bath. For example, the nickel layer changes from crystalline to amorphous as the P content increases, and its resistance to oxidation increases as a consequence. The chemistry of the immersion gold plating bath, the plating time in the immersion gold bath (which can be correlated to the gold thickness), and the layout of the PWB, together with the susceptibility of the nickel layer to oxidation, are some of the important factors which influence the extent of the occurrence of black pad. For example, a BGA pad which is connected to a large copper pad on the board often suffers more severe black pad attack than the other pads for the same ball grid array (BGA) package, due to the higher current density (and higher electrochemical potential) for the small pad which is connected to a large pad, as a result of relatively faster depletion of the reactant (i.e., gold ions) in the vicinity of the large pad, as compared with the small pad in electrical connection during the electrochemical reaction (Ref 50).

When black pads are present, proper metallurgical bonding between the solder and the pad will not form during soldering. The occurrence of black pad is difficult to detect from the visual appearance of the PWB and solder joints. Boards with black pad can easily pass visual inspections and functional tests during board assembly, but will lead to catastrophic failures in the field due to the separation of the solder from the pad. More detailed discussions can be found in Chapter 2.

For higher end applications, electrolytic nickel and gold (more expensive than ENIG) provides a more reliable surface finish. Care must be taken to limit the thickness of gold, as too much gold dissolved in the solder joint may cause "Au embrittlement," while noting that the Sn-Ag-Cu alloy is less sensitive to gold embrittlement than the tin-lead alloy. Use of the electrolytic nickel-gold may also be constrained by the layout and available "real estate" of the board. Another consideration for the nickel-gold finish (including ENIG and electrolytic nickel-gold) is that the interface between the tin and the nickel layer is in general more brittle than joints between tin and copper (which is formed for tin-based solders on boards with OSP and immersion silver finishes). This is a particularly important consideration for products (such as handheld devices) for which dynamic mechanical reliability (such as drop) is important.

Immersion silver (I-Ag) is a more recent and less costly alternative. The immersion process is a self-limiting process and can only provide a very thin layer of coating (several hundred atomic layers). The solderability, ICT probe ability, aluminum wire bondability, and contact/switch pad performance of I-Ag are not as good as nickel-gold, but are adequate for most applications. For I-Ag, the exact chemistry, thickness, surface topography, as well as the distribution of organic constituents within the silver layer, must be carefully selected and specified (Ref 47). X-ray fluorescence (XRF) is the technique often used to monitor the thickness of the surface finish for process control purposes. Careful calibration of the XRF instrument is very critical, especially for I-Ag. Handling and storage also need to be carefully controlled for I-Ag.

In terms of solderability for lead-free soldering (Ref 51–54), immersion tin (I-Sn) and ENIG surface finishes provide the best wetting results on fresh boards, followed by I-Ag and OSP. However, after storage and heat exposures, the wetting of the I-Sn finish degrades the fastest, with less wetting degradation for the I-Ag and OSP finishes, while the wetting of the ENIG finish remains excellent through various pre-conditioning treatments and heat exposures. Fresh I-Ag boards can withstand up to four lead-free reflow cycles before the final reflow soldering process, and at least two reflow cycles before the wave soldering process. By contrast, I-Sn finished boards cannot withstand multiple lead-free reflow cycles or a reflow cycle prior to wave soldering process without significant degradation in wetting, unless the I-Sn thickness is significantly increased.

The impact of different surface finishes on press fit connectors is another factor to be considered when selecting PWB surface finishes for products such as backplanes. The plastic deformation of the Sn plating, the hardness of the surface finishes, along with the dimensional tolerances of the pin and through-holes, the design of the connector pins (and their mechanical compliance), and the mechanical properties of the PWB, all affect the insertion force required (and consequently the retention force) for the press fit connectors. Typically, the difference in the insertion force among the different surface finishes (Ni/Au, I-Ag, I-Sn) is not significant (<10%).

Components

Termination Metallurgy. Traditionally, most components have leads (QFP, SOT, etc.), balls (BGA, CSP), or end terminations (chip capacitors/resistors, LGA, QFN, etc.), plated or coated with the tin-lead alloy, to provide solderability. Although the tin-lead plating provides adequate solderability with lead-free solder as well, it will not meet the environmental legislation requirements. Some tin-lead-plated components may be used with the lead-free solder during the transition; the impact of lead in lead-free solder joints has been discussed previously.

A number of lead-free component termination finishes have been evaluated (Ref 13, 55) and used over the years. For passive components (such as chip capacitors and resistors), matte Sn plating has been used for many years with the tin-lead solder, and can be used with lead-free solder as well. For leaded components (e.g., quad flat pack or QFP), plating of matte tin or tin alloys may be used with lead-free solders (forward compatible). The tin whisker concern will be discussed in a later section. Nickel-lead has been used with the tin-lead solder for many years, and Ni-Pd-Au is currently an alternative for leaded components for lead-free soldering; Ni-Pd typically does not provide as good wettability as tin. Area array packages with SAC balls are available and work well with the SAC solder.

Temperature Requirements. Because of the temperature impact on components, efforts have been made to minimize the lead-free soldering temperature, but the soldering temperature must still be high enough to offer a robust process window to enable good yields for large volume production, for a large variety of products. For reflow soldering, assuming the minimum peak temperature to be 235 °C (455 °F) (to be discussed in a later section), the maximum temperature depends on the temperature delta across the board, which, in turn, depends on the board size, thickness, layer count, layout, copper distribution, component size and thermal mass, thermal capacity of the oven, and certain unavoidable process variations and measurement tolerances. Large thick boards with large complex components (such as CBGA, CCGA, etc.) typically have temperature delta as high as 20 to 25 °C (68 to 77 °F) (Ref 56). Rework is another process which contributes to elevated temperature exposure for the components.

For wave soldering, since the solder pot temperature is typically already quite high for the tin-lead solder, the increase needed for the lead-free solder is generally less as compared with reflow, and the dwell time with the solder pot is also generally very short (a few seconds).

When all of the application requirements are taken into consideration, 260 °C (500 °F) peak temperature has been proposed as the temperature required for components for lead-free soldering (Ref 57). The requirements (including soldering peak temperature and tolerances) are captured in the IPC/JEDEC Standard 020C (Ref 58), according to the component volume and thickness, and process conditions (such as rework). The actual component body temperature may be different from the temperature measured on the board, and different components may have different temperatures depending on component thermal characteristics and locations on the board. For thin packages (<1.6 mm), 260 °C (500 °F) reflow temperature is specified, whereas for thick components (≥2.5 mm or 0.098 in.), the reflow temperature specified is only 245 to 250 °C (473 to 482 °F). However, IPC/JEDEC standard 020 also specifies that a lead-free component shall be capable of being reworked at 260 °C (500 °F) within eight hours of removal from dry storage or bake. On the other hand, it is recognized that for certain products (such as small/medium form factor boards), which typically have a smaller temperature delta, a lower maximum temperature may be adequate.

The IPC/JEDEC Standard 020 is for nonhermetic solid state surface mount devices (Ref 58), but is generally referenced for other components as well for reflow applications. For wave soldering, the peak temperature may be as high as 270 °C (518 °F), but the duration at the peak temperature is only a few seconds. In this situation, heat is transferred to the components on the bottom side of the board through direct contact, and to the components on the topside of the board through convection and conduction through the leads, where the component temperature is generally different (lower in most cases) from the soldering temperature. The thermal shock (from the pre-heat temperature to the peak temperature) is also an important parameter for components (especially ceramic capacitors) for wave soldering.

Internal Materials. The internal materials for the components must also meet the requirements of environmental regulations. Each of the constituent materials, such as the molding compound, lead-frame alloy, plating material on the lead-frame, die attach material, bonding wires, underfill, and the various materials of the substrate, etc., must all be examined to ensure that the environmental requirements, especially for the six substances prescribed under the European Union RoHS legislation, are met (to be discussed in more detail in a later section). In this regard, particular attention needs to be paid to the flame retardants, e.g., PBB and PBDE, in the molding compound. Careful evaluation needs to be carried out with alternative flame retardants, in view of the recent failures reported in connection with certain alternative flame retardant materials in the molding compound.

Processes, Equipment, and Quality For Lead-Free Soldering

SMT Reflow

Reflow Process. The key parameter for the reflow profile is the peak temperature. Adequate reflow temperature is needed for the solder to melt, flow and wet, interact with the copper on the pad and the component termination, and form sound intermetallic bond when cooled and solidified. Typically, 30 °C (55 °F) superheat (above the melting temperature) is desired. For lead-free soldering, because of concerns about the thermal stability of the components, efforts are needed to minimize the soldering temperature. For SAC alloy with the eutectic temperature at 217 °C (422 °F), the minimum reflow peak temperature should be 235 °C (455 °F) for large volume manufacturing (Ref 61), taking into account process robustness, yield, variety of component finishes, oven thermal stability and tolerance, etc. The dwell time (or time above liquidus, TAL) is typically 40–90 s. The reflow profile may be straight ramp, or with a pre-heat plateau for the purpose of homogenizing the temperature distribution across the board and/or minimizing voiding in the solder joints. Reflow profile development to minimize the temperature delta across the board, especially for large complex boards, is an important issue. The conveyor speed, board orientation, and zone settings are important parameters. For example, having a slower conveyor speed and orienting the short side of the board parallel to the conveyor direction help reduce the temperature delta.

In terms of printability, tack, slump, and solder balling, there is no clear and consistent difference between the Sn-Pb and lead-free solder pastes (Ref 59–60), because these performances depend on the solder paste formulation, not directly on the solder alloy. Very clear and consistent differences have been observed, however, in wettability between the tin-lead and lead-free solder pastes. In general, the wettability of lead-free solder paste is not as good as the tin-lead solder paste. For example, lead-free solder paste exhibits very limited spreading on OSP during reflow, and exposed corners after reflow are quite common, unless overprint or round corner pads are used. The difference in wettability between OSP and ENIG surface finishes, which is already evident for the tin-lead solder, becomes even more pronounced for lead-free solders. This has been observed with a variety of solder paste and flux formulations from a number of vendors.

Lead-free reflow may be done in air or N_2. Typically N_2 is not required, and the use of N_2 may even increase certain defects (such as tombstoning) especially for small passive components. In certain situations, N_2 may help improve wetting (which in turn may help reduce the amount of voids in the solder joints). For flip chip applications, where flux is used instead of solder paste, N_2 becomes necessary to form reliable solder interconnects. An N_2 atmosphere with O_2 level below 1000 ppm has been found to be effective (Ref 65, 67).

Case studies. Lead-free soldering processes have recently been developed for various advanced packages, including 0201 (0.02 in. × 0.01 in.) components, 01005 (0.01 in. × 0.005 in.) components, flip chip, CSP, BGA, CCGA, pin-in-paste (PIP), package stacking, etc. (Ref 62–68).

0201 assembly processes have been developed through a systematic study (Ref 63), including pad design, pick-place equipment evaluation, component qualification, and process optimization, and the overall yield is slightly lower for the lead-free solder as compared with the Sn-Pb solder, for reflow in air. The overall process for 01005 (Ref 66) is similar to 0201; however, reflow soldering for 01005 may require nitrogen, due to the small size of the solder deposits, to protect the activity of the flux; further work is underway to optimize the process for reflow in air.

Lead-free solder flip chip assembly on FR-4 (a standard glass epoxy substrate) has been demonstrated to be feasible (Ref 65, 67) both on ENIG and OSP surface finishes with reflow in nitrogen. The self-alignment properties with Sn-Ag-Cu is the same as or similar to tin-lead, and the different surface finishes (ENIG and OSP) showed no difference with both showing self-centering with misplacement up to 50% off pad, using dip fluxing or flux jetting as the fluxing method.

Sn-A4.0g-0.5Cu balled, 0.4mm (0 0157 in.) pitch CSP components on Sn-3.9Ag-0.6Cu solder paste, placed up to ~50% off-pad, can self-align to the pad after reflow in air (Ref 62). By considering the PWB pad size/location tolerances and solder paste alignment tolerances, the required alignment accuracy for the pick and place equipment was calculated to be ±50 μm at 3σ (standard deviation).

Large CCGA devices could self-align to the pad while placed up to 25% off the pad, with the SAC solder (Ref 75). The main concerns are with the thermal profiling for reflow and rework, and the impact of the temperature on the surrounding components.

Package stacking, or package-on-package (POP), has been developed for Sn-Pb and SAC (Ref 68). In this process, two or more CSPs (each may contain one or multiple dice), are stacked together using the pick-place machine, and only a single reflow (along with the rest of the surface mount devices, or SMD, on the board) is needed for the SMT process. This can be done on either or both sides of the board.

For PIP, various PWB variables, stencil apertures, solder paste printing methods and parameters, and component types and insertion methods, have been evaluated for both tin-lead and lead-free solders (Ref 64), as an alternative to wave soldering. PWBs with thicknesses of 1.0 mm (0.039 in.) and 1.6 mm (0.063 in.), with both ENIG and OSP surface finishes, have been assembled successfully with the PIP process using metal squeegees.

Second reflow. Component fall-off during the second reflow process has also been studied as a function of component weight and solder surface tension and contact area (Ref 69). The component weight to contact/pad area ratio has been found to be slightly higher for eutectic tin-lead than SAC; however, the difference has been found to be statistically insignificant. Therefore, the same rule of thumb (g/in^2) can be used.

Quality. The IPC 610 standards (Ref 70–71) have recently been revised to include appropriate workmanship criteria for lead-free solders. Operator and AOI (automatic optical inspection) training is needed for lead-free solder because of the different appearance of the solder joints. Lead-free solder joints are generally more dull and grainy than tin-lead solder joints. This difference in appearance is determined by the metallurgy of the solder alloys and is generally not a reflection of the workmanship (Ref 5). Whereas the tin-lead solder solidifies as a typical eutectic microstructure, the SAC alloy, even though it is a ternary eutectic alloy, solidifies as an off-eutectic microstructure, under typical soldering conditions, due to non-equilibrium solidification. Tin dendrites, which are formed as a result of the non-equilibrium solidification, create the grainy and dull appearance of lead-free solder joints. The issue of "hot tearing," associated with dendritic solidification and shrinkage leading to surface cracks, is another issue under discussion.

Even though the types of defects for lead-free solders are similar for tin-lead, lead-free solder has generally been found to generate more defects of voiding, tomb-stoning, solder beading, bridging, and misalignment. It takes considerable efforts to achieve the same yield for the lead-free solder as for the tin-lead solder, especially for wave soldering. This is believed to be related to the inferior wettability of the lead-free solder and the consequently reduced wetting force. For example, self-alignment (or self-centering) has been found to be less with the lead-free solder than with the Sn-Pb solder (Ref 72), especially for passive components (such as 0201). Therefore, process optimization is critical for volume manufacturing with lead-free solder.

Wave Soldering

For wave soldering with lead-free solder, a higher solder pot temperature, typically 255 to 270 °C (491 to 518 °F), will be required (Ref 73). Flux application (spray, foaming, etc.) and amount, and preheat temperature and time, must be optimized for each particular flux. A longer pre-heat may be needed in order to keep the thermal shock (difference between the preheat and peak temperatures) below 100 °C (180 °F) to protect ceramic components (especially ceramic chip capacitors). Dual wave soldering, already popular for tin-lead soldering, will still be used for lead-free, and inert (N_2) atmosphere may be used to improve yield, for example, by reducing bridging and improving wetting and hole filling.

The amounts of dross formed with SAC and tin-lead solders have been found to be very similar (Ref 73), at the same temperature, for the same duration, and under the same atmosphere, while the tin-copper solder forms considerably more dross. Just as with the tin-lead solder, N_2 helps reduce dross formation for lead-free solders.

As with reflow, lead-free solder wetting during wave soldering is generally not as good as with the tin-lead solder, leading to reduced hole filling. However, the overall influences of other process parameters, such as conveyor speed, dwell time and contact length, component orientation and soldering direction (parallel or perpendicular), etc., on the yield and quality for wave soldering, are generally similar for the lead-free and tin-lead solders. For example, increasing the conveyor speed (i.e., reducing the contact time) helps to reduce solder balls and bridging, but may reduce hole filling at the same time. The use of thieving pads can effectively reduce bridging for SMD components, for both Sn-Pb and lead-free solders.

For lead-free wave soldering, the chemical composition of the molten solder in the pot needs to be closely monitored, especially for lead and copper. For the tin-copper solder, the liquidus temperature will change by as much as 6 °C (11 °F) when the copper composition changes by 0.2%; such a change may cause significant change in the wave dynamics and the soldering quality (such as bridging). The buildup of copper in the molten solder can also cause sluggishness of the solder, disturbing the wave dynamics. The density of the tin-copper intermetallic compound or IMC (8.3 g/cm^3) is greater than that of the SAC (7.4 g/cm^3) and tin-copper solder (7.3 g/cm^3), causing the IMC dispersion in the molten solder, as compared with that of the tin-lead solder (8.7), where the IMC tends to float and can therefore be more easily removed.

Rework and Repair

Rework for lead-free solders has been found to be more difficult, because the lead-free solder alloys typically do not wet or wick as easily as the Sn-Pb solder due to their difference in wettability. This can be easily seen with QFP packages. In spite of these differences, successful rework methods (both manual and semi-automatic) have been developed (Ref 74–75) with lead-free solders (Sn-Ag-Cu, or Sn-Ag), for many different types of components. Most of the rework equipment for tin-lead can still be used for lead-free solder. For area array packages, it is helpful to use a rework system with split vision and temperature profiling features. The soldering parameters must be adjusted to accommodate the higher melting temperature and reduced wettability of the lead-free solder. The other precautions for tin-lead rework (such as board baking) still apply to lead-free rework.

Studies have shown that reliable lead-free solder joints, with proper grain structures and intermetallics formation, can be produced using appropriate rework processes. Care must be taken to minimize any potential negative impact of the rework process on the reliability of the components and the PWB. Surface insulation resistance (SIR) tests must be performed to ensure the compatibility between the reflow/wave solder flux and the rework flux, i.e., to ensure that the rework flux and any products of reaction between the reflow/wave solder flux and the rework flux do not pose any unacceptable risk for electromigration and dendritic growth for no-clean applications.

The issue of component mixing, or cross-contamination, warrants special concern, especially during the industry-wide transition to lead-free. If a tin-lead solder board is to be repaired (for example for warranty repair at some future time) with a lead-free solder, the reliability of homogeneously mixed lead-free solder and tin-lead solder is probably not inferior to the tin-lead solder in most cases; however, the temperature impact on the components (especially plastics package parts) could be a concern. Careful consideration must be given to the use of area array packages with lead-free balls to repair a tin-lead solder board. In this case, if the temperature is not high enough, reliability concerns may arise; this is the backward compatibility issue to be discussed later in this chapter. On the other hand, repairing a lead-free soldered board with the tin-lead solder could create solder joints which are not as reliable as the lead-free solder joints on the rest of the board.

Equipment

For reflow ovens, the ability to minimize the temperature delta for large complex boards is a key differentiator for lead-free soldering. In a recent study (Ref 56), the temperature distribution of PWBs with various sizes and component types was investigated for lead-free reflow soldering, in comparison with Sn-Pb soldering. It was found that the temperature delta for a 400 × 350 mm (16 × 14 in.) board was as high as 18 to 20 °C (32 to 36 °F) for both the Sn-Pb and the lead-free profiles, using several different state-of-the-art convection reflow ovens. The temperature delta will be even greater for larger and more complex boards with large components (such as CCGA). The energy consumption is also investigated for lead-free and Sn-Pb soldering processes, and an almost 20% increase in energy consumption was seen for the 400 × 350 mm (16 × 14 in.) board for the lead-free profile, as compared with the tin-lead profile.

The desire to minimize the temperature delta has also generated renewed interest in certain less popular soldering methods, such as vapor phase soldering. Throughput, maintenance, handling of chemicals, tombstoning of passive components, and thermal shock of the board, are some of the concerns with the vapor phase soldering process. Localized heating may also be used for temperature sensitive components, but the throughout will be fairly low.

Equipment for lead-free wave soldering must have adequate pre-heating capacity in order to keep the thermal shock below 100 °C (180 °F) between the pre-heat temperature and the soldering peak temperature. Fluxers with uniform flux application are another important consideration for lead-free wave soldering. Special materials (such as titanium alloys) for the solder pot and the other components (such as the pump impeller) in contact with the molten solder are needed to prevent erosion due to the higher soldering temperature and higher Sn content of the lead-free solder.

AOI, X-ray, and ICT will need to be recalibrated and reprogrammed for lead-free solder boards. A recent study involving both 2D and 3D X-ray has found that (Ref 76) the difference in the grayscale level between the Sn-Pb and lead-free solder interconnects for the various components examined ranged from 6% to 12%, in line with what would be expected from the density difference (approximately 12%) between the tin-lead and lead-free solder alloys. Overall, there is no significant difference in the x-ray images of lead-free and tin-lead solder joints using the same equipment, and the same settings can generate images that can be analyzed with no major difficulties, even though certain fine-tuning can be done to further optimize the x-ray images.

Lead-Free Solder Interconnect Reliability

The reliability of the interconnection for a PCB assembly depends on the reliability of the components, the PCB, and the solder joints, as illustrated in Fig. 2.

Component Reliability

Tin Whisker Growth and Tin Pest. Whiskers have been observed over the past 60 years on the surfaces of tin and other metals. A tin whisker is a spontaneous columnar or cylindrical filament of monocrystalline Sn emanating from the surface of a plating finish, with a high aspect ratio (length/width >2), and may be kinked, bent, or twisted, but rarely branch and generally has consistent cross-sectional shape and may have striations/rings around it (Ref 77). Tin whiskering is a localized growth of Sn at weak spots of the Sn oxide layer. Tin whiskers can potentially cause shorts between the adjacent metallic elements (such as leads), and/or adversely affect the RF performance of the circuit. Tin whiskers are completely different from tin dendrites as a result of electromigration of an ionic species through an electrochemical reaction, details of which can be found in Chapter 5. Tin whiskers can grow with different tin alloys (including tin-lead) under a variety of conditions, but pure tin is most prone to whiskering as many alloying elements (such as lead and bis-

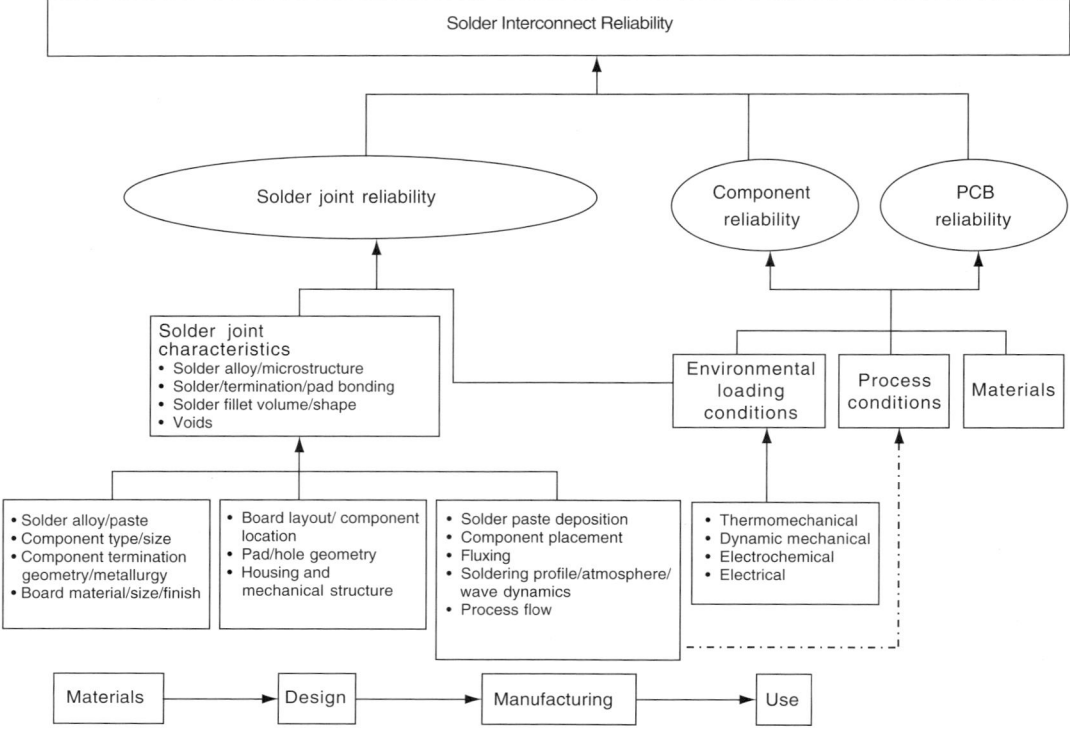

Fig. 2 Solder interconnect reliability

muth) can inhibit or retard tin whisker growth, whereas certain elements (most notably copper, which is insoluble in tin), may actually promote tin whiskering.

The issue of tin whisker growth has generated a great deal of concern, especially for fine pitch (<0.5mm or 0.020 in.) QFP and for products with a long service life greater than five years. Although the fundamental mechanism of tin whisker growth is still under intense study and is not yet completely clear, the root cause of tin whisker growth is believed to be compressive stresses in the tin plating. For example, compressive stresses can build up in the tin plating by large, irregular intermetallic Cu_6Sn_5 growth at the interface between the substrate (e.g. the copper leadframe) and the tin plating layer. Compressive stresses can also build up in the tin plating by mechanical deformation (such as lead forming), coefficient of thermal expansion (CTE) mismatch and other mechanical processes. More detailed discussions can be found in Chapter 6.

There are several factors which can help mitigate the risk of tin whisker growth (Ref 5, 78–79), especially for copper leadframes. The mitigation techniques include:

- the use of *matte* tin plating (approximately 2 μm over a Ni barrier layer, or >10 μm if Ni barrier layer is not used) with large grains (>1 μm) and low carbon content (C < 0.05% wt), instead of *bright* Sn;
- the use of tin alloys (even with a small percentage of soluble alloying elements, such as bismuth), instead of 100% tin (while copper is not effective in retarding tin whisker growth)
- the use of a porosity-free nickel barrier layer (1.3 μm minimum, preferably >2 μm thick) or silver underlayer, underneath the tin plating
- the use of immersion tin (<20 μin)
- the use of copper lead-frame, instead of Alloy 42 (FeNi42)
- reflowing or fusing the tin plating, or tin hot dipping
- annealing (150 °C, or 302 °F for 1 h or 170 °C, or 338 °F for 10 min)
- the use of a conformal coating

Mitigating tin whisker growth on Alloy 42 (Ni-Fe alloy) leadframes have so far proven to be more difficult, due to the CTE mismatch between the leadframe and the plating, which generates stresses in the tin plating during thermal cycling. The mitigation techniques discussed above are much less effective for Alloy 42 leadframes.

As the mechanism of tin whisker growth is not yet completely understood and agreed upon, and the acceleration mechanism for tin whisker growth appears to be quite complex (including temperature and relative humidity, temperature cycling, and even ambient conditions), the JEDEC standard for measuring whisker growth (Ref 77, 80) includes the following:

1. Temperature cycling from between −55 and −40 (+0/−10) °C to 85 (+10/−0) °C, for minimum 1000 cycles, air to air, with 5- to 10-minute soak, typically 3 cycles/h;
2. Ambient temperature/humidity storage at 30 (±2) °C and 60 (±3)% RH, for 3000 h
3. High temperature/humidity storage at 60 (±5) °C and 87 (+3/−2)% relative humidity (RH), for minimum 4000 h

All three tests are required and each test condition is to be performed independently. The sample size, pre-conditioning, and inspection procedure are also defined in this test method. As of this writing, no industry standard is yet available for the acceptance criterion for tin whiskers. Acceptance criteria have been proposed by several industry consortia (for example, whisker length <40–67 μm depending on product category), and some component vendors have proposed the maximum whisker length of 50 μm after two years of storage as the criterion.

Reliability requirements are product application dependent; this is true for whisker growth as well. The mitigation techniques and the testing parameters discussed previously may be excessive for certain applications while inadequate for others. Therefore, they should be tailored, refined and optimized for each specific application. It is also noted that in many applications, the tin plating (Sn melting point 232 °C, or 449 °F) will be reflowed during the lead-free soldering process of the PWB, and this will substantially alter the characteristics of the tin plating (such as thickness distribution, microstructure, and stress state) and thereby its risk of tin whiskering.

At the present time, tin whisker growth remains an active field for research. Overall, a rational approach should be taken in managing the risk of tin whiskering. Towards this end, predictive models for assessing the risk of Sn whisker growth are urgently needed.

Tin pest is a phenomenon primarily associated with the spontaneous allotropic transformation of pure tin, from white Sn (β) with a body-centered tetragonal structure (density 7.3 g/cm^3) to gray tin (α) with a diamond face-centered cubic structure (density 5.77 g/cm^3), at temperatures below 13 °C (55 °F), with the maximum phase transformation rate around −40 °C (−40 °F). This phase transformation is accompanied by a significant volumetric increase, leading to the catastrophic loss of mechanical strength and integrity of tin. Again, a number of soluble alloying elements (such as lead, antimony, and bismuth) can inhibit or significantly retard the occurrence of tin pest (similar to the retardation effect of tin whisker growth), whereas silver is less effective and copper is not effective in retarding the occurrence of tin pest. Mechanical deformation may increase the risk of occurrence for tin pest as well. In theory, the tin pest phenomenon poses a potential reliability risk, but it is rarely observed under practical application conditions (due to the presence of impurities in tin), so its practical reliability implications are still unclear.

Thermal Stability. Lead-free soldering generally requires higher soldering temperatures, as discussed previously. Different components have different sensitivities to the soldering temperature. For example, ceramic resistors, and especially capacitors, are very sensitive to the ramp rate but are not so sensitive to the actual temperature. A fast ramp (for example, >3 °C/s, or 37 °F/s heating rate) or a large thermal shock (for example, >100 °C, or 180 °F) can cause cracking of a ceramic chip capacitor. Aluminum electrolytic capacitors, on the other hand, are extremely sensitive to temperatures. Connectors and some of the plastic package parts may also exhibit increased failures (such as delamination, popcorning, deformation and warpage, etc.) at higher temperatures. Popcorning, for example, can further cause cracking in the molding, component package warpage (which can further result in solder ball shorts for fine pitch devices), and bond wire damage.

Roughly, for every 10 °C (18 °F) increase in temperature, the moisture sensitivity level (MSL) degrades by one level. As such, new molding compound may have to be used for components intended for lead-free soldering in order to maintain the same MSL. Components must now be qualified to the higher soldering temperature for lead-free soldering, as defined by IPC/JEDEC J-STD-020 (Ref 58).

Forward/Backward Compatibility. Overall, the transition to lead-free soldering needs to be completed soon. However, for certain product categories, the transition may be prolonged due to exemption status, technological and economics complexities, supply chain readiness, market requirements, etc. Therefore, tin-lead and lead-free components may be used with tin-lead and lead-free solders in various combinations, during the transition, as illustrated in Fig. 3. Forward compatibility refers to the compatibility of current components (qualified for tin-lead soldering) with the lead-free solder, and backward compatibility refers to the compatibility of the lead-free components with the tin-lead solder (in terms of the metallurgy, soldering process, and reliability).

Assuming that the components can meet the temperature requirements as discussed above, tin and Ni-Pd-Au platings are generally considered to be forward compatible with lead-free solder, as well as backward compatible with the tin-lead solder. This makes it much easier to manage production lines with the tin-lead solder and lead-free solder in co-existence within the same factory during the transition to lead-free.

The backward compatibility of area array packages (such as CSP and BGA with SAC balls) with the tin-lead solder, however, is very much questionable. This is primarily due to the fact that the SAC alloy, with a melting temperature of 217 °C (422 °F), will not always completely melt during reflow with the tin-lead solder, typically at reflow peak temperatures between 205 to 225 °C, or 401 to 437 °F (or even as low as 200 °C, or 392 °F in extreme cases). As such, there will be little or no self-alignment, which is critical, especially for finer pitch area array packages, with coplanarity issues further aggravating the situation due to the lack of collapse. Further, very little mixing takes place leading to grossly segregated microstructures. Poor interfacial bonding and increased voids are some of the issues which have been observed leading to poor solder interconnect reliability, which in combination with the process issues, render the SAC area array packages incompatible with the tin-lead solder (Ref 57). Such issues have been reported in detail in Ref 5, 81–82, with reliability test data including thermal cycling and drop test.

When the reflow temperature is high enough (>225 °C, or 437 °F or preferably >235 °C, or 455 °F) and self-alignment does occur, mixing takes place between the SAC balls and the tin-lead solder, and the reliability of the interconnect using the tin-lead solder paste and SAC balls is not inferior to that using the tin-lead solder paste and tin-lead balls, for area array packages. Such a scenario may be technically feasible in practice, but in theory may present significant difficulties for operations and logistics on the factory floor, and the other components on the board may not be able to survive the higher soldering temperature. As such, area array packages (such as CSP and BGA) with SAC balls are considered to be backward incompatible with the tin-lead solder (Ref 57).

Other component reliability issues related to lead-free solders include flip chips and wafer level CSPs with lead-free solder bumps and balls, where the higher soldering temperature and higher stiffness of the lead-free solder can adversely effect the reliability of the low-k dielectric layer on the die. Low k dielectric is needed for high speed applications, but is typically more fragile and prone to cracking.

The higher temperature for soldering lead-free flip chips to the substrate within the component package (1st level interconnection) can also have an adverse impact on the reliability of the substrate, similar to the soldering process for the 2nd level interconnection, as discussed in the following section.

PWB Reliability

Other than the solderability issues discussed above, the higher temperature of lead-free soldering presents reliability concerns for PWBs, such as discoloration, warpage, delamination, outgassing and blistering, pad lifting, laminate fracturing, conductive anodic filament (CAF), cracking of copper barrels and foils, and interconnect separation, etc. Some of these issues become apparent immediately after the soldering process, whereas others may cause the occurrence of a latent failure, during the use of the product.

As the soldering temperature increases, the Z-axis coefficient of thermal expansion (CTE) mismatch between the laminate material, the glass fiber, and the copper, will exert greater stresses on the copper. This stress may potentially cause failures by cracking the copper of the plated vias, for example, depending on the thickness of the PWB, the laminate material, the soldering profile, the copper distribution, the via geometry, and the copper plating thickness, etc. In a recent study (Ref 83), accelerated testing was done using air-to-air thermal cycling from −40 to 125 °C (−40 to 257 °F) for 3000 cycles (40 min per cycle) to evaluate the effect of lead-free SAC soldering temperatures on the long-term reli-

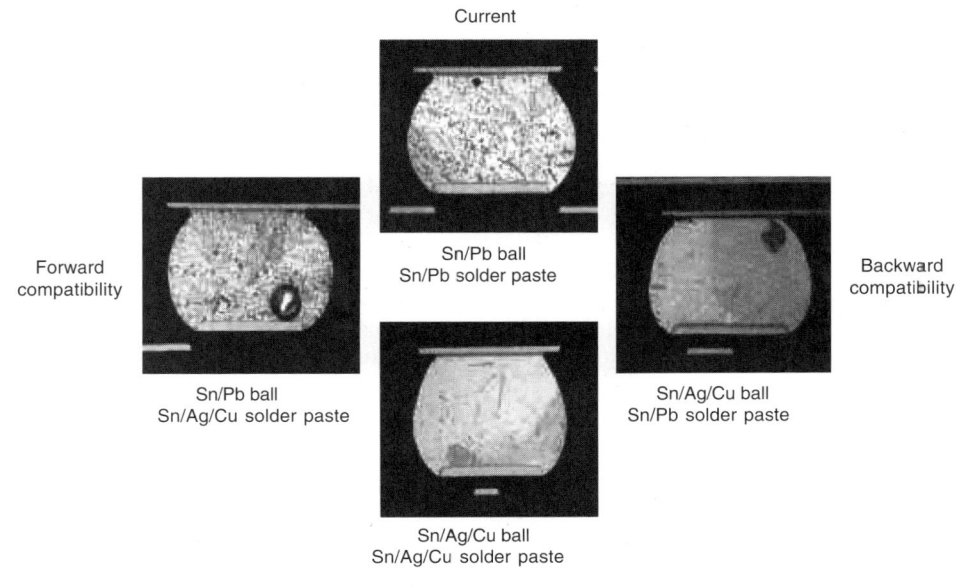

Fig. 3 Solder and component combinations

ability of plated through hole (PTH) vias (with 7 to 1 aspect ratio), and connections were continuously monitored during the testing. The effect of copper weight, aspect ratio, and the use of non-functional pads were also studied. It was found that, with standard T_g (glass transition temperature) (135 °C, or 275 °F) materials, PTH reliability was substantially reduced by lead-free soldering (with two reflow cycles at 255 °C, or 491 °F peak temperature) as compared with tin-lead soldering (at 215 °C, or 419 °F reflow peak temperature), due to two failure modes: barrel cracking and foil cracking (particularly on ½ oz copper foil). This becomes a particular concern for high reliability applications, which demand less than 100 ppm failure rates (Ref 83) for complex boards which may have over 10,000 holes per board.

Another study (Ref 84) used the IST (interconnect stress test) to evaluate the PTH performance on 18 layer, 3.12 mm (0.125 in.) thick boards with 0.34 mm (0.0135 in.) diameter drill holes, and T_g values of 140 °C (284 °F), 170 °C (338 °F), and 180 °C (356 °F). Up to 5 simulated reflow cycles were done for tin-lead soldering (215 °C, or 419 °F peak temperature) and lead-free SAC soldering (259 °C, or 498 °F). PWB discoloration was observed due to oxidation of the epoxy matrix during lead-free soldering. The IST test heats the coupon internally by applying a DC current to built-in resistive traces. Switching the current on for 3 min heats the coupon from room temperature to 150 °C (302 °F), and switching the current off and using forced air cooling for 2 min returns the coupons to room temperature. The results indicated that the lead-free soldering significantly reduced the PTH reliability, due to barrel cracking, near the center of the board. Time-to-delamination tests were also done at 260 °C (500 °F) and 288 °C (550 °F), by ramping the sample from room temperature to the peak temperature at 10 °C/minute, or 18 °F/minute and holding at temperature until catastrophic failure occurred. The evaluation studied the loss of the ability of the dielectric bond line to absorb stress, due to polymer embrittlement and oxide layer damage, causing the separation or delamination between the resin and copper and/or between the resin and the reinforcement. The 288 °C (550 °F) test was used as an accelerated test for lead-free soldering, similar to the 260 °C (500 °F) test used today for tin-lead soldering. The results show that the time to delamination degraded very significantly when held at 288 °C (550 °F), as compared with 260 °C (500 °F).

In an experimental study (Ref 85), different substrate materials, with T_g at 155 °C (311 °F), 170 °C (338 °F), and 210 °C (410 °F) (by DSC) were tested through IST using a 22-layer, 120 mil thick, PTH/post interconnect test vehicle, including vias with aspect ratios of 3 to 1 and 8 to 1. The test vehicles were pre-conditioned through three reflows at 255 °C (491 °F) peak temperature. For the IST test, the test coupon was heated to 150 °C (302 °F) for 3 min, allowed to cool to ambient temperature for 2 min, and reheated to 150 °C (302 °F), for 1000 cycles. Barrel cracking in the central zone of the plated through-hole vias due to fatigue was found to be the dominant failure mechanism. No coupons measured resistance degradation in the inner layer to PTH barrel (post) interface. This study demonstrated the effect of copper plating thickness and integrity on the results of the IST testing, and recommended that any tests to compare PWB substrate materials be performed with identical copper plating processes.

Several important materials parameters were included in another study (Ref 86), such as the T_g, T_d, and CTE. T_g is the temperature at which the material transforms from a relatively rigid or glassy state to a more deformable or softened state, and the properties of the resin are different (such as higher CTE) above the T_g as compared with below the T_g. The decomposition temperature T_d (at 5% mass loss), measures the actual chemical and physical degradation of the resin system. This study suggests that the different material properties (T_g, T_d, and CTE) can vary independently of one another, and they are all important for the reliability of the PWB. Balancing these properties (high T_g, high T_d, and low Z-axis CTE) is important in achieving PWB reliability for lead-free soldering.

Much work is still needed to determine under what conditions of PWB layout, thickness and via geometry, alternative laminate materials may be needed for lead-free soldering. Multiple materials parameters, including T_g, CTE, T_d, and T_{288} (time to delamination at 288 °C, or 550 °F), for the PWB, must be considered, especially for thick boards with high aspect ratio vias. Of particular concern are "dicy-cured" laminate materials, which are generally more prone to delamination (and consequently CAF failures and laminate failures), since the dicyandiamide ("dicy"), which is used as a chemical bonding agent through cross-linking to enhance the resin/glass bonding, is hydrophilic. This is not to say that lower cost materials cannot be used with lead-free soldering. In fact, a variety of PWB laminate materials (such as CEM3, FR2, FR4, halogen-free, flexible circuits, etc.) have been used with lead-free solder in volume manufacturing, depending on the application. Testing requirements for PWBs for lead-free soldering have been proposed (Ref 87), and application-specific solutions are needed. For PWB testing, both failures after pre-conditioning (6 to 10 reflows at 260 °C, or 500 °F, peak temperature) and latent failures (after thermal cycling test) need to be examined.

Electrochemical Reliability

Electrochemical reliability is determined by the resistance of the flux residue to electromigration and dendritic growth in no-clean applications. In no-clean soldering, flux residues are left on the PWB after the soldering process. During the lifetime of the product, when flux residues are dissolved in moisture condensation on the board, electrochemical reactions will take place between conductor traces under an electrical bias, causing reduction in the SIR. If electromigration and dendritic growth occurs, even more fatal failures can occur due to the formation of "shorts" between the conductor traces. This needs to be assessed for lead-free solders because of the different flux systems and the higher soldering temperatures (Ref 5). More details can be found in Chapter 5.

CAF is another failure mechanism of the PWB due to electrochemical reactions (Ref 88). The higher temperature for lead-free soldering can damage the bond between the glass fibers and the epoxy resin in a glass-fiber reinforced (FR) epoxy laminate, causing physical degradation of the bond and delamination (as discussed in the previous section). Upon delamination, moisture and ionic contamination can permeate and migrate along the gaps between the glass fiber and the epoxy resin, providing an electrochemical pathway. Under an applied voltage, an electrochemical reaction occurs, and a filament grows connecting the cathode and the anode leading to a short, which can cause catastrophic field failures. As the board density increases in miniaturized electronics products, the distance between vias decreases and the risk of CAF failures increases. Moisture absorption, ionic contamination, adhesion between the glass fiber and the epoxy, are some of the key characteristics which need to be controlled in order to reduce the risk of CAF failures. The higher soldering temperature for lead-free solder can have a significant impact on the occurrence of CAF failures (Ref 88–90). Additional details will be discussed in Chapter 5.

Thermomechanical and Dynamic Mechanical Reliability

The mechanical failures of the solder joints may be due to mechanical overload or fatigue. Overload failures occur when the stress in the solder joint exceeds the capacity/strength of the solder alloy (for example, in mechanical tests such as pull, drop, static bending, impact, etc.), causing it to fail in a single event. By contrast, fatigue failures happen even at stress levels far below the strength of the solder alloy, under cyclic loading, through a wear-out mechanism over a certain period of time. The cyclic loading may be imposed by cyclic temperature excursions, and/or vibration, and the solder joint fails through crack initiation and propagation until the crack becomes a complete "open" through the solder joint.

The thermomechanical reliability of solder interconnects, including both tin-lead and lead-free alloys, is determined by creep and fatigue interaction of the solder alloy. The CTE mismatch in the system (component, solder, and substrate) imposes cyclic strains (most notably shear strains) in the solder under varying temperatures. The spatial and temporal distribution of shear strains in the solder joint is dependent on the geometrical and materials parameters of the interconnect system, including the CTE mismatch, the range of the temperature variation, the component configuration and solder joint distribution, the solder joint geometry, and the solder alloy elastic-plastic and creep constitutive relationships.

The deformation and fracture behavior of a solder alloy, in turn, depends to a large extent on the microstructural characteristics of the alloy (to be further discussed in Chapters 2 and 3). For example, the creep resistance of the SAC alloy has been related to finely dispersed Ag_3Sn phases in the matrix (Ref 41, 43). As another example, voids (such as Kirkendall voids between the Cu-Sn intermetallic layer and the copper) can propagate into cracks under cyclic mechanical or thermomechanical loading, and cracks from multiple void locations eventually link up, developing into a continuous crack and causing a solder joint failure (Ref 12). The higher soldering temperature and the higher Sn content associated with the SAC alloy may exacerbate this phenomenon. The occurrence of this phenomenon is also dependent on the surface finish condition, which in turn depends on the plating chemistry and process control, making its occurrence somewhat random, causing anxiety in the industry. The complex creep/fatigue interactions and damage mechanics will be discussed in detail in later chapters as well.

There have been many experimental studies on the reliability of lead-free solders for various components, including BGA, CCGA, CSP, PLCC, QFP, TSOP, PTH, 0201, and 2512 (for example, Ref 91–98), on a variety of PWB surface finishes (such as HASL, OSP, and ENIG). Under typical conditions, the SAC alloy is significantly more creep resistant than the tin-lead alloy due to differences in microstructure (such as fine Ag_3Sn phases in the matrix), and the SAC solder interconnect has been found to offer greater reliability than the tin-lead solder. However, this may not always be the case. The National Center for Manufacturing Sciences study in 1997 (Ref 16) has found that the ranking of alloys in terms of reliability relative to tin-lead varies with the thermal cycle conditions and component type. For example, using 1206 ceramic chip resistors on FR4 boards, when thermal-cycled between 0 and 100 °C (32 and 212 °F) (with 10 °C /min., or 50 °F/min. ramp rate and 5 min. dwell), the reliability of tin-lead was found to be similar to Sn-3.5Ag. However, when thermal-cycled between −55 and 125 °C, or 67 and 257 °F (with 11 °C /min., or 20 °F/min., ramp rate and 20 min. dwell), the reliability of tin-lead was found to be greater than Sn-3.5Ag. A recent study in 2002 (Ref 99) noted that below a certain stress threshold, the steady state creep strain rate is much lower for the SAC alloy than for the tin-lead alloy; however, above the threshold, the trend is reversed. This is believed to be the reason for the different reliability comparison between the SAC and tin-lead solders, under different thermal cycling conditions and for different components (Ref 5, 100). In most typical use conditions, the SAC solder is more reliable than the tin-lead solder. However, for large components and components with a large CTE mismatch with the substrate, and/or under severe thermal cycling conditions, the SAC alloy may not always be as reliable as the tin-lead solder. More details will be discussed in later chapters.

The increased use of handheld electronics products has posed new reliability concerns (Ref 1). Increased portability increases the risk of products being dropped. Reduction in product weight and size has led to smaller, thinner, and more compact housings, offering less protection to the components and solder interconnects on the PWB. Use of stacked dice and stacked packages increases the package mass and results in greater mechanical loading on the solder joints when the product is dropped. Smaller solder joints for miniaturization are more vulnerable to failures under dynamic mechanical loading. Based on these considerations, the dynamic mechanical reliability, especially for bending and drop, is considered a predominant failure mechanism for handheld products (Ref 101–103). This is particularly true for large area array packages such as CSP and BGA, and the failure modes from the drop test include cracking of the solder balls and of the copper trace. For lead-free solder interconnects, the dynamic mechanical reliability warrants further study.

The effect of macro-voids (especially in BGA/CSP solder joints) on solder joint reliability is a complex subject, as it depends on the size, location, and distribution of the voids as well as the mechanical loading conditions. For example, voids near the solder/component interface are believed to be more detrimental to solder joint reliability than those near the solder/PWB interface. The volume and size of the voids depend on the solder paste, BGA/CSP solder ball alloy, PWB surface finish and solderability, solder mask, micro-vias in pad, PWB moisture condition, and reflow profile. For example, better wetting generally results in less voids, which can explain the difference in voiding on different surface finishes. Furthermore, in terms of void location, PWBs with ENIG finish can more readily release the voids, which then float to the solder/component interface, whereas with the OSP finish, voids tend to remain at the solder/PWB interface due to the difference in wettability. The location of the voids changes during the second reflow, due to buoyancy. These complexities often cause large scatter in the reliability test data. A recent study by IPC suggests that neither the size or the number of macro-voids in a solder-joint appears to have any significant effect on its thermomechanical reliability under the test conditions (Ref 21). The impact of macro-voids on dynamic mechanical reliability, however, may be more pronounced.

Accelerated reliability tests are often used for predicting the useful life of the solder interconnects. The primary consideration for designing an accelerated testing profile is to maximize the damage to the solder interconnects per unit time (i.e., testing efficiency), while maintaining the failure mode and mechanism that are relevant to the actual application. The acceleration factor is needed to relate the results of the accelerated life tests to the actual field use reliability. The acceleration factor is different for SAC and tin-lead, because the acceleration factor depends on materials properties. Acceleration factors also depend on the test conditions (especially the dwell time), because of the different creep rates of SAC and tin-lead alloys.

Complicating the picture further is the use of different solder alloys on the same board (such as SAC for reflow and tin-copper for wave soldering). Detailed discussions on accelerated reliability testing methodology can be found in Chapter 7.

Failure analysis is useful for providing the failure locations, modes and mechanisms, verifying the reliability test data and reliability prediction, and providing insights into the physical, chemical, electrical, mechanical, and thermal behaviors of solder joints (Ref 104–105). The failure modes of lead-free solder interconnects are similar to those for the tin-lead solder (Ref 104), under various loading conditions. Techniques for failure analysis for lead-free solder joints will be presented in Chapter 10.

Design For Lead-Free Soldering and Environmental Compliance

Design for lead-free soldering and environmental compliance includes several aspects, such as: PWB layout design rules for lead-free solders; design for lead-free reliability; design to environmental requirements regarding hazardous substances; and design best practices for recycling.

No major changes in design rules are anticipated for the switch to lead-free soldering. For wave soldering of through-hole components (Ref 73, 96), some changes in design rules may be necessary to accommodate the difference in the physical properties (such as surface tension) between the lead-free and tin-lead solder alloys. The general guidelines, such as board orientation relative to the soldering direction, still apply to lead-free wave soldering.

For lead-free soldering, it is important to optimize board layout, component distribution and copper distribution in the PWB in order to minimize the temperature delta across the board, in order to keep the soldering temperature as low as possible, and to minimize the impact on the components. This is particularly important for large complex boards.

For reliability life prediction of solder interconnects through creep fatigue interactions under thermomechanical loading conditions, time and path dependent creep models are needed. A modified Coffin-Mason type equation is typically assumed to relate the number of cycles to failure to the creep strain energy density. Work is ongoing to evaluate the materials constants in the Coffin-Mason equation for SAC (Ref 106–107).

Time dependent inelastic finite element analysis (FEA) methodology provides an effective tool for design for reliability (DFR) of lead-free solder interconnects. Such DFR methodology through solder joint reliability prediction (Ref 93, 98, 108–109) must take into account the intrinsic (such as the resistance of the solder alloy to creep and fatigue) and extrinsic (such as the solder joint geometry, environmental and loading conditions, as well as board and package dimensions and elastic/thermal properties) factors of the interconnect system (Ref 100). Typically, the analysis is achieved through macro modeling of the entire assembly to identify the critical solder joint, and then micro modeling of the critical solder joint with prescribed displacement to determine the distribution of stress, strain, and/or strain energy density in the critical joint, and life prediction through the fatigue life models. Detailed discussions can be found in forthcoming chapters.

For environmental compliance, the selection of components which can meet the environmental regulation requirements (to be discussed in more detail in the following section) is a critical design consideration. Product design best practices for service, disassembly and recycling are also of great value to the industry.

Environmental Compliance Outlook

Environmental Compliance Requirements

Environmental compliance is becoming a global effort in the electronics industry. Lead-free is only part of RoHS requirements, and environmental compliance encompasses RoHS and WEEE requirements, as illustrated in Fig. 4.

Article 4 of European Union RoHS Directive (Ref 4) states that "Member States shall ensure that, from 1 July 2006, new electrical and electronic equipment put on the market does not contain lead, mercury, cadmium, hexavalent chromium, polybrominated biphenyls (PBB) or polybrominated diphenyl ethers (PBDE)." The European Commission has further proposed a draft whereby (Ref 110) "A maximum concentration value of 0.1% by weight in homogeneous materials for Pb, Hg, Cr^{+6}, PBB, and PBDE and of 0.01% by weight in homogeneous materials for Cd shall be allowed. Homogeneous material means a material that can not be mechanically disjointed into different materials." The term "homogeneous" is understood as of uniform composition throughout. Examples of "homogeneous materials" are individual types of plastics, ceramics, glass, metals, alloys, paper, board, resins, and coatings. The term "mechanically disjointed" means that the materials can be, in principle, separated by mechanical actions such as unscrewing, cutting, crushing, grinding, and abrasive processes. For example, a plastic cover is a "homogeneous material" if it consists of one type of plastic that is not coated with or has attached to it or inside it any other kinds of materials. In this case the limit values of the Directive would apply to the plastic. An electric cable that consists of metal wires surrounded by non-metallic insulation materials is an example of a "non-homogeneous material" because the different materials could be separated by mechanical processes. In this case, the limit values of the Directive would apply to each of the separated materials individually. A semiconductor package contains many "homogeneous materials" which include plastic molding material, Sn-electroplating coating on the leadframe, the leadframe alloy, gold bonding wires, etc.

The European Union WEEE (waste electrical and electronic equipment) legislation (Ref 111), several years in the making, mandates that, effective 2006, producers (i.e., companies whose names appear on the product) must be responsible for the take-back and disposition of their products at the end-of-life (EOL) of the product. The legislation is designed to tackle the fast increasing waste stream of electrical and electronic equipment. In Japan, the "Home Appliance Recycle Law" of 2001 mandates that 60% of the e-waste must be recycled. In the U.S., recycling programs are in place in many states and localities.

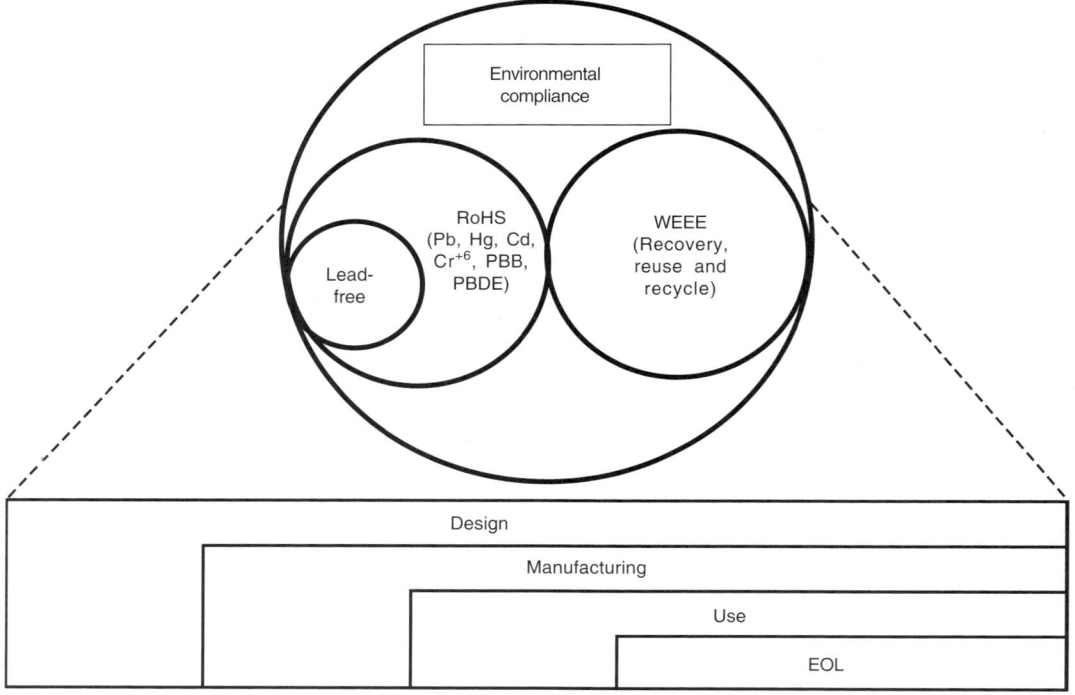

Fig. 4 Environmental compliance requirements

Electronics Recycling and End-of-Life Management

It is anticipated that WEEE will have a great impact on the industry. WEEE covers "equipment dependent on electric currents or electromagnetic field and equipment for generation, transfer, and measurement," including "components, sub-assemblies, and consumables, which are part of the product at the time of discarding." This is, indeed, very broad in scope. Producers are responsible for the take-back and disposition of products, including collection, treatment, recovery, and environmentally sound disposal. On average, no less than 70 to 75% (by weight) per appliance must be recovered, including no less than 50 to 65% through re-use and recycling. The financial impact on the industry is estimated to be billions of dollars per year.

Recycling and environmentally responsible EOL management are considered the ultimate solution to the environmental concerns posed by the ever increasing volume of electronic products (Ref 112). At the same time, increased recognition of the potential economic value of the "from cradle to cradle" approach to product EOL management through the reuse and repurposing of components and subassemblies through multiple product life cycles, provides additional impetus to accelerate efforts in this important business arena. OEMs, EMS companies, and recycling companies, along with consumers and national/local governments, all have important roles to play in the process.

The functional signature analysis methodology (Ref 113–116) offers a powerful tool for managing EOL. From the user perspective, failure is represented as an inability of the product, or system, to properly perform its function, or sub-functions, and a product of a roughly fixed life may in fact be composed of parts with varying lives. A "signature" means a recognizable function in the analysis that can be attributed to a particular performance deterioration. This means that there is a signature under normal operation, and a signature when a deterioration is present. The aim of the signature analysis method is to find an aggregate signature representing the functional performance during the life of the module, to have a single measure of the reliability performance of the whole module/function including the performance limits that can be used to identify the source of failures or performance deterioration, and to aggregate the measures from modules to produce a measure for the whole higher-level system, in a hierarchical way. Therefore, the functional signature analysis methodology aims at defining methods to extract the indicators (or signatures) by analyzing the functional performance over life and to have a single measure of the reliability performance to identify the source of failures and performance deterioration by signals available for external measurement, and predict the function performance degradation over life. By assessing the remaining life of the components and modules, the functional signature analysis methodology enables the conscious and managed re-use and re-purposing of functional components and modules (including electrical, mechanical, electronic, electromechanical, etc.) and their reintroduction into the product lifecycle.

This "from cradle-to-cradle" approach promises to fully exploit the total economic lifecycle of products. Often, a product is retired from the market, either because of functional failures of the product caused by one or more nonfunctional components, or because the product is no longer valued by the customer for whatever reasons, such as the introduction of a newer model. Hence, the valued lifetime of a product is often much shorter than its viable physical lifetime (or life expectancy). Even if the product is retired from the market because of true functional failures, a large number of components and modules in the product may still have considerable remaining lifetime. By assessing the remaining lifetime of the components and modules of a product at the end of its first life through the functional signature analysis, certain components and modules of the product can be appropriately recovered for new use (second life); such new use may include assembly into new products, knowing the remaining lifetime of the components and the target life of the new product.

Environmental Compliance Challenges

With the RoHS compliance deadline fast approaching for EEE products put on the market in the EU, the challenges ahead fall primarily in these areas: requirements, infrastructure, technology, and economics (Ref 117–120).

The enforcement and verification mechanisms of the RoHS requirements are far from being clear. There are still uncertainties regarding exemptions. Other countries (such as China) are in the process of establishing similar, but not necessarily identical, requirements. It is therefore difficult for the industry to comply with requirements which have not been fully defined. As such, original equipment manufacturers or OEMs (who, as producers, directly own the legal obligation for compliance) are coming up with their own, often different, interpretations and requirements.

In terms of the knowledge infrastructure, whisker growth acceptance criteria, as well as standards for component and product numbering and marking, materials declaration and verification for RoHS compliance (such as scope, measurement methodology, reporting, and compliance verification and certification), need to be established. Industry materials database and B2B (business-to-business) data exchange mechanisms for materials declaration, best practices, and design for environment tools are also needed to facilitate the transition to environmental compliance.

There are several challenges in terms of the physical infrastructure. Certain equipment (such as short reflow ovens, wave soldering machines, etc.) needs to be upgraded for lead-free soldering, as discussed in previous sections. Analytical instruments for environmental compliance verification need to be evaluated. This is no trivial issue, as the RoHS definition (currently pending final approval) will be on the level of homogeneous materials. As such, a component typically comprises a number of homogeneous materials, and RoHS compliance needs to be verified for each homogeneous material for each of the six substances, at levels as low as 100 ppm. It is likely that different instruments may be needed for different substances. The accuracy and repeatability, as well as cycle time and cost, need to be studied for different instruments, for shop floor applications (inspection), as well as laboratory applications (calibration and verification). Considering the number of components and the number of supply tiers involved in the entire manufacturing process of each component, this may become an overwhelming endeavor for the industry. As such, it is hoped that the industry effort for compliance assurance management will be based primarily on due diligence and certification for suppliers involved in each tier of the supply chain, instead of massive analytical verification on individual components and products.

Another infrastructural challenge for the industry is the readiness of the component supply chain. There has been great awareness of RoHS compliance requirements in the electronic component supply community. However, the RoHS compliance covers everything (with exemptions) that ends up on the EEE product, including mechanical components. Often, mechanical components are custom-made by small local vendors. The awareness and readiness for RoHS compliance needs to be raised in these areas. Managing component traceability, obsolescence and the gray market is also a challenging task.

There is also much work to be done in the technical arena. Design rules need to be further refined for lead-free soldering. Process optimization for high-yield volume manufacturing with lead-free solder is also a critical capability, considering the variety of designs, component types, PWB finishes, and process complexities. PWB material requirements for lead-free soldering need to be further studied, as materials with different characteristics (T_g, T_d, and CTE) may be needed, especially for large, thick, and complex boards. Simply changing to a high T_g laminate material may not always be the right answer, from economic and reliability considerations. The reliability of lead-free solders warrants further study as well, in terms of creep and fatigue properties, acceleration factors, under different loading conditions (thermomechanical, dynamic mechanical, temperature/relative humidity, etc.). Compliance verification methodology for environmental compliance is yet another technical area to be worked on. The practical reliability implications of the tin pest phenomenon (Ref 121) also remain a field for further research.

As one may expect, all of these challenges have economical implications. Our ultimate challenge is, therefore, to manage these challenges at the lowest cost for the industry. For example, in terms of cost, there is no simple formula to assess the precise cost impact for converting to lead-free soldering, because the field is still dynamic and fluid, and volume dependency is an important factor. Lead-free solder paste will cost more than the tin-lead solder paste; however, the differential is expected to decrease as the volume increases, because the metal cost is only a fraction of the solder paste cost. SAC bar solder and wire-core solder will cost several times as much as the tin-lead solder because of the higher metal cost for tin and silver, and the tin-copper solder is much cheaper than the SAC alloy. Components and possibly PWBs for higher soldering temperatures may also increase cost. Any special handling or baking for components will also add to cost. Yield is another factor which can significantly affect cost. Equipment upgrading, if needed, also contributes to the initial cost increase. Other cost factors are due to the complexity in managing the transition, training, materials handling and tracking.

Summary

A holistic approach is needed for environmental compliance in the electronics industry. Supply chain management, design for the environment (including design for environmental compliance, design for recycling, design for lead-free soldering, and design for lead-free reliability), and recycling and product end-of-life management, are all important components of the entire environmental compliance efforts.

Managing the compatibility issues is critical to lead-free transition. These include materials compatibility (solder, components and PWB), process compatibility (reflow, wave soldering, rework, equipment, and yield), design compatibility, reliability compatibility, and business compatibility (cost, supply chain, and operations).

Lead-free solder interconnect reliability is a critically important, yet complex, issue, for the industry. The entire interconnect system must be considered for lead-free reliability, including the component (thermal stability and Sn whisker growth), PWB, and the solder joints, under different environmental conditions (electrochemical, thermo-mechanical, and dynamic mechanical). Due to the complex creep and fatigue interactions (which are related to microstructural evolution), the reliability comparison between the tin-lead and lead-free solder interconnects is not straightforward. Accelerated testing methodology, failure analysis methodology, and reliability prediction methodology, for various alternatives to the tin-lead solder, are important tools which are urgently needed by the industry.

ACKNOWLEDGMENTS

The author would like to thank his colleagues across the industry worldwide, especially his past and current colleagues at Ford Motor Company, Visteon Corporation, and Flextronics International, for their support, collaboration, and insightful discussions in the area of lead-free soldering and environmental compliance, since 1991.

REFERENCES

1. D. Shangguan, Packaging & Board Assembly Technology Trend and Impact on the Supply Chain, (Keynote) *Proceedings of the 6th IEEE CPMT Conference on High Density Microsystem Design and Packaging and Component Failure Analysis (HDP'04),* June 2004, p 14–17
2. U.S. Department of Health and Human Services, Toxicology Profile for Lead, Clement International Corporation, April 1993
3. M. Costic, J. Sullivan, B. Bryant, and D. Shangguan, LCI for Automotive Electronic Systems: Substitution Assessment of Ag-Sn for Pb-Sn Solder, *Proceedings of the International Symposium on Electronics and the Environment,* May 1996, p 58–63
4. Directive 2002/95/EC of the European Parliament and of the Council of 27 Jan 2003 on the Restriction of the Use of Certain Hazardous Substances in Electrical and Electronic Equipment, *Off. J. Eur. Union,* 13,2,2003, L37/19
5. D. Shangguan, Study of Compatibility for Lead-free Solder PCB Assembly, *Proceedings of the IPC and Soldertec First International Conference on Lead-Free Electronics,* June 2003, p 298–308
6. D. Shangguan, Key Topics in Lead-Free Solder PCB Assembly, *Proceedings of NEPCON* 2003, p 111–121
7. D. Shangguan, Managing Compatibility Issues For Transitioning To Volume Manufacturing For Lead-Free PCB Assembly, *Proceedings of 2003 International Printed Circuit & Electronics Assembly Fair Technical Conference and Exhibition,* Dec 10–12, 2003
8. D. Shangguan, Managing Lead-Free Compatibility, *Circuits Assem.,* Nov 2003, p 30–33
9. D. Shangguan, Challenges & Opportunities in Lead-Free Solder PCB Assembly, *Global SMT & Packaging,* Oct 2002, p 18–24
10. D. Shangguan, Leading the Lead-Free Transition, *Circuits Assem.,* March 2004
11. D. Shangguan, Understanding Compatibility and Clarifying Issues in Lead-Free

Transition, *Electronics Manufacturing China,* April 2004, p 20–24
12. D. Shangguan and A. Achari, Evaluation of Lead-Free Eutectic Sn-Ag Solder for Automotive Electronics Packaging Applications, *Proceedings of the International Electronics Manufacturing Technology Symposium,* Sept 1994, p 25–37
13. D. Shangguan and A. Achari, Lead-Free Solder Development for Automotive Electronics Packaging Applications, *Proceedings of the Surface Mount International Conference,* Aug 1995, p 423–428
14. D. Shangguan and G. Gao, Environmentally Conscious Manufacturing Technologies for Automotive Electronics, *Proceedings of the Second International Symposium on Electronic Packaging Technology (ISEPT '96),* Dec 1996, p 391–402
15. D. Shangguan and G. Gao, Lead-Free & No-Clean Soldering for Automotive Electronics, *Solder. Surf. Mt. Technol.,* 26 (1997), p 5–8
16. National Center for Manufacturing Sciences (NCMS), Lead-Free Solder Project–Final report, Aug 1997
17. A. Rae and C. Handwerker, NEMI's Lead-Free Alloy: Still on Target, *Circuits Assem.,* April 2004, p 20–25
18. J.S. Hwang, Environment-Friendly Electronics: Lead-Free Technology, Electrochemical Publications Ltd., 2001
19. J. Lau, C.P. Wong, N.C. Lee, and S.W.R. Lee, Electronics Manufacturing With Lead-Free, Halogen-Free and Conductive Adhesive Materials, McGraw-Hill, 2003
20. S. Ganesan and M. Pecht, Lead-Free Electronics, CALCE EPSC Press, 2003
21. IPC Solder Products Value Council White Paper, Round Robin Testing and Analysis of Lead Free Alloys: Tin, Silver, Copper, http://leadfree.ipc.org/LeadFreeWP006.asp, 2005
22. A. Achari, M.R. Paruchuri, and D. Shangguan, Lead-Free Solder Compositions, U.S. patent 5,863,493 (1999), and European Patent 0847829
23. J. Oliver, M. Nylen, O. Rod, and C. Markou, Fatigue Properties of Sn3.5Ag0.7Cu Solder Joints and Effects of Pb-Contamination, Journal of SMT, Vol 15, Issue 4, 2002, p 23–29
24. D. Shangguan and A. Achari, Lead-Free Solder Alloys, U.S. Patent 5,429,689 (1995), and Mexico Patent 196053
25. M. Paruchuri and D. Shangguan, Lead-Free Electrical Solder and Method of Manufacturing. U.S. Patent 6,360,939 (2002)
26. M.R. Paruchuri and D. Shangguan, Low Temperature Lead-Free Solder Compositions, U.S. Patent 5,755,896 (1998)
27. M.R. Paruchuri and D. Shangguan, Lead-Free, Low Temperature Solder Compositions, U.S. Patent 5,833,921 (1998)
28. W. Ren, M. Lu, S. Liu, and D. Shangguan, Thermal Mechanical Property Testing of Lead-Free Solder Joints, *Sold. Surf. Mt. Technol.,* 9(3) (1997), p 37–40
29. W. Ren, M. Lu, S. Liu, and D. Shangguan, A Study of a Lead-Free Solder Joint, Extended Abstract, *Proceedings of 1997 Society of Experimental Mechanics (SEM) Spring Conference on Experimental Mechanics & Experimental/Numerical Mechanics in Electronic Packaging,* Bellevue, WA, June 1997, p 30–31
30. W. Ren, Z. Qian, M. Lu, S. Liu, and D. Shangguan, Thermal Mechanical Properties of Two Solder Alloys, *Proceedings of the 1997 ASME International Mechanical Engineering Congress & Exposition,* EEP-Vol 22, AMD-Vol 226, Nov 1997, p 125–130
31. M. Lu, W. Ren, S. Liu, and D. Shangguan, A Unified Multi-Axial Sub-Micron Fatigue Tester with Applications to Electronic Packaging Materials, *Proceedings of the 47th Electronic Components & Technology Conference,* May 1997, p 144–148
32. M. Lu, Z. Qian, W. Ren, S. Liu, and D. Shangguan, Investigation of Electronic Packaging Materials by Using a 6-Axis Mini Thermo-Mechanical Tester, *Int. J. Solids Struct.,* 36(1) (1999), p 65–78
33. W. Ren, Z. Qian, M. Liu, S. Liu, and D. Shangguan, Investigation of a New Lead-Free Solder Alloy Using Thin Strip Specimen, *ASME Transactions Journal of Electronic Packaging, 121* (1999), p 271–274
34. Visteon Announces Production of the Automotive Industry's First Lead-Free Solder Electronic Module, Visteon Press Release, Dec 18, 2000
35. S. Chada, A. Herrmann, W. Laub, R. Fournelle, D. Shangguan, and A. Achari, Microstructural Investigation of Sn-Ag and Sn-Pb-Ag Solder Joints, *Solder. Surf. Mt. Technol.,* 26 (1997), p 9–13

36. D. Shangguan, A. Achari, and W. Green, Application of Lead-Free Eutectic Sn-Ag Solder in No-Clean Thick Film Electronic Modules, *IEEE Transactions on Components, Packaging, and Manufacturing Technology—Part B*, 17(4) (1994), p 603–611
37. S. Chada, W. Laub, R. Fournelle, and D. Shangguan, Microstructural Evolution of Sn-Ag Solder Joints Resulting from Substrate Copper Dissolution, *Proceedings of SMTA Conference*, Sept 1999, p 412–418
38. S. Chada, A. Hermann, W. Laub, R. Fournelle, D. Shangguan, and A. Achari, Microstructural Investigation of Sn-Ag and Sn-Pb-Ag Solder Joints, *Proceedings of NEPCON West '96*, Feb 1996, p 195–205
39. J.R. Lloyd, C. Zhang, H.L. Tan, D. Shangguan, and A. Achari, Measurement of Thermal Conductivity and Specific Heat of Lead-Free Solder, *Proceedings of the 1995 IEEE/CPMT International Electronics Manufacturing Technology Symposium*, Oct 1995, p 252–262
40. S. Chada, R.A. Fournelle, and D. Shangguan, An Improved Numerical Method for Predicting Intermetallic Layer Thickness Developed During the Formation of Solder Joints on Cu Substrates, *J. Electron. Mater.*, 28(11) (1999), p 1194–1202
41. V. Igoshev, J. Kleiman, D. Shangguan, C. Lock, S. Wong, and M. Wiseman, Microstructural Changes in Sn-3.5Ag Solder Alloy During Creep, *J. Electron. Mater.*, 27(12) (1998), p 1367–1371
42. S. Chada and D. Shangguan, Microstructural Evolution of Sn-Ag Based Lead-Free Solders, Chapter 22 in: "Environment-Friendly Electronics: Lead-Free Technology" (by J.S. Hwang), Electrochemical Publications Ltd., 2001, p 513–565
43. V.I. Igoshev, J.I. Kleiman, D. Shangguan, S. Wong, and U. Michon, Fracture in Sn-3.5Ag Solder Alloy Under Creep, *J. Electron. Mater.*, Vol 29(12) (2000), p 1356–1361
44. S. Chada, R.A. Fournelle, W. Laub, and D. Shangguan, Copper Substrate Dissolution in Eutectic Sn-Ag Solder and Its Effect on Microstructure, *J. Electron. Mater.*, Vol 29(10) (2000), p 1214–1221
45. D. Shangguan, Conductor Systems for Thick Film Electronic Circuits, U.S. Patent 6,476,332 (2002)
46. D. Geiger, M. Arra, and D. Shangguan, Evaluation of Sn-Zn Solder Paste, *Proceedings of IPC/JEDEC 2004 International Conference on Lead-Free Electronic Assemblies and Components*, March 2004
47. D. Shangguan, M.R. Paruchuri, and A. Achari, Method of Forming Interconnections on Electronic Modules, U.S. Patent 6,082,610 (2000)
48. D. Shangguan and R.J. Gordon, Solder Composition and a Method For Making the Same, U.S. Patent 6,416,597 (2002)
49. D.A. Geiger and D. Shangguan, Investigation of Solder/Flux Residue Effect on RF Signal Integrity Using Real Circuits, *Proceedings of the SMTA International Conference*, 2003. Also to appear in *Solder. Surf. Mt. Technol.*
50. G. Gao, Private communication (2004)
51. R. Gordon, S. Marr, and D. Shangguan, Evaluation of Immersion Silver PWB Finish for Automotive Electronics, *Proceedings of SMTA International Conference*, Sept 2000, p 583–591
52. M. Arra, D. Shangguan, and D. Xie, Wetting of Fresh and Aged Lead-Free PCB Surface Finishes by Sn-Ag-Cu Solder, *Proceedings of APEX 2003*, March 2003, p S12-2-1/7. Also, *Journal of the Hong Kong Printed Circuit Association*, 2004, p 22–29
53. M. Arra, D. Shangguan, J. Sundelin, T. Lepistö, and E. Ristolainen, Aging Mechanisms of Immersion Tin and Silver PCB Surface Finishes in Lead-Free Solder Applications, *Proceedings of the 3rd IPC/JEDEC Annual Conference on Lead-Free Electronic Assemblies and Components*, 2003, p 170–179
54. M. Arra, D. Shangguan, D. Xie, J. Sundelin, T. Lepistö, and E. Ristolainen, Study of Immersion Silver and Tin PCB Surface Finishes in Lead-Free Solder Applications, *Proceedings of the IPC and Soldertec First International Conference on Lead-Free Electronics*, June 2003, p 423–446. Also in *J. Electron. Mater.*, Vol 33 (No. 9), 2004, p 977–990
55. M. Arra, T. Castello, D. Shangguan, and E. Ristolainen, Characterization of Mechanical Performance of Sn/Ag/Cu Solder Joints With Different Component Lead Coatings, *Solder. Surf. Mt. Technol.*, Vol 16(1) 2004, p 35–43. Also, *Proceedings of SMTA International Conference*, 2003
56. D. Geiger, D. Shangguan, and S. Y., Ther-

mal Study of Lead-Free Reflow Soldering Processes, *Proceedings of the 3rd IPC/JEDEC Annual Conference on Lead-Free Electronic Assemblies and Components,* 2003, p 95–98
57. White Paper, Guidelines for Suppliers Transitioning to RoHS Compliant Components (Rev. 1.1), Electronics Manufacturing Services (EMS) Forum on Lead-Free PCB Assembly, 2004
58. IPC/JEDEC J-STD-020C, Moisture/Reflow Sensitivity Classification for Nonhermetic Solid State Surface Mount Devices, July 2004
59. M. Arra, D. Geiger, D. Shangguan, S. Yi, F. Grebenstein, H. Fockenberger, K.H. Kerk, and H. Wong, Performance Evaluation of Lead-Free Solder Paste, *Proceedings of SMTA Conference,* Oct. 2001, p 850–857
60. M. Arra, D. Shangguan, E. Ristolainen, and T. Lepisto, Solder Balling of Lead-Free Solder Paste, *J. Electron. Mater.,* Vol 31 (No. 11), 2002, p 1130–1138
61. M. Arra, D. Shangguan, E. Ristolainen, and T. Lepistö, Effect of Reflow Profile on Wetting and Intermetallic Formation Between Sn/Ag/Cu Solder, Components and Printed Circuit Boards, *Solder. Surf. Mt. Technol.,* Vol 14(2), 2002, p 18–25
62. M. Arra, D. Geiger, D. Shangguan, and J. Sjoberg, Study of SMT Assembly Processes for Fine Pitch CSP Packages, *Solder. Surf. Mt. Technol.,* Vol *16(3),* 2004. Also, *Proceedings of APEX 2004*
63. D. Geiger, F. Mattsson, D. Shangguan, M. Ong, P. Wong, M. Wang, T. Castello, and S. Yi, Process Characterization of PCB Assembly Using 0201 Packages With Lead-Free Solder, *Solder. Surf. Mt. Technol.,* 2003, p 22–27. Also, *Proceedings of NEPCON West,* Dec 2002
64. G. Pfennich, H. Fockenberger, L.C. Tat, T. Ho, and D. Shangguan, Board Design and Process Optimization for Paste-in-Hole Using Lead-Free Solder, *Proceedings of the SMTA International Conference,* Sept 2004
65. J. Sjoberg, D.A. Geiger, D. Shangguan, and T. Castello, Alternative Assembly Methods for Lead-Free Solder Flip Chips on FR-4 Substrates, *Proceedings of IMAPS Nordic,* 2004
66. F. Mattsson, D.A. Geiger, D. Shangguan, and T. Castello, PCB Design and Assembly Process Study of 01005 Size Passive Components Using Lead-Free Solder, *Proceedings of SMTA International Conference,* Sept 2004
67. D.A. Geiger, J. Sjoberg, D. Shangguan, and T. Castello, Lead-Free Solder Flip Chips on FR-4 Substrates with Different PCB Surface Finishes, Underfills and Fluxes, *Proceedings of Semicon West 2004,* p 31–36
68. D. Geiger, D. Shangguan, S. Tam, and D. Rooney, Package Stacking in SMT for 3D PCB Assembly, *Proceedings of IEEE IEMT 2003,* Paper S207P6, July 2003
69. Y. Liu, D.A. Geiger, and D. Shangguan, Determination of Components' Candidacy for Second Side Reflow With Lead-Free Solder. *Proceedings of ECTC 2005,* May 2005, p 970–976
70. IPC-A-610D, Acceptability of Electronic Assemblies, Feb 2005
71. IPC-A-610C, Acceptability of Electronic Assemblies, Amendment 1, Nov 2001
72. M. Arra, D. Geiger, D. Shangguan, and J. Sjoberg, Study of Component Self-Alignment During Reflow Using Sn/Pb And Sn-Ag-Cu Solders, *Proceedings of IMAPS Nordic 2004*
73. M. Arra, D. Shangguan, S. Yi, R. Thalhammer, and H. Fockenberger, Development of Lead-Free Wave Soldering Process, *IEEE CPMT Transactions on Electronics Packaging Manufacturing,* Vol 25(4) (2002), p 289–299. Also, *Proceedings of APEX 2002,* Jan 2002, p 4(1)–4(10)
74. D. Shangguan, Reworking Lead-Free Solder in PCB Assembly, *SMT,* June 2003, p 38
75. D. Geiger, J. Yu, and D. Shangguan, Development of Assembly & Rework Processes for Large and Complex PCBs Using Lead-Free Solder, *Global SMT & Packaging,* April 2004, p 16–19. Also, *Proceedings of APEX 2004*
76. D. Geiger, T. Castello, and D. Shangguan, X-Ray Inspection of Area Array Packages Using Tin-Lead and Lead-Free Solders, *Proceedings of SMTA International Conference,* Sept 2004
77. JEDEC Standard JESD22A121, Measuring Whisker Growth on Tin and Tin Alloy Surface Finishes, May 2005
78. G.T. Galyon and R. Gedney, Avoiding Tin Whisker Reliability Problems, *Circuits Assem.,* Aug 2004, p 26

79. J.H. Lau and S.H. Pan, 3D Nonlinear Stress Analysis of Tin Whisker Initiation on Lead-Free Components, *Journal of Electronics Packaging,* Vol 125 (2003), p 621–624
80. N. Vo, I. Boguslavsky, and P. Bush, NEMI Recommends Standard Test Methods to Assess Propensity for Tin Whisker Growth, *SMT,* Nov 2003, p 36–41
81. F. Hua, R. Aspandiar, T. Rothman, C. Anderson, G. Clemons, and M. Klier, Solder Joint Reliability of Sn-Ag-Cu BGA Components Attached With Eutectic Pb-Sn Solder Paste, *Proceedings of SMTA International Conference,* Sept 2002
82. T. Gregorich and P. Holmes, Low-Temperature, High-Reliability Assembly of Lead-Free CSPs, *Proceedings of IPC/Soldertec Lead Free Electronic Components and Assemblies,* Oct 2003
83. J. Smetana, Plated Through Hole Reliability With High Temperature Lead Free Soldering, *"The Board Authority"* (Technical Supplement to *CircuiTree* magazine), Vol 4, No. 1, April 2002, p 50–64
84. J.J. Davignon and R. Reed, Effects of NEMI Sn/Ag/Cu Alloy Assembly Reflow on Plated Through Hole Performance, *Journal of SMT,* Oct 2000, p 23–30
85. D. Leys and S.P. Schaefer, PWB Dielectric Substrates for Lead-Free Electronics Manufacturing, *CircuiTree,* Aug 1, 2003 http://www.circuitree.com/CDA/Article-Information/features/BNP_Features_Item/
86. E. Kelley, An Assessment of the Impact of Lead-Free Assembly Processes on Base Material and PCB Reliability, *Proceedings of APEX 2004,* Feb 2004
87. White Paper, Guidelines for PCB Suppliers Transitioning to Lead-Free Soldering (Rev 1.0), Electronics Manufacturing Services (EMS) Forum on Lead-Free PCB Assembly, Oct 2003
88. L. Turbini, W.R. Bent, and W.J. Ready, Impact of Higher Melting Lead-Free Solders on the Reliability of Printed Wiring Assemblies, *Journal of SMT,* Oct 2000, p 10–14
89. A. Brewin, L. Zou, and C. Hunt, Susceptibility of Glass-Reinforced Epoxy Laminates to Conductive Anodic Filamentation. *NPL Report* MATC(A)155, Jan 2004 http://libsvr.npl.co.uk/npl_web/pdf/matc155.pdf
90. K. Karavakis and S. Bertling, Conductive Anodic Filament (CAF): The Threat to Miniaturization of the Electronics Industry, *CircuiTree,* Dec 2004, p 70–73
91. D. Shangguan, Analysis of Crack Growth in Solder Joints, *J. Solder. Surf. Mt. Technol., 11(3)* (1999), p 27–32
92. D. Xie, D. Geiger, M. Arra, D. Shangguan, and H. Phan, Reliability of CSP/Lead-Free Solder Joints with Different PCB Surface Finishes and Reflow Profiles, *Proceedings of SEMICON West/IEMP Conference,* July 2002, p 323–328
93. J. Smetana, R. Horsley, J. Lau, K. Snowdon, D. Shangguan, J. Gleason, I. Memis, D. Love, and B. Sullivan, Design, Materials and Process for Lead-Free Assembly of High Density Packages, *Solder. Surf. Mt. Technol.,* Vol 16(1) 2004, p 53–62
94. J. Lau, N. Hoo, R. Horsley, J. Smetana, D. Shangguan, W. Dauksher, D. Love, I. Memis, and B. Sullivan, Reliability Testing and Data Analysis of High-Density Packages' Lead-Free Solder Joints, *Solder. Surf. Mt. Technol.,* Vol 16(2) 2004, p 46–68. Also, *Proceedings of APEX 2003,* March 2003, p S42-3-1/24
95. J. Lau, W. Dauksher, E. Ott, D. Shangguan, T. Castello, J. Smetana, R. Horsley, D. Love, I. Memis, and B. Sullivan, Reliability Testing and Data Analysis of an 1657CCGA Package with Lead-Free Solder Paste on Lead-Free PCBs, *Proceedings of the 54th IEEE ECTC,* June 2004 p 718–725
96. J. Lau, D. Shangguan, W. Dauksher, D. Khoo, G. Fan, W. Loong-Fee, and M. Sanciaume, Lead-Free Wave-Soldering and Reliability of a Light-Emitting Diode (LED) Display, *Proceedings of the IPC and Soldertec First International Conference on Lead-Free Electronics,* June 2003, p 116–124
97. D.A. Geiger, M. Wang, D. Shangguan, T. Castello, and F. Mattson, Reliability Study of Solder Joints for 0201 Components, *Proceedings of SMTA International Conference,* 2003
98. J. Smetana, R. Horsley, J. Lau, K. Snowdon, D. Shangguan, J. Gleason, I. Memis, D. Love, and B. Sullivan, Lead-Free Design, Materials and Process of High Density Packages, *Proceedings of APEX 2003,* Anaheim, CA, March 2003, p S42-1-1/13
99. D. Xie, M. Arra, H. Phan, D. Shangguan, D. Geiger, and S. Yi, Life Prediction of Lead-free Solder Joints for Handheld

Products, *Proceedings of the Telecomm Hardware Solutions Conference & Exhibition*, SMTA/IMAPS, May 2002, p 83–88
100. J. Liang, D. Shangguan, and S. Downes, Effects of Load and Thermal Conditions on Lead-Free Solder Joint Reliability, *Symposium Proceedings of Lead-Free Solders and Processing Issues Relevant to Microelectronic Packaging,* TMS Annual Meeting, 2004. Also in *J. Electron. Mater.,* Special Issue, Vol 33 (No. 12), Dec 2004, p 1507–1515
101. M. Arra, D. Xie, and D. Shangguan, Performance of QFP Lead-Free Solder Joints Under Dynamic Mechanical Loading, *International Journal of Microcircuits and Electronic Packaging,* Vol 24(4), 2001, p 346–365
102. M. Arra, D. Xie, and D. Shangguan, Performance of Lead-Free Solder Joints Under Dynamic Mechanical Loading, *Proceedings of ECTC 2002,* May 2002
103. D. Xie, D. Geiger, D. Shangguan, D. Rooney, and L. Gullo, Characterization of Fine Pitch CSP Solder Joints Under Board-Level Free Fall Drop. *Proceedings of ASME InterPack 2005,* San Francisco, CA, May 2005
104. J. Lau, D. Shangguan, T. Castello, Ro Horsley, J. Smetana, W. Dauksher, D. Love, I. Memis, and B. Sullivan, Failure Analysis of Lead-Free Solder Joints for High-Density Packages, *Solder. Surf. Mt. Technol.,* Vol *16(2)* 2004, p 69–76. Also, *Proceedings of APEX 2003,* March 2003, p S42-4-1/11
105. J. Lau, T. Castello, D. Shangguan, W. Dauksher, J. Smetana, R. Horsley, D. Love, I. Memis, and B. Sullivan, Failure Analysis of Lead-Free Solder Joints for a 1657 CCGA (Ceramic Column Grid Array) Package, *Proceedings of IMAPS 2004,* Nov 2004
106. J.H. Lau, D. Shangguan, D.C.Y. Lau, T.T.W. Kung, S.W.R. Lee, Thermal-Fatigue Life Prediction Equation for Wafer-Level Chip Scale Package (WLCSP) Lead-Free Solder Joints on Lead-Free Printed Circuit Board (PCB), *Proceedings of the 54th IEEE ECTC,* June 2004, p 1563–1569
107. J.H. Lau, S.W. Ricky Lee, D. Shangguan, D. Lau, and T. Kung, Thermal-Fatigue Life Prediction Equations for Lead-Free Solder Joints on Lead-Free Printed Circuit Board (PCB), *Proceedings of IMAPS 2004,* Nov 2004
108. J. Lau, W. Dauksher, J. Smetana, R. Horsley, D. Shangguan, T. Castello, I. Memis, D. Love, and R. Sullivan, Design for Lead-Free Solder Joint Reliability of High-Density Packages, *Solder. Surf. Mt. Technol.,* Vol *16(1)* 2004, p 12–26. Also, *Proceedings of APEX 2003,* March 2003, p S42-2-1/13
109. J. Lau, R. Lee and D. Shangguan, Thermal Fatigue Life Prediction of Lead-Free Solder Joints. *Proceedings of ASME International Mechanical Engineering Conference & Exposition,* Nov 13–19, 2004
110. European Commission, Frequently Asked Questions on RoHS and WEEE, May 2005
111. Directive 2002/96/EC of The European Parliament and of The Council of 27 Jan 2003 on Waste Electrical and Electronic Equipment (WEEE), *Off. J. Eur. Union,* 13.2.2003, L 37/24
112. D. Shangguan, Electronics Recycling and End of Life Management, *SMT,* Nov 2003, p 48
113. J.A. van den Bogaard, G. Hulsken, D. Shangguan, A.C. Brombacher, J.S.J. Jayaram, and R.A. Ion, Using Dynamic Reliability Models to Extend the Economic Life of Strongly Innovative Products Through Re-Use, *Proceedings of 2004 IEEE International Symposium on Electronics and the Environment*
114. G. Hulsken, A.C. Brombacher, J.A. van den Bogaard, H.P. Wynn, A. Di Bucchianico, B. Peeters, T. Theunissen, and D. Shangguan, Functional Signature Analysis for Product End-of-Life Management, *Proceedings of Recycling Electrical and Electronic Equipment Conference,* Nov 2003
115. G. Hulsken, A.C. Brombacher , J.A. van den Bogaard, H.P. Wynn, A. Di Bucchianico, B. Peeters, T. Theunissen, and D. Shangguan, End-of-Life Management of Electronics Products Through Functional Signature Analysis, *Proceedings of APEX 2004,* Feb 2004
116. G. Hulsken, H.P. Wynn, A. Di Bucchianico, J.A. van den Bogaard, N.A. Mushkudiani, and D. Shangguan, Functional Signature Analysis for Product End-of-Life Management. *Proceedings of the 3rd IPC/JEDEC Annual Conference on Lead-Free Electronic Assemblies and Components,* 2003, p 233–238

117. D. Shangguan, A Holistic Approach to Lead-Free Transition and Environmental Compliance, *Proceedings of SMTA 2004,* Chicago, IL
118. D. Shangguan, Environmental Leadership in Electronics Manufacturing: Lead-Free and Beyond, *Proceedings of 2004 IEEE International Symposium on Electronics and the Environment,* May 2004
119. D. Shangguan, Strategy for Environmental Leadership in Electronics Manufacturing: Lead-Free and Beyond, *Proceedings of the High Level SMT Conference,* April 2004
120. D. Shangguan, Lead-Free Transition and Environmental Compliance: Challenges Ahead for the Industry, *Global SMT & Packaging,* Sept 2004
121. Y. Kariya, C. Gagg, and W.J. Plumbridge, Tin Pest in Lead-Free Solders, *Solder. Surf. Mt. Technol.,* Vol *13(1)* (2000), p 39–40

CHAPTER 2

Microstructural Evolution and Interfacial Interactions in Lead-Free Solder Interconnects

Zequn Mei, Cisco Systems, Inc.

Introduction

The microstructure inside a solder joint and at the interface between the solder joint and the substrate determines the mechanical properties of the solder joint. The soldering process and subsequent solid state aging and thermal cycling determine the initial microstructure and its evolution. The reaction between the solder and the substrate at their interface is important, because it affects wetting, joint strength, and reliability. While the formation of an intermetallic layer is desired for wetting and metallurgical bonding, excessive interfacial reaction may degrade the solder joint integrity and reliability.

This chapter reviews literature on the microstructure of solder joints and the interactions of solders with substrates and related solder joint reliability issues. The substrates here are limited to Cu, Ni-coated Cu, electroless nickel/immersion gold (ENIG), and hot air solder leveled (HASL) Sn-Pb. The solders are mainly Sn, eutectic or near eutectic Sn-Ag, and Sn-Ag-Cu alloys. Specific reliability issues discussed here include "black pads" of ENIG, gold embrittlement, compatibility of Pb-free solders with Pb-containing surface finish, and Kirkendall voids.

Microstructural Evolution in Pb-Free Solders

Phase Diagram and Equilibrium Solidification

Several studies have been reported on the phase equilibria in the Sn-Ag-Cu system. The ternary eutectic has been determined to be at a composition of Sn-3.5Ag-0.9Cu (all percentages by weight) and a temperature of 217.2 °C (423 °F) (Ref 1). The equilibrium solidification of a Sn-Ag-Cu alloy near the ternary eutectic composition consists of three stages: primary, secondary, and tertiary, and involves the liquid, L, and three solid phases, β-Sn, Ag_3Sn, and Cu_6Sn_5. If the three solid phases are denoted generically as S1, S2, and S3, the primary stage is $L \Rightarrow S1$, and the temperature at which the primary solid phase S1 starts to form, is the liquidus temperature. S1 can be any of the three solid phases, β-Sn, Ag_3Sn, or Cu_6Sn_5, depending on the composition of the alloy relative to the ternary eutectic composition. As the temperature decreases, the secondary solidification starts, $L \Rightarrow S1 + S2$, followed by the tertiary stage of the solidification, $L \Rightarrow S1 + S2 + S3$, at the ternary eutectic temperature, 217.2 °C (423 °F).

The liquidus surface, i.e., the temperature versus the composition for the primary stage of the solidification, has been determined through experiments and thermodynamic calculations (Ref 1–3). The surface of the secondary solidification, i.e., the temperature versus the composition for the $L \Rightarrow S1 + S2$ reaction, has also been presented (Ref 1). The isothermal sections of the Sn-Ag-Cu phase diagram have been determined for the temperatures at 270, 240, 223, and 219 °C (518, 464, 433, and 426 °F) (Ref 1), for the temperature at 400 and 600 °C (752 and 1112 °F) (Ref 2), and for the temperatures at 240 and 450 °C (404 and 842 °F) (Ref 3).

Nucleation and Growth

The phase diagrams described previously tell us only the phases formed during the equilibrium solidification. The microstructure, defined here as the size, shape, and distribution of the solid phases, is determined by the kinetics, i.e., the nucleation and growth mechanisms during solidification.

Nucleation. The nucleation mechanism in the eutectic Sn-Pb is different from that in the Sn-Ag-Cu eutectic. In the eutectic Sn-Pb, the solidification starts with the homogeneous nucleation of the Pb-rich phase (Ref 4) or heterogeneous nucleation of Pb on Cu or Ni in the case of a solder joint (Ref 5, p 65–67), on the eutectic solidification process that the required undercooling for Pb to nucleate on a Cu or Ni surface was only 2 °C (35 °F). Subsequently, the Sn-rich phase nucleates heterogeneously on the previously solidified Pb-rich phase. The Sn-rich phase wets on the Pb-rich phase with the wetting angle of approximately 61 to 66°. The catalytic effect of the Pb-rich phase for the nucleation of the Sn-rich phase reduces the undercooling for the solidification, and provides a starting point for the two phases to grow side-by-side.

On the other hand, in the eutectic Sn-Ag-Cu system, the undercooling for the nucleation of the solid Ag_3Sn phase is small, 7.2 °C (45 °F) (Ref 6) with differential scanning calorimetry (DSC) in the Sn-3.8Ag-0.7Cu alloy. But the formation of the Ag_3Sn does not facilitate the nucleation of the β-Sn phase. The β-Sn does not wet on the previously formed Ag_3Sn, therefore; it will not nucleate heterogeneously on the Ag_3Sn. The undercooling for the β-Sn phase is 29 °C (84 °F) in the same alloy as found in Ref 6. It was observed in Ref 1 that the large undercooling of the β-Sn phase in the presence of large Ag_3Sn intermetallic particles, and the same conclusion was reached, that the intermetallics particles are ineffective as heterogeneous nucleation substrates for Sn.

Growth. The growth mechanisms of eutectic Sn-Pb and eutectic Sn-Ag-Cu are also different. In the eutectic solidification process (Ref 5), the eutectic alloys are divided into two types: normal and anomalous; 63Sn-37Pb is a normal eutectic, while the Pb-free eutectic alloys (eutectic Sn-Ag, Sn-Cu, and Sn-Ag-Cu) are anomalous.

Normal alloys are associated with regular microstructures of a lamellar or rod form. During their solidification process, the constituent phases of a normal eutectic alloy grow into the liquid phase in a coupled or coordinated mechanism, which usually requires that the growth rates of the constituent phases are the same, and their growth mechanisms are non-faceting. For a normal binary eutectic alloy, two metals usually have similar melting points, and the two constituent phases have the same volume fraction.

In the case of eutectic Sn-Pb, the melting points of Sn and Pb are 232 and 327 °C (450 and 621 °F), respectively. The volume fractions of the Sn-rich and Pb-rich phases are approximately 70 and 30%. Both phases are non-faceting. During their solidification process, the Pb-rich and Sn-rich phases grow into the liquid phase side by side. The Pb atoms rejected by the solidified Sn-rich phase, or the Sn atoms rejected by the solidified Pb-rich phase, diffuse at the front of the solid/liquid interface, toward the Pb-rich phase or the Sn-rich phase, respectively. This side-by-side advance of the two individual eutectic phases and the short range redistribution of the solutes between the two phases leads to lamellar or rod arrays with a fine spacing. The two phases in their lamellar array maintain certain crystallographic orientation with respect to each other (Ref 5), e.g., (010) Sn // (111) Pb // lamellar interface plane, and [211] Sn // [211] Pb // growth direction. The epitaxial crystallographic relationship between the Sn-rich and Pb-rich lamellar phases is probably the result of the minimization of the interfacial energy and maximization of the growth rate. The advance of lamellar arrays into the liquid phase continues until hindered by other lamellae. An area with the same lamellar orientation is called a colony. A colony is actually a grain, even if it contains two phases.

The eutectic Sn-3.5Ag, Sn-0.7Cu, and Sn-3.5Ag-0.9Cu alloys, on the other hand, are anomalous. The difference in the melting point between Sn and Ag (or Cu) is large; the difference in the volume fraction of the constituent phases, β-Sn and Ag_3Sn (or Cu_6Sn_5), is also large. More importantly, the Ag_3Sn and Cu_6Sn_5 are faceting phases, while β-Sn is a non-faceting phase. During solidification, the faceting phase, Ag_3Sn or Cu_6Sn_5, grows by layer deposition involving the lateral propagation of a step across the liquid/solid interface. On the other hand, the non-faceting phase, β-Sn, advances into the liquid phase by tree-like, non-faceted dendrites. Because of the different growth mechanisms, the growth rates of the β-Sn and Ag_3Sn (or Cu_6Sn_5) are quite different. Therefore, the growths of the β-Sn, and Ag_3Sn (or Cu_6Sn_5) are independent or only loosely coupled.

Solidification Microstructure

The microstructures resulting from the loosely coupled growth of the constituent phases in the Sn-Ag-Cu eutectic alloys are different from the lamellar/colony of the eutectic Sn-Pb. The microstructures shown in most of the literature on eutectic or near eutectic Sn-Ag and Sn-Ag-Cu solder joints are composed of dendrites of β-Sn in the volume fractions varying between 20 and 80%, depending on the cooling rates, and the time and temperature above the liquidus temperature. Between the dendrites, there exist lamellar arrays of β-Sn, Ag_3Sn, and Cu_6Sn_5, as shown in Fig. 1. In some studies, large Ag_3Sn plates also appeared in the solder joints.

Sn Dendrites. The dendrites of the β-Sn phase are the result of the primary solidification. The presence of the primary β-Sn in the eutectic Sn-Ag or the eutectic Sn-Ag-Cu solder joints is probably due to (a) the large undercooling below the eutectic temperature prior to the formation of the β-Sn phase, and (b) the shift of solder composition away from the eutectic as the result of Cu pick-up by the solder.

As stated before, the required large undercooling for the nucleation of the β-Sn phase is due to the fact that the β-Sn phase does not nucleate heterogeneously on the $AgSn_3$ or Cu_6Sn_5 phases that formed at the smaller undercoolings, so the β-Sn phase has to nucleate homogeneously. Since the solidification temperature of the β-Sn phase is much lower than the eutectic temperature, the original composition of Sn-3.5Ag or Sn-3.5Ag-0.9Cu no longer represents the eutectic composition at the lowered solidification temperature; rather, the eutectic composition is now at the intersection of the extended L/β liquidus and the solidification temperature. This predicts a microstructure consisting of primary β-Sn phase and eutectic mixture of Ag_3Sn (and Cu_6Sn_5) and the β-Sn phase.

As the cooling rate increases, the undercooling increases, which leads to a larger volume fraction of the β-Sn dendrites, and a smaller volume fraction of the eutectic lamellar region (Ref 7). The spacing between the dendrites becomes finer with increasing cooling rates.

The second reason for the formation of the primary β-Sn phase in the eutectic Sn-Ag solder joints is the Cu dissolution from the soldering pad into the solder joints, which shifts the composition of the solder joint from its eutectic point. As shown in Ref 8, the volume percentages of both the Cu_6Sn_5 phase and the β-Sn dendrites increase with increasing reflow temperature and reflow time.

The nucleation of the β-Sn phase is facilitated when the eutectic Sn-Ag-Cu is soldered to a Au-Ni(P) surface. A study (Ref 9) of the eutectic Sn-Ag-Cu BGA solder joints between two Cu pads, between a Cu pad and an Au-Ni(P) pad, and between two Au-Ni(P) pads, and observed a higher volume density and finer spacing of the β-Sn dendrites in the solder joints bounded by two Au-Ni(P) pads. Also, Ref 10 compared the microstructures of the Sn-3.5Ag-0.5Cu and Sn-3.5Ag-0.5Ni solder joints, and a higher density of the β-Sn dendrites with finer spacing was observed in the Ni-containing solder joints.

Large Ag_3Sn Plates. As stated previously, the Ag_3Sn phase nucleates with minimal undercooling, while the β-Sn phase requires an un-

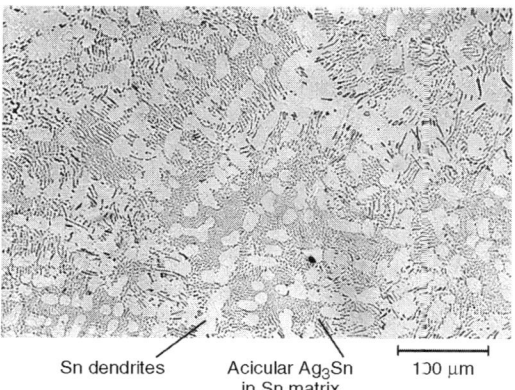

Fig. 1 Typical microstructure of a 95Sn-3.5Ag solder resulting from non-equilibrium cooling showing primary dendrites of Sn-rich solid solution and interdendritic space filled with acicular Ag_3Sn in a Sn-rich solid solution matrix.

dercooling up to 30 °C (86 °F) for nucleation. During the solidification process, the Ag_3Sn nucleates first and may grow into large plates before the β-Sn matrix is solidified. Sn-3.8Ag-0.7Cu BGA solder balls solidified at a slower cooling rate of 0.02 °C/s (0.036 °F/s) and the large Ag_3Sn plates could cross the whole solder joint as demonstrated in Ref The large Ag_3Sn plates pose adverse effects on ductility and fatigue resistance of solder joints as reported by (Ref 11 to 13).

The formation of the large Ag_3Sn was reported to depend on three factors: (a) Ag content, (b) cooling rate, and (c) Cu content. Large Ag content promotes the Ag_3Sn formation. Silver content is recommended to be less than 3 wt% (Ref 6). A slower cooling rate allows more time for Ag_3Sn to grow, which requires the long-range diffusion of Ag and Sn atoms in the liquid phase. It was also determined (Ref 6) that the cooling rate of BGA solder joints should be faster than 1 °C/s (60 °C/min.) or 2 °C/s (108 °F/min.). The typical cooling rate in a surface mount reflow oven is in the range of 50 to 120 °C/min. (90 to 216 °F/min.). Therefore, large Ag_3Sn plates are not usually present in the solder joints. But the risk exists for some applications, especially with large components or thick substrates or PCBs. The effect of cooling rates on the formation of large Ag_3Sn in eutectic Sn-Ag solder joints was also reported (Ref 14).

The Cu content in the solder joints is shown in the literature to promote the formation of large Ag_3Sn plates. It was reported that the increased percentage of the large Ag_3Sn plates with increasing Cu content as the microstructures of Sn-3.8Ag-0.7Cu, Sn-3.8Ag-0.35Cu, and Sn-3.5Ag (Ref 6). Large Ag_3Sn plates adjacent to the Cu substrate where the Cu atom dissolved into the molten solder joint were reported (Ref 11). The formation of large Ag_3Sn next to the Cu/solder interface was observed (Ref 10).

Cu_6Sn_5. In a eutectic Sn-Ag-Cu solder joint, the Cu_6Sn_5 phase appears as more blocky particles, while the Ag_3Sn is always recognized as elongated plates. In the secondary solidification of L $\Rightarrow$ Cu_6Sn_5 + β-Sn, or tertiary L $\Rightarrow$ Cu_6Sn_5 + β-Sn + Ag_3Sn, the faceted Cu_6Sn_5 do not grow in good coordination with the nonfaceted β-Sn. The two faceted phases, Cu_6Sn_5 and Ag_3Sn, do not exhibit coupled growth either, as shown in Ref 1. Under certain special conditions, the Cu_6Sn_5 phase exhibits as a nonfacet phase. For example, the Cu_6Sn_5 dendrites in the eutectic Sn-Ag solder joints were formed by a laser soldering process (Ref 14). The laser soldering process superheated the Sn-Ag solder to 400 to 600 °C (752 to 1112 °F). The great superheating leads to more rapid dissolution of Cu from the Cu substrate. The fast cooling rate may also have an effect on the non-faceting of the Cu_6Sn_5 phase. In laser soldering, the cooling rate is high as is the undercooling at the interface between liquid and Cu_6Sn_5 phase. The tendency for the transition from faceting to non-faceting increases as the interface undercooling increases. A minor addition of Co (<1%) into the eutectic Sn-Ag-Cu solder reduced the faceting of the Cu_6Sn_5 particles and enhanced the nucleation of the Cu_6Sn_5 phase in the liquid phase, and resulted in a greater volume fraction of the eutectic region with a finer lamellar spacing (Ref 15).

The amount of Cu_6Sn_5 particles in eutectic Sn-Ag and Sn-Ag-Cu solder joints on Cu substrate increases with increased reflow time and reflow temperature because of the increased Cu dissolution (Ref 7, 14). Also, there are more Cu_6Sn_5 particles adjacent to the Cu substrate than inside the solder joint.

Microstructural Evolution in Solid State Aging

Like other materials, grain growth occurs for the eutectic Sn-Ag-Cu solder joints in solid state annealing or aging. The sizes of the β-Sn, Ag_3Sn, and Cu_6Sn_5 increase with aging time at the temperatures between ambient and 190 °C (374 °F) (Ref 7, 10, 14, 16–19). Because of the low melting temperature, the grain growth may occur even at ambient temperature.

The grain growth rate depends on the initial microstructure. For the initial microstructure produced by quenching, the sizes of all phases are fine, and the growth of the phase occurs quickly even at the ambient temperature (Ref 14, 17). For the initial microstructure produced by a slow casting, the grain growth was not significant even at 150 °C (302 °F).

The density of the Cu_6Sn_5 may increase with the aging time, especially in the quenched solder joints from high reflow temperatures and long reflow time, because the fast cooling rate suppressed the precipitation of the Cu_6Sn_5 phase during solidification. The subsequent solid state aging provided an opportunity for the supersaturated Cu to precipitate (Ref 7). However, if the eutectic Sn-Ag-Cu solder joints are attached to a Ni surface, the density of the Cu_6Sn_5 decreases with the aging time. The Cu_6Sn_5 particles inside the solder joint moved toward the Ni/solder interface and reappeared as the Cu-Ni-Sn intermetallic phase (Ref 16). Similar observations were reported in solder joints involving Sn-3.5Ag solder, Cu pad on the substrate, and a 1206 ceramic chip capacitor with Ni-Sn termination plating after 1000 cycles of thermal shock testing between −40 and 150 °C (−40 and 302 °F) with 18 min. dwell at each temperature extreme (Ref 20). The driving force for the migration of the Cu_6Sn_5 phase from the interior of the solder joint to the solder/Ni interface is that the Cu-Ni-Sn intermetallic is a more preferred phase than the Cu_6Sn_5 intermetallic from the thermodynamics consideration.

Interactions between Substrates and Solders: Introduction

The interactions between solders and substrates are described in two classes: liquid solder/substrate reaction during soldering and solid solder/substrate reaction during subsequent aging. During soldering, two processes occur simultaneously: (a) the substrate metal dissolves into the molten metal and (b) the active constituents in the solder combine with the substrate metal to form intermetallic compounds (IMCs) at the substrate/solder interface.

Molten Solder-Substrate Interactions

When a substrate contacts a molten solder, it may dissolve. At the same time, Sn in the solders interacts with the substrate to form IMC at their interface. The relative amount of the substrate that may dissolve into the molten solder is determined by its solubility limit at the temperature, and the dissolution rate of the substrate into the molten solder depends on the substrate material and temperature. In general, the temperature dependence follows an Arrhenius behavior.

Phase diagrams may be used as a guide to identify which intermetallic phase forms at the liquid solder/substrate interface. However, an equilibrium phase diagram merely indicates what phases are thermodynamically stable, given a particular composition and temperature. The kinetics of solder-substrate reactions determine the structure of the solder/substrate interface. There may be metastable phases that are not present on the equilibrium phase diagram that can form as a result of soldering, as described later. The kinetics of intermetallic growth at the solder/substrate interface depends on the nucleation rate of the particular intermetallic phase and transportation rate (by diffusion or convention) of the constituent elements to the reaction interface.

Dissolution during Soldering

Solubility of Cu in Molten Solders. The solubility limit of a substrate metal in a molten solder determines the maximum thickness of the substrate that may dissolve into the molten solder, if the solder joint volume is known, and it also affects the dissolution rates.

The solubility of Cu in molten Sn is measured and calculated in the temperature range of 232 to 260 °C (450 to 500 °F) in the Sn-rich portion of the Sn-Cu binary phase diagram (Ref 1). The solubility limit increases from 1.75 to 2.5 at.% (0.94 to 1.35 wt%) at temperatures from 232 to 260 °C (450 to 500 °F). The solubility limit in the temperature range is almost linear.

The solubility limits of Cu in molten eutectic Sn-Pb solder reported in the literature are not consistent. The solubility limits measured between 225 and 505 °C (437 and 941 °F) (Ref 21). The solubility increases from 0.88 to 14 at.% (0.4 to 8.1 wt%) in that temperature range. The experimental data were curve-fitted to an Arrhenius relation:

$$C_s = C_{s0} \exp\left(-\frac{Q}{RT}\right) \quad \text{(Eq 1)}$$

The values obtained from the curve fitting are: C_{s0} (at.%) = 3111, and Q = 33.8 kJ/mol. The solubility values at 255 and 310 °C (491 and 590 °F) calculated from the empirical formula from Ref 21 are 1.4 to 2.9 at.% (0.65 to 1.35 wt%), respectively. The solubility of Cu in eutectic Sn-Pb was also measured (Ref 22) at 255 and 310 °C (491 and 590 °F), and the solubility limits are 2.3 to 4.75 at.% (1.05 to 2.2 wt%), respectively. The measured values recorded in Ref 22 are larger than those in Ref 21 and 23, which recorded the maximum content of Cu dissolved in the Sn-37Pb solder at 230 and 250 °C (446 and 482 °F) were 2.2 to 2.84 at.% (1.0 to 1.3 wt%), respectively. The Cu-Pb-Sn isothermal section was calculated at 235 °C (455 °F) (Ref 24). The calculated solubility of Cu at 235 °C (455 °F) is

0.55 at.% (0.25 wt%). In Ref 25, a Cu-Pb-Sn isothermal section at 200 °C (392 °F) was presented, and the solubility of Cu in 63Sn-37Pb was approximately 0.05 wt%. The experimentally measured values (Ref 22, 23) are significantly larger than either curve-fitted values (Ref 21) or thermodynamically calculated values.

The solubility limit of Cu in molten eutectic Sn-Ag was determined experimentally (Ref 8). Solder joints of eutectic Sn-Ag were reflowed on Cu for 5 to 10 h, to achieve the Cu saturation. The Cu concentration in the solder matrix was determined by energy dispersive x-ray (EDX). The measurement was conducted at temperatures between 232 and 300 °C (450 and 572 °F). The test data were curve-fitted to an Arrhenius relationship such as Eq 1, and the results are: C_{s0} (at.%) = 9870, and Q = 35.3 kJ/mol.

Table 1 lists the solubility of Cu in molten Sn, eutectic Sn-Pb, and eutectic Sn-Cu in the temperature range between 232 and 260 °C (450 and 500 °F). It is seen that the solubility increases in the order of eutectic Sn-Pb, pure Sn, and eutectic Sn-Ag. The addition of Ag in molten Sn enhances the solubility of Cu. The 240 °C (464 °F) isothermal section of Sn-Ag-Cu ternary phase diagram (Ref 3) demonstrates that with the addition of Ag, the Cu solubility in molten Sn may increase up to 5 wt% (9 at.%) at 240 °C (404 °F). The phase diagram was built based on the equilibrium phase diagrams of its three constituent binary systems and the information that no ternary compound has been found.

The solubility limit (C_s, at.%) of a substrate material in a molten solder may be used to determine the minimum thickness (Δh_{sub}) of the substrate or under bump metallization (UBM) for a given solder bump dimension (the ratio of solder volume (V) to contact area (A), or the thickness of the solder joint, h_{solder}):

$$C_s = \frac{\Delta h_{sub} A}{v_{sub}} \bigg/ \frac{V}{v_{sol}} \quad \text{(Eq 2)}$$

Rearranging Eq 2 gives:

$$\Delta h_{sub} = \frac{v_{sub}}{v_{sol}} C_s \frac{V}{A} \approx \frac{v_{sub}}{v_{sol}} C_s h_{solder} \quad \text{(Eq 3)}$$

where v_{sub} and v_{sol} are the mole volumes of the substrate metal and the solder. The substrate metal here is Cu, and its mole volume is 7.12 cm^3. The mole volumes of Sn, Pb, and Ag are 16.2 cm^3, 18.3 cm^3, and 10.3 cm^3, respectively and 63wt%Sn-37wt%Pb is, in at.%, 74.8at.%Sn-25.2at.%Pb, and 96.5wt%Sn-3.5wt%Ag is 96.2at.%Sn-3.8at.%Ag. The mole volume of 74.8at.%Sn-25.2at.%Pb is approximately equal to 16.75 cm^3, and that of 96.5at.%Sn-3.8at.%Ag is 16.01 cm^3.

To get a feeling of the thickness of Cu substrate that may dissolve into a molten solder joint, assuming that the solder joint thickness is 0.1 mm (0.0039 in.) in a flip chip and 0.5 mm (0.020 in.) in a BGA package, respectively, and using the solubility limits listed in Table 1 and Eq 3, we obtain that the thickness of Cu substrate that may dissolve into the flip chip joint of Sn-37Pb, Sn, or Sn-3.5Ag at 232 °C (450 °F) is 0.4, 0.8, and 1 µm, respectively, and the Cu thickness that may dissolve into the molten joints of the three solders for the BGA package at 232 °C (450 °F) is 2, 4, and 5 µm, respectively.

Solubility of Ni in Molten Solders. The solubility limit of Ni in molten Sn and eutectic Sn-Pb is extremely small at temperatures lower than 400 °C (752 °F), as shown in the Ni-Sn binary phase diagram (Ref 26), the binary section of Ni-63Sn37Pb of the Ni-Sn-Pb ternary phase diagram (Ref 27), and the isothermal sections at 170, 240, and 400 °C (338, 464, and 752 °F) of the ternary Sn-Pb-Ni phase diagrams (Ref 25). It was estimated in Ref 27 that the Ni solubility limit in molten eutectic Sn-Pb is approximately 10^{-5} at.%.

Dissolution in Solder Bath. The dissolution rates of Cu, Au, Ag, Pd, Pt, and Ni in molten 60Sn-40Pb were determined by immersing metal wires into liquid solder bath and measuring the wire diameter changes as a function of immersion time and solder bath temperature (Ref 28). The curves of wire diameter change versus immersion time are linear; therefore, dissolution rates can be calculated from the slopes. The wires used were thin, 0.5 mm (0.020 in.), and solder pots were large, 2.9 kg (39 lb); therefore, the dissolution rates thus determined represent the upper limits for substrate metals dissolving into 60Sn-40Pb solder joint.

Table 1 Solubility of Cu (at.%) in three molten solders

Temperature		Cu in Sn-37Pb(a)	Cu in Sn(b)	Cu in Sn-3.5Ag(c)
°C	°F			
232	450	1	1.75	2.2
260	500	1.5	2.5	3.4

(a) Source: Ref 21. (b) Source: Ref 1. (c) Source: Ref 8

The temperature dependence of the radial dissolution rate follows the Arrhenius behavior as described by:

$$r = A \exp(-B/T) = A \exp(-Q/RT) \quad \text{(Eq 4)}$$

where r is the rate of radial dissolution, A and B are constants, T is absolute temperature, Q is the activation energy, and R is the gas constant. The values of the parameters for various metals are listed in Table 2, and data from Ref 28 are plotted in Fig. 2.

The Cu dissolution rates were measured between 232 and 482 °C (450 and 900 °F). The dissolution rates at temperatures in the 232 to 250 °C (450 to 482 °F) range are between 0.1 and 0.15 µm/s, and the dissolution rate at temperatures outside the testing range may be extrapolated if the dissolution process is the same. In other words, if at a temperature outside the testing range there is a different intermetallic formed at the solder/substrate interface, the empirical relation may not be valid.

As shown in Fig. 2, the dissolving rate of Ni in liquid 60Sn-40Pb solder pot is approximately 20 to 30 times slower than that of Cu.

Cu Dissolution in Solder Joints. Different from a solder bath, the volume of a solder joint is limited or comparable with the substrate. Initially, a substrate dissolves into a molten solder joint linearly with time, as if dissolving into a large solder bath. The dissolving rate slows down as the concentration of substrate metal in the molten joint approaches its solubility limit.

The studies in Ref 22 and 29 presented an analytical method for modeling the dissolution kinetics under the transient condition. The dissolution rate may be modeled with the Nernst-Brunner relationship (Ref 30), which gives the following time dependence of substrate metal concentration in a solder system:

$$C - C_0 = (C_s - C_0)\left[1 - \exp\left(-\frac{KA}{V}t\right)\right] \quad \text{(Eq 5)}$$

where C_0 and C (at.%) are the substrate metal concentration at time 0 and time t, C_s is the solubility concentration of substrate metal, V and A are the solder volume and the contact area between the solder and the substrate, and K is the dissolution rate constant. Equation 5 shows that the process for a substrate metal to saturate a molten solder is similar to the process for electric charge to saturate a capacitor.

The dissolution rate constant, K, follows an Arrhenius relationship:

$$K = K_0 \exp\left(-\frac{Q}{RT}\right) \quad \text{(Eq 6)}$$

The values of the constant K_0 and activation energy Q were obtained for the Cu/liquid Sn system by the least square curve fitting of the experimental data (Ref 29). The activation energy Q is 21 kJ/mol below 415 °C (779 °F), and 17 kJ/mol above 415 °C (779 °F). This change is related to a change in the morphology of the Cu-Sn intermetallic layer. Below 415 °C (779 °F), the observed layer consists of ε-phase (Cu_3Sn) adjacent to the copper and η-phase (Cu_6Sn_5) in contact with the liquid. According to the Cu-Sn phase diagram, the η-phase undergoes a peritectic decomposition above 415 °C (779 °F). Correspondingly, layers formed above this temperature consisted of δ-phase (Cu_4Sn) adjacent to the copper and ε-phase adjacent to the liquid. The fact that the surface intermetallic influences the dissolution rate is evidence that the surface reaction is involved in the rate limiting mechanism for dissolution.

The values of the constant K_0 and activation energy Q were obtained, $Q = 13$ kJ/mol, and $K_0 = 2.9$ mm/min., for the Cu/eutectic Sn-Pb system, by the least square curve fitting of the experimental data at temperatures of 255 and 310 °C (491 and 590 °F) for the ratio of solder volume to contact area of 0.33 to 0.56 mm (0.012 to 0.022 in.) (Ref 22). The ratio of the solder volume to contact area (approximately equal to the solder joint height) between 0.33 and 0.56 mm (0.012 and 0.022 in.) is within a range that would be typical for some surface mount ball grid array (BGA) packages.

The values of the constant K_0 and activation energy Q were obtained, $Q = 14.9$ kJ/mole, and $K_0 = 14.2$ mm/min., for the Cu/eutectic Sn-Ag system by the least square curve fitting of the experimental data at temperatures of 254, 272, and 302 °C (489, 521, and 575 °F), for the ratio

Table 2 Values of the parameters in Equation 4

Metal	A (in./s)	B (1/T(Rankine))	A (µm/s)	Q (J/mole)	Q (eV)
Au	308	13,600	7,823,200	62,821	0.65
Ag	1.51	9,500	38,354	43,883	0.45
Cu	0.99	11,300	25,146	52,197	0.54
Pd(<525 °F)	524	18,000	13,309,600	83,146	0.86
Pd(>600 °F)	14.2	16,100	360,680	74,369	0.77
Ni	133	15,900	33,782	73,446	0.76
Pt	346	22,800	8,788,400	105,318	1.09

Source: Ref 28

of solder volume to contact area of 0.2 to 0.4 mm (0.007 to 0.015 in.) (Ref 8).

Ni Dissolution in Solder Joints. Reference 31 studied the dissolution of Ni-coated Cu into molten solders of 57Bi-43Sn, 62Sn-38Pb, 96.5Sn-3.5Ag, 100Sn, and 95Sn-5Sb, at 30 or 40 °C (86 or 104 °F) above the melting temperatures of the respective solders at 2, 6, and 20 min. The Cu layer was approximately 2 to 4 μm, and the Ni coating was approximately 0.5 to 2 μm. It was observed that the Cu and Cu-Ni substrate metal dissolution rates increased as the Sn content of the solders and the reflow temperature increased. For the Sn-rich solders, 95Sn-5Sb and 96.5Sn-3.5Ag, and 100Sn, a Ni layer up to 2 μm thick is not a good reaction barrier for the extended reaction time of 20 min.

The kinetics were measured at 250 °C (482 °F) of the dissolution of surface finishes and intermetallic compound growth (Ref 32). The surface finishes investigated included Cu, Au-Ni(P), Au-Pd-Ni(P), and Au-Ni (electroplated). The Pb-free solders investigated include Sn-3.5Ag, Sn-3.8Ag-0.7Cu, and Sn-3.5Ag-3.0Bi. They observed that the dissolution rate of Au-Ni(P) and Au-Pd-Ni(P) is approximately one-half that of Cu at 250 °C (482 °F). Electroplated Ni dissolved much less in Sn-Ag and Sn-Ag-Bi than electroless Ni(P) did. The presence of Cu in solders significantly reduces the dissolving rate of the Au-Ni(P) and Au-Pd-Ni(P), as shown by the comparison of the dissolving rates in Sn-3.5Ag versus Sn-3.8Ag-0.5Cu. The reduction of the Ni dissolution rate by the small Cu content in the molten solders was also observed (Ref 35), the reason being that the Cu presence changes the interfacial intermetallic from $(Ni,Cu)_3Sn_4$ to $(Cu,Ni)_6Sn_5$.

Experimental research was conducted on the dissolution rate of Cu-Ni alloy substrates into Sn-3.5Ag and Sn-3.8Ag-0.7Cu solders as a function of time and Cu-Ni ratio of the substrate at 250 °C (482 °F) from 15 s to 5 min (Ref 33). The dissolution of Cu, Ni, and Cu-Ni alloys into high Sn solders was significantly faster than for the eutectic Sn-Pb solder, and even small Ni additions into Cu significantly reduced the dissolution rate. In the solder alloyed with Cu, Sn-3.8Ag-0.7Cu, dissolution into the solder was slower than for the Sn-3.5Ag solder.

The Characteristic Time. One of the obvious features of all these experimental studies is that liquid solders (used in all studies listed previously) are not saturated with Cu within several seconds at the soldering temperatures. If we define the characteristic time as the time when the

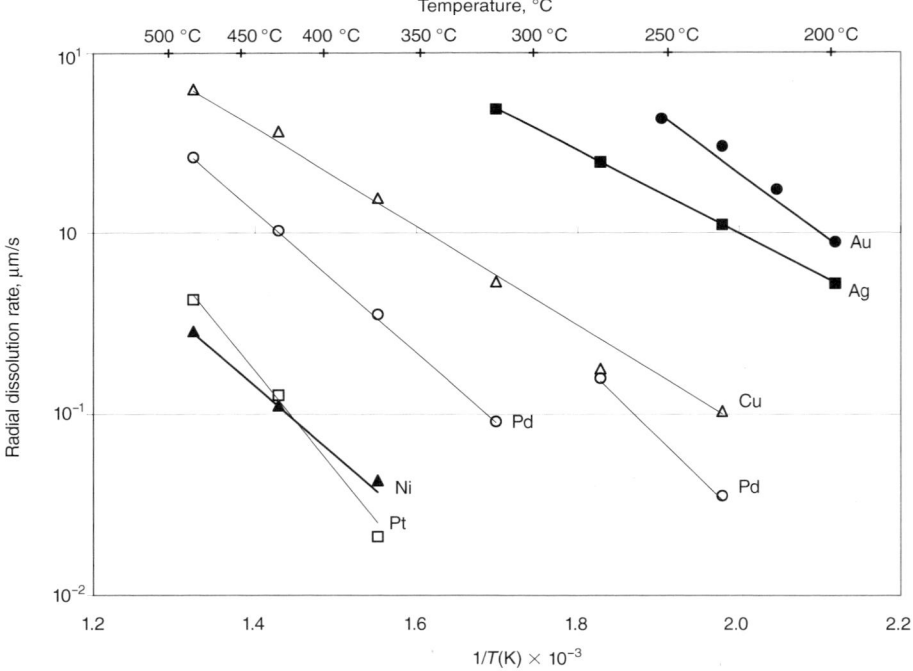

Fig. 2 Temperature dependence of the dissolution rates of Au, Ag, Pd, Cu, Pt, and Ni in molten 60Sn-40Pb solder, from Bader (Ref 28). The dots are data points from Bader's article, and the lines are calculated from the parameters in Table 2.

dissolving substrate metal reaches 63% (1/e) of the saturation limit, most data listed previously show that the characteristic times are between 2 and 4 min. For example, the measurement of eutectic Sn-Pb, and eutectic Sn-Ag on Cu and on Ni, showed that the characteristic times are approximately 3 min. (Ref 31). The shortest time in the studies listed previously, 0.5 min. at approximately 300 °C (572 °F) was reported in Ref 8, of eutectic Sn-Ag on Cu.

The characteristic time for the substrate dissolution, t_0, may be calculated:

$$t_0 = \frac{V}{A} K = hK \qquad \text{(Eq 7)}$$

where h is the ratio of the solder volume to the contact area, and roughly the thickness of the solder joint. Therefore, the characteristic time depends on the soldering temperature (Eq 6), and the solder joint thickness.

Comparing the characteristic times between eutectic Sn-Pb and eutectic Sn-Ag at the same temperatures, it can be shown that Cu saturates in eutectic Sn-Ag much faster than in eutectic Sn-Pb. The characteristic time at 255 °C (491 °F) for a solder joint thickness of 0.5 mm (0.019 in.), is 3.3 min. for the Cu/eutectic Sn-Pb (Ref 22), and 1.1 min. for the Cu/eutectic Sn-Ag (Ref 8).

Using Eq 7 and experimentally obtained K values from Ref 8 and 22, we may estimate the characteristic times for Cu dissolution into typical solder joints. Currently, the sizes of solder joints in flip chip, CSP, and BGA packages are approximately 0.1, 0.3, and 0.5 mm (0.003, 0.011, and 0.019 in.), respectively. If those solder joints are made of 63Sn-37Pb, and reflowed at 220 °C (428 °F), the characteristic times for Cu dissolution are 4.4, 2.6, and 0.9 min., respectively. If those solder joints are made of Sn-3.5Ag and reflowed at 250 °C (482 °F), the characteristic times are 1.2, 0.7, and 0.2 min., respectively. As can be seen later, the interfacial IMC grows much faster after Cu saturates the solder joint. Before Cu saturates the solder joint, the IMC grows and dissolves simultaneously. Therefore, for a flip chip of Pb-free solder joints, the IMC will be exceptionally thick.

The dissolution from the preformed IMCs into the liquid Sn has little effect on the growth behavior of IMCs since the liquid Sn is saturated with Cu within several seconds at the soldering temperature of 280 °C (536 °F) (Ref 34). Using Eq 7 and K value from Ref 29, the characteristic time is less than 1 s.

Formation of Cu-Sn Interfacial Intermetallic

Nucleation. As a molten solder contacts a Cu substrate, Cu_6Sn_5 (η) forms instantaneously. The nucleation kinetics of the η-phase IMC were investigated in Ref 35 by hot dipping copper coupons into a molten tin bath at temperatures varying from 240 to 300 °C (464 to 572 °F). The formation of η nuclei as small, round bumps at the molten Sn/Cu interface was observed after just 1 s at temperatures between 240 and 300 °C (464 and 572 °F). After as little as 10 s, experimental observations of Sn and eutectic Sn-Pb indicated that the η phase is continuous over the Cu substrate surface (Ref 34, 36, 37). Once a continuous layer is established, further growth requires diffusion of the reacting species (Sn and/or Cu) through the intermetallic layer.

The density of nuclei versus temperature exhibited a typical inverse-C type behavior with a maximum nucleation rate occurring at an intermediate temperature (Ref 35). As explained in the latter section, the interfacial η-phase growth is controlled by the diffusion along its scallop or grain boundaries. The grain size formed during soldering is mostly determined by the nucleation density.

Phase Diagram. The reactions between the Cu substrate and liquid solders of pure tin, eutectic Sn-Pb, Sn-Ag, and Sn-Ag-Cu can often be explained with the Cu-Sn binary equilibrium phase diagram, because Pb and Ag are found not to participate in the interfacial IMC structure. Below 415 °C (779 °F), there are two equilibrium phases: η-Cu_6Sn_5 and ε-Cu_3Sn. Above 415 °C (779 °F), the two equilibrium phases are ε-Cu_3Sn and δ-Cu_4Sn. However, the addition of Pb or Ag into the Cu-Sn system may change the relative stabilities of Cu-Sn intermetallics. For example, when Cu reacts with liquid 63Sn-37Pb, η-Cu_6Sn_5 forms; when Cu reacts with liquid 95Pb-5Sn, ε-Cu_3Sn forms. Ternary phase diagrams are helpful in these cases.

For the Sn-Pb-Cu system, the calculated liquid surfaces and isothermal sections at 170, 200, 283, 350, and 400 °C (338, 392, 541, 662, and 752 °F) are presented in Ref 25. The calculated liquid surface and isothermal sections are at 150 and 235 °C (302 and 455 °F) (Ref 24).

Identification of Interfacial Intermetallic Phases. The primary IMC formed at the interface between high-Pb Sn-Pb solder, e.g., 90Pb-10Sn or 95Pb-5Sn, and the Cu substrate, is ε-phase (Cu_3Sn) (Ref 38). The ε-phase is also reported in Ref 39 as the primary IMC to form at the interface between Cu and 27Sn-73Pb at temperatures above 290 °C (554 °F).

The primary IMC formed at the interface between a Cu substrate and eutectic or near eutectic Sn-Pb, Sn-Ag, Sn-Ag-Cu, and pure Sn is η-phase (Cu_6Sn_5). Less certain is if another stable IMC, ε-phase (Cu_3Sn), forms or not, at the interface between the Cu substrate and η-phase during soldering. The reason for the uncertainty is because the ε-phase is very thin, if it exists, and its identification requires a transmission electron microscope (TEM) study, as an ordinary scanning electron microscope (SEM) cannot clearly identify the thin ε-phase in the as-soldered condition.

A study of IMC at a liquid tin/copper interface (Ref 40) found η-Cu_6Sn_5 formed almost immediately upon contact of the liquid solder with solid copper and developed with a scalloped morphology. The ε-phase grew at much earlier reaction times at higher temperatures. For example, after a 2-min. reaction time, the ε-phase was clearly discernible at 325 °C (617 °F), but it was not evident at 250 °C (482 °F) after 5 min. They plotted the square of the thickness of the ε phase with respect to the reaction time at each reaction temperature, assuming that the ε-layer growth is controlled by the bulk diffusion and obeys $t^{1/2}$ kinetics. The incubation time for the nucleation of the ε-phase is given by the intercept of the plot on the time axis. The incubation times generally decreased with increased reaction temperatures; they were 219.3, 271.3, 74.5, and 84.8 s for 250, 275, 300, and 325 °C (482, 527, 572, and 617 °F), respectively.

A study of bulk Sn/Cu and near eutectic Sn-Ag-Cu/Cu samples reflowed for approximately 60 s at 238 and 230 °C (460 and 446 °F) (Ref 41), respectively. Wavelength dispersive spectrometry (WDS) analysis showed Cu_6Sn_5 to be growing, just as in eutectic Sn-Pb/Cu. Also, between this layer and the Cu substrate, a layer of Cu_3Sn had formed during the reflow process.

A Sn-3.5Ag/Cu couple reacted at 240 °C (464 °F) for 48 h, and two IMCs, η and ε, were observed at the interface (Ref 3). A Sn3.5Ag/Cu couple reacted at 450 °C (842 °F) and showed three IMCs at the interface, δ-Cu_4Sn, ε-Cu_3Sn, and η-Cu_6Sn_5. The δ and ε are stable phases, but η is not at 450 °C (842 °F). It was concluded that η did not form during the 450 °C (842 °F) aging, but formed during the subsequent solidification, because the η phase did not thicken with increasing aging time at 450 °C (842 °F) as the δ and ε phases did.

Most studies show that Ag atoms do not participate in the interfacial IMC formation, except for Ref 42. They did energy dispersive x-ray (EDX) analyses of interfacial IMC between Sn-3.5Ag and Cu formed at 250 to 375 °C (482 to 707 °F) for 10 to 60 min. and found that the atomic composition of the scallop-shaped intermetallic is Cu:Sn:Ag = 54.4:45.39:0.31, which corresponds to the η-$Cu_6(Sn_{0.993}Ag_{0.007})_5$ phase. Following prolonged soldering reactions, a thin layer of IMC appears at the $Cu_6(Sn_{0.993}Ag_{0.007})_5$/Cu interface. The atomic composition of this thin layer of IMC is Cu:Sn:Ag = 75.21:24.70:0.09, which corresponds to the ε-$Cu_3(Sn_{0.996}Ag_{0.004})$ phase.

A detailed TEM work was reported on the IMC at the interface between Cu substrate and Sn-3Ag-(0,3,6)Bi solders formed at 250 °C (482 °F) for 1 min. (Ref 43). TEM micrograph clearly shows two reaction layers. The first layer is much thicker than the second layer and grows up to approximately a few hundred nm in thickness. The second layer adjacent to the Cu substrate is thinner than 100 nm and consists of small grains. A microbeam of electrons was used for selected area diffraction (SAD). The first layer is identified as η-Cu_6Sn_5, a high temperature Cu_6Sn_5 phase. The Cu_6Sn_5 can take two different phases; one is the high-temperature stable phase η-Cu_6Sn_5, which is a hexagonal close-packed structure, and the other is the low-temperature stable phase η'-Cu_6Sn_5 phase with a superstructure type, having a phase transformation approximately in the range of 168 to 189 °C (334 to 372 °F). In the diffraction patterns obtained from the first layer, the superlattice of the η'-Cu_6Sn_5 phase cannot be found. The second layer was identified as ε-Cu_3Sn, with a periodic antiphase-domain structure based on the Cu_3Ti-type ordered lattice. Bismuth contents (0, 3, 6 wt%) in Sn-3Ag do not have any influence on the interfacial IMC structure.

Formation of Ni-Sn Interfacial Intermetallics

Reaction Between Ni and Sn, Eutectic Sn-Pb, and Sn-Ag. The reaction products between the Ni substrate and solders of pure Sn, eutectic Sn-Pb, and Sn-Ag can be understood using the binary Ni-Sn phase diagram, because neither Pb nor Ag is involved in the interfacial IMCs. There are three stable IMCs, Ni_3Sn_4, Ni_3Sn_2, and Ni_3Sn. The primary IMC formed at the interface of Ni solders is scallop-shaped Ni_3Sn_4. Sometimes, a thin layer of Ni_3Sn_2 or Ni_3Sn was also observed between Ni and Ni_3Sn_4. Interestingly enough, a metastable phase, $NiSn_3$, was reported to exist at the Ni_3Sn_4 solder boundary.

A study was conducted of the IMC at the interface between Ni and molten Sn-3Ag-(0,3,6)Bi solders (Ref 43). The solder joints were reflowed at 250 °C (482 °F) for 1 min. The primary interfacial IMC phase is Ni_3Sn_4, which has a scalloped edge of approximately 0.5 µm, much thinner than the scallops of Cu_6Sn_5 formed at the Cu substrate of 1 to 2 µm. A thin layer is observed at the interface between the Ni_3Sn_4 and Ni layers. TEM micrograph shows that the layer consists of small grains, and the layer width is approximately 40 nm. Microbeam diffraction patterns indicate that the layer is η-Ni_3Sn_2 of the high-temperature stable phase. The Ni_3Sn_2 phase can exist in two structure types. The crystallographic data is listed in Ref 44 and 45. The η-Ni_3Sn_2 of the high-temperature phase is based on the hexagonal close-packed Ni-As structure, and it undergoes a transformation to λ-Ni_3Sn_2 of orthorhombic symmetry as the temperature decreases below 600 °C (1112 °F). The existence of η-Ni_3Sn_2 may decrease the interfacial fatigue resistance (Ref 46).

The presence of Ni_3Sn_2 and Ni_3Sn was detected at the interface between Sn-3.5Ag and Ni in the as-reflowed condition (Ref 47). The reflow was done at 250 °C (482 °F) for 30 to 120 s at intervals of 30 s. SEM cross sections showed a single layer of IMC at the solder/Ni interface. However, x-ray diffraction (XRD) analysis identified the presence of Ni_3Sn_4 and Ni_3Sn at the interfacial IMC layer, after 30 s of reflow. After 60 s of reflow, Ni_3Sn_2 was identified, in addition to Ni_3Sn_4 and Ni_3Sn that were already present at 30 s. However, this additional phase of Ni_3Sn_2 disappeared after 120 s reflow. The specimens for XRD were prepared by mechanically removing the solder and etching away the remaining solder.

Metastable $NiSn_3$ Phase. $NiSn_3$ is not one of the stable phases in the Ni-Sn system. Its existence at the interface between Sn-3.5Ag and Ni(2 µm)/Cu plate in the as-reflowed condition was reported in Ref 47. Soldering Sn-3.5Ag on a Ni-Cu substrate was performed at 250 °C (482 °F) for 30 to 120 s, and the XRD analysis detected the presence of metastable $NiSn_3$.

Another example of the formation of the metastable phase $NiSn_3$ at the interface between electroplated 5 µm Sn (or 90Sn-10Pb) and 2.5 µm electrolytic Ni during long-term aging can be found in Ref 48. This showed that on the top of the thin continuous layer of Ni_3Sn_4, there were discontinuous large platelets of $NiSn_3$, which grew in significant amounts at temperatures between 75 and 165 °C (167 and 329 °F). The fast growth of the $NiSn_3$ was identified as the cause for the solderability deterioration during aging. The growth of the $NiSn_3$ depends on the Sn plating condition (matte Sn versus bright Sn) and may be suppressed by the addition of Pb in the plated Sn. Solderability could be restored by thermally decomposing $NiSn_3$ into Ni_3Sn_4 and Sn. The decomposition kinetics of $NiSn_3$ was analyzed, and the results indicated that the decomposition reaction occurs within 3 s at 250 °C (482 °F), while the decomposition reaction required about 10^5 s at 215 °C (419 °F). Thus, higher temperature soldering processes may not be impacted by the presence of $NiSn_3$, while lower temperature soldering processes would require the removal of this phase prior to soldering.

Reaction Between Ni and Sn When Cu is Present. If Cu is present at the interface between Sn and Ni, the interfacial reaction products are complicated, because three elements, Sn, Cu, and Ni, are involved in the interfacial IMCs. Experimentally, in general, when the Cu concentration is small, $(Ni,Cu)_3Sn_4$ phase forms. The Cu atoms substitute partially the Ni positions in Ni_3Sn_4. When the Cu concentration is high, $(Cu,Ni)_6Sn_5$ phase forms; the Ni atoms take some positions of Cu in the Cu_6Sn_5 phase.

Three scenarios are observed when Cu atoms get to the Ni-Sn interface: (a) Eutectic Sn-Ag-Cu solder contacts a Ni surface, (b) Sn-containing solders contact Ni-coated Cu substrate, and the Cu atoms diffuse through the Ni layer during soldering, and (c) A solder joint contacts a Ni substrate on one side and a Cu substrate on the other; and the Cu atoms diffuse through the molten solder joint and reach the solder/Ni interface.

While both the Cu-Sn and Ni-Sn phase diagrams have been well characterized, only sections of the ternary phase diagram of the Cu-Ni-Sn system are available at 235 °C (455 °F) (Ref 33, 49) and 240 °C (464 °F) (Ref 50, 51), based on experimental data and thermodynamic modeling. The version of the diagram from Ref 50 is presented as Fig. 3. It is seen that the $(Cu,Ni)_6Sn_5$ and $(Ni,Cu)_3Sn_4$ compounds are presented in this diagram as narrow composition range compounds that extend the domain of the respective binaries Cu_6Sn_5 and Ni_3Sn_4 to a ternary domain. The Cu-Sn-Ni phase diagram also displays a large two-phase region between the $(Cu,Ni)_6Sn_5$ and the Sn-Cu-Ni solid solution, and a large two-phase region between the $(Ni,Cu)_3Sn_4$ and the Sn-Cu-Ni solid solution, and a three-phase region among $(Ni,Cu)_3Sn_4$, $(Cu,Ni)_6Sn_5$, and the Sn-Cu-Ni solid solution. These features support the idea of a substitutional mechanism where atoms such as Cu and Ni or Ni and Cu substitute for each other in their binary compounds with Sn.

Sn-Ag-Cu on Ni. The reaction between the Sn-Ag-Cu solders and Ni substrate at 250 °C (482 °F) for 10 min. and 25 h was studied (Ref 50). Nine different Sn-Ag-Cu solders, with the Ag concentration fixed at 3.9 wt% and Cu concentration varied between 0 and 3.0 wt%, were prepared. When the reaction time was 10 min., at low-Cu concentration (≤0.2 wt%), only a continuous $(Ni_{1-x}Cu_x)_3Sn_4$ layer formed at the interface. When the Cu concentration increased to 0.4 wt%, a continuous $(Ni_{1-x}Cu_x)_3Sn_4$ layer and a small number of discontinuous $(Cu_{1-y}Ni_y)_6Sn_5$ particles formed at the interface. When the Cu concentration increased to 0.5 wt%, the $(Cu_{1-y}Ni_y)_6Sn_5$ became a continuous layer. At higher Cu concentrations (0.6 to 3.0 wt%), the $(Ni_{1-x}Cu_x)_3Sn_4$ layer disappeared, and only the $(Cu_{1-y}Ni_y)_6Sn_5$ was present.

The result in Ref 41 of 93.6Sn-4.7Ag-1.7Cu is consistent with the observation made in Ref 50; both observed $(Cu,Ni)_6Sn_5$, instead of Ni_3Sn_4. The formation of the $(Cu,Ni)_6Sn_5$ at the interface drained the Cu concentration in the solder matrix. In Ref 41, a simple calculation of the total Cu content in the observed $(Cu,Ni)_6Sn_5$ layer indicated that after reflow the interfacial IMC already contained approximately 72% of the Cu originally present in the solder joint. When the solder joints were subjected to solid state aging at 150 °C (302 °F), the further growth of the $(Cu,Ni)_6Sn_5$ would eventually consume essentially all of the available Cu in the solder joint. Because the mechanical properties of Sn-Ag-Cu alloys depend upon the Cu content, this consumption can be expected to alter the mechanical properties of these Pb-free solder joints. After depletion of the Cu from the solder, further annealing then gradually transformed the $(Cu,Ni)_6Sn_5$ phase into the $(Ni,Cu)_3Sn_4$ phase.

A study was conducted (Ref 52) of the interface between 95.5Sn-4.0Ag-0.5Cu and electroless Ni. It was observed that when the reflow temperature was below 300 °C (572 °F), $(Cu,Ni)_6Sn_5$ was observed. However, above 300 °C (572 °F), two different IMCs were found, bulky and faceted $(Cu,Ni)_6Sn_5$ and needle-like $(Cu,Ni)_3Sn_4$.

Different results were reported in Ref 9 and 32. The IMC product at the interface between Sn-3.8Ag-0.7Cu and Au-Ni(P)-Cu was $(Ni,Cu)_3Sn_4$ after reflow at 250 °C (482 °F) for 2, 6, and 20 min. (Ref 32), or reflow for 1, 6, and 11 times with the peak temperature at 260 °C (500 °F) (Ref 9).

Sn-Ag and Sn-Pb on Ni/Cu. The IMC formation between molten solders (eutectic Sn-Pb and Sn-Ag) and Ni(3 μm)/Cu(5 μm) was explored (Ref 53). The reflow temperatures were 225 and 260 °C (437 and 500 °F) for Sn-Pb and Sn-Ag, respectively. After only one reflow cycle at 260 °C (500 °F), Cu of approximately 1.5 at.% was observed in the interfacial $(Ni,Cu)_3Sn_4$ intermetallic. Since there were no Cu atoms in the solder matrix, the Cu in the interfacial IMC must have come from the diffusion of Cu through Ni to form the Ni_3Sn_4 IMC. The Cu content remained at approximately 1.5 at.% after ten reflow times, and the interfacial IMC remains Ni_3Sn_4 type. The grain size of $(Ni,Cu)_3Sn_4$ IMC increased with the number of reflow cycles.

For Sn-Pb on Ni-Cu, after one reflow at 225 °C (437 °F), only one layered-type $(Ni,Cu)_3Sn_4$ with 1 μm thickness formed between the solder and the Ni-Cu substrate. However, after three reflow cycles, another island-like $(Cu,Ni)_6Sn_5$ IMC laying on top of the layered-type $(Ni,Cu)_3Sn_4$ was seen. On the other hand, the thickness and the grain size of the $(Ni,Cu)_3Sn_4$

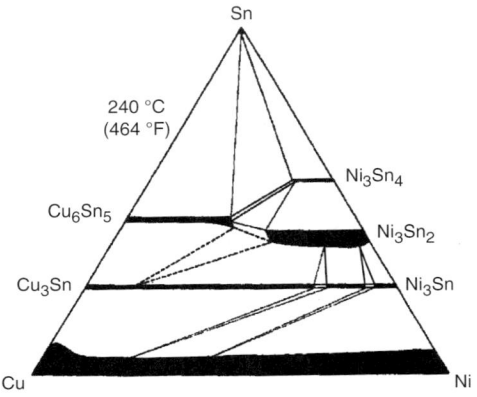

Fig. 3 The Cu-Ni-Sn isotherm at 240 °C (464 °F) Source: Ref 50

are nearly identical, even after ten cycles of reflow. Nevertheless, the amount of the $(Cu,Ni)_6Sn_5$ increased with the reflow cycles.

The different interfacial structures, $(Ni,Cu)_3Sn_4$ for Sn-Ag versus $(Ni,Cu)_3Sn_4$ + $(Cu,Ni)_6Sn_5$ for Sn-Pb, is thought to be due to the grain growth behavior. For the Sn-Ag sample, the grain size of $(Ni,Cu)_3Sn_4$ increased significantly with the reflow cycle, which suppressed the grain boundary diffusion, and therefore the formation of the $(Cu,Ni)_6Sn_5$.

The IMC formation at 250 °C (482 °F) at the interface between molten Sn-3.5Ag and four substrates: pure Cu, Ni(2 μm)/Cu, Ni(4 μm)/Cu, and pure Ni was investigated (Ref 47). After 30 s, Ni_3Sn_4 + metastable $NiSn_3$ formed on the Ni(2 μm)/Cu, Ni_3Sn_4 on Ni(4 μm)/Cu, and Ni_3Sn_4 + Ni_3Sn on bare Ni. Approximately 10 at.% of Cu was detected in the interfacial IMC layers on both (2 μm)/Cu and Ni(4 μm)/Cu. It is interesting that the Cu atoms penetrated 4 μm of Ni in only 30 s at 250 °C (482 °F).

When the Ni plating is thin, the interfacial IMC undergoes evolution during soldering. There was a study of the reaction at 260 °C (500 °F) between the eutectic Sn-Ag and Au(500 Å)/Ni(1000Å)/Cu(7500 Å) for various times ranging from 10 s to 5 min. (Ref 54). Two interfacial IMCs with three morphologies were observed. The first phase, formed at the initial stage of the reaction, is predominantly $(Ni,Cu)_3Sn_4$. At the longer time, the initially formed $(Ni,Cu)_3Sn_4$ started to transform to $(Ni,Cu)_6Sn_5$. At the same time, the underlying Cu layer also reacts with Sn to form another $(Ni,Cu)_6Sn_5$ with a different morphology from the one transformed from $(Ni,Cu)_3Sn_4$.

Solder Joint in Contact with Cu on One Side and Ni on the Other. A eutectic Sn-Pb or Sn-Ag solder joint may contact Cu on one side and Ni on the other side. For example, solder balls are first attached to a BGA substrate with ENIG finish. The BGA package with solder balls is then attached to a motherboard with Cu pads. When the solder balls are attached to the BGA substrate of ENIG finish, a layer of Ni_3Sn_4 forms at the interface between the molten solder and the ENIG finish. When the BGA package is soldered to the motherboard, Cu atoms from the Cu pads dissolve into the molten solder, move cross the solder joint, and react with Sn and Ni to form $(Cu,Ni)_6Sn_5$ intermetallic on the interface between the solder and the ENIG finish, as shown in Fig. 4(a). This layer of $(Cu,Ni)_6Sn_5$ is located on the top of the Ni_3Sn_4 layer formed during solder ball attachment on the BGA substrate, as shown in Fig. 4(b). As shown in Ref 55, this additional layer of Cu-Ni-Sn ternary intermetallic makes the interface brittle.

The diffusion coefficient of Cu in liquid Sn has been reported in Ref 56, $D(Cu/Sn(l))$ $(cm^2/s) = 0.00018 \exp(-2113.6/T)$, where T is the absolute temperature in Kelvin. The time for Cu to cross a liquid Sn joint of size (h) may be estimated as Dh^2 (Ref 57). For a BGA joint of 0.5 mm (0.019 in.) diameter, Cu atoms need approximately 15 min. to cross; for a CSP joint of 0.3 mm (0.011 in.), Cu atoms need approximately 7 min.; and for a flip chip of 0.1 mm (0.003 in.), Cu atoms need approximately 0.6 min. This analytical estimation indicates that the formation of the second layer Cu-Ni-Sn intermetallic is probably a sensitive function of the reflow temperature and time, and the formation

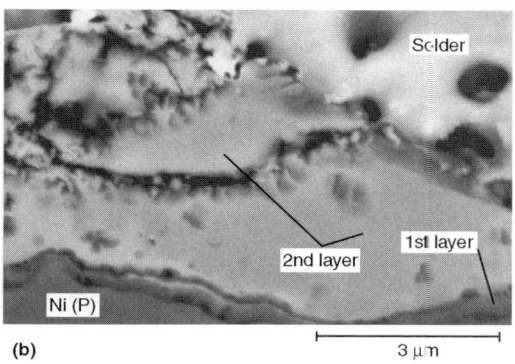

Fig. 4 Movement of copper atoms to form $(Cu,Ni)_6Sn_5$ intermetallic. (a) During reflow, Cu atoms from the other side of the solder joint move across the solder joint and form $(Cu,Ni)_6Sn_5$ intermetallic layer on the interface between the Ni(P) and solder. (b) Higher magnification view of the interface in (a). The interfacial intermetallic consists of two layers: the first layer, next to the Ni(P), is Ni_3Sn_4, formed during solder ball attachment to BGA substrate. The second layer is $(Cu,Ni)_6Sn_5$, formed as Cu atoms move across the molten solder joint and react with the Ni and Sn atoms.

of the ternary IMC is more an issue for smaller solder joints.

Growth of Cu-Sn Interfacial Intermetallic

During soldering, initially, the liquid solder alloy is unsaturated with respect to the substrate metal Cu, and the interfacial IMC layer will tend to dissolve into the solder. Thus the IMC/solder interface would retreat toward the substrate. On the other hand, either the Sn atoms in the liquid solder diffuse though the interfacial IMC and react with the Cu substrate to form additional IMC at the IMC/Cu interface, or the Cu atoms diffuse through the interfacial IMC layer and react with Sn atoms in the liquid solder, adding the IMC at the solder/IMC interface. Net growth in the interfacial IMC layer would occur as long as the growth reaction outpaced the dissolution reaction. Eventually, the Cu will approach its solubility limit in the liquid solder and the dissolution reaction will stop. Afterwards, the interfacial IMC grows at a higher speed because of the absence of dissolution. Finally, in cooling from the maximum reflow temperature, the solder becomes supersaturated with respect to Cu atoms. This can result in the formation of additional interfacial IMC on solidification. Of course, this additional growth of interfacial IMC depends on the cooling rate.

The measured layer thickness, X, versus time, t, at a certain temperature, T, may be adequately modeled with a power law:

$$X(t,T) = k(t)^n \quad \text{(Eq 8)}$$

$$k = k_0 \exp(-Q/RT)$$

It is useful to distinguish among the following three types of IMC growth mechanism based on the value of the time exponent, n:

- $n = 1$: Linear growth implies that the growth rate is limited only by the reaction rate at the growth site (i.e., solder/substrate interface). In other words, the growth is not limited by the rate at which the constituents of the IMC are able to diffuse to the reaction site.
- $n = 1/2$: Parabolic growth kinetics applies when the layer growth is controlled by the volumetric diffusion of elements to the reaction interface. Growth of the IMC layer becomes increasingly difficult as the layer grows because diffusion of one or more of the IMC constituent elements must diffuse through the existing IMC layer to reach the reaction site.
- $n = 1/3$: Subparabolic growth kinetics applies when the layer growth is controlled by grain boundary diffusion of elements to the reaction site.

Experimental Data of the Kinetics of IMC Growth in Liquid Solders. Data addressed are for IMC growth in eutectic Sn-Pb, liquid Sn, and liquid Sn-Ag.

IMC Growth in Eutectic Sn-Pb. There are some data on the kinetics of interfacial IMC growth between liquid eutectic Sn-Pb and Cu (Ref 22, 23, 31, 58). It is intended here to compare these data to see consistency among them. It should be noted that the determination of the growth kinetics of the interfacial IMC through cross-sectional analysis presents difficulties because of the scallop morphology. The height and width of the scallop traces in a cross section depends on the location of the cross section within the scallop. Also, scallops growing normal to the copper surface (and the cross section) will be taller on the section surface than if they were in some off-normal orientation.

The interfacial IMC growth was measured at 189, 255, and 310 °C (372, 491, and 590 °F) in two separate conditions: concurrent IMC growth and dissolution, and IMC growth in Cu-saturated liquid solder in the absence of dissolution (Ref 22, 58). In both cases, after specified times for IMC growth, the specimens were cooled on a large, room temperature steel block. The cooling rate was believed to be fast enough to suppress any additional IMC layer growth during solidification. For the measurement of IMC growth in Cu-saturated solder, the tests were done in two steps: first, solder was put on a Cu plate and reflowed for 15 min. to saturate the liquid solder with Cu; next, a second Cu piece was placed on top of the solder in a "sandwiched" configuration, adjacent to the Cu-saturated solder from the initial contact. Therefore, the IMC layer growth on this Cu piece should reflect only the growth effect and not the effects of dissolution.

For the interfacial IMC growth in the liquid Sn-Pb that is not saturated with Cu:

$$n = 0.37, k_0 = 7.75 \text{ }\mu\text{m/min}^{0.37},$$
$$\text{and } Q = 8.0 \text{ kJ/mol}$$

For the interfacial IMC growth in the Cu saturated Sn-Pb liquid:

$$n = 0.25, k_0 = 17.5 \text{ }\mu\text{m/min}^{0.25},$$
$$\text{and } Q = 9.0 \text{ kJ/mol}$$

Figure 5 shows the significant difference in the IMC growth at 215 °C (419 °F) for the two growth conditions.

Measured IMC layer growth was at 215 °C (419 °F) for 2, 6, and 20 min. (Ref 31). The results are plotted in Fig. 5 in comparison with (Ref 22, 58). Data (Ref 31) are enveloped by curves of the unsaturated and saturated solders. Growth data were collected (Ref 31) in an unsaturated solder, so the growth rate should be slower than (Ref 22, 58) data of the saturated solder. Data (Ref 31) are faster than (Ref 22, 58) unsaturated data probably because of the different rates of IMC dissolution. The dissolution rate, as seen previously in Eq 5, depends on the ratio of the solder volume to the contact area. The ratio in (Ref 22, 58) test was approximately 0.3 to 0.5 mm (0.011 to 0.019 in.), while the ratio in (Ref 31) test was not specified.

The IMC growth determined in unsaturated Sn-Pb solder at 230 and 250 °C (446 and 482 °F) for up to 5 min (Ref 23). Their data are compared with (Ref 22, 58) data in Fig. 6 and a good agreement is observed.

IMC Growth in Liquid Sn. The thickening behavior of η-phase and ε-phase in liquid Sn-Cu reaction couples over reaction times from 30 s to more than 4,000 min. and temperatures from 250 to 325 °C (482 to 617 °F) was studied (Ref 40). The thickness versus time and temperature data are fitted to Eq 8 with the least-square regression and values of the parameters:

For thickening of the η-phase,

$$n = 0.3, \ k_0 = 2.52 \ \mu m/s^{0.3},$$
$$\text{and } Q = 13.4 \ kJ/mol$$

For thickening of the ε-phase,

$$n = 0.57, \ k_0 = 2.26 \ \mu m/s^{0.57},$$
$$\text{and } Q = 29.2 \ kJ/mol$$

IMC Growth in Liquid Sn-Ag. Kinetics of interfacial IMC between liquid eutectic Sn-Ag and Cu substrate were reported in several studies. Extensive data was reported (Ref 9, 31, 32) of several prominent Pb-free solders (eutectic or near eutectic Sn-Ag, Sn-Ag-Cu, Sn-Ag-Bi, Sn-Bi, Sn-Sb, and pure Sn) and eutectic Sn-Pb, on several metal finishes (Cu, Ni-Cu of different thickness, Au-electroless Ni(P), Au-Pd-Ni(P), and Au-electrolytic Ni). References 3 and 59 measured the interfacial IMC layer over long times, up to 70 h at 240 and 250 °C (464 and 482 °F). Both data showed that in the long time range, for example, beyond 1 h, the IMC thickness varied with time in parabolic kinetics, implying that the bulk diffusion through the IMC layer is the limiting factor for the IMC growth.

Kinetics data was reported at 226, 251, and 292 °C (438, 483, and 557 °F), over 8 min. (Ref 7, 8). The isothermal data were fitted to Eq 8, and the parameters obtained are:

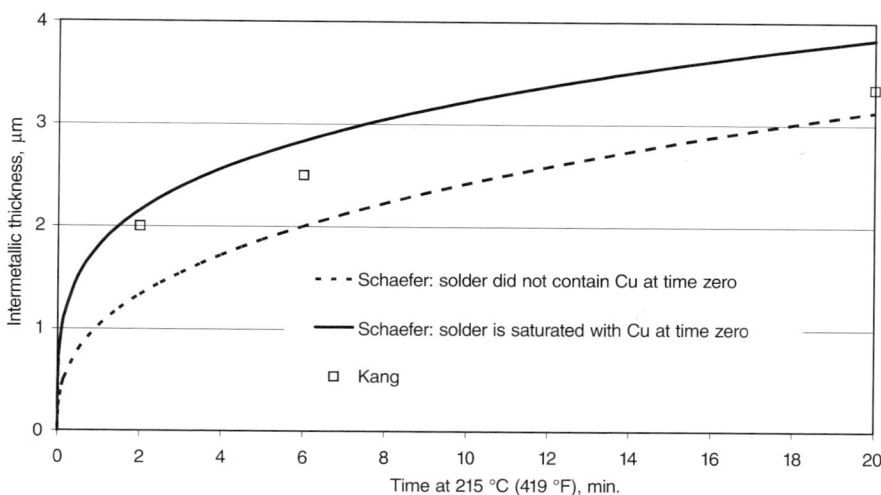

Fig. 5 Intermetallic layer thickness vs. time at 215 °C (419 °F) reflow, from Schaefer et al. (Ref 22) and Kang et al. (Ref 31). For Schaefer's data, in one case, the solder did not contain Cu at time zero, and in the other case, the solder is saturated with Cu at time zero. The plots are produced from the curve-fitted parameters in Ref 22.

$n = 0.31$, $k_0 = 11.6$ μm/s$^{0.31}$, and $Q = 7.6$ kJ/mol, (226 °C $\leq T \leq$ 251 °C)

$n = 0.37$, $k_0 = 391.2$ μm/s$^{0.31}$, and $Q = 23.2$ kJ/mol, (251 °C $\leq T \leq$ 292 °C)

The activation energy values obtained here are in close agreement with that reported in Ref 60.

At higher temperatures, 250 to 375 °C (482 to 707 °F), and longer time, up to 60 min., the IMC growth at the interface between Cu and eutectic Sn-Ag is different (Ref 42). The n value for the η-phase is approximately 0.75, and the n value for the ε-phase varies between 0.59 and 1.23. The increased n values at the higher temperatures and longer times were attributed to the unclosed liquid grooves between intermetallic scallops, which provides fast diffusion paths for Cu atoms to move to the intermetallic fronts. Therefore, the IMC growth is controlled more by the IMC reaction rate and less by the diffusion rate.

Theories for Interfacial IMC Growth. As shown in the previous section, when growth data are fitted to a growth law of Eq 8, the time exponent, n, is approximately 1/3. Two models were proposed to explain the 1/3 power dependence on time. One model assumes that the scalloped grains of the η-phase coarsen by a ripening reaction driven by the Gibbs-Thompson effect, and that Cu dissolves into the molten solder along liquid channels between the scallops (Ref 41). This model predicts a 1/3 power dependence of the mean scallop radius, hence, the layer thickness, on time. The second model assumes that the grain boundary diffusion is the predominant transport mechanism (Ref 58). The model includes the geometric effects caused by grain boundary grooving. The model predicts layer growth which follows a $t^{1/3}$ dependence on time, t. The difference between the two models is in the assumption of the dominant diffusion element. In Ref 61, the Cu atoms are assumed to diffuse faster through the grain boundaries of the interfacial IMC layer than the Sn atoms. In Ref 58, the Sn atoms are the faster diffusion species than Cu.

Schaefer's Model. The main assumption that leads to the $t^{1/3}$ dependence of the IMC thickness X is that the grain boundary diffusion is dominant and the volumetric diffusion is negligible. If the grain boundary has a constant width (δ) and grains are hexagonal with diagonal distance (d), the fraction of the total area available for grain boundary diffusion is proportional to δ/d. From the First Fick's Law of diffusion:

$$\frac{dX}{dt} \propto J = -D \frac{\delta}{d} \frac{dC}{dX} \quad \text{(Eq 9a)}$$

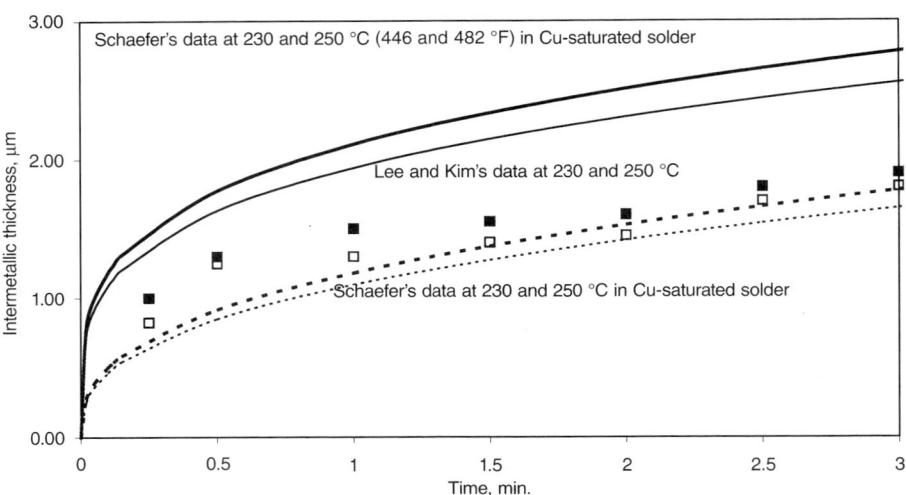

Fig. 6 Intermetallic layer thickness vs. time at 230 and 250 °C (446 and 482 °F) reflow, from (Ref 22) and (Ref 23). For (Ref 22) data, in one case, the solder did not contain Cu at time zero, and in the other case, the solder is saturated with Cu at time zero. The plots are produced from the curve-fitted parameters in Ref 23.

Here is another important assumption; $d = X$, meaning the equiaxed grain. Also, $dC/dX = \Delta C/X$, where ΔC is the Sn concentration at the η/Cu boundary minus that at the solder/η boundary, which is determined from the equilibrium phase diagram. Therefore,

$$\frac{dX}{dt} \propto -D \frac{\delta}{X} \frac{\Delta C}{X} = \frac{A}{X^2} \quad \text{(Eq 9b)}$$

where A is a constant, independent of time. Integration of 9(b) leads to $X = At^{1/3}$.

The assumption of $d = X$ is close to the reality. Experimental data from Ref 62 showed that the thickness to base ratio changed from roughly 0.8 to 1.2 as η-scallops grew from 1.3 to 10 μm. If the assumption of $d = X$ is not true, the dependence of X on t will be different from 1/3. In fact, Ref 63 replaces the $d = X$ with $d = kt^n$, and obtains different time dependence of X on t.

It is also interesting to note that the derivation that leads to the $t^{1/3}$ dependence of X does not have to assume that the Sn is the faster diffusing element. A similar derivation in Ref 23 is applied to grain boundary diffusion, but with the assumption that the Cu is the faster diffusion species.

Interdiffusion between Sn and Cu in Interfacial IMC Growth. There is disagreement in the referenced literature in the dominant diffusion species, Cu or Sn, for the interfacial IMC formation.

The η-phase has a scalloped shape at the η/Sn (liquid) interface, indicating that the reaction forming the η-phase occurs at the η/Sn (liquid) interface. This is consistent with the assumption that the Cu atoms diffuse through the η-phase and react with Sn atoms at the η/Sn (liquid) interface. If the Sn atoms diffused through the η-phase, the reaction would have occurred at the η/Cu or the η/ε interface, which could not explain the scalloped grain morphology.

It is known from much experimental data that Cu atoms dissolve into molten solder upon the initial contact of the Cu substrate with the liquid solder, and the dissolution of Cu into liquid Sn continues until the Cu content in the liquid solder reaches the solubility limit.

A study of the Cu-Sn reaction couple at temperatures between 250 and 325 °C (482 and 617 °F), observed that the ε-layer always grew as a somewhat undulated planar layer in phase with the η-phase scallops (Ref 40). This morphology indicated that the ε grows by reaction of Cu with η at the ε/η interface. This is consistent with the assumption that Cu is the dominant diffusion species.

A study was conducted of the interdiffusion process for soldering of Sn-3.5Ag alloy on Cu by using the Cr-evaporated surface as a reference line (Ref 64). At the beginning of soldering, Cu was observed to dissolve into the molten Sn-3.5Ag until saturation. The Sn-Ag solder dissolved with Cu collapsed, crossing the reference line into the Cu substrate side. As a result, the scallop-shaped Cu_6Sn_5 IMC was formed at the newly formed Sn-Ag-Cu solder/Cu interface at the Cu substrate side relative to the reference line. A sample that had been reflowed at 350 °C (662 °F) for 3 min. and water quenched was reflowed again at 250 °C (482 °F) for 6 min. Because the solder matrix was saturated with Cu when reflowed at 350 °C (662 °F), the η-phase growth at 250 °C (482 °F) took place without further Cu dissolution. The SEM cross section showed that the Cu/η interface moved into the Cu substrate. The result suggested that Sn is the dominant diffusing species for the IMC formation.

It was concluded that Sn is the faster diffusion species than Cu, based on the phase formation relationships and morphological development of the interface between Cu and Ag_3Sn reflowed at 450 °C (842 °F) for 576 h (Ref 3). The interfacial structure consists of $Cu/Cu_4Sn/(Ag_4Sn, Cu_3Sn)/Ag_3Sn$. The layer between Cu_4Sn and Ag_3Sn, labeled here by (Ag_4Sn, Cu_3Sn), is a two-phase mixture where the Ag_4Sn phase is the matrix mixed with the Cu_3Sn phase. The existence of the Cu_4Sn layer between the two-phase region and the Cu substrate indicates that Sn is the fastest moving species among the three elements. In this process, Sn diffused toward the Cu side reacted with Cu and the Cu_4Sn formed, with the growth front of the Cu_4Sn phase at the Cu/Cu_4Sn interface. Sn diffused out of the Ag_3Sn, leaving an Ag-rich Ag_3Sn phase, and possibly leading to the formation of the Ag_4Sn phase. By contrast, if Cu moved faster than Sn, the Cu would be in the matrix of Ag_3Sn, and a two-phase region rather than the Cu_4Sn would exist adjacent to Cu.

Numerical Analysis of Interfacial IMC Growth. Isothermal growth data were used to predict the IMC layer thickness ensuing a non-isothermal reflow (Ref 22, 58). The temperature-time reflow profile was divided into a differential time interval Δt, during which the temperature was considered as constant. The layer thickening during each Δt can be calculated from the isothermal growth data. The isothermal growth data was collected as a function of time and temperature, and curve fitted into empirical equations. When using the empirical equations for calculating the IMC layer thickening during each differential time interval Δt, an "equivalent time" needs to be used. The concept of an equivalent time assumes that, irrespective of the prior thermal history, an existing layer of given thickness, x', at a certain temperature, T', grows at the same rate as a layer of thickness, x', formed isothermally at T'.

Schaefer's method was improved, taking into consideration the dissolution of the η-layer into the molten solder until the copper concentration in the solder reaches the saturation limit, as well as the additional growth due to the reprecipitation of Cu as η after the Cu reaches it saturation limit (Ref 7). In consideration of Cu dissolution into the molten solder to reach the Cu solubility limit, they assumed that Cu dissolution into the molten solder is entirely from the dissolved η IMC. They also assumed that during cooling from the reflow temperature, Cu becomes incrementally supersaturated. The excess Cu atoms in the solder react with readily available Sn to form η in the bulk solder, or migrate to the interface and precipitate on the interfacial η layer. The portion that precipitated at the interface is calculated based on the growth controlled by diffusion, and the diffusion coefficient of Cu in liquid Sn was used (Ref 56).

Kinetics of Ni_3Sn_4 in Liquid Solder

The growth kinetics of the Ni_3Sn_4 IMC layer was measured in Ref 63 at 200, 225, and 250 °C (392, 437, and 482 °F) for 30, 60, 150, and 300 s, at the interface between eutectic Sn-Pb and Pd-Ni-Cu substrate. The best fit of the data to Eq 8 yields:

$$n = 0.31, Q = 9.26 \text{ kJ/mol}$$

$$k_0 \approx 1 \text{ μm/s}^{0.31} \text{ (225 and 250 °C)}$$

$$k_0 = 0.61 \text{ μm/s}^{0.31} \text{ (200 °C)}$$

The difference in k_0 values was attributed to the fact that at 225 and 250 °C (437 and 482 °F), only the liquid phase was in contact with the IMC layer, while at 200 °C (392 °F), $PdSn_4$ grains formed very close to the interface. The presence of $PdSn_4$ in contact with the IMC layer caused the decreased IMC layer growth.

Interfacial IMC Microstructures

Morphology. Three morphologies of the interfacial η-Cu_6Sn_5 layer were observed (Ref 39):

- *Cellular layer with a rugged interface.* This appears similar to columnar grains in plane view, but the cross section shows branches with substantial intergranular space. Hence, the layer is not dense. The interface with the solder is rugged.
- *Dense layer with scalloped interface.* This appears similar to the cellular grains in plane view, but the layer is dense beneath the surface. The interface with the solder appears similar to scallops.
- *Dense layer with planar interface.* The morphology of the η-layer varies gradually from a cellular film with a rugged interface to a dense film with a scalloped interface as the Pb content, temperature, and reaction time increased. The ε-phase is always dense and nearly planar.

Fast cooling rates resulted in a relatively planar Cu_6Sn_5 layer, while a nodular Cu_6Sn_5 morphology was present for slower cooling (Ref 65). The reflow time also had an effect on the intermetallic morphology, with shorter reflow times leading to a relatively planar, η-phase morphology, while longer reflow times produced a more nodular or scalloped η-phase. The ε-phase appeared to exhibit planar growth irrespective of the reflow time.

Spalling. The IMC, initially formed at the substrate/liquid solder interface, sometimes starts to move away from the interface as a result of prolonged or multiple reflows. This phenomenon often seems related to Ni, especially electroless Ni(P) substrate.

A study was conducted of interfacial IMC structures between a few Pb-free solders (Sn-3.5Ag, Sn-3.8Ag-0.7Cu, and Sn-3.5Ag-3.0Bi) and several substrates (Cu, Au-Ni(P), and Au-Pd-Ni(P)) reflowed at 250 °C (482 °F) for up to 20 min. (Ref 32). For Sn-3.5Ag and Sn-3.5Ag-3.0Bi on Au-Ni(P) or Au-Pd-Ni(P), the majority of the Ni_3Sn_4 IMC formed were separated or spalled away from the interface, and only a thin layer of the IMC remained attached to the interface, while for Sn-3.8Ag-0.7Cu on Au-Ni(P) or Au-Pd-Ni(P), the interfacial IMC of $(Cu,Ni)_6Sn_5$ adhered well to the interface. In the electroplated

Ni, all three Pb-free solders produced well-adhered IMC of Ni_3Sn_4.

Similar results were reported in Ref 66 of 99.3Sn-0.7Cu, 95.5Sn-3.8Ag-0.7Cu, 95Sn-3.5Ag, and 96Sn-2Ag-2Bi, on electroless Ni(P) formed at 260 °C (500 °F) for approximately 60 s. After reflow, the interfacial IMC for the solders containing Cu is $(Cu,Ni)_6Sn_5$ and had good adhesion with electroless Ni(P), while interfacial IMC for the solders that do not contain Cu is Ni_3Sn_4 with a needle-shaped morphology that spalled off the surface of electroless Ni(P).

The phenomenon of IMC spalling at the interface between eutectic Sn-Bi and Sn-Pb and electroless Ni with different P contents was studied (Ref 67). It was observed that the spalling was promoted by longer time and related to the P content in electroless Ni.

Sn-3.5Ag-1.0Cu was used to solder flip chip packages with Al-Ni(V)-Cu under bump metallization (UBM)/on electroless Ni(P) (5 µm)/immersion Au (0.125 µm) substrate (Ref 68). They observed that after three reflows at 250 °C (482 °F), the $(Cu,Ni)_6Sn_5$ formed at the UBM/solder interface started to separate from the interface. After 10 reflows, an IMC layer was found to "float" in the center of the solder joint. The IMC grains, which had drifted to the center of the solder, were still in a layered structure, although not a continuous form. A high magnification SEM and EDX analysis showed that after 5 or 10 reflows, Cu and Ni were no longer detected at the UBM layer, and Sn and V were the major compositions on top of the Al layer. A ductile to brittle transition was observed as the reflow times increased. For the same UBM structure and solder joint, when they were not assembled to the Ni-Au substrate, spalling did not occur until 20 reflows. A possible explanation was that the Ni from the substrate side dissolved into the molten solder during reflow, moved to the IMC at the UBM/solder interface, thereby increasing the grooving of $(Cu,Ni)_6Sn_5$.

Texture. A study was conducted using x-ray diffraction to observe the IMCs at the interface between molten solders and the Cu substrate (Ref 39). The solders were pure Sn, 82Sn-18Pb, 63Sn-37Pb, and 27Sn-73Pb. A strong texture was detected in both the hexagonal η-phase and the orthogonal ε-phase in (101) and (002) pole figures, respectively. The growth directions were identified to be $\langle 101 \rangle$ and $\langle 102 \rangle$ for the η-phase and $\langle 102 \rangle$ and $\langle 031 \rangle$ for the ε-phase, normal to the Cu surface. The growth direction does not change with the morphology and the thickness of the IMC layer.

Intermetallic Joints. Thin (30 µm) joints were made consisting of Cu-Sn intermetallics formed by isothermal solidification of Cu-Sn diffusion couples (Ref 69). During the initial stages of isothermal solidification, both Cu_6Sn_5 and Cu_3Sn phases grew, even though the former is the dominant one. After consumption of all available Sn, the Cu_3Sn phase grew reactively at the expense of Cu and Cu_6Sn_5. Finally, they obtained solder joints that consisted of only Cu_3Sn, for 5 h at 325 °C (617 °F). Five µm thick intermetallic joints consisting of Cu_6Sn_5 and Cu_3Sn can be made by stacking and compressing two Sn-coated Cu plates together at 0.05 to 0.1 MPa (0.007 to 0.001 ksi) at 280 °C (536 °F) for approximately 4 min. (Ref 34). No flux was used and the sample was reacted in ambient air.

Effects of Microelements on IMC between Sn-Ag-Cu Solder and Cu Substrate. Copper plates were dipped into baths of Sn-3.5Ag-0.7Cu and Sn-3.5Ag-0.5Cu-0.07Ni-0.01Ge at 250 °C (482 °F) for 15 s, respectively (Ref 70). SEM cross sections showed that the thickness of the interfacial IMC layer formed with Sn-3.5Ag-0.5Cu-0.07Ni-0.01Ge was 5 µm, which is several times that formed with Sn-3.5Ag-0.7Cu. The intermetallic morphology formed between Cu and Sn-3.5Ag-0.7Cu is of pebble shape. The IMC morphology with Sn-3.5Ag-0.5Cu-0.07Ni-0.01Ge tended to be worm-shaped, and they were seen to drift or spall into the solder matrix. X-ray diffraction study shows that the IMC formed between Cu and the two alloys are the same crystalline structure, $(Cu,Ni)_6Sn_5$. Electron probe microanalysis (EPMA) showed that Ni was and Ge was not present at the interface.

Sn-3.5Ag-0.07Ni was investigated to clarify the microelement effects on the growth and morphology of interfacial IMC. The shape and thickness of the IMC formed between Sn-3.5Ag-0.07Ni and the Cu substrate are the same as those formed between 3.5Ag-0.5Cu-0.07Ni-0.01Ge and the Cu substrate. Notice that Sn-3.5Ag-0.07Ni is Cu free, and the source of the Cu atoms in the interfacial IMC is the Cu substrate. The contention is that during soldering, Ni aggregates at the interface of Cu/Sn-3.5Ag-0.07Ni and Cu/Sn-3.5Ag-0.5Cu-0.07Ni-0.01Ge and enhances diffusion of Cu atoms from the Cu substrate to form a thicker IMC layer than that formed at the Cu/Sn-3.5Ag-0.7Cu interface.

The base alloy of Sn-3.8Ag-0.7Cu was added with Au ranging from 0.1 to 5 wt% (Ref 71).

DSC study showed that Au addition promotes the formation of a quaternary-eutectic phase (AuSn$_4$, Au$_3$Sn, β-Sn, and Cu$_6$Sn$_5$) at 204.5 °C (400.1 °F). The addition of the Au to Sn-3.8Ag-0.7Cu increases the liquidus temperature and the temperature ranges of the phase equilibrium field for the primary phases. Such effect from Au addition was less pronounced when the alloys were reacted with a Cu substrate. Because of the formation of the Au-Cu-Sn ternary interface intermetallic, the majority of the Au in the solder matrix was drained to the interface. The drainage of Au reduced the impact of Au on the phase equilibria of the solder alloys in the joint. It was also observed that the involvement of Au in the interface reaction results in a change of the interface morphology from the conventional scallop structure to a composite-like structure consisting of (Au,Cu)$_6$Sn$_5$ grains and finely dispersed β-Sn islands.

Solid Solder-Substrate Reactions

During solid state aging, the interfacial IMC thickness increases and its morphology evolves from a scalloped shape to a flat and uniform shape. Also, upon solidification, the solid solder is either supersaturated with the substrate metal that dissolved during soldering or filled with precipitated IMCs. The substrate metal atoms in either the supersaturated solid solution or the precipitated IMCs may diffuse back to the solder/substrate interface, resulting in an additional interfacial IMC growth. Excessive interfacial IMC formation during solid state aging produces segregation of chemical species that do not take part in the IMC formation. Excessive solid state aging also produce a large quantity of porosities at the solder/substrate interface due to either the Kirkendall effect or the material density reduction in the IMC formation.

Experimental Data

Solders/Cu Interface. There is a large body of literature on the solid state intermetallic growth at the Cu/solder interface (Ref 72). In terms of the range of time and temperature, the most comprehensive investigation of intermetallic growth kinetics for samples of Cu-Sn and Cu-60Sn-40Pb was conducted during the early 1970s (Ref 73, 74).

The thickness, X, of the intermetallic layer can be represented by an equation of the usual form:

$$X(t,T) = X_0 + k(t)^n \qquad \text{(Eq 8a)}$$
$$k = k_0 \exp(-Q/RT)$$

where X_0 is the thickness of the intermetallic in the as-soldered condition (at $t = 0$). The exponent, n, was determined to be 0.5 for intermetallic growth in a semi-infinite Cu-Sn diffusion couple (Ref 75), and ~0.33 for intermetallic growth in a thin solder layer on Cu (Ref 73, 74). The parameters obtained from the curve fitting to the data from Ref 73 and 74 of the total Cu-Sn intermetallic thickness (Cu$_6$Sn$_5$ and Cu$_3$Sn) versus time and temperature are:

For pure Sn on wrought Cu:
$n = 0.347$, $k_0 = 7.18 \times 10^{-3}$,
$Q = 523$ (kcal/mol)

For 60Sn-40Pb on wrought Cu:
$n = 0.372$, $k_0 = 3.56 \times 10^{-3}$,
$Q = 7.941$ (kcal/mol)

Activation energies for the Cu$_6$Sn$_5$ is lower than that in the Cu$_3$Sn phase (Ref 75, 76). It follows that the proportion of Cu$_6$Sn$_5$ in the total intermetallic thickness increases as the temperature decreases.

Solders/Ni Interface. The growth kinetics of the IMC Ni$_3$Sn$_4$ among 100 Sn, 95Sn-3.5Ag, and 63Sn-37Pb solders on electroplated Ni during solid state aging at 160 °C (320 °F) up to 36 days were studied (Ref 77). The growth rates were 0.12 μm/h$^{0.5}$ for 100 Sn, 0.17 μm/h$^{0.5}$ for Sn-3.5Ag, and 0.19 μm/h$^{0.5}$ for Sn-37Pb. The faster growth rate for Sn-37Pb solder was attributed to a relatively high homologous aging temperature. At the end of the 36 days of aging, metallographic analysis revealed cracks and separation in the Ni$_3$Sn$_4$ IMC layer, which may have a detrimental effect on the mechanical strength of the solder joint.

Interface Reliability

Pb-rich Phase Band

The effect of isothermal aging on the fatigue crack growth behavior at the Sn-Pb solder/Cu interface with emphasis on the role of the interfacial microstructure was tested (Ref 78). Fatigue crack growth rates along the solder/Cu interface versus strain energy release were measured as a function of aging time from 7 to

30 days at 140 °C (284 °F). The aging produced a continuous Pb-rich phase band next to the interfacial IMC, resulting in an easy path for fatigue crack propagation. The threshold strain energy release rate for fatigue crack growth was changed from 25 J/m^2 at the as-reflowed condition to 10 J/m^2 after 30 days of aging.

Blocky Ag$_3$Sn

Fatigue crack propagation in 95Sn-5Ag solder joint between Cu plates formed at 280 °C (536 °F) for 15 min. was tested (Ref 11) Next to the interfacial Cu$_6$Sn$_5$ IMC layer, there were large, nodular Ag$_3$Sn particles. This interfacial microstructure was shown to result in inferior fatigue resistance, with the fatigue crack path following the interfacial Ag$_3$Sn IMC phase.

Critical IMC Thickness

The shear strength of the BGA solder joint of three alloys on Cu formed at 270 °C (518 °F) for different times were studied (Ref 19). The three alloys are pure Sn, Sn-1.5Cu, and Sn-2.5Cu. The Cu additions into the Sn solder enhanced the interfacial IMC layer growth during soldering. There exists a critical thickness for the IMC layer at which the shear strength is a maximum. For all three solders, the critical IMC layer thickness is approximately 1.2 µm, corresponding to the reflow times of approximately 60 s for the pure Sn and approximately 15 s for Cu-containing solders. The critical IMC thickness is closely related to the changes in the fracture mode. When the IMC thickness is thinner than the critical thickness, the shear fracture occurs inside the solder, and the shear strength increases with the reflow time because of the increased Cu$_6$Sn$_5$ precipitates in the solder matrix. As the interfacial IMC grows beyond the critical thickness during soldering, there is an onset of a brittle fracture in the IMC layer, resulting in the decreased shear strength with reflow time.

Kirkendall Voids in IMC

At least four separate studies have reported void formation in the ε-Cu$_3$Sn phase and degradation of the joint strength after long-term solid state aging. An SEM photo of a cross section of a Cu/eutectic Sn-Pb interface after baking for 20 days at 125 °C (257 °F) is shown in Fig. 7.

The interface microstructure and shear strength of 96.5Sn-3.5Ag and 62Sn-36Pb-2Ag on Cu substrate after aging at 150 °C (302 °F) for up to 0, 50, 250, 500, and 1000 h were studied (Ref 18). At the solder/Cu interface, the duplex structure with Cu$_6$Sn$_5$ next to the solder and Cu$_3$Sn next to the Cu substrate was observed in all aged samples. As the aging time increased, voids form in the Cu$_3$Sn phase. The shear strength of both Sn-Ag and Sn-Pb-Ag decreased with the aging time, which is related to a fracture mode change from the mixture of solder and IMC at the zero aging time, to the complete fracture within the IMC layer after 1000 h aging.

The void formation in the Cu$_3$Sn layer is related to the form of the Cu substrate used, e.g., voids were observed in the eutectic Sn-Ag solder joint on electroplated Cu after aging at 190 °C (374 °F) for 3 days (Ref 14). It is interesting that they also aged the same solder on a rolled Cu at 190 °C (374 °F) for 12 days. Though the ε-phase exists, no voids were found in either the ε-phase or the η-phase. During the formation of Cu$_3$Sn, the mass imbalance, due to the different diffusion rates of Sn and Cu, induces the formation of vacancies or fine Kirkendall voids, which may be accelerated by the hydrogen introduced during the electroplating process.

Miniature Charpy tests were conducted on four different solder balls (eutectic Sn-Pb, near eutectic Sn-Ag-Cu, eutectic Sn-Zn, and Sn-Zn-Bi) bonded to Cu or electroless Ni(P)-immersion Au, and subsequently aged at 150 °C (302 °F) for up to 1000 h (Ref 79). A large number of voids were observed in the ε-phase after 500 h

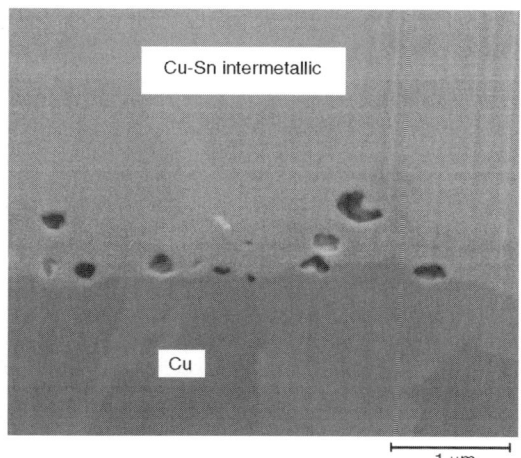

Fig. 7 Voids at the interface between 63Sn-37Pb solder joint and Cu substrate. The sample is solder joint in a BGA package attached to a motherboard with Cu pads. After attachment, the assembly was baked at 125 °C (257 °F) for 20 days.

aging at the interface between Sn-Pb or Sn-Ag-Cu and Cu substrate, and in the vicinity of the γ-Cu_5Zn_8 segment after 1000 h aging at the interface between Sn-Zn or Sn-Zn-Bi and Cu substrate. Ductile to brittle transition associated with the fracture inside the solder to within the IMC phase was seen as the aging time increased.

Drop and shear tests were conducted of BGAs with near eutectic Sn-Ag-Cu solder balls on Cu pads after thermal aging at 100, 125, 150, and 175 °C (212, 257, 302, and 347 °F) for 3, 10, 20, and 40 days (Ref 80). Kirkendall voids were observed at the Cu/Cu_3Sn interface. Voids occupied 25% of the pad/solder interface after only 3 days of 125 °C (257 °F) aging. The void density increased with the aging time and temperature. The drop performance degraded 80% from time 0 to 10 days of 125 °C (257 °F) aging.

Black Pad

ENIG is being used as a metal finish on PCB and BGA package substrates, as well as under bump metallization in flip chips, to protect the copper pads from oxidation and present a solderable surface. "Black pads" refer to the failures of solder joints attached to the ENIG finish. There appear to be two distinctive types of failures. The first occurs mostly in the component assembly process. Solder joints do not wet well or de-wet on soldering pads. The unwetted or dewetted soldering pads are strikingly dark, which is why the failure is called a black pad. Figure 8(a) is an optical photo of black pads (Ref 81). Figure 8(b) is an SEM photo of the pad surface, showing corrosion cracks. High contents of carbon and oxygen, sometimes phosphorus, are seen on the black pad surface by an SEM/EDX, as shown in Fig. 8(c). The failure occurred during a "toothpick" test. As a toothpick was used to touch lightly the leads of a PQFP package, the leads lifted off the solder pads.

The second type of failure is often observed in mechanical tests, such as bending, shocking, and vibration. In these tests, the solder joints break at low strength and show extremely flat fracture surface, as shown in Fig. 9(a) to (c) (Ref 82). However, the fracture surface is not black, and there is no abnormal concentration of either carbon or oxygen, as shown in Fig. 9(d) to (e). There is little or no trace of corrosion cracks, or "mud" cracks, on the fracture surface. There may be a relationship between the two types of failures; the first type of failure may be the worst case of the second type of failure. It appears that the first type of failure is related to poor control of the plating parameters of electroless Ni and immersion Au plating baths, while the second type of failure is related to the interfacial microstructure between ENIG and solder.

Electroless Ni/Immersion Au (ENIG). Electroless Ni is actually a nickel-phosphorus alloy. Phosphorus ions (P^{2+}) are present as reducing agent ($P^{+2} \Rightarrow P^{4+}$) in the plating bath, but some of the phosphorus ions are reduced ($P^{2+} \Rightarrow P^0$) themselves and co-deposited with Ni (Ref 83). The electroless Ni plating baths are typically divided as low P ($\sim$5 wt% P), middle P (8–10 wt% P), and high P (13–15 wt% P). For most of the applications in the electronic packaging industry, the middle P bath is being used. At the low P concentration, the electroless Ni is crystalline, while at the middle and high P concentrations, the electroless Ni is amorphous or crystalline with extremely small grain size (Ref 84). At ambient temperature, electroless Ni with the low phosphorus content is magnetic, while electroless Ni with the middle and high phosphorus contents are non-magnetic. In the vicinity of 7 wt% P, both the strength and ductility of the electroless Ni rise abruptly (Ref 84). The corrosion resistance of electroless Ni increases with its phosphorus content, whereas its wettability degrades as the phosphorus content increases (Ref 85, 86) or improves (Ref 87). The solubility of phosphorus in nickel is nearly zero; the stable phases are Ni and Ni_3P, according to the Ni-P phase diagram. Therefore, the electroless Ni is a super-saturated solid solution of phosphorus. When it is subjected to a high temperature excursion, for example >240 °C (464 °F) for 30 min., electroless Ni becomes kinetically unstable, leading to Ni_3P precipitation. Therefore, the mechanical properties of electroless Ni are sensitive to thermal treatment (Ref 84).

Immersion Au is a displacement deposition. In an immersion Au bath, Au ions receive electrons supplied from Ni atoms on the substrate. As a result, Ni atoms become ions and dissolve into the bath, and Au ions become atoms and deposit on the substrate. The electron exchange is determined by the standard electrode potential. In theory, the deposition by an immersion process is self-limiting; once the Ni surface is covered with an atomic layer of Au, the deposition stops because the Au ions in the bath and the Ni atoms on the substrate are no longer in contact. In reality, the thickness of immersion Au is approximately 0.1 μm. When the immer-

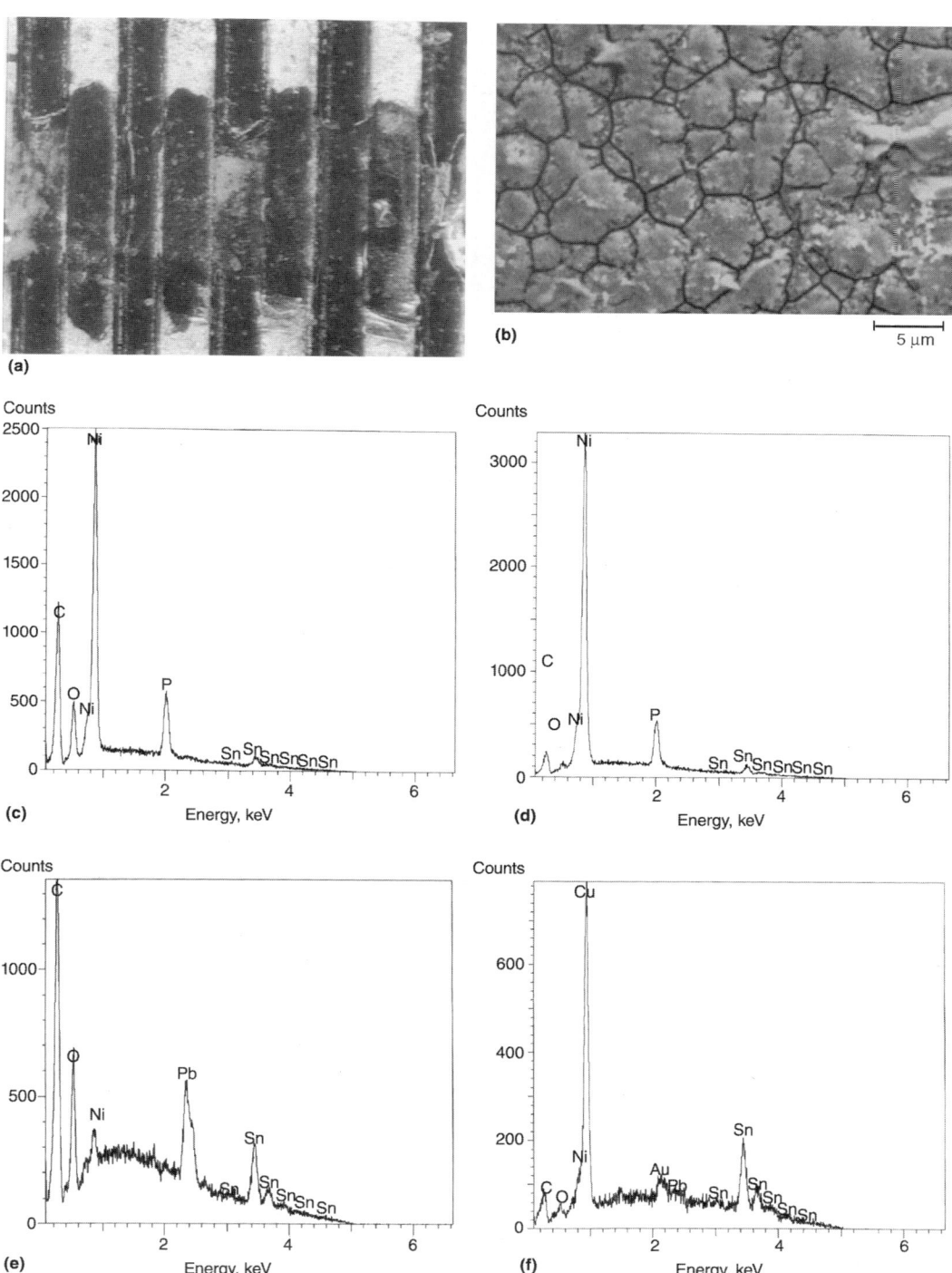

Fig. 8 "Black pad" failure. (a) Optical photo of top view of "black pads." After components with weak joint strength were removed, the pads underneath were exposed, showing strikingly dark color. (b) SEM photo of a black pad, showing mud crack at the electroless Ni nodule boundaries and corrosion trace around the nodule boundaries. (c) EDX plot of a black pad, showing high contents of carbon and oxygen. (d) EDX plot of a regular (non-black) pad, showing low contents of carbon and oxygen. (e) EDX plot of the surface of a component lead that was pulled off from a black pad, showing high contents of carbon and oxygen. (f) EDX plot of the surface of a component lead that was pulled off from a non-black pad, showing low contents of carbon and oxygen.

sion time is prolonged, the Au thickness increases slightly; the electron exchange continues through the surface defects, e.g., pin-holes. The displacement reaction between Ni atoms and Au ions results in a phosphorus enriched layer next to the immersion Au coating, since the phosphorus atoms in electroless Ni do not participate in the displacement reaction.

General Features of Black Pads

Sporadic Failures. The failure of solder joints attached to ENIG is occasional. In the reference literature, there are numerous reports showing that solder joints on the ENIG finish are reliable (e.g., Ref 88, 89). ENIG seems less of a reliability problem for solder joints in flip chip packages (the first-level interconnect) than in

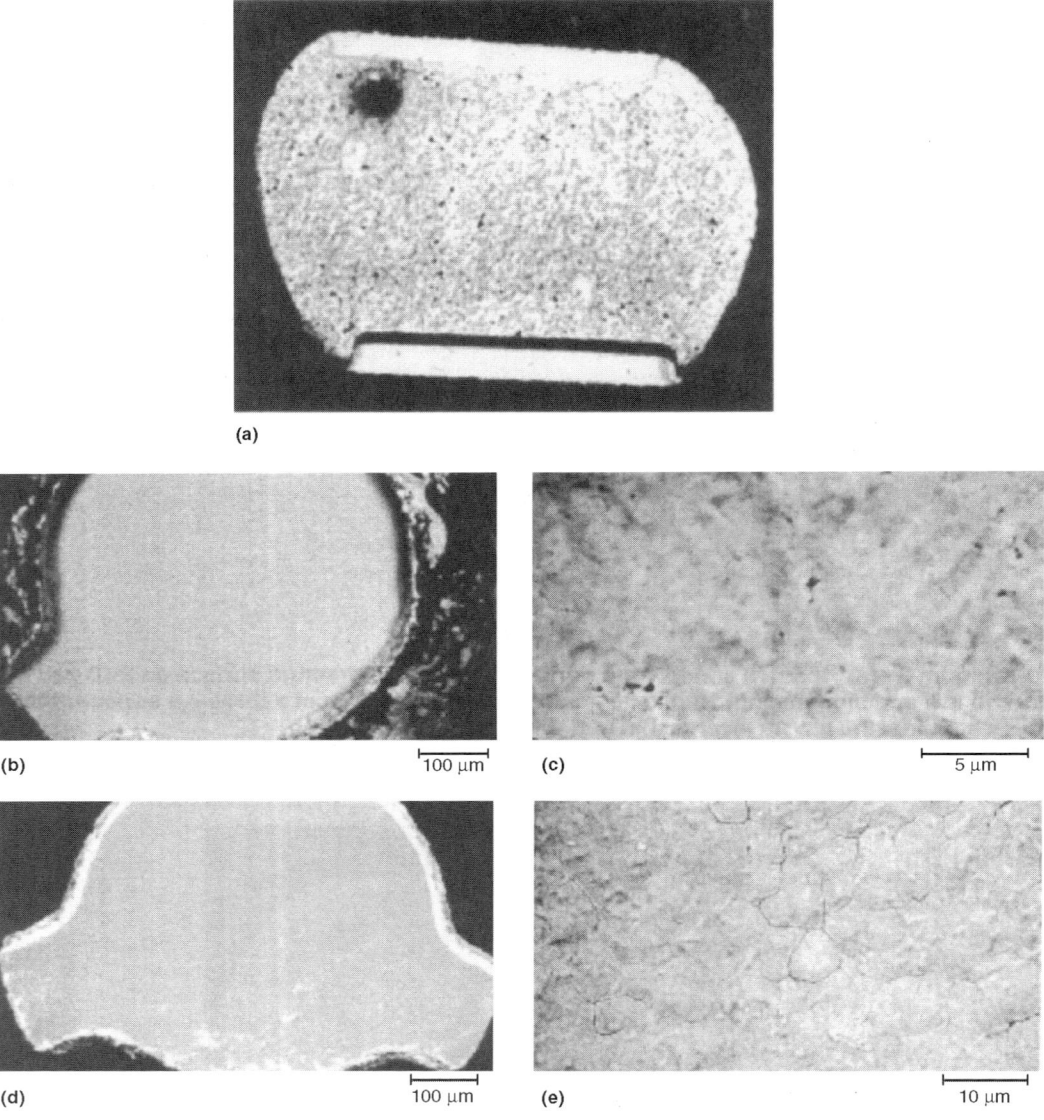

Fig. 9 "Black pad" brittle interfacial failure in which the surface does not look black. (a) Pad/solder failure mode typical of failed BGAs on E-Ni/I-Au boards. (b) Low-magnification SEM photograph of fracture surface on PBGA solder ball side. (c) High-magnification SEM view of (b). (d) Low-magnification SEM photograph of the fracture surface on the pad side; there is little or no trace of "mud" crack or corrosion trace. (e) High-magnification SEM view of (d). Source: Ref 82

BGA packages or leaded components (the second-level interconnect).

The poor wetting or de-wetting failure occurs often during a particular time period or for a particular product lot. During that time period or product lot, a certain percentage of assembled products fail. In the failed assembled parts, only some pads are unwetted or dewetted. These failed pads are often at the same locations for all the failed assemblies. Later, it will be seen that the wetting problem is caused by the galvanic corrosion in the immersion Au bath. Occasionally, various plating process parameters could be traced as the origin of the failure, for example, prolonged time in the immersion Au bath (Ref 90, 91). However, at most times, the failure cannot be traced to particular plating parameters, and cannot be "turned on" or "turned off" by controlling the plating parameters.

The brittle fracture during mechanical tests does not occur for the solder joints attached to all ENIG pads. ENIG pads made with different plating parameters or produced by different vendors exhibit different solder joint strength and percentage of brittle fracture. But so far, a recipe of ENIG plating for eliminating brittle interfacial fracture has not been published in the reference literature.

Interface Microstructure. Figure 10 shows a cross section of a Sn-37Pb solder ball attached on an ENIG finish. A dark layer (denoted as Ni(P)$^+$ layer) is visible between Ni-Sn intermetallic and electroless Ni (denoted as Ni(P)). Chemical analysis indicates that the phosphorus content in the Ni(P)$^+$ is roughly twice that in the Ni(P) substrate. When molten Sn-Pb solder contacts Ni(P), only Ni-Sn intermetallic is observed; Neither P nor Pb was detected with EDX in the Ni-Sn intermetallic. As a result, P is expelled from the Ni-Sn intermetallic layer and segregates at the interface between the Ni-Sn intermetallic and Ni(P), just like Pb that segregate at the interface between solder and the Ni-Sn intermetallic. The Ni(P)$^+$ layer was made of Ni$_3$P crystalline phase by TEM analysis (Ref 11, 92, 93), and possibly of Ni$_5$P$_4$, also by TEM analysis (Ref 49).

The interfacial intermetallic is mainly Ni$_3$Sn$_4$. There may be others; thin layers of both Ni$_3$Sn and Ni$_3$Sn$_2$ were reported in TEM and x-ray diffraction studies (Ref 43, 90, 94, 95). Most studies conclude that phosphorus does not take part in the interfacial intermetallic. However Ref 43 observed the phase of Ni$_3$SnP, where P replaces Sn in the Ni$_3$Sn$_2$ structure.

Interfacial Fracture. Figures 8(a) and (b) are the SEM images of two sides of the fracture surface; Figures 8(c) through (f) are EDX plots for various locations on the fracture surfaces. The solder ball side of the fracture surface is flat and featureless; there are Ni, Sn, Pb, and a very small amount of P. It is deduced that there must be a thin layer of Ni-Sn intermetallic on the solder ball side of the fracture surface.

The PCB side of the fracture surface is also very flat. At higher magnifications, some surface

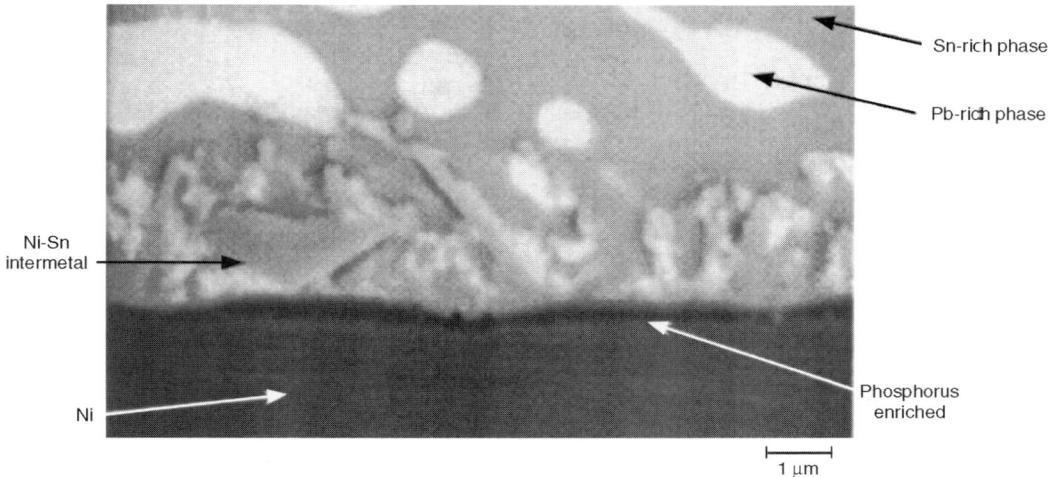

Fig. 10 SEM cross section of the interface between Sn-37Pb solder and ENIG finish, showing Ni-Sn intermetallic phase and a phosphorus enriched layer

cracks are visible at electroless Ni nodule boundaries. There are Ni, Sn, and P. The presence of Sn at the fracture surface indicates that there must be some residues of Ni-Sn intermetallic phase. Similar observations were also reported in Ref 96.

Based on the fracture surface analysis, the brittle interfacial fracture occurs at the interface between the Ni(P)$^+$ and Ni$_3$Sn$_4$, possibly in the Ni$_3$Sn$_2$ or Ni$_3$Sn. It was thought initially that the brittle interfacial intermetallic was due to the phosphorus segregation. As shown later, this hypothesis does not seem to be the root cause.

Sensitivity to Testing Methods. In the product qualification tests, the brittle interfacial fracture of solder joints attached to ENIG pads occurs mostly during shock and vibration tests, and less during thermal cycling tests (Ref 97). This is because the strength of solder materials is highly sensitive to strain rate and temperature, while the strength of the solder/ENIG pad interface is not. So, at higher temperatures and lower strain rates, the solders become much softer, accommodate most of the deformation, and become the weakest link. At lower temperatures or high strain rates, the solders are stronger than the solder/ENIG pad interface, resulting in the brittle fracture at the interface.

In Ref 98, two testing methods were used, i.e., four-point bending and pull. In the four-point bending test, a rectangular plate of PCB with one BGA attached at the center is bent to failure. The failure occurs in one of the two modes: brittle fracture at the solder/ENIG pad interface, or pads being pulled out. There is a good correspondence between the fracture mode and fracture strength; the fracture load for pulling out the pad is about twice that for the interfacial brittle fracture.

The pull test was intended as a process monitoring test (PMT) because the dewetting or brittle fracture problem could not be controlled by the plating parameters. The test vehicle was designed as a single FR4 layer with a solid copper plane on one side of FR4. Solder mask was used to create a 5 by 5 array of soldermask-defined pads where the pads were simply exposed regions of the Cu plane. In this design, the "pads" have essentially infinite adhesion strength. Two rectangular plates with a 5 by 5 array of pads in the center were soldered together with standard surface mount assembly processes. The test results, however, showed that the test vehicle was not adequate to differentiate the brittle fracture at the solder/pad interface from the failure within the solder joints. The correlation between the fracture load and the percentage of the interface fracture area on the fracture surface was not obvious. An improvement of the process monitoring test was described in Ref 99. The brittle fracture was more susceptible to bending than to pull tests (Ref 100).

Thermal cycling, tensile, shear, and mechanical fatigue tests of solder joints of several alloys, including Sn-37Pb, Sn-0.7Cu, and Sn-3.5Ag-0.5Cu, attached to ENIG and Cu pads, in a flip chip test vehicle without underfill were conducted (Ref 101, 102). The thermal fatigue temperature profiles were 0 to 100 °C (32 to 212 °F), and -40 to 125 °C (-40 to 257 °F), both at 1 cycle/h. The thermal cycling, tensile, and shear tests did not reveal any failure at the solder/pad interface. In fact, the thermal fatigue lives and tensile and shear strength of solder joints were the same on both Cu and ENIG. Only in the mechanical fatigue test was brittle interfacial fracture observed at the solder/ENIG interface.

Defect Mapping. It was found in failure analysis that for any particular PCB design, there was a repetitive pattern to the locations that exhibited the black pads (Ref 103, 104). A black pad can be located immediately beside a pad or between two pads that are perfectly acceptable. It was found that the affected pads were electrically connected to some other feature on the board that tended to have a larger ENIG surface area that was not affected. Some affected pads were electrically connected to larger plated-through-hole sites used for connectors. It was suggested that in the immersion Au bath, there is some electric potential difference between the affected pads and the large connecting ENIG area. The potential difference results in a galvanic cell action that permitted the attack of the smaller pads in preference to the larger pad.

Dependence on Solder Composition and Flux. Most failures of ENIG finish reported in the reference literature relate to eutectic Sn-Pb solder joints. When solder joints of other alloys are attached to ENIG finish, the fracture behaviors are different. A ceramic BGA with 90Pb-10Sn solder balls attached to an ENIG finish using 63Sn-37Pb solder paste was tested (Ref 82). The assembled package was four-point bent to fracture. The failure mode was peeling of the PCB pads. Another test is more convincing. The outmost row of 63Sn-37Pb solder balls in a PBGA was removed and replaced with 90Pb-10Sn balls. The modified PBGA package was then attached to a PCB with ENIG finish using 63Sn-37Pb solder paste. The assembled package was four-point bent to fracture. The fracture mode was interfacial fracture for all solder balls, except the outmost row that was 90Pb-10Sn. Those solder balls fractured in mixed modes; i.e., some failed inside the solder, and others at the interface between the solder balls and package pads which were electrolytic Ni/electrolytic Au. The elimination of the brittle interfacial fracture with high-Pb solder balls attached to ENIG is probably related to the different IMCs between high-Pb solder and ENIG, in comparison with the IMCs formed between eutectic Sn-Pb and ENIG.

Mechanical fatigue tests of solder joints of several alloys, Sn-37Pb, Sn-0.7Cu, and Sn-3.5Ag-0.5Cu, attached to ENIG and Cu pads in a flip chip test vehicle without underfill were conducted (Ref 102). The fatigue lives of Sn-0.7Cu solder joints attached to ENIG and Cu are the same, and in both cases the fatigue cracks extended through the solder joints, not at the solder/pad interface. On the contrary, the fatigue lives of Sn-37Pb and Sn-3.5Ag-0.5Cu attached to ENIG finish are shorter than those attached to Cu, and the difference was attributed to the interfacial fracture mode in the case of ENIG versus the solder joint failure in the case of Cu.

It was found that when indium-containing solders, e.g., 62Sn-37Pb-1In, 61Sn-36Pb-2In, 60Sn-35Pb-5In, and 40In-40Sn-20Pb, were used to attach PBGA with 63Sn-37Pb solder balls to ENIG finish, the tendency of the interfacial brittle fracture was significantly reduced (Ref 98).

Several fluxes were used to attach BGAs to test coupons. Based on the four-point bending test results, one flux increased the interfacial strength of the joint by approximately 15%, as compared with the other fluxes. However, the fracture mode remained the same (interfacial fracture).

Failure mechanisms include phosphorus segregation, corrosion in immersion gold bath, brittle Ni-Sn intermetallics, and Kirkendall voids.

Phosphorus Segregation. As shown previously, there is a phosphorus enriched layer, $Ni(P)^+$, next to the Ni-Sn intermetallic layer, and brittle fracture occurs at the interface between the $Ni(P)^+$ and Ni-Sn intermetallic. It was proposed that the phosphorus segregation caused a poor adhesion of the Sn-Ni intermetallic (Ref 82).

To verify that the phosphorus segregation is the root cause, PCBs with ENIG finishes of various P contents were prepared by varying the pH value and temperature of the electroless Ni plating bath. In general, the P content increases when the pH is reduced and the temperature is increased. After receiving the PCBs, PBGAs were assembled to the boards. The metallization on the PBGA package side was electroplated Ni/electroplated Au. Therefore, the failure should not occur on this side of the solder joint. A four-point bending test was used to measure the fracture load and the results are plotted in Fig. 11 as a function of the phosphorus content. It is seen that at the lower P contents, the solder joints failed by the interfacial brittle fracture and the fracture strength was approximately 40 to 60 lb. At higher P contents (except for the point at 10.5%), the failure mode is pulling out the pads and the strength is approximately 90 lb.

Based on the results, it is concluded that the phosphorus segregation is not the root cause for the brittle interfacial fracture. Actually, a relatively high P content in electroless Ni is beneficial because of the improved corrosion resistance of Ni in the immersion Au bath.

Corrosion in Immersion Au Bath. Solder joints attached to bare electroless Ni (no immersion Au) have high joint strength, and the failure mode in the four-point bending was pulling out pads (Ref 81). It was also found that an ENIG finish that had prolonged time in the immersion Au bath resulted in solder joints of almost zero strength, as shown in Fig. 11 (the circle at 9 wt% P). These two observations suggest that some impurities (carbon and oxygen, for example) or damage might have been introduced in the immersion Au bath. This suggestion is supported by the rework experiments. By reattaching a new BGA package to previously fractured pads, the solder joint strength increases significantly, and the percentage of the interfacial fracture reduces. It seems that the first interfacial fracture took away "garbage," and made the next package attached to the Ni surface much stronger.

The hyperactive corrosion effect of the Ni surface during the immersion Au plating is demonstrated in Ref 103. Figure 12(a) is "picture of the top view of electroless Ni after the immersion Au was etched off, showing "mud cracks." A cross section of the mud cracks is shown in Fig. 12(b). A small electrical bias (0.1 to 0.5 V) was applied on two pads of electroless Ni immersed in an immersion Au bath. It was observed that at the pad with the higher electrical potential, Ni was heavily corroded; at the pad with the lower potential, thicker than normal Au was deposited. It was suggested that if two pads were connected through a long connection, there may be some potential difference between the two pads, resulting in one pad being corroded. This model explains the phenomenon of "defect map" as described previously.

Brittle Ni-Sn Intermetallic. Ni_3Sn_4 was determined to be very brittle by hardness and fracture toughness tests (Ref 105). It is likely that Ni_3Sn and Ni_3SnP, the other possible phases present at the solder/ENIG interface, are also brittle. This mechanism is supported by the fracture surface analysis described previously; both sides of the fracture surface contain traces of Ni-Sn intermetallic. The actual failure process could start from mud cracks and Ni-Sn intermetallic fragmentation followed, resulting in the fracture along the interface.

Kirkendall Voids. Micro-voids at the interface between $Ni(P)^+$ and the Ni-Sn intermetallic were observed (Fig. 13) (Ref 93). These voids were thought to have been introduced by Kirkendall effects. It was proposed that the presence of high tensile stresses in the $Ni-P^+$ layer due to the volume reduction as a result of the phase change from the amorphous Ni-P to crystalline Ni_3P in $Ni-P^+$, creates mud cracks in the $Ni-P^+$ layer. Propagation of these mud cracks along the Kirkendall voids results in the brittle fracture along this interface. It was found that the higher the density of these voids, the lower the strength of the joint.

Au Embrittlement

It is known from early literature that Au has a propensity to embrittle solder joints within a range of concentration that extends from 2 to 7 wt%. That is one of the reasons that Au is being used in the electronic industry as a thin film on soldering pads. During soldering, the Au film dissolves into the molten solder; upon solidification, $AuSn_4$ precipitates and evenly distributes within the solder joints. The Au concentration in solder joints of BGA, CSP, and flip chip packages is typically less than 1 wt%; therefore, these joints are usually not brittle. Recently, it was observed that during solid state aging, the $AuSn_4$ precipitate particles move from the interior of the solder joint to the interface between the solder and the substrate, resulting in a brittle fracture at the interface (Ref 82).

Previous Literature on Au Embrittlement. Figures 14(a) through (c) plot the tensile strength, shear strength, and elongation as a function of the Au content in eutectic Sn-Pb alloy (Ref 106–108). Although the test data scatter, it is seen that both the tensile and the shear strength increase initially as the Au content in-

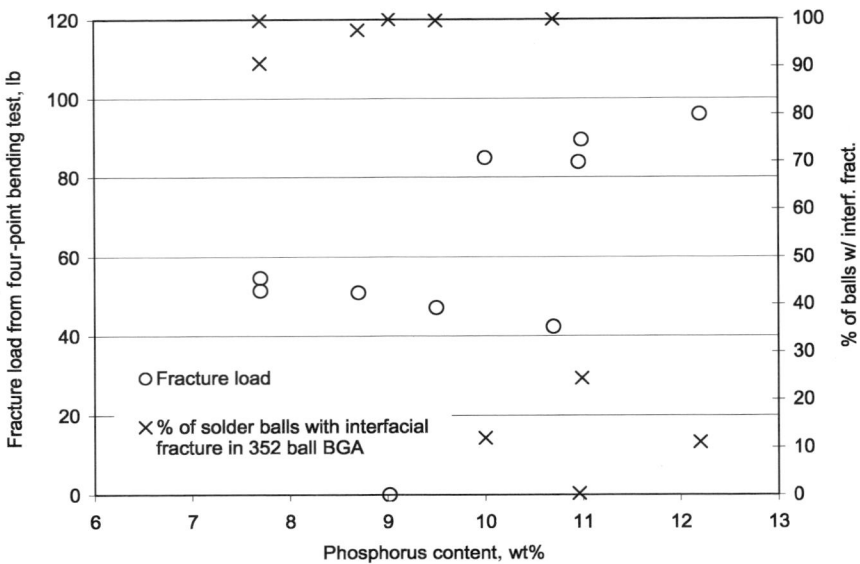

Fig. 11 Fracture load and percentage of solder balls with interfacial fracture. The data were obtained from four-point bending tests of 352 ball BGA packages with electrolytic Ni-Au finish attached to motherboards with ENIG finishes of different phosphorus contents.

creases from zero, peak around 1 to 6 wt%, then decrease as the Au content increases further. The elongation also increases from 0% Au, reaches the peak at approximately 3 wt%, and deteriorates sharply at 6 wt%.

Figure 14(d) plots the fatigue lives at room temperature of eutectic Sn-Pb solder joints as a function of their Au contents (Ref 109). The test pieces are solder joints of 3 by 3 arrays between two Au-coated Cu plates. A constant cyclic strain of 13% is applied in shear. As fatigue cracks propagate, the strength of the solder joints decreases. The fatigue life of a test piece is defined as the cycle reaching the ½ load drop. It is seen that with just 1 wt% of Au addition, the fatigue life decreases significantly.

For fine pitch surface mount components, such as BGA packages and CSPs, the Au content in a solder joint is less than 1%. The effects of Au content in fine pitch surface mount solder joints with respect to mechanical shock, vibration, and thermal cycling can be found in Ref 110. Au contents up to 10.1 wt% were tested. The testing was not severe enough to cause a significant portion of the solder joints to fail. Based on metallurgical examination of solder joints with concentration more than 4 wt%, they recommend that the Au concentration should be limited to 3 wt%.

Recent Observation of Au Embrittlement After Baking. As shown in Ref 82, PBGA packages were soldered on an ENIG finish as usual. After the assembly, packages were baked at 150 °C (302 °F) for 2 weeks. Some of the packages were reflowed the second time, after being baked. Figures 15(a) through (c) are SEM cross sections of solder joints under the conditions of (a) as-reflowed, (b) baked, and (c) baked, then reflowed. In the as-reflowed sample, there was only a thin Ni_3Sn_4 layer at the interface between the solder and the substrate, and the $AuSn_4$ par-

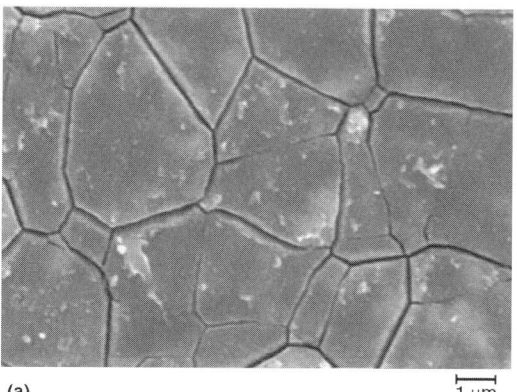

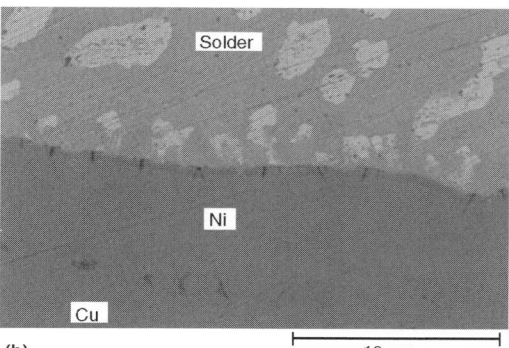

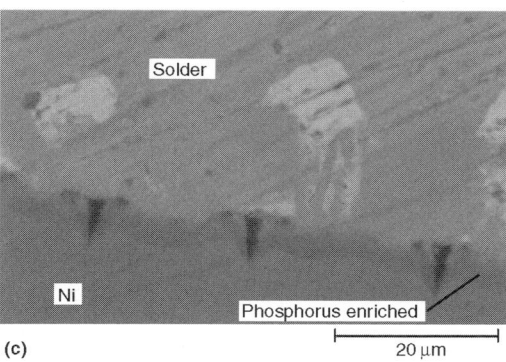

Fig. 12 Hyperactive corrosion effect of the nickel surface during immersion sold plating. (a) BGA package on this ENIG pad on motherboard fell off due to poor wetting. Mud-like cracks are shown on the pad. (b) and (c): Cross sections of mud-like cracks on ENIG pads

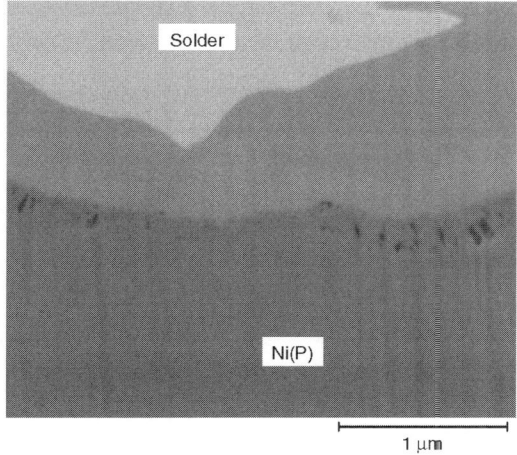

Fig. 13 Voids at the interface between 63Sn-37Pb solder and ENIG pad

ticles resided inside the solder joint. In the baked sample, a coarse intermetallic layer developed above the Ni_3Sn_4 layer, and $AuSn_4$ migrated from the solder interior onto the interface between the solder and the substrate. It must be emphasized that a relatively small amount of Au can result in a thick $AuSn_4$ layer, because of the ratio between Au and Sn being 1 to 4. In the baked, then reflowed sample, the $AuSn_4$ intermetallic layer dissolved from the interface into the solder joint.

Four-point bending tests were conducted on the assembled packages of the three conditions: (a) as-reflowed, (b) baked, and (c) baked, then reflowed. The fracture loads were 45, 30, and 45 lb, for conditions (a), (b), and (c), respectively. The fracture in conditions (a) and (c) was the same, i.e., cleavage between the solder joint and PCB pads as described in Fig. 9(a). Specifically, the fracture was at the interface between the Ni_3Sn_4 intermetallic layer and the phosphorus enriched layer, $Ni(P)^+$, the brittle interfacial fracture related to ENIG finish. The fracture in condition (b) was different; the fracture occurred at both the solder ball/PCB pad interface, and the solder ball/BGA pad interface. The metal finish on the BGA pad was electrolytic Ni/electrolytic Au; the brittle fracture at this interface cannot be related to ENIG. Figures 16(a) to (e) show that the main crack extended between the Au-Sn intermetallic and Ni-Sn intermetallic, and minor cracks went through the Ni_3Sn_4 intermetallic and the $Ni(P)^+$ layer.

The Au concentration in the solder joints was estimated to be approximately 0.1 wt%, based on the following dimensions: the Au film on PCB pad is 5 μin., the Au coating on BGA pad is 15 μin., and the solder joint is a column with the same diameter of the PCB pad and the height of 0.03 in. This concentration is much smaller than 3 wt%, the limit concentration based on the previous literature on the Au embrittlement. The 3 wt% limit is based on the brittle fracture of the solder joint. The brittle fracture observed here is a new fracture mode, a delamination at the interface, instead of brittle fracture of the solder.

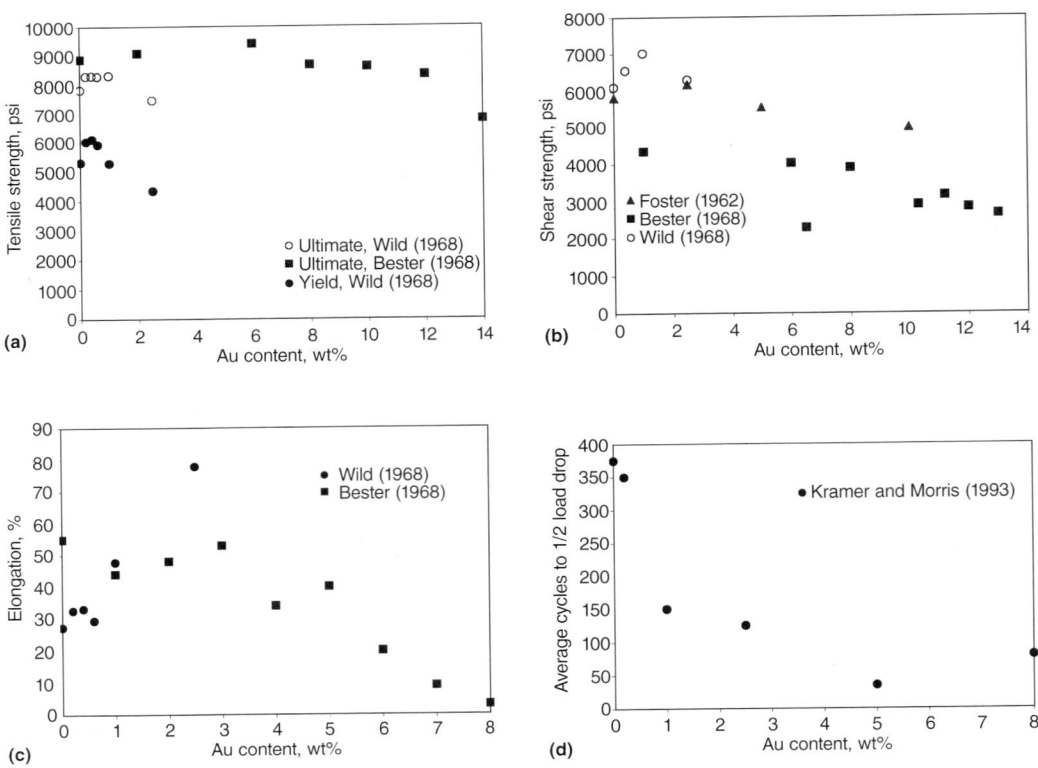

Fig. 14 Mechanical properties as a function of gold content in eutectic tin-lead solder joints. (a) tensile strength, (b) shear strength, (c) elongation, as function of Au content in eutectic Sn-Pb alloy. Source: Wild (Ref 106), Bester (Ref 107), and Foster (Ref 108), and (d) fatigue lives at room temperature of eutectic Sn-Pb solder joints as function of their Au contents. Source: Ref 109

This brittle interfacial fracture due to the redeposition of the Au-Sn intermetallic during baking has been confirmed in shear tests (Ref 27, 111–114).

Metallurgical Mechanism

The First Reflow. When eutectic Sn-Pb is reflowed on Ni-Au coated Cu, Au and Ni dissolve into the liquid solder. The dissolution is controlled by the temperature and the solubility of Ni and Au in the liquid solder. The dissolution rate of thin wires of Ni and Au in a rather large volume of molten solder was measured (Ref 28) (Eq 4) (Table 2). By extrapolating the data toward lower temperatures, the Ni and Au dissolution rates are estimated to be 0.04 and 1.25 µm/s, respectively.

The solubility limits of Au and Ni in liquid eutectic Sn-Pb are estimated to be 3 to 4 at.% and 10^{-5} at.% at 209 °C (408 °F) (Ref 27). Solubility limits of Au in Pb-free alloys are presented in Ref 115. At 50 °C (122 °F) superheat, the solubility of Au is 4 wt% in 57Bi-43Sn, 13 wt% in 60Sn-40Pb, 30 wt% in 96Sn-4Ag, and 30 wt% in pure Sn.

A typical Au finish of 0.1 µm thickness should dissolve into molten solder in less than a second. One µm of Au layer was consumed in less than 10 s at 250 °C (482 °F) by eutectic Sn-

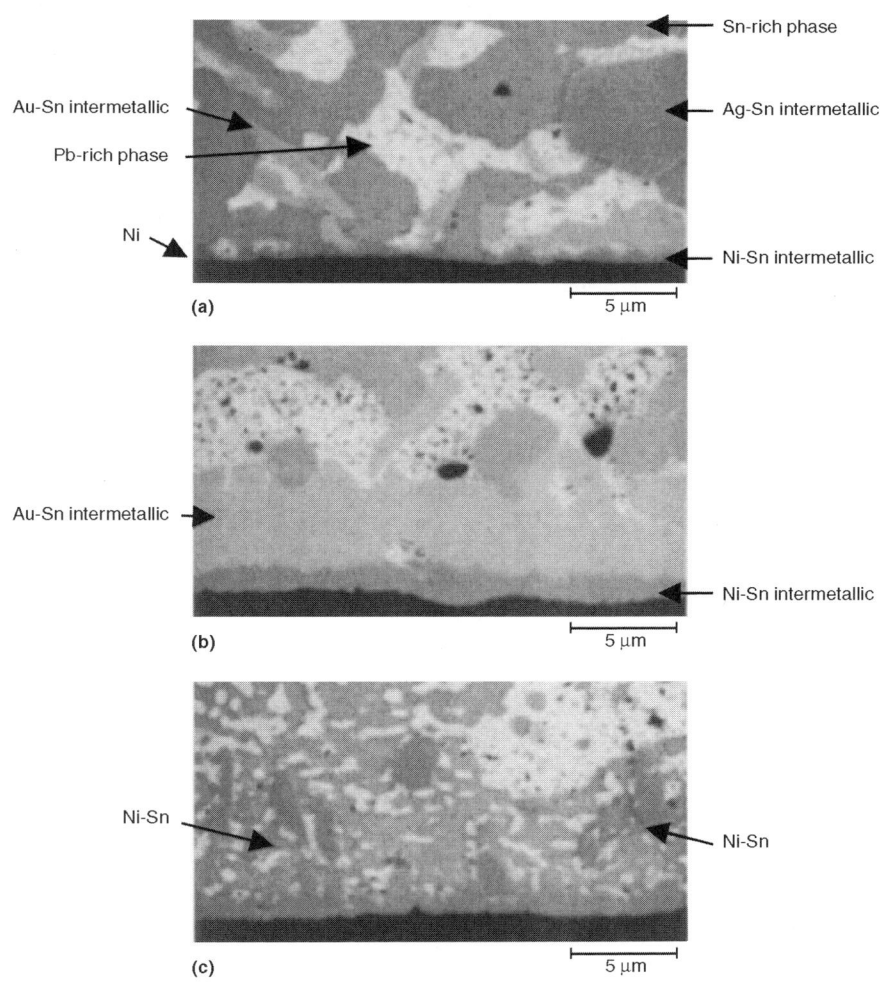

Fig. 15 SEM cross sections of solder joints under the conditions of (a) as-reflowed, (b) baked, and (c) baked, then reflowed. In the as-reflowed sample, there was only a thin Ni_3Sn_4 layer at the interface between the solder and the substrate, and the $AuSn_4$ particles resided inside the solder joint. In the baked sample, a coarse intermetallic layer developed above the Ni_3Sn_4 layer, and $AuSn_4$ migrated from the solder interior onto the interface between the solder and the substrate.

Ag in an experiment (Ref 94). A very thin layer of Ni is expected to dissolve, given its extremely low solubility limit in molten solder, and the dissolution process will go on for only a few seconds to reach the solubility limit. For the rest of the reflow time, typically 60 s, the exposed Ni reacts with the liquid solder to form Ni-Sn intermetallic, mainly Ni_3Sn_4.

On cooling from the soldering temperature, needle-like $AuSn_4$ forms and is evenly dispersed throughout the bulk of the solder joint, if the level of Au in the solder exceeds the solubility limit of 0.3 wt% at the freezing temperature (Ref 116).

Baking. The redeposition layer has been characterized, and it is not a binary $AuSn_4$ IMC, but rather $Au_{0.5}Ni_{0.5}Sn_4$ (Ref 27, 117), or $Au_{0.45}Ni_{0.55}Sn_4$, using SEM/EDX. X-ray diffraction and TEM diffraction identified that $Au_{0.5}Ni_{0.5}Sn_4$ is an $AuSn_4$-based ternary com-

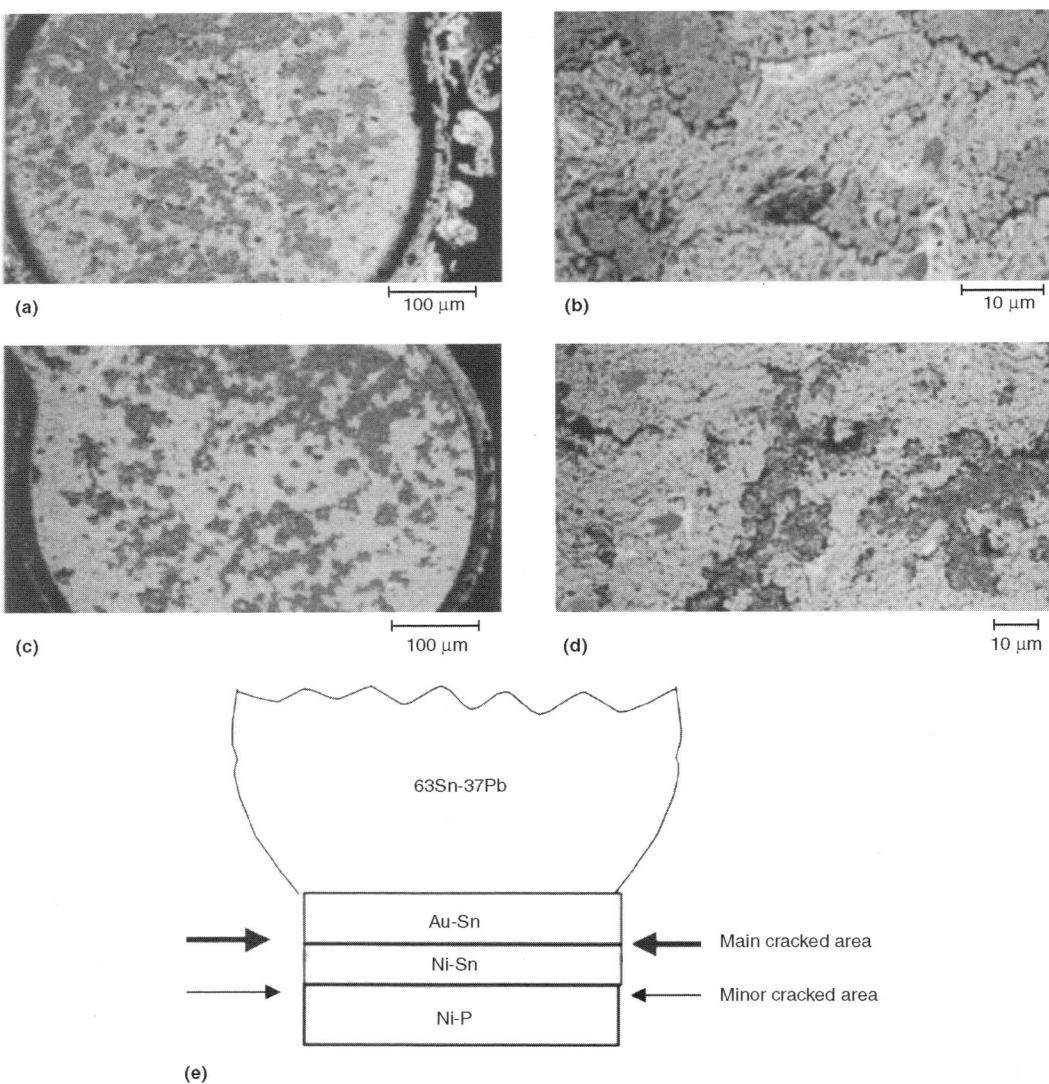

Fig. 16 Views of gold embrittlement in a PBGA assembly after aging at 150 °C (300 °F) for two weeks, showing that the main crack extended between the Au-Sn intermetallic and Ni-Sn intermetallic, and minor cracks went through the Ni_3Sn_4 intermetallic and the $Ni(P)^+$ layer. (a) and (b) SEM views of the PBGA package side. Light areas are $AuSn_4$, dark areas are Ni_3Sn_4, and bright spots are lead-rich solder. (c) and (d) SEM views of the fracture surface on the PCB side. Light areas are Ni_3Sn_4, dark areas are Ni-P, and bright spots are lead-rich solder. (e) Schematic drawing showing the fracture locations on the PBGA assembly after aging.

pound in which Ni substitutes for Au in the AuSn$_4$ phase, and the composition of the Au$_{0.5}$Ni$_{0.5}$Sn$_4$ corresponds to the solubility limit of Ni in AuSn$_4$ (Ref 118). Indeed, Au$_x$Ni$_{1-x}$Sn$_4$ with various values of x were observed (Ref 119).

TEM identified that the Ni$_3$Sn$_4$ layer is composed of relatively large, single crystal grains (Ref 116). On the other hand, the Au$_x$Ni$_{1-x}$Sn$_4$ layer shows a nanocrystalline structure with many fine pores, especially along the grain boundaries. It is believed that the negative volume change associated with the solid state reaction between the Au$_x$Ni$_{1-x}$Sn$_4$ compound and its constituents produces these fine pores. The presence of the fine pores explains the brittle fracture observed in bending and shear tests.

The simultaneous depletion of AuSn$_4$ from the bulk and the redeposition of (Au,Ni)Sn$_4$ at the interface suggests that the basic driving force for the reconfiguration is that the ternary phase is an energetically preferred phase at 150 °C (302 °F), and the relocation of the AuSn$_4$ onto the interface is to seek Ni at the interface. Under this driving force, the details of the kinetic process of the reconfiguration are proposed in Ref 118 as follows. At 150 °C (302 °F), the solubility limit of Au in eutectic Sn-Pb is approximately 0.2 to 0.3 at.%. In comparison, the Ni solubility in the eutectic Sn-Pb at 150 °C (302 °F) is negligible. During baking at 150 °C (302 °F), the Au in bulk AuSn$_4$ dissolves into the eutectic Sn-Pb solder up to its solubility limit. Au diffusion in Pb-Sn is known to be fast (Table 3), so the Au can easily reach the interface to form the preferred Au$_x$Ni$_{1-x}$Sn$_4$ phase where Ni is available from the Ni$_3$Sn$_4$. The drainage of Au at the interface causes the diffusion of Au from the bulk of the solder to the interface.

The intermetallic layer in Table 3 was observed after as little as 3 h of baking at 150 °C (302 °F) (Ref 117). It thickens roughly as $t^{1/2}$, where t is the baking time, as expected for diffusion-controlled growth. Empirical relations of the ternary IMC thickness versus baking time at 150 °C (302 °F) were obtained in Ref 41 and 117 and their results were consistent.

The Second Reflow. If the baked solder joints are reflowed once again, the redeposited Au$_x$Ni$_{1-x}$Sn$_4$ layer at the interface dissolves back into the molten solder rapidly. This is because the Au$_x$Ni$_{1-x}$Sn$_4$ is not stable at the temperatures above the melting point. The solubility limits of Au in molten eutectic Sn-Pb, Sn-Ag, or pure Sn are greater than 10 at.%. When the process of baking, then reflow is repeated, the Au$_x$Ni$_{1-x}$Sn$_4$ phase deposits, then dissolves, but the Ni$_3$Sn$_4$ layer thickens monotonically. The redeposition of Au$_x$Ni$_{1-x}$Sn$_4$ during baking becomes slower and less pronounced after each cycle of baking, then reflow (Ref 117). This is probably due to two reasons. The thickened Ni$_3$Sn$_4$ becomes a diffusion barrier for the Ni to diffuse through to form Au$_x$Ni$_{1-x}$Sn$_4$. Also, the dissolved Au$_x$Ni$_{1-x}$Sn$_4$ made Ni available in the solder matrix, and Au and Sn do not need to travel to the interface to react with Ni, but rather react with Ni and form Au$_x$Ni$_{1-x}$Sn$_4$ in the solder matrix.

Control of the Au Embrittlement. A solder joint attached to Ni-Au finish becomes more and more brittle with time because of the gradual deposition of the Au$_x$Ni$_{1-x}$Sn$_4$ at the solder joint/pad interface. One way to restore the reliability of the solder joint is to reflow it every 5 years or so, to dissolve the brittle Au$_x$Ni$_{1-x}$Sn$_4$ back to the solder matrix. This solution, however, may not be practical.

It is suggested that the thickness of the Au metallization be reduced until its concentration in the solder falls below the solubility limit of ~0.3 wt% (Ref 117, 120). However, as the solubility limit of Au reduces with temperature, the limit of 0.3 wt% does not eliminate the driving force for the redeposition of Au$_x$Ni$_{1-x}$Sn$_4$ at lower temperatures.

To introduce Ni or Ag into the solder matrix to trap the Au$_x$Ni$_{1-x}$Sn$_4$ phase, a small amount of Ni can be added to the solder to avoid the resettlement of Au$_x$Ni$_{1-x}$Sn$_4$ (Ref 119). These Ni atoms react with Sn to become Ni$_3$Sn$_4$ particles inside the solder joint, which serve as sinks for Au$_x$Ni$_{1-x}$Sn$_4$. Different behaviors of Sn-3.5Ag solder on Ni-Au pads were observed (Ref 94). During the isothermal aging at 160 °C (320 °F) for 875 h, only a small amount of Au$_x$Ni$_{1-x}$Sn$_4$ resettled back to the interface, and a continuous Au$_x$Ni$_{1-x}$Sn$_4$ layer did not form at the interface. It was observed that the many Au$_x$Ni$_{1-x}$Sn$_4$ particles were trapped by the Ag$_3$Sn particles inside the solder joint, and were hindered from resettling back to the interface. Slower and thinner deposition of Au$_x$Ni$_{1-x}$Sn$_4$

Table 3 Diffusion coefficients of Au in polycrystalline Pb, and single crystal Sn along "a" and "c" directions

	D_0, cm^2/s	Q, cal/mol
Au in Pb	4.1×10^{-3}	9,350
Au in Sn (c)	5.8×10^{-3}	11,000
Au in Sn (a)	1.6×10^{-1}	17,700

at the interface in Sn-3.5Ag than in Sn-37Pb is reported in Ref 112.

Similarly, an addition of Cu into the solder was found effective in retarding the redeposition and in preventing the Au-embrittlement (Ref 112, 121). When Cu is present in the solder, the interfacial intermetallic phases are different. Instead of the Ni_3Sn_4 and $(Au,Ni)Sn_4$ in the case of eutectic Sn-Pb on Ni-Au, the interfacial IMC is quaternary intermetallic $(Au,Ni)_2Cu_3Sn_5$ in the case of Sn-3.5Ag-0.7Cu on Ni-Au. After 45 days aging at 155 °C (311 °F), the quaternary intermetallic layer is only 4 µm for the Sn-3.5Ag-0.7Cu on Ni/Au, while the ternary $(Au,Ni)Sn_4$ is 20 µm in the case of eutectic Sn-Pb on Ni-Au.

Reflowing solder joint at high temperatures (greater than 240 °C, or 464 °F) was demonstrated to be effective in suppressing the $Au_xNi_{1-x}Sn_4$ at the solder/pad interface during subsequent baking (Ref 122). The high temperature reflow introduces Ni and Cu into the solder substrate and traps the $Au_xNi_{1-x}Sn_4$ phase.

ACKNOWLEDGMENTS

The author is grateful to Mason Hu and Mark Brillhart for technical discussions and support.

REFERENCES

1. K.-W. Moon, W.J. Boettinger, U.R. Kattner, F.S. Biancaniello, and C.A. Handwerker, *J. Electron. Mater.*, Vol 29 (No. 10), 2000, p 1122
2. I. Ohnuma, M. Miyashita, K. Anzai, X.J. Liu, H. Ohtani, R. Kainuma, and K. Ishida, *J. Electron. Mater.*, Vol 29, (No. 10), 2000, p.1137
3. S.-W. Chen and Y.-W. Yen, *J. Electron. Mater.*, Vol 28, (No. 11), 1999, p 1203
4. F. Hua, Ph.D. thesis, Vanderbilt University, 1995
5. R. Elliott, *Eutectic Solidification Processing Crystalline and Glassy Alloys*, Butterworths & Co Ltd., 1983
6. S.K. Kang, W.K. Choi, D.-Y. Shih, D.W. Henderson, T. Gosselin A. Sarkhel, C. Goldsmith, and K.J. Puttlitz, *Electronic Component and Technology Conference Proc.*, IEEE/CPMT, 2003, p 64
7. S. Chada, and D. Shangguan, *Environment-Friendly Electronics: Lead-Free Technology*, J.S. Hwang, Ed., Electrochemical Publications Ltd, 2001, p, 513
8. S. Chada, R.A. Fournelle, W. Laub, and D. Shangguan, *J. Electron. Mater.*, Vol 29, (No. 10), 2000, p 1214
9. S.K. Kang, W.K. Choi, P. Lauro, D.W. Henderson, T. Gosselin, and D.N. Leonard, *Electronic Component and Technology Conference Proc.*, IEEE/CPMT, 2002, p 146
10. C.-M. Chuang, P.-C. Shih, and K.-L. Lin, *J. Electron. Mater.*, Vol 33, (No. 1), 2004, p 1
11. P.L. Liu, and J.K. Shang, *J. Electron. Mater.*, Vol 29, (No. 5), 2000, p 622
12. D.R. Frear, J.W. Jang, J.K. Lin, and C. Zhang, *JOM*, Vol 53, (No. 6), June 2001, p 28
13. D.W. Henderson, T. Gosselin, A. Sarkhel, S.K. Kang, W.K. Choi, D.Y. Shih, C. Goldsmith, and K.J. Puttlitz, *J. Mater. Res.*, Vol 17, (No. 11), Nov 2002, p 2775
14. W. Yang, and R.W. Messler, *J. Electron. Mater.*, Vol 23, 1994, p 765
15. I.E. Anderson, J.C. Foley, B.A. Cook, J. Harringa, R.L. Terpstra, and O. Unal, *J. Electron. Mater.*, Vol 30, (No. 9), 2001, p 1050
16. M.D. Cheng, S.F. Yen, and T.H. Chuang, *J. Electron. Mater.*, Vol 33, (No. 3), 2004, p 171
17. Q. Xiao, H.J. Bailey, and W.D. Armstrong, *Transactions of the ASME*, Vol 126, June 2004, p 208
18. S. Ahat, M. Sheng, and L. Luo, *J. Electron. Mater.*, Vol 30, (No. 10), 2001, p 1317
19. C.K. Shin, Y.-J. Baik, and J.Y. Huh, *J. Electron. Mater.*, Vol 30, (No. 10), 2001, p 1323
20. D. Shangguan, and A. Achari, Evaluation of Lead-Free Eutectic Sn-Ag Solder for Automotive Electronics Packaging Applications, *Proceedings of the International Electronics Manufacturing Technology Symposium*, (La Jolla, CA), Sept 1994, p 25–37
21. I. Okamoto, and T. Yasuda, *Trans. JWRI*, Vol 15, (No. 2), 1986, p 73
22. M. Schaefer, W. Laub, R.A. Fournelle, and J. Liang, *Design and Reliability of Solders and Solder Interconnections*, R.K. Mahidhara, D.R. Frear, S.M. Sastry, K.L. Murty, P.K. Liaw, and W. Winterbottom, Ed., The Minerals, Metals & Materials Society, 1997, p 247

23. J.-H. Lee, and Y.-S. Kim, *J. Electron. Mater.*, Vol 31, (No. 6), 2002, p 576
24. W.J. Boettinger, C.A. Handwerker, and U.R. Kattner, *The Mechanics of Solder Alloy Wetting & Spreading,* F.G. Yost, F.M. Hosking, and D.R. Frear, Ed., Van Nostrand Reinhold, New York, 1993, p 103
25. K.N. Tu, and K. Zeng, *Mater. Sci. Eng. R34,* 2001, p 1
26. T.B. Massalski, *Binary Alloy Phase Diagrams,* H. Okamoto, Ed., Vol 2, ASM International, Dec 1, 1990
27. A. Zribi, R.R. Chromik, R. Presthus, J. Clum, K. Teed, L. Zavalij, J. DeVita, J. Tova, and E.J. Cotts, *Electronic Components and Technology Conference,* IEEE CPMT, 1999, p 451
28. W.G. Bader, Dissolution of Au, Ag, Pd, Pt, Cu and Ni in a Molten Tin-Lead Solder, *Welding Journal,* Research Supplement, Dec 1969, p 551s
29. Y. Shoji, S. Uchida, and T. Ariga, *Welding Journal,* Research Supplement, Vol 60, Jan 1981, p 19s
30. E.A. Moelwyn-Hughes, *The Kinetics of Reactions in Solution,* Oxford University Press, London, 1947
31. S.K. Kang, R.S. Rai, and S. Purushothaman, *J. Electron. Mater.,* Vol 25, (No. 7), 1996, p 1113
32. S.K. Kang, D.Y. Shih, K. Fogel, P. Lauro, M.J. Yim, G. Advocate, M. Girffin, C. Goldsmith, D.W. Henderson, T. Gosselin, and D. King, *Electronic Component and Technology Conference Proc.,* IEEE/CPMT, 2001, p 448
33. T.M. Korhonen, P. Su, S.J. Hong, M.A. Korhonen, and C.-Y. Li, *J. Electron. Mater.,* Vol 29, (No. 10), 2000, p 1194
34. F. Bartels, J.W. Morris, Jr., G. Dalke, and W. Gust, *J. Electron. Mater.,* Vol 23, (No. 8), 1994, p 78
35. R.A. Gagliano, G. Ghosh, and M.E. Fine, *J. Electron. Mater.,* Vol 31, (No. 11), 2002, p 1195
36. Y. Wu, J.A. Sees, C. Pouraghabagher, L.A. Foster, J.L. Marshall, E.G. Jacobs, and R.F. Pinizotto, *J. Electron. Mater.,* Vol 22, (No. 7), 1993, p 769
37. S. Bader, W. Gust, and H. Hieber, *Acta Metall.,* Vol 43, (No. 1), 1995, p 329
38. D. Grivas, D. Frear, L. Quan, and J.W. Morris, Jr., *J. Electron. Mater.,* Vol 15, (No. 6), 1986, p 355
39. K.H. Prakash, and T. Sritharan, *J. Electron. Mater.,* Vol 31, (No. 11), 2002, p 1250
40. R.A. Gagliano, and M.F. Fine, *J. Electron. Mater.,* Vol 32, (No. 12), 2003, p 1441
41. A. Zribi, Clark, A.L. Zavalij, P. Borgesen, and E.J. Cotts, *J. Electron. Mater.,* Vol 30, (No. 9), 2001, p 1157
42. T.H. Chuang, H.M. Wu, M.D. Cheng, S.Y. Chang, and S.F. Yen, *J. Electron. Mater.,* Vol 33, (No. 1), 2004, p 22
43. C.-W. Hwang, J.-G. Lee, K. Suganuma, and H. Mori, *J. Electron. Mater.,* Vol 32, (No. 2), 2003, p 52
44. P. Villars, *Pearson's Handbook of Crystallographic Data for Intermetallic Phases,* ASM International, 1997, p 1593, 2549
45. C. Smithells, E.A. Brandes, E. Adolph, and G. Brook, *Smithells Metals Reference Book,* Butterworth-Heinemann, London, 1992, p 1–79
46. P.L. Liu, and J.K. Shang, *Metall. Trans. A,* Vol 31A, 2000, p 2867
47. W.K. Choi, and H.M. Lee, *J. Electron. Mater.,* Vol 28, (No. 11), 1999, p 1251
48. J. Haimovich, *Welding Journal,* Research Supplement, Vol 8, (No. 3), 1989, p 102s
49. K. Zeng, V. Vuorinen, and J.K. Kivilahti, *Electronic Components and Technology Conference,* IEEE CPMT, 2002, p 693
50. C.E. Ho, R.Y. Tsai, Y.L. Lin, and C.R. Kao, *J. Electron. Mater.,* Vol 31, (No. 6), 2002, p 584
51. C.-H. Lin, S.-W. Chen, and C.-H Wang, *J. Electron. Mater.,* Vol 31, (No. 9), 2002, p 907
52. Y.-D. Jeon, S. Nieland, A. Ostmann, H. Reighl, and K.-W. Paik, *J. Electron. Mater.,* Vol 32, (No. 6), 2003, p 548
53. C.-S. Huang, J.-G. Duh, Y.-M. Chen, *J. Electron. Mater.,* Vol 32, (No. 12), 2003, p 1509
54. J.Y. Park, C.W. Yang, J.S. Ha, C.-U. Kim, E.J. Kwon, S.B. Jung, and C.S. Kang, *J. Electron. Mater.,* Vol 30, (No. 9), 2001, p 1165
55. S. Yee, J. Zeng, and R. Jay, *Surface Mount International,* 2003
56. C.H. Ma, and R.A. Swalin, *Acta Metall.,* Vol 8, (No. 6), 1960, p 388
57. P.G. Shewmon, *Transformations in Metals,* Chap. 2, J. Williams Book Company, 1983
58. M. Schaefer, R.A. Fournelle, J. Liang, *J. Electron. Mater.,* Vol 27, (No. 11), 1998, p 1167

59. W.K. Choi, and H.M. Lee, *J. Electron. Mater.*, Vol 29, (No. 10), 2000, p 1207
60. J. London, and D.W. Ashall, *Brazing Soldering,* Autumn 1986, p 49
61. H.K. Kim, and K.N. Tu, *Phys. Rev. B53,* 1996, p 16027
62. H.K. Kim, H.K. Lioun, and K.N. Tu, *Appl. Phys. Lett.*, Vol 66, (No. 18), 1995, p 2337
63. G. Ghosh, *J. Electron. Mater.,* Vol 28, (No. 11), 1999, p 1238
64. K.-S. Bae, and S.-J. Kim, *J. Electron. Mater.,* Vol 30, (No. 11), 2001, p 1452
65. X. Deng, G. Piotrowski, J.J. Williams, and N. Chawla, *J. Electron. Mater.,* Vol 32, (No. 12), 2003, p 1403
66. J.W. Jang, D.R. Frear, T.Y. Lee, and K.N. Tu, *J. Appl. Phys.*, Vol 88, (No. 11), 2000, p 6359
67. C.-S. Huang, J.-H. Yeh, and B.-L. Young, *J. Electron. Mater.,* Vol 31, (No. 11), 2002, p 1230
68. F. Zhang, M. Li, C.C. Chum, and Z.C. Shao, *J. Electron. Mater.,* Vol 32, (No. 3), 2000, p 123
69. J.S. Kang, R.A. Gagliano, G. Ghosh, and M.E. Fine, *J. Electron. Mater.,* Vol 31, (No. 11), 2003, p 1238
70. C.-M. Chuang, and K.-L. Lin, *J. Electron. Mater.,* Vol 32, (No. 12), 2003, p 1426
71. J.-Y. Park, R. Kabade, C.-U. Kim, T. Carper, S. Dunford, and V. Puligandla, *J. Electron. Mater.,* Vol 32, (No. 12), 2003, p 1474
72. A.D. Romig, Jr., Y.A. Chang, J.J. Stephens, D.R. Frear, V. Marcotte, and C. Lea, Physical Metallurgy of Solder-Substrate Reactions, *Solder Mechanics—A State of the Art Assessment,* D.R. Frear, W.B. Jones, and K.R. Kinsman, Ed., TMS, 1991
73. P.J. Kay, and C.A. Mackay, *Transactions of the Institute of Metal Finishing,* Vol 54, 1976, p 68
74. D.A. Unsworth, and C.A. Mackay, *Transactions of the Institute of Metal Finishing,* Vol 51, 1973, p 85
75. M. Onishi, and H. Fujibuchi, *Trans. Jpn. Inst. Met.,* Vol 16, 1975, p 539
76. Z. Mei, A.J. Sunwoo, and J.W. Morris, Jr., *Metall. Trans. A,* Vol 23A, 1992, p 857
77. T.-Y. Pan, H.D. Blair, J.M. Nicholson, and S.-W. Oh, E. Suhir, Y.C. Lee, M. Shiratori, and G. Subbarayan, Ed., Advances in Electronic Packaging, *Pacific Rim/ASME International Intersociety Electronic & Photonic Packaging Conference Proc.,* Vol 2, Interpack '97, p 1347
78. D. Yao, and J.K. Shang, *Metall. Mater. Trans. A,* Vol 26A, 1995, p 2677
79. M. Date, T. Shoji, M. Fujiyoshi, K. Sato, and K.N. Tu, *Electronic Component and Technology Conference Proceeding,* 2004, p 668
80. T.-C. Chiu, K. Zeng, R. Stierman, D. Edwards, and K. Ano, *Electronic Component Technology Conference,* 2004, p 1256
81. Z. Mei, A Failure Analysis and Rework Method of Electronic Assembly on Electroless Ni/Immersion Au Surface Finish, *Surface Mount Technologies,* 1999
82. Z. Mei, M. Kaufmann, A. Eslambolchi, and P. Johnson, *Electronic Components and Technology Conference,* IEEE CPMT, 1998
83. W. Reidel, *Electroless Nickel Plating,* ASM International, 1991
84. A.H. Graham, R.W. Lindsay, and H.J. Read, *J. Electrochemical Society,* April 1965, p 401
85. J.L. Fang, X.R. Ye, and J. Fang, *Plat. Surf. Finish.,* July 1992, p 44
86. K.-L. Lin, and J.-M. Jang, *Mater. Chem. Phys.,* Vol 38, 1994, p 33
87. B.-L. Young, J.-G. Duh, and B.-S. Chiou, *J. Electron. Mater.,* Vol 30, (No. 5), 2001, p 543
88. S. Wiegele, P. Thompson, R. Lee, and E. Ramslan "Reliability and Process Characterization of Electroless Nickel-Gold/Solder Flip Chip Interconnect Technology," Motorola, Inc.
89. N.S. Kim, S.Y. Jung, and S.Y. Oh, *Electronic Components and Technology Conference,* IEEE CPMT, 2003, p 77
90. P. Snugovsky, P. Arrowsmith, and M. Romansky, *J. Electron. Mater.,* Vol 30, (No. 9), 2001, p 1262
91. Y. Tomita, Q. Wu, A. Maeda, S. Baba, and N. Ueda, *Electronic Components and Technology Conference,* IEEE CPMT, 2000, p 861
92. J.W. Jang, P.G. Kim, K.N. Tu, D.R. Frear, and P. Thompson, *J. Appl. Phys.,* Vol 85, (No. 12), 1999, p 8456–8463
93. D. Goyal, T. Lane, P. Kinzie, C. Panichas, K. Chong, and O. Villalobos, *Electronic Component and Technology Conference Proceeding,* IEEE/CPMT, 2002
94. C.M. Liu, C.E. Ho, W.T. Chen, and C.R. Kao, *J. Electron. Mater.,* Vol 30, (No. 9), 2001, p 1152

95. C.-Y. Lee, and K.-L. Lin, *Thin Solid Films,* Vol 249, 1994, p 201
96. E. Bradley, and K. Banerji, *Electronic Components and Technology Conference,* IEEE CPMT, 1995, p 1028
97. P.J. Callery, "Mechanical Reliability of Mid-Range Ball Grid Array Packages," HP Internal Report, 1998
98. A. Eslambolchi, P. Johnson, M. Kaufmann, and Z. Mei, "Electroless Ni/Immersion Au Evaluation—Final Program," HP Internal Report, Aug 1998
99. P. Johnson, Z. Mei, A. Eslambolchi, M. Kaufmann, *Proceedings of SemiCon West,* G1-9, July 13–14, 1999
100. E. Bradley, *IPC/ITRI Technical Council on Electroless Ni/Immersion Au,* Presentation, Dec 1997, San Antonio, TX
101. C. Zhang, J.-K. Lin, and L. Li, *Electronic Components and Technology Conference,* IEEE CPMT, 2001, p 463
102. J.-K. Lin, A. De Silva, D. Frear, Y. Guo, J.-W. Jang, L. Li, D. Mitchell, B. Yeung, and C. Zhang, *Electronic Components and Technology Conference,* IEEE CPMT, 2001, p 455
103. N. Biunno, A Root Cause Failure Mechanism for Solder Joint Integrity of Electroless Nickel/Immersion Gold Surface Finishes, Proceedings of the IPC Printed Circuit Expo, Paper S18-5, March 14–18, 1999
104. F.D. Houghton, *Future Circuits International,* Vol 6, 2000, p 121
105. D.R. Frear, F.M. Hosking, and P.T. Vianco, *Materials Developments in Microelectronic Packaging Conference Proc.,* (Montreal), Aug 19–22, 1991, p 229
106. R.N. Wild, *Proceeding of Nepcon,* June 1968, p 198
107. M.H. Bester, Proceeding of Internepcon, Oct 1968, p 211
108. G.F. Foster, *Papers on Soldering,* ASTM Standard Technical Publication, 1962, p 13
109. P.A. Kramer, J. Glazer, J.W. Morris, Jr., *Metall. Mater. Trans. A,* Vol 25A, June 1994, p 1249
110. J. Glazer, P.A. Kramer, and J.W. Morris, Jr., *J. Surf. Mt. Technol.,* Oct 1991
111. C.H. Zhong, S. Yi, Y.C. Mui, C.P. Howe, D. Olsen, and W.T. Chen, *Electronic Components and Technology Conference,* IEEE CPMT, 2000, p 151
112. J.-H. Lee, J.-H. Park, D.-H. Shin, Y.-H. Lee, and Y.-S. Kim, *J. Electron. Mater.,* Vol 30, (No. 9), 2001, p 1138
113. M.R. Marks, *J. Electron. Mater.,* Vol 31, (No. 4), 2002, p 265
114. T. Taguchi, R. Kato, S. Akita, A. Okuno, H. SuZuki, T. Okuno, *Electronic Components and Technology Conference,* IEEE CPMT, 2001, p 675
115. D.M. Jacobson and G. Humpston. *Gold Bull.,* Vol 22, (No. 1), 1989, p 9
116. A. Prince, The Au-Pb-Sn Ternary System, *J. Less-Common Met.,* Vol. 12, 1967, p 107
117. A.M. Minor, and J.W. Morris, Jr., *Metall. Mater. Trans.,* Vol 31A, March 2000, p 798
118. H.G. Song, J.P. Ahn, A.M. Minor, J.W. Morris, Jr., *J. Electron. Mater.,* Vol 30, (No. 4), 2001, p 409
119. C.E. Ho, R. Zheng, G.L. Luo, A.H. Lin, and C.R. Kao, *J. Electron. Mater.,* Vol 29, (No. 10), 2000, p 1175
120. P.G. Kim, and K.N. Tu, *J. Appl. Phys.,* Vol 80, 1996, p 3822
121. C.-S. Chi, H.-S. Chang, K.-C. Hsieh, and C.L. Chung, *J. Electron. Mater.,* Vol 31, (No. 11), 2002, p 1203
122. R.K. Kinyanjui, A. Zribi, and E.J. Cotts, *Electronic Components and Technology Conference,* IEEE CPMT, 2002, p 161

CHAPTER 3

Fatigue and Creep of Lead-Free Solder Alloys: Fundamental Properties

Paul T. Vianco, Sandia National Laboratories

Introduction

Surface mount technology changed the function of soldered interconnections on printed wiring assemblies (PWAs) from that of simply providing electrical continuity, to including the expressed role of mechanical attachment between the device or component and the circuit board substrate. As a consequence, the mechanical properties of the circuit board solder joint and specifically, those of the solder material, have become critical to the reliable operation of the PWA. Of course, the solder joints must withstand catastrophic damage due to overload stresses and mechanical shock, the effects of which are exemplified in Fig. 1(a). Secondly, the solder joints must not degrade when subjected to lower stresses that are sustained at a constant level, or that cycle during the service life of the product. Cyclic loads can arise from mechanical vibrations or temperature fluctuations; both conditions slowly degrade the integrity of a soldered interconnection until it becomes an electrically open path. The progression of cyclic or fatigue damage in a 63Sn-37Pb solder joint caused by temperature variations is shown in Fig. 1(b).

In order to predict the mechanical properties of the solder joint, it is necessary to understand the various structures and manner in which their properties effect those of the whole joint. The solder joint is comprised of the solder alloy, the substrate materials that are being joined together, and the interfaces between the substrate materials and the solder that is typically comprised of one or more intermetallic compound (IMC) layers. In some cases, the base materials have a solderable finish to which the solder has been joined. Then, there is a solder/protective finish interface that typically also generates an IMC layer. There is also an interface between the substrate material and the solderable finish. A solderable finish is covered with a protective finish before soldering. The protective finish is dissolved into the molten solder during the assembly process. The dissolved protective finish can impact the mechanical properties of the solder, the extent of which depends upon the thickness of the coating and the volume of the solder. In summary, the mechanical performance of the solder joint is a function of the mechanical properties of the solder, the base materials (including solderable finishes, if present), and the mutual interfaces.

In addition, there are two nonmaterial factors that affect the mechanical performance of a soldered interconnection. Those factors are (a) the *joint geometry*, in particular, the gap thickness and the footprint area of the joint, and (b) the *rate* at which a stress is applied to the joint. First, the joint geometry is discussed. Generally, a small gap thickness and/or large footprint places the joint under a *plane strain* condition that limits the ability of the solder to readily deform. As a result, the solder and the joint appear to be stronger than would be calculated strictly from the bulk strength of the solder. Conversely, a larger gap and/or smaller footprint causes a *plane stress* condition to prevail in the joint. The solder can more easily deform, resulting in a joint strength that is more commensurate with calculations based upon the bulk strength of the solder. The second factor is the loading rate. The rate at which a stress is applied to the joint effects its apparent strength because of the high strain rate sensitivity of most tin (Sn)-based solder alloys. A high strain rate sensitivity indicates that the yield and ultimate tensile strengths increase rapidly with loading rate. Thus, the solder material and associated joint will appear to be stronger (and less ductile) as the speed at which the stress is applied, increases.

The mechanical performance of the solder joint is determined by the mechanical properties of the materials system—they are the solder, substrate materials, and the interface IMC layer(s)—combined with the joint geometry and the loading rate. Because of these many factors, it is often difficult to quantitatively predict the deformation response of a solder joint without the use of finite element analysis. However, it is possible to establish several generalized, qualitative trends. For example, factors that increase the apparent strength of the joint, which include (a) a thinner gap and/or large footprint (i.e., plane strain condition), (b) higher applied stresses, (c) a faster loading rate, and (d) a higher bulk solder strength, will increase the likelihood of failure occurring near the solder/substrate interface. The crack path may be in the solder immediately adjacent to the IMC layer or in the IMC layer itself. In some instances, fracture may occur at the IMC layer/base material interface or, in the case of brittle substrate materials, within the latter structures.

On the other hand, relatively low stresses (e.g., below the yield stress) that are applied at reduced loading rates cause the deformation to shift into the solder material. These conditions characterize the service life environment of most solder interconnections as well as accelerated aging tests. As a consequence, the solder deformation will include time-dependent or *creep* deformation. In the case of cyclic loading environments, the solder deforms by a combination of creep and fatigue responses. Temperature vari-

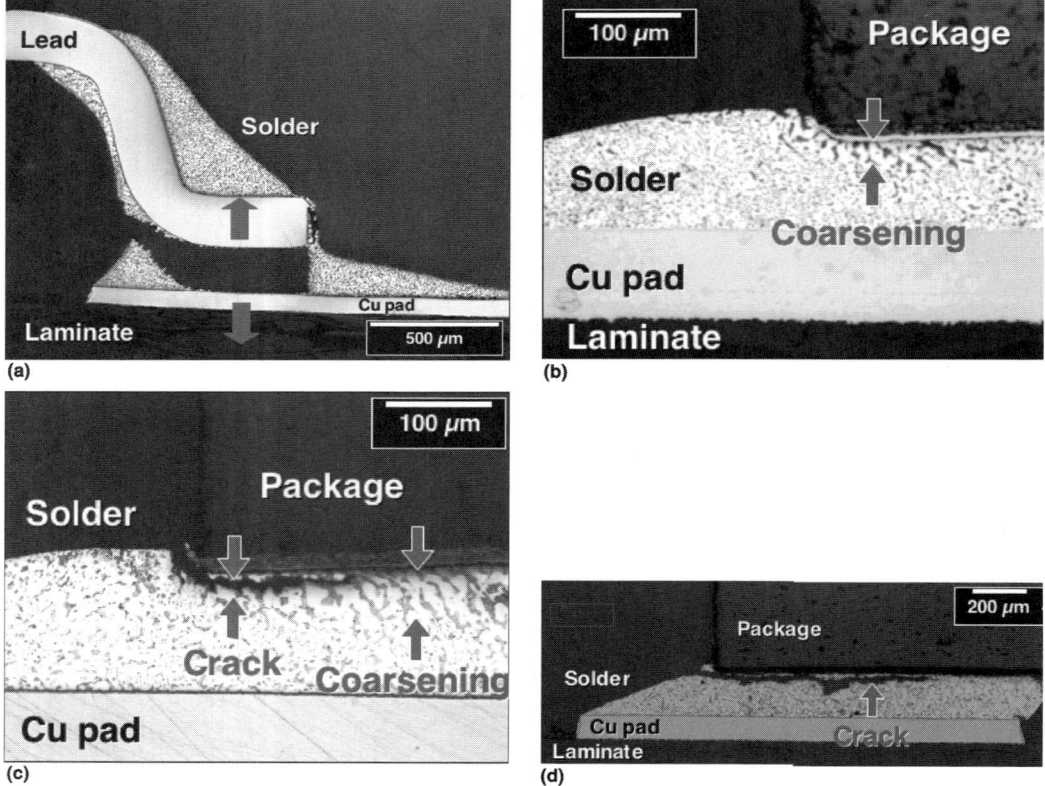

Fig. 1 Optical micrographs that illustrate damage to surface mount solder joints due to (a) mechanical shock and (b to d) fatigue. Note the coarsening of the Pb-rich phase in latter case that is the signature of thermal mechanical fatigue in Sn-Pb solders.

ations, in combination with thermal expansion mismatch of materials comprising the interconnection structure, can result in a form of slow loading-rate fatigue for which creep becomes a primary contributor to the material deformation; this condition is referred to as thermal mechanical fatigue (TMF).

The case of mechanical vibration is less well defined in terms of the predominance of creep versus time-independent (plastic) deformation. Although the applied stresses are often quite low, the strain rates can be sufficiently high, so that a mixture of creep and plastic deformation modes define the fatigue response of the material. This sensitivity of fatigue behavior to cyclic loading frequency has been widely studied for tin-lead (Sn-Pb) solders and is recognized as an important variable in the fatigue response of lead-free alloys (Ref 1–3).

In general, the long-term reliability of electronic solder joints is determined by the creep and fatigue properties of the solder alloy. Interestingly, as solder joints become smaller, to accommodate the increased miniaturization of electronic products, the joint size will have a greater impact on interconnection performance than simply the plane stress/plane strain condition. The so-called *size or length-scale effect* refers to a change of bulk mechanical properties for the solder, including those of creep and fatigue, as the joint dimensions become commensurate with the size of microstructural features (e.g., grain size, colony size, etc.) that characterize the alloy.

This chapter will focus on the fundamental aspects of fatigue and creep deformation modes in Pb-free solders. Particular emphasis will be placed upon the Sn-Ag-Cu solders because these compositions are the leading candidates to replace the eutectic Sn-Pb alloy for reflow soldering. Where appropriate, the properties of the 96.5Sn-3.5Ag solder can provide an approximation of the behavior of the Sn-Ag-Cu materials. A brief synopsis is made of time-independent or plastic deformation. The purpose of that synopsis is to describe yield stress and elastic modulus properties. The yield stress is often used to distinguish the regimes of time-dependent and time-independent deformation. For example, the stress used in a creep test is often designed to be 20%, 40%, and so on, of the yield stress at test temperature. The stress levels may be later adjusted, depending upon the early findings of the creep experiments. Also, the plastic deformation properties are used in some quantitative analyses of creep and fatigue data.

The two remaining sections describe the fundamental fatigue and creep properties of the Pb-free solders. The properties of the bulk solder, as well as those of solder joints, are highlighted. The relevance of bulk solder properties stems from the need to develop constitutive models that predict the long-term reliability of the solder interconnections. Each section will summarize quantitative data, as well as address microstructural considerations pertaining to deformation in Pb-free solder compositions.

Material Deformation

Time-Independent Deformation

Time-independent or plastic deformation refers to a material performance that results from relatively fast loading rates. In the laboratory, time-independent deformation is typically generated by the stress-strain experiments. The tests are carried out under either strain-rate control or stress-rate control, but most often, the experiments are performed under strain-rate control. An approximate boundary between time-independent deformation and time-dependent deformation for solders are strain rates of 10^{-4}–10^{-5} s^{-1}. The test sample dimensions are typically large, relative to the microstructural features of the material. However, there is a growing need to understand size or length-scale effects on these properties as solder interconnections become increasingly smaller, particularly solder joint dimensions less than 100 μm.

The evaluation of bulk solder properties can be carried out in tension, compression, or in shear. Shear test techniques include a simple shear sample or torsion test. Output data include the true stress-true strain curve, the static elastic moduli, the yield strength (0.2% offset), ultimate (tensile, compressive, or shear) strength, and ductility properties such as reduction in area (RA) or percent elongation. Test procedures and data analysis regiments have been established in the American Society for Testing and Materials (ASTM) specifications (Ref 4–7). Multiple extensometers or strain gages can be used to measure deformation along several directions with respect to the stress axis, thereby providing the means to determine the bulk (or volume) moduli and Poisson's ratio of the material (Ref 8). Complex triaxial stress states can be introduced into test specimens by the simultaneous application of both tension/compression and torsion; however, the data analysis can be very involved.

Test procedures have been modified in order to assess the performance of actual solder joints. It is important to recognize that standardized test methodologies (e.g., ASTM) are not in place to address specific solder joint geometries. (There are ASTM standards available for the testing of organic adhesive joints and American Welding Society specifications for testing braze joints. Unfortunately, neither adhesive nor braze joints exhibit the same performance characteristics as do solder joints due to differences in material properties.) Solder joint testing is preferred to the testing of bulk solder material, when it is desired to assess the properties of a specific interconnection structure (i.e., substrate materials, interfaces, solder, and the geometry). Solder joints may be tested in tension or compression using the butt joint configuration. Shear test specimen configurations include the simple-lap, double-lap, and ring-and-plug specimen types. The butt joint geometry can be used to obtain torsion-shear test data. Because the solder joint mechanical properties are sensitive to the geometry, a complete description of the joint dimensions should always accompany the test data.

The time-independent mechanical properties of all metals and alloys are sensitive to the applied strain rate (or stress rate) to some degree. The strain rate sensitivity parameter, m, is expressed in Eq 1:

$$\sigma = c(d\varepsilon/dt)^m \quad \text{(Eq 1)}$$

where σ is the post-yield stress; c is a constant; and $(d\varepsilon/dt)$ is the strain rate (Ref 9). Solders and, in particular, the Sn-based, Pb-free alloys are highly strain rate sensitive. A listing of the yield strength, ultimate strength, and ductility properties of solders must also be accompanied by the strain rate used in the test procedure.

Microstructure

The mechanical properties of a metal or alloy are determined by the microstructure of the material. This is the so-called *microstructure-properties relationship*. Microstructural features that affect mechanical performance include: grain size, sub-grain structure (low-angle boundaries, cells, etc.), the distribution of matrix and particle phases, the atomic structure of the matrix phase (e.g., single element or solid solution), dislocation line defects (density and mobility), and point defects such as vacancies and interstitial atoms (density and mobility). Rest assured, these defects are present in Pb-free solders and are as important to their mechanical properties as they are to the properties of aluminum (Al) alloys, steels, and advanced superalloys! The transmission electron micrographs in Fig. 2 illustrate several features in the microstructure of the Pb-free 91.84Sn-3.33Ag-4.83Bi and traditional 63Sn-37Pb solders. Although it is beyond the scope of this text to provide a complete dissertation on the role of defects and microstructure on the mechanical properties of materials, several generalizations are described in the following text that can be applied to the behavior of the Sn-based, Pb-free solders.

The solidus temperatures (T_s) of nearly all Pb-free solders for PWAs are in the range of 215 to 225 °C (419 to 437 °F). The homologous tem-

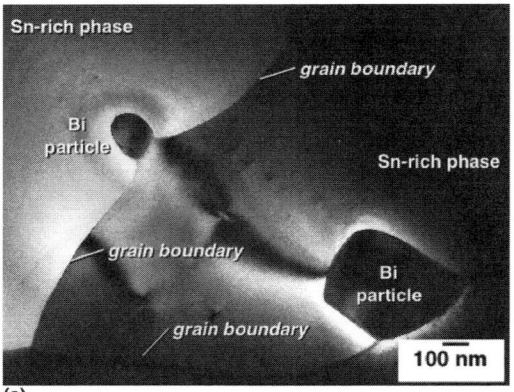

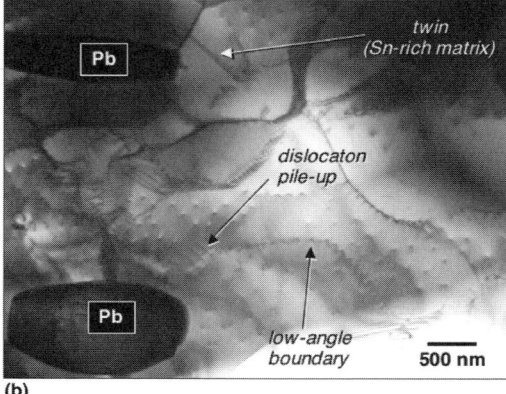

Fig. 2 Transmission electron micrographs showing grain boundaries, dislocations, and other features in the microstructure of (a) 91.84Sn-3.33Ag-4.83Bi lead-free solder, and (b) 63Sn-37Pb solder

perature, T_h, defined as T/T_s where T is the "use" temperature, will be in the range of $0.60 < T_h < 0.87$ for use temperatures of 25 to 150 °C (77 to 302 °F). By comparison, these temperatures would be equivalent to those experienced by advanced superalloys in the hot-section of an operating jet aircraft engine (700 to 900 °C, or 1292 to 1652 °F). *Therefore, even at room temperature, solder alloys are considered to be in high-temperature service.*

The microstructure has a unique effect on the high-temperature mechanical properties of an alloy. At elevated temperatures, time-dependent deformation and, in particular, diffusion-based processes dominate, as compared to the traditional dislocation glide mechanism. Because grain boundaries can enhance diffusion processes by serving as so-called *fast diffusion* or *short-circuit diffusion* pathways, high temperature strength is improved by increasing the grain size in the material, thereby reducing the number of such boundaries. This trend is opposite that of low-temperature plastic (time-independent) deformation for which the yield strength decreases with and increase of grain size, as described by the Hall-Petch relationship (Ref 10). The Hall-Petch relationship would describe deformation in solder when the dislocation glide mechanism dominates, that is, at fast strain rates and/or at very low temperatures ($< \approx -100$ °C, or -148 °F). It is apparent from these limiting conditions that, for most service and laboratory testing environments, only a very weak Hall-Petch relationship will prevail between grain size and yield strength in solder alloys.

The Sn-based, Pb-free solders have second and third phase particles distributed in the microstructure. In the case of the Sn-Ag-Cu alloys, those particles are comprised of the Cu_6Sn_5 and Ag_3Sn intermetallic or covalent compounds (Ref 11, 12). Precipitate particles serve as barriers to the motion of dislocations, whether the dislocations are gliding (low temperatures) under a critical resolved shear stress or climbing by means of thermally-activated, vacancy diffusion (high temperatures). Also, such particles hinder the movement of grain boundaries. In each of these roles, the particle phases can strengthen the material; this mechanism is called *precipitation strengthening or hardening.* That strengthening process is optimized for a large number of small, homogeneously distributed particles. In the case of the Sn-Ag-Cu alloys, only a nominal strengthening effect is to be expected from the Ag_3Sn particle phase. First of all, those particles are non-homogeneously distributed in the microstructure. Secondly, given the typical cooling rates of assembly processes, the Ag_3Sn particles precipitate with a relatively large size and, as such, will be *incoherent* with the matrix phase, thereby limiting the extent to which the surrounding mismatch stress field will impact the movement of point and line defects.

The Sn-Ag-Cu alloys experience a significant undercooling (25 to 30 °C, or 77 to 86 °F) upon solidification, even at relatively slow cooling rates of 0.02 °C/s, or 32 °F/s (Ref 13). The undercooling allows for the formation of pro-eutectic Ag_3Sn phase in the microstructure. The morphology of the pro-eutectic Ag_3Sn is that of isolated plates that, when viewed on-side, appear as long, needle-shaped particles. There is no data to indicate that the Ag_3Sn particles are detrimental to either the time-independent or time-dependent deformation of the Sn-Ag-Cu solders. Also, accelerated aging studies of Sn-Ag-Cu solder joints have not been able to determine, conclusively, whether or not the Ag_3Sn particles degrade the fatigue resistance of the interconnections. At this time, test results have indicated that particles oriented in the direction of crack development may enhance the rate of crack propagation, and those particles oriented at right angles to the crack growth direction may actually serve as barriers to crack growth. In this regard, it appears that the Ag_3Sn particles likely increase the uncertainty of the fatigue resistance of potential interconnections.

At elevated temperatures, the precipitated particles will coarsen with time. However, because the Cu_6Sn_5 and Ag_3Sn particles of Sn-Ag-Cu solders are covalently bonded compounds, the coarsening rate is relatively slow. This point has been confirmed in aging experiments (Ref 14). The slow coarsening rate is caused by the high activation energy required to break the covalent bonds in the particles in order to allow copper (Cu) and silver (Ag) to diffuse towards growing particles. As such, particle coarsening during aging is not expected to cause a significant change to the strength properties of the Pb-free solders.

Grain boundaries can serve as fast diffusion pathways for vacancies and interstitial atoms when deformation takes place at elevated temperature. The same point can be made with regards to the interface between the matrix and particle phases. However, the particle interfaces that are widely distributed lack interconnectivity with each other, causing them to be less effective as short-circuit diffusion paths than are the more extensive lamellae structures of alloys having a higher concentration of solute phase (e.g., 58Bi-42Sn and the traditional 63Sn-37Pb solders).

An alternative mechanism that can improve both the time-independent and time-dependent strength properties of a metal alloy is *solid-solution strengthening*. Solute atoms in the solvent phase lattice distort the latter, thereby causing local stress fields that act as barriers to the mobility of dislocations. Solid-solution strengthening tends to be more stable than precipitation hardening, particularly at elevated temperatures, because the solute can remain in solution by design, per the binary and ternary (equilibrium) phase diagrams. The solute atoms will not precipitate out into large particles, which can reduce the strengthening effect. Also, the wider dispersion of solute atoms within the solvent phase tends to be more effective at limiting the motion of dislocations. The Sn-Ag-Cu alloys do not exhibit a solid-solution strengthening mechanism because Cu and Ag have a negligible solubility in the Sn-rich matrix (solvent) phase. On the other hand, the Sn-Ag-Bi ternary Pb-free solders benefit from the solid-solution strengthening mechanism. As a result of a 3 to 5 wt% solubility of bismuth (Bi) in Sn, these alloys and several other higher-order compositions exhibit very high yield strengths (Ref 15–17).

Deformation processes at elevated temperatures necessarily raise the issue of microstructural stability. It is known that dislocation creation, motion, and annihilation will form features in the microstructure, e.g., dislocation pile-ups, dislocation tangles, sub-grain development, and small cavities caused by the associated movement of vacancies. However, high temperatures can also initiate other changes to the microstructure. The simultaneous occurrence of recovery processes can lead to the accelerated annihilation of point defects. Deformation at elevated temperatures can also initiate *dynamic recrystallization* processes that cause grain growth and a loss of grain boundaries. In the case of time-independent deformation, the consequence of dynamic recrystallization would cause a softening of the material during stress-strain testing, a phenomenon that has been observed in the Sn-Ag-Cu solders (Ref 18). These microstructural changes would also be expected to affect, in real time, the time-dependent deformation component.

The time-independent mechanical properties of Sn-Ag-X solders reflect largely the deformation of the Sn-rich matrix. In the case of the Sn-Ag-Cu alloys, the properties of elemental Sn will predominate the alloy behaviors. Elemental Sn has two crystal structures. The common structure, β-Sn, occurs at temperatures T > 13 °C (55 °F); it is body centered tetragonal (BCT). At temperatures T < 13 °C (55 °F), the α-Sn occurs that has the diamond cubic crystal structure. The allotropic transformation of β-Sn to α-Sn as the temperature drops lower than 13 °C, or 55 °F—the source of the so-called "tin pest" phenomenon—has drawn recent attention as a possible degradation mechanism of Pb-free solders as illustrated by Y. Kariya, et al. (Ref 19). The BCT structure of the more common β-Sn phase has a limited number of available slip planes, causing the crystal structure to have a greater strength, but also lower ductility, than do the more common face-centered cubic (FCC) and body-centered cubic (BCC) structures. The β-Sn structure will develop deformation twins. It is the formation of these twins that result in the so-called "crying" sound of Sn when it is deformed. Stacking faults are not commonly observed, due to the high stacking fault energy of Sn. Therefore, dislocation glide, deformation twinning, and diffusion-based processes are the microstructural processes that control time-independent deformation in the Pb-free solders.

Pb-Free Solders

The time-independent (stress-strain or plastic) deformation properties of the Pb-free solders will be examined. The discussion will be limited primarily to the 96.5Sn-3.5Ag eutectic and Sn-Ag-Cu alloys.

The values of the strain rate exponent, m, in Eq 1 are in the range of 0.06 to 0.08 for the alloys 96.5Sn-3.0Ag-0.5Cu, 95.5Sn-3.9Ag-0.6Cu, and 95.8Sn-3.5Ag-0.7Cu (Ref 20).

Shown in Fig. 3 are yield stress data for 95.5Sn-3.9Ag-0.6Cu (wt%) solder that were obtained by compression testing, using several strain rates (Ref 18, 21). The nominal sample dimensions were 19 mm (0.74 in.) length and 10 mm (0.39 in.) diameter and were tested in the as-cast condition. Minimum and maximum values of the duplicate tests were within ±5% about the mean value. The plot shows the expected decrease of yield stress with increased test temperature. The strain rate sensitivity of the yield stress is most distinct at the lowest temperatures, and then diminishes somewhat at the higher test temperatures. The yield stresses that were measured at the lowest strain rates, provide a benchmark for determining stresses to be used in constant load creep tests.

Often, it is preferred to perform mechanical tests and, in particular, creep tests on samples that were exposed to an annealing treatment beforehand. The annealing treatment stabilizes the microstructure and improves the consistency and reproducibility of the data. Time-dependent deformation is more sensitive to microstructural variations than time-independent deformation that occurs at faster strain rates. The compression yield stress as a function of test temperature and annealing treatments of 125 °C (257 °F) for 24 h and 150 °C (302 °F) for 24 h are shown in Fig. 4(a) and 4(b) for the strain rates of 4.2×10^{-5} s^{-1} and 8.3×10^{-4} s^{-1}, respectively (Ref 21). Although there was very little difference between the two annealing conditions, the introduction of either annealing treatment caused a significant drop in yield stress at the slower strain rate (and generally lower temperatures). This trend suggests that these treatments will have an impact on the creep performance of this Sn-Ag-Cu alloy.

The effect of Sn-Ag-Cu composition was also investigated. Shown in Fig. 5 are compression yield stress data for the following solders: 95.5Sn-4.3Ag-0.2Cu, 95.5Sn-3.9Ag-0.6Cu, 95.5Sn-3.8Ag-0.7Cu (Ref 22). The data obtained in Fig. 5 were generated at a strain rate of 4.2×10^{-5} s^{-1}. The samples were tested in the as-cast condition or subsequent to an annealing treatment of 125 °C (257 °F) for 24 h. The sensitivity of the yield stress to the solder composition was most pronounced at the lowest test temperature; that sensitivity diminished with increasing temperature. The annealing treatment reduced the yield stress to the greatest degree at test temperatures between 25 and 75 °C (77 and 167 °F) for each of the alloys. The largest effect was observed for 95.5Sn-3.9Ag-0.6Cu alloy.

Similar trends of yield stress as a function of solder composition were observed at the faster strain rate of 8.3×10^{-4} s^{-1} (Ref 22). The impact of annealing treatment on the yield stress values was of a lesser degree for all of the solders. It was observed that the compression yield stress values obtained in Ref 22 for the 95.5Sn-3.8Ag-0.7Cu alloy, using a strain rate of 8.3×10^{-4} s^{-1} and test temperatures of 25, 75, and 125 °C (77, 167 and 257 °F), were in agreement with the trend line obtained by Pang and co-

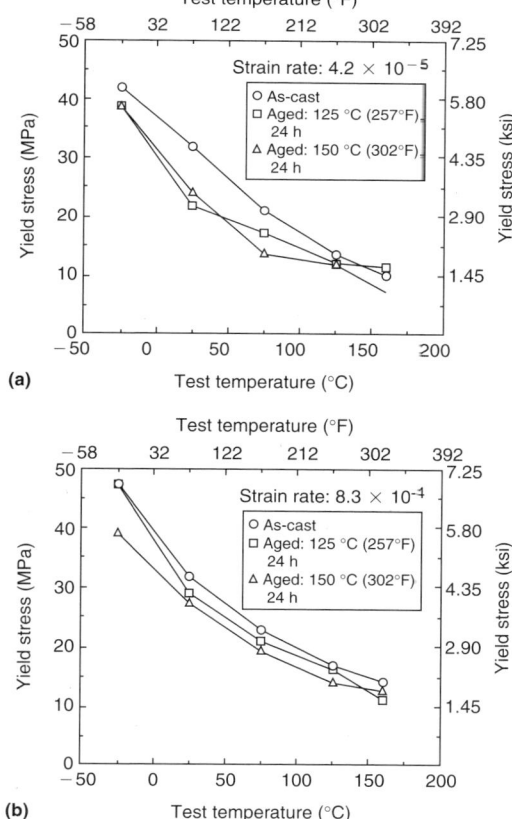

Fig. 4 Compression yield stress as a function of test temperature for the 95.5Sn-3.9Ag-0.6Cu, using strain rates of (a) 4.2×10^{-5} s^{-1} and (b) 8.3×10^{-4} s^{-1}. The specimens were tested in the following conditions: as-cast; annealed 125 °C (257 °F), 24 h; and annealed: 150 °C (302 °F), 24 h. Minimum and maximum values are within ±5% about the mean.

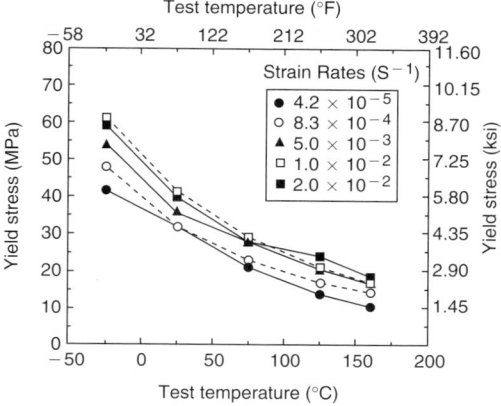

Fig. 3 Compression yield stress as a function of test temperature for the as-cast 95.5Sn-3.9Ag-0.6Cu, using strain rates of 4.2×10^{-5} s^{-1} 8.3×10^{-4} s^{-1}, 5.0×10^{-3} s^{-1}, 1.0×10^{-2} s^{-1}, and 2.0×10^{-2} s^{-1}. Minimum and maximum values are within ±5% about the mean.

workers in a study of the tensile yield strength of the same alloy and similar test conditions (Ref 23). In the latter tests, the test samples had a gage section with dimensions of 15 (0.59 in.) mm length and 3.0 (0.11 in.) mm diameter.

The authors F. Ochoa, et al. investigated the effect of controlled cooling rates on the time-independent deformation properties of the eutectic 96.5Sn-3.5Ag solder (Ref 24). The specimens were cooled at either 24 °C/s, 0.5 °C/s, or 0.08 °C/s (75 °F/s, 32.9 °F/s, or 32.14 °F/s). Bulk samples were tested in tension at room temperature, using a strain rate of 1×10^{-3} s^{-1}. The tensile yield strength decreased with a slower cooling rate; the values were 23.3 ± 0.4 MPa (3.38 ± 0.06 ksi), 18.0 ± 1.4 MPa (2.61 ± 0.20 ksi), and 17.0 MPa (2.47 ksi), respectively. A similar trend would be expected for the yield strength property of Sn-Ag-Cu alloys (Ref 20).

Frequently, the interpretation of fatigue data is based upon the extent of *plastic* strain only. The plastic strain can be obtained by subtracting the time-independent, or elastic strain component, from the total strain. Also, the more fundamental analyses of experimental creep data, in particular, the steady-state strain rate behavior, uses the *normalized stress* parameter. The normalized stress is the ratio of the applied stress (σ, normal stress, τ, shear stress) to the respective modulus value (E, Young's modulus and G, shear modulus) at that test temperature (T), that is, $\sigma/E(T)$ and $\tau/G(T)$ for normal and shear stress conditions, respectively. Therefore, it is important to have a record of the elastic moduli of the Pb-free solders.

There are two types of elastic moduli. First, there is the *static elastic modulus* that is measured from the stress-strain response of the solder when subjected to tension or compression testing (Ref 25). The second type is referred to as the *dynamic elastic modulus* and is measured by the passage of sound waves through the material (Ref 26). In the latter case, because sound wave propagation in a solid is based upon atomic vibrations that are very rapid, inelastic deformation is largely eliminated from the material response. Therefore, the modulus is determined from nearly pure elastic deformation. On the other hand, the static modulus is sometimes preferred when calculating plastic strain because it accounts for all deformation leading up to the yield stress as defined by the 0.2% offset criterion.

The static elastic modulus was determined for the Sn-Ag-Cu ternary solders. Shown in Fig. 6(a) are static elastic modulus values as a function of temperature for the 95.5Sn-3.9Ag-0.6Cu solder (Ref 18). These data were measured in tests performed at 4.2×10^{-5} s^{-1}. Samples were evaluated in the as-cast condition or subsequent to one of two annealing treatments: 125 °C (257 °F) for 24 h or 150 °C (302 °F) for 24 h. The ranges between the maximum and minimum values used to determine the means in Fig. 6(a) were generally ±(500 to 1000 MPa, or 72.52 to 145.04 ksi), causing any differences observed between the three sample conditions to be insignificant. This degree of scatter is not uncommon for static modulus data. The modulus values exhibit a repeatable trend that included a maximum for tests performed at 25 °C (77 °F) rather than an expected, monotonic decrease of values over the temperature range. There were no visible indications that the β-Sn to α-Sn allotropic transformation was responsible for the decrease of modulus for T < 25 °C (77 °F).

Most obvious was that fact that the moduli values are nearly an order of magnitude less than expected, based upon previous data for 100Sn, 96.5Sn-3.5Ag, and 95.5Sn-5Sb alloys (Ref 27). The role of inelastic deformation was examined by measuring the static elastic modulus at faster strain rates. The results are shown in Fig. 6(b). A general trend was observed, in which the modulus increased with strain rate, particularly at the lower temperatures, thereby implying that inelastic deformation likely had a role in the static modulus measurements at low strain rates. However, the faster strain rates did not bring the moduli to within range of the expected values.

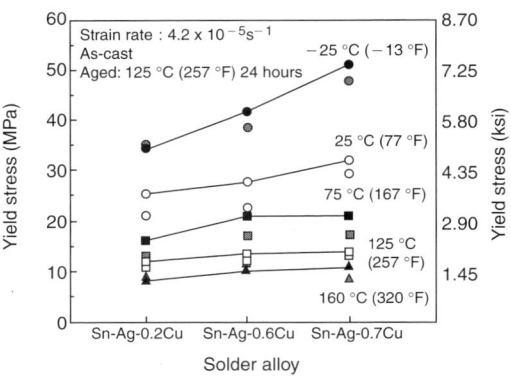

Fig. 5 Compression yield stress data for solders: 95.5Sn-4.3Ag-0.2Cu, 95.5Sn-3.9Ag-0.6Cu, 95.5Sn-3.8Ag-0.7Cu (wt%) from tests performed at 4.2×10^{-5} s^{-1}. The samples were in the as-cast condition (black symbols) or following the annealing treatment of 125 °C (257 °F), 24 h (corresponding gray symbols). Minimum and maximum values are within ±5% about the mean. Source: Ref 2

It was interesting to note that the same dependency of modulus on temperature (i.e., the maximum at 25 °C, or 77 °F) was reproduced at each of the strain rates in Fig. 6(b).

Shown in Fig. 7 are the static elastic moduli as a function of temperature for the three solders: 95.5Sn-4.3Ag-0.2Cu, 95.5Sn-3.9Ag-0.6Cu, 95.5Sn-3.8Ag-0.7Cu (Ref 22). The strain rate was 4.2×10^{-5} s^{-1} and the samples were in the as-cast condition. (The trends and data magnitudes were similar at a strain rate of 8.3×10^{-4} s^{-1}.) In all three cases, the values were lower by nearly an order of magnitude than expected. Also, the same test temperature dependency—that is, there being a maximum value at 25 °C (77 °F)—was observed for all three alloys. By comparison, a study was conducted by Pang and co-workers in which the static elastic modulus of bulk 95.5Sn-3.8Ag-0.7Cu alloy was measured by the tension testing of dog-bone samples (Ref 28). Those experiments were performed at temperatures of 25, 75, and 125 °C (77, 167, and 257 °F). The strain rates ranged from 5.6×10^{-4} s^{-1} to 5.6×10^{-2} s^{-1}. The static elastic moduli, which are listed in Table 1, were commensurate with expected values per Ref 27.

The dynamic modulus was determined for the 95.5Sn-3.9Ag-0.6Cu solder, using acoustic wave propagation through a cylindrical sample (Ref 18). The data appear in Fig. 8. It can be inferred from the sample-to-sample scatter between the two as-cast tests that the aging treatment had no significant effect on the modulus. The data in Fig. 8 confirmed that the static elastic moduli values in Fig. 6 were nearly an order of magnitude less than the corresponding dynamic moduli, even when the former were measured at the fastest strain rates. This comparison suggests that there is a component of linear inelastic deformation in the initial loading of the

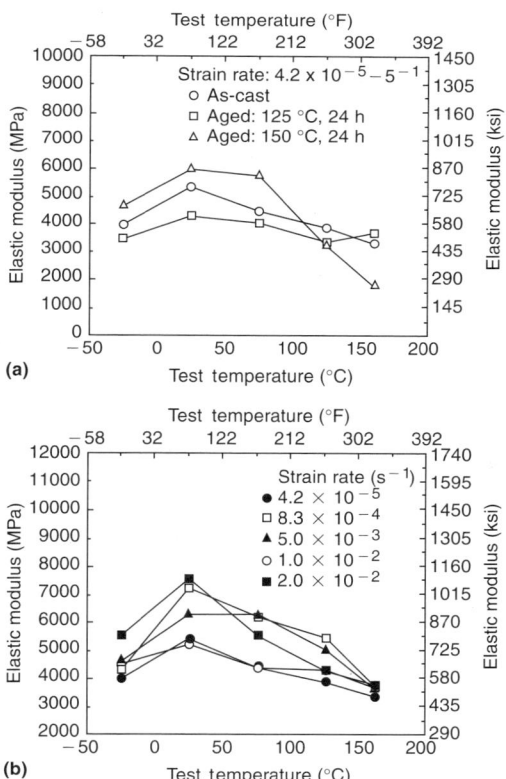

Fig. 6 Static elastic modulus as a function of temperature for the 95.5Sn-3.9Ag-0.6Cu composition: (a) strain rate was 4.2×10^{-5} s^{-1} and sample conditions were: as-cast; aged: 125 °C (257 °F), 24 h; and aged: 150 °C (302 °F), 24 h and (b) samples were in the as-cast condition and exposed to strain rates of 4.2×10^{-5} s^{-1}, 8.3×10^{-4} s^{-1}, 5.0×10^{-3} s^{-1}, 1.0×10^{-2} s^{-1}, and 2.0×10^{-2} s^{-1}. Minimum and maximum values are within a range of ±(500–1000 MPa, or 72.52–145.04 ksi) about the mean.

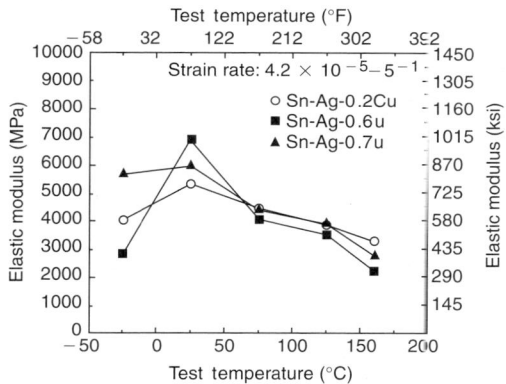

Fig. 7 The static elastic moduli as a function of temperature for the three solders: 95.5Sn-4.3Ag-0.2Cu, 95.5Sn-3.9Ag-0.6Cu, 95.5Sn-3.8Ag-0.7Cu. The strain rate was 4.2×10^{-5} s^{-1} and the samples were in the as-cast condition. Minimum and maximum values are within a range of ±(500–1000 MPa, or 72.52–145.04 ksi) about the mean.

Table 1 Static Young's modulus as a function of test temperature and strain rate measured for the bulk 95.5Sn-3.8Ag-0.7Cu solder

Strain rate (s^{-1})	Static elastic modulus (MPa)		
	25 °C (77 °F)	75 °C (167 °F)	125 °C (257 °F)
5.6×10^{-4}	44400	30700	18800
5.6×10^{-3}	50300	36000	25700
5.6×10^{-2}	58000	42400	32500

Source: Ref 28

stress-strain curve. It is not understood if this linear inelastic deformation is based upon a mechanism that is distinct from that which underlies post-yield plastic deformation. In some circumstances, the static modulus may be preferred to the dynamic modulus because the former captures the linear inelastic deformation behavior. Finally, mathematical expressions were developed to calculate the dynamic moduli and Poisson's ratio values as a function of temperature for the 95.5Sn-3.9Ag-0.6Cu solder; the parameters for those equations appear in Table 2 (Ref 18).

Pb-Free Solder Microstructures

The studies in Ref 18 and 22, which examined the time-independent properties of the Sn-Ag-Cu solders, included extensive reviews of the microstructure of the specimens after testing. In spite of compressive strains that reached 0.12 in some cases, as well as the occurrence of softening attributed to dynamic recovery and recrystallization processes, there was no crack damage to the microstructures of any of the samples. There was evidence of grain boundary development within the larger Sn-rich phase (dendrite) regions. This phenomenon is shown by the micrograph in Fig. 9 that was taken of a 95.5Sn-3.9Ag-0.6Cu solder exposed to stress-strain testing at 160 °C (320 °F) and a strain rate of 8.3×10^{-4} s^{-1}. The material was in the as-cast condition. Grain boundary formation was present for all test temperatures at the fast strain rate (8.3×10^{-4} s^{-1}); however, it was only observed at the 160 °C (320 °F) test temperature under the slower strain rate (4.2×10^{-5} s^{-1}) tests. Because (a) the grain boundaries developed within the Sn-rich dendrites; (b) it occurred at both high *and* low test temperatures; and (c) it was observed primarily under the faster strain rate, their source was likely the low-angle boundary formation during work hardening rather than dynamic recrystallization of pre-existing grain boundaries. The latter were not observed in similar areas prior to testing. Once formed, sliding may have subsequently accentuated those grain boundaries.

Fatigue Deformation

A brief synopsis will be presented of fatigue deformation; the reader is referred to more detailed treatises on this subject in Ref 29–31. Fatigue deformation results when a cyclic stress or cyclic strain is applied to the material. When the deformation is caused by a defined range of applied stress, then fatigue is said to be *stress controlled*. On the other hand, fatigue that results from a repeated strain displacement is said to be *strain controlled*. Nearly all fatigue conditions that are applicable to soldered interconnections, whether it be TMF that is generated by the combination of temperature variations and thermal expansion mismatch or fatigue that is caused by mechanical vibration, result in strain controlled, cyclic deformation.

Irrespective of whether the conditions are stress controlled or strain controlled, there are several parameters that describe the fatigue cycle. The case of strain control is exemplified by the sinusoidal fatigue cycle shown in Fig. 10(a). The list of parameters includes: (a) the strain amplitude, ε_a; (b) the strain range, $\Delta\varepsilon$; (c) the mean strain, ε_m; (d) the maximum strain, ε_{max}; and (e) the minimum strain, ε_{min}. Another parameter associated with fatigue tests is the strain ratio, $R = |\varepsilon_{max}/\varepsilon_{min}|$. A symmetric strain cycle would have ε_m equal to zero and $R = 1$. Because of the difficulty with gripping (soft) solder specimens and subsequent machine chatter that often accompanies the tension-to-compression cycle transition, fatigue tests are often performed in tension-tension. Referring to Fig. 10(a), tension-tension fatigue tests are characterized by a min-

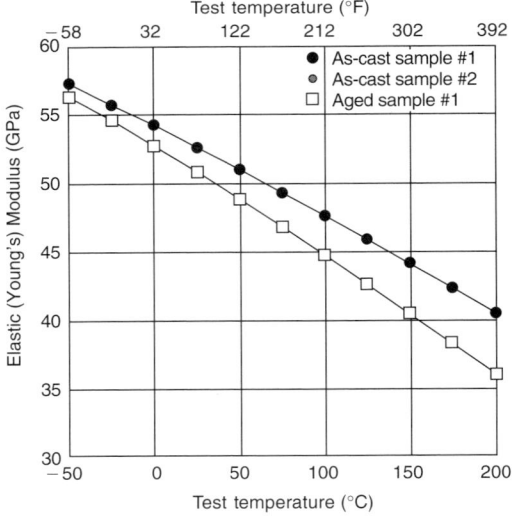

Fig. 8 Dynamic Young's modulus of the Sn-Ag-Cu solder as a function of temperature as determined by acoustic wave propagation techniques. The data were obtained from two as-cast specimens and one aged (125 °C, or 257 °F for 24 h) specimen. Source: Ref 18

imum strain that does not drop below zero so that the material is always under a tension stress. The strain amplitude, $\Delta\varepsilon_a$ or strain range, $\Delta\varepsilon$, define the extent of fatigue deformation in the solder.

A second important parameter of the fatigue is the frequency, ν, of the cycle. The frequency, which is the reciprocal of the period, τ, and is shown in Fig. 10(a), describes the number of strain reversals that occur per unit time—typically, per second. The fatigue process is often categorized as being either *low-cycle fatigue* or *high-cycle fatigue*. Because of the range of fatigue behaviors exhibited by metals and alloys, the demarcation point between low-cycle and high-cycle fatigue is not well defined by a single frequency value. Rather, a better parameter to distinguish between the two fatigue regimes is the number of cycles-to-failure, N_f. Low-cycle fatigue is characterized by $N_f < 10^4$ and high-cycle fatigue occurs when $N_f > 10^4$ (Ref 29). Thus, low-cycle fatigue has a relatively large

Table 2 Parameters to calculate the dynamic moduli and Poisson's ratio of as-cast 95.5Sn-3.9Ag-0.6Cu solder as a function of temperature

Material	Parameter	Quadratic function coefficients (a + bT + cT²)		
		a	b	c
As-cast #1	Young's modulus	54.21	-6.358×10^{-2}	-2.685×10^{-5}
	Shear modulus	20.24	-2.635×10^{-2}	-6.503×10^{-5}
	Bulk modulus	56.19	-5.461×10^{-3}	-4.223×10^{-5}
	Poisson's ratio	0.3392	1.722×10^{-4}	4.879×10^{-9}
As-cast #2	Young's modulus	52.77	-7.637×10^{-2}	-3.885×10^{-5}
	Shear modulus	19.59	-3.110×10^{-2}	-1.077×10^{-5}
	Bulk modulus	57.39	-1.258×10^{-2}	-2.703×10^{-5}
	Poisson's ratio	0.3467	1.874×10^{-4}	1.096×10^{-7}

Source: Ref 18

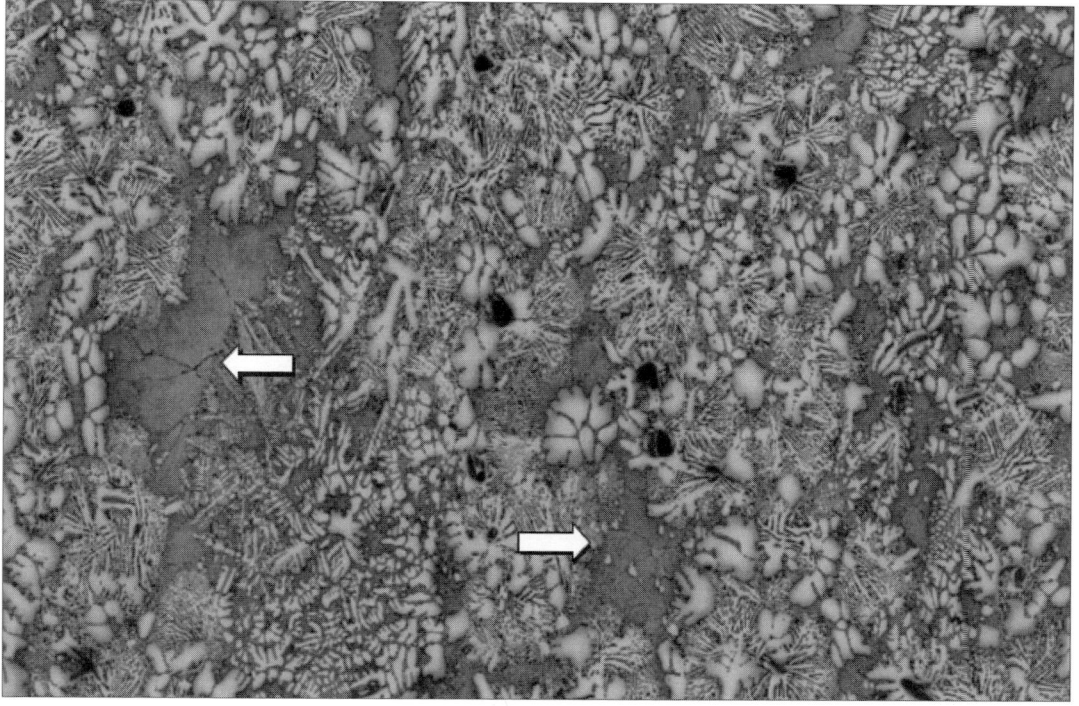

100 μm

Fig. 9 Optical micrograph (bright field) of a 95.5Sn-3.9Ag-0.6Cu sample tested at 160 °C (320 °F) and a strain rate of 8.3×10^{-4} s^{-1}. The material was in the as-cast condition. The arrows indicate areas of grain boundary development in the larger Sn-rich phase dendrites.

strain range so that the resulting stresses typically exceed the yield stress of the material on each cycle. Reduced strain ranges characterize high-cycle fatigue in which stresses remain less than the yield stress of the material. High cycle fatigue data is represented by the familiar S-N plot (displacement control). The "S" is the engineering stress and is traditionally plotted as the dependent variable while "N" is cycles-to-failure (N_f) and serves as the independent variable in the S-N plot. Low-cycle fatigue performance is typically plotted as cycles-to-failure (independent variable) versus plastic strain range, $\Delta\varepsilon$ (dependent variable).

It is important to recognize that fatigue data is prone to scatter. A 95% confidence interval band on cycles-to-failure can span a range of ±10% or more about the mean value. The variability increases as the strain or stress range decreases in extent. Variations in starting microstructure are often responsible for the test data scatter.

Solder can be exposed to either type of fatigue regime. Low-cycle fatigue most often underlies the failure of solder interconnections exposed to cyclic temperature environments (see discussion that follows). Solder joints exposed to mechanical vibration—for example, those due to the vibration of the printed circuit board—undergo high-cycle fatigue. Discussions of low-cycle fatigue by solders will often include the aspect of *frequency effect*. The frequency effect refers to the sensitivity of low-cycle fatigue deformation to the cyclic frequency. This effect was studied extensively by Solomon for the 60Sn-40Pb solder (Ref 1, 32). Solomon observed that the fatigue degradation of 60Sn-40Pb solder at 35 °C, or 95 °F (as measured by N_f, with a 50% load drop criterion) became insensitive to cycle frequency that exceeded 10^{-3} s^{-1}. At this time, similar tests have not been performed on the candidate Pb-free solders to determine the existence of such a frequency effect.

Solder interconnects are most often exposed to TMF while in service. Thermal mechanical fatigue takes place when temperature fluctuations combine with thermal expansion mismatch between different substrate materials, the *global mismatch*, or between the solder and the substrate materials, the *local mismatch*, to generate a cycle strain in the interconnections. The complicating factor of TMF, when compared to isothermal fatigue, is that the temperature changes *concurrently* with the strain cycle so that the entire strain event ($\Delta\varepsilon$) does not occur at the same temperature. More specifically, the fatigue deformation in a TMF cycle depends upon not only the temperature difference, ΔT, which determines the strain range, but also upon the minimum and maximum temperature values that define the limits of ΔT. Those temperature limits, and the time spent at them, contribute a significant amount of time-dependent deformation to the solder.

Thermal mechanical fatigue is a low-cycle fatigue process that is described, explicitly, by the temperature limits (Fig. 10b). The cycle is characterized by (a) relatively slow ramp rates between the maximum and minimum temperatures; and (b) hold times at the minimum and maximum temperatures. Both parameters determine the cycle frequency. It has been demonstrated with the Pb-bearing solders that the ramp

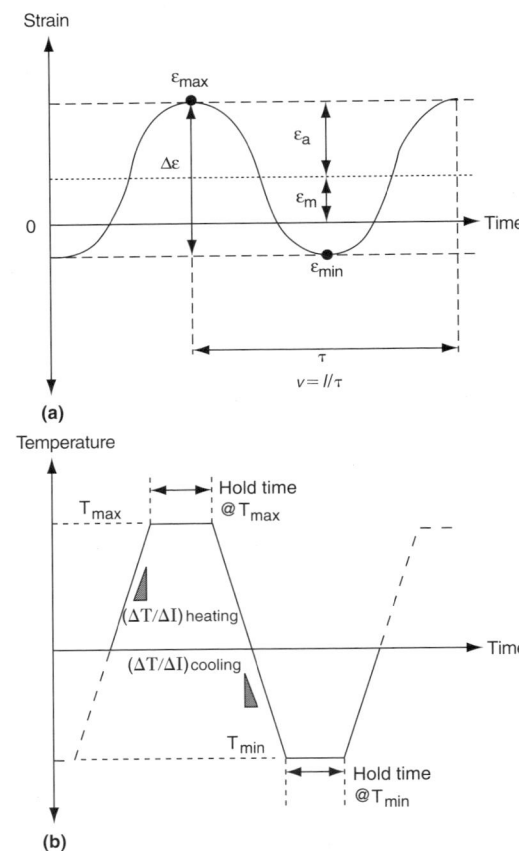

Fig. 10 (a) The parameters of a sinusoidal fatigue cycle: the strain amplitude, ε_a; the strain range, ε_r; the mean strain, ε_m; the maximum strain, ε_{max}; and the minimum strain, ε_{min}. (b) Temperature cycle responsible for TMF of soldered interconnections in service environments or during accelerated testing.

time between temperature extremes, as well as the hold times at those extremes, have a significant impact on the fatigue life of the solder. It is expected that the Pb-free solders will exhibit a similar effect, the magnitude of which is currently being documented in several studies.

Thermal mechanical fatigue strain can reach extremely high values in common electronic packaging applications. A worst-case scenario can be drawn from the placement of the ceramic package on a laminate circuit board. For example, a quarter-inch long ceramic chip component (coefficient of thermal expansion, CTE, = 6×10^{-6} C^{-1}) is soldered to an FR4 circuit board (CTE = 15×10^{-6} C^{-1}). When exposed to an accelerated aging condition that has the temperature extremes, -55 and 150 °C (-67 and 302 °F), the resulting shear strain, γ, in a 0.002 in. (3 mm) thick solder joint reaches 0.10 (10%). In appropriately designed solder interconnections, TMF strains will generally range from 0.001 to 0.01 for most service and accelerated aging applications. The deformation is primarily as a shear strain.

The ramp rates and the hold times of the TMF cycle (Fig. 10b) have a significant impact on the extent of fatigue deformation in the solder. Because the solder is always functioning at an elevated service temperature as defined by T_h, time-dependent deformation, or creep, occurs simultaneously in the solder. This phenomenon is the so-called *creep-fatigue interaction* that has been studied extensively for the Sn-Pb and high-Pb solders (Ref 33). The ramp rate determines the strain rate imposed on the solder between the temperature extremes. Under a very rapid strain rate (e.g., thermal shock testing), there is less opportunity for creep deformation to occur. On the other hand, a lower temperature ramp rate increases the opportunity for creep deformation to occur in the solder. Similarly, longer hold times allow for an increased proportion of time-dependent deformation, particularly at the higher temperature limits. Since the solder is usually under strain-controlled fatigue, the creep deformation is recorded as a drop in the stress; this mechanism is referred to as *stress relaxation*.

The reliability of an interconnection is determined by the accumulation of deformation in the solder. Several theoretical approaches have been proposed to quantitatively describe the accumulation of deformation during the creep-fatigue process. Those theories range from empirical approaches based upon such metrics as strain range, stress range, or strain energy density, to more complex computational approaches that use constitutive equations and finite element analysis, combined with fracture mechanics techniques. (Ref 34–37). (Of course, the spatial distribution of the deformation within the joint geometry has an important role in the failure of the interconnection.) In turn, the extent of accumulated deformation is dependent upon the parameters of the fatigue cycle. In the case of isothermal low-cycle fatigue, the factors that increase cyclic deformation in the solders are: (a) higher temperatures; (b) a larger strain range; and (c) a reduced cycle frequency or strain rate. In the case of TMF, which is a subset of low-cycle fatigue, the factors that raise the rate of deformation are: (a) a wider temperature range which increases the effective strain range; (b) higher maximum and minimum temperatures that accelerate creep deformation; (c) slower temperature ramps that provide a greater opportunity for creep to take place; and (d) longer hold times that likewise increase the extent of creep deformation. Isothermal fatigue and TMF tests performed on the 63Sn-37Pb and 60Sn-40Pb alloys have confirmed the above premises (Ref 38, 39). It is expected that similar trends will describe the fatigue of the Pb-free solders; the differences will lie in the magnitude with which those factors affect the performance of the new alloys.

The fatigue deformation of a polycrystalline material can be broken down into two processes. First, there are the microstructural changes associated with the annihilation, creation, and movement of point, line, and surface defects. Second, those aforementioned microstructural changes lead to the initiation of discontinuities, namely voids and cracks, which are often referred to as fatigue damage in order to distinguish the latter from the (deformation) activities of defects beforehand. The fatigue life begins with a stabilization of the microstructure. In a strain controlled experiment, the stress will increase or decrease, depending upon whether the material exhibits *cyclic strain hardening* or *cyclic strain softening* properties, respectively. In general, crystalline materials that are in an annealed (soft) state will harden. Conversely, cyclic softening will occur in already work hardened materials as an excess of defects are annihilated by dynamic recovery and/or recrystallization processes. Most Sn-based solders cyclically soften in strain-controlled fatigue (Ref 40). Upon completion of the softening or hardening stage (which is typically 20% of the pre-crack fatigue life), a nearly steady-state stress value will prevail until crack initiation (damage) begins. Crack initiation and propagation are indicated by a marked drop in the stress that is commensurate with the loss of load-bearing cross section in the material.

Microstructure

Fatigue deformation in polycrystalline materials begins with the development of a microstructure as a result of the creation, annihilation, and movement of point and line defects within individual grains of the phases in the solder (see Fig. 2). Due to the high homologous temperature at which solders must operate, grain boundaries can move as part of dynamic recrystallization processes. The dislocations that are created as a consequence of the cyclic deformation combine to form dislocation networks and tangles that often develop into shear bands, which are literally pathways of high dislocation densities that facilitate deformation. Also, the dislocations can generate low-angle boundaries that develop into high-angle grain boundaries, creating a subgrain or cell structure within the original grains of the material. Eventually, the dislocation structures create localized stresses that exceed the yield and tensile strength of the material, resulting in the initiation of a microcrack. Microcracks can also form at the intersections of slip planes or at the intersection of deformation twins. In this regard, Sn is susceptible to the formation of deformation twins. Once initiated, these microcracks become the sites of increased stress concentration so that crack propagation becomes the dominating mode of further fatigue degradation, and finally, failure of the material.

There is an absence of extensive transmission electron microscopy (TEM) studies that address fatigue deformation in the Pb-free solders (or, for that matter, in the traditional Sn-Pb alloys). However, as polycrystalline materials, it is likely that the mechanisms described previously can all contribute to the fatigue of the Sn-based solders. Deformation within the pro-eutectic Sn-rich dendrites will have a significant effect on fatigue deformation. It is expected that the Ag_3Sn and finer Cu_6Sn_5 particles will provide obstacles to the motion of dislocations in the ternary eutectic regions, thereby increasing the apparent fatigue strength of these materials.

However, it must be appreciated that the fatigue of Pb-free solders takes place at high homologous temperatures. As such, mechanisms that are specific to creep deformation, as well as ancillary processes such as recovery and (dynamic) recrystallization, will contribute, if not dominate, the fatigue process, particularly as the temperature becomes higher. For example, the diffusion of vacancies can occur under the cyclic stress or strain. The vacancies accumulate at other defects, often grain boundaries, resulting in the formation of pores, which then serve as sites for crack initiation. Grain and phase boundaries are susceptible to migration at elevated temperatures. Also, neighboring grains (or phases) can literally slide past one another; this phenomenon is referred to as *grain (phase) boundary sliding* and has been observed extensively in the high temperature deformation of metals, including Sn-Pb solders at room temperature (Ref 41, 42). A predominance of diffusion-based deformation processes will alter the effectiveness that precipitated particles have on fatigue performance with respect to their interactions with dislocations.

As noted previously, high homologous temperature can activate recovery and recrystallization processes. The recovery processes can serve to annihilate point and line defects, thereby stalling the initiation and propagation of cracks. It is for this reason that fatigue deformation that occurs prior to crack development contributes a significant proportion to the overall fatigue life of Sn-based solders. In a similar manner, recrystallization processes can remove accumulated defects in the microstructure by the sweeping action of migrating grain boundaries, resulting in a concurrent softening of the material strength.

Pb-Free Solders

Next, the fatigue behavior of Pb-free solders will be examined, targeting primarily the Sn-Ag-Cu compositions. As expected, the availability of bulk sample fatigue data—either isothermal fatigue or TMF—is relatively limited for Pb-free solders at the present time. Experiments by Pang and co-workers examined the isothermal fatigue of bulk 95.5Sn3.8Ag-0.7Cu solder (Ref 23). Tests were performed at two temperature/frequency combinations: 25 °C (77 °F)/1 Hz and 125 °C (257 °F)/10^{-3} Hz. Four total strain ranges ($\Delta\varepsilon_t$) were used: 0.02, 0.035, 0.05, and 0.075. The cycles-to-failure (N_f)—failure being defined at 50% load drop point—were correlated to both plastic strain range ($\Delta\varepsilon_p$) as well as to inelastic strain energy (ΔW), the latter being the area within the hysteresis loop, using the following Coffin-Manson-like relationships:

$$(N_f)^m \Delta\varepsilon_p = C \quad \text{(Eq 2)}$$

$$(N_f)^n \Delta W = D \quad \text{(Eq 3)}$$

where m and n are the fatigue exponents, C and D are constants, $\Delta\varepsilon_p$ is the plastic strain range, and N_f is the number of cycles-to-failure. The corresponding equations that were compiled by Pang and co-workers for their isothermal fatigue data were:

Plastic strain range, $\Delta\varepsilon_p$:

25 °C (77 °F)/1 Hz: $N_f^{0.91}\Delta\varepsilon_p = 26.3$ (Eq 4)

125 °C (257 °F)/10^{-3} Hz: $N_f^{0.85}\Delta\varepsilon_p = 9.2$ (Eq 5)

Inelastic strain energy density, ΔW:

25 °C (77 °F)/1 Hz: $N_f^{1.1}\Delta W = 1.5 \times 10^4$ (Eq 6)

125 °C (257 °F)/10^{-3} Hz: $N_f^{0.90}\Delta W = 310$ (Eq 7)

Two generalizations can be made from Eq 4 to 7. First of all, irrespective of whether deformation was recorded as $\Delta\varepsilon_p$ or ΔW, the fatigue exponent, which is the exponent of N_f, changed very little between the two test conditions, suggesting that the isothermal fatigue of the 95.5Sn-3.8Ag-0.7Cu solder does not have a particularly strong frequency dependence for these particular values. Secondly, a cross comparison of Eq 4 versus 6, as well as Eq 5 versus 7, indicated similar fatigue exponent values within each pair. Consequently, the plastic strain range parameter, $\Delta\varepsilon_p$, and the strain energy density, ΔW, differ only by a constant, albeit, that constant was not the same between the two temperature/frequency conditions. Thus, the plastic strain range, which is generally much easier to extract from experimental data, can serve as a suitable metric for measuring the isothermal fatigue damage of the Pb-free solder under these conditions.

There is a considerable amount of fatigue data for the eutectic and off-eutectic binary Sn-Ag solders. Those properties can be used to estimate the performance of the Sn-Ag-X and Sn-Ag-X-Y solders, keeping in mind that the ternary and quaternary additions can affect the fatigue properties of the alloy, even if only indirectly by altering the solidification microstructure. Shown in Fig. 11(a) is a plot of the total strain range versus cycles-to-failure (defined with the 50% load drop criteria) for the 96.5Sn-3.5Ag solder tested at 25 and 80 °C (77 to 176 °F) (Ref 43). The frequency of the tests was 0.5 Hz; there was a 0 s hold time at the maximum and minimum strain. These data can serve as a benchmark of the fatigue performance of the 96.5Sn-3.5Ag solder as a function of strain range and test temperature. The plot in Fig. 11(b) compares the fatigue performances of the 96.5Sn-3.5Ag and 63Sn-37Pb solders under the same test conditions (25 °C, or 77 °F). The Pb-free solder had a higher fatigue resistance than did the Sn-Pb alloy.

Frequency and hold time effects have also been evaluated for the isothermal fatigue of 96.5Sn-3.5Ag solder (Ref 43). The number of cycles-to-failure, N_f, does not appear to be strongly dependent upon frequencies that exceed approximately 10^{-3} Hz. A similar trend was observed with the 60Sn-40Pb solder; however, in the latter case, a loss of fatigue life was observed when frequencies were less than approximately 10^{-4} Hz where the material entered the low-cycle fatigue regime (Ref 1, 32). Unfortunately, such a frequency benchmark has not been determined for the Sn-Ag Sn-Ag-Cu solders.

The effect of cycle hold time was compared between the 96.5Sn-3.5Ag and 63Sn-37Pb solders in Fig. 11(c) (Ref 43). Those tests were performed at 25 °C (77 °F), using a strain range of 0.006, and a frequency of 10^{-2} Hz. The results indicated that hold times in excess of approximately 120 s (2 min.) do not affect the fatigue behavior of 96.5Sn-3.5Ag solder. A similar conclusion can be drawn for the 63Sn-37Pb solder. Finally, it is important to recognize that the minimum hold time parameter (120 s in the previous tests) would likely decrease at higher temperatures and the reduced frequencies for both low-cycle fatigue or TMF as creep (relaxation) deformation occurs more rapidly in the material.

Isothermal fatigue tests have been performed on solder joint configurations. Nearly all such studies have subjected the solder joint to shear deformation. Solomon performed studies on 60Sn-40Pb solder joints that resulted in a Coffin-Manson relationship between $\Delta\gamma_p$, the plastic shear strain range, and N_f (Ref 1). Lau and coworkers performed isothermal fatigue experiments on the 0.17 mm (0.006 in.) diameter, 95.5Sn-4.0Ag-0.5Cu area array (chip scale package) solder joints using load control (Ref 44). Those tests were performed at room temperature and at a frequency of 0.2 Hz. The pad finish on the latter was electroless Ni-immersion Au (ENIG). An empirical relationship was developed between the cycles-to-failure, N_f, and the strain energy, ΔW, which is the area of the fatigue hysteresis loop. The data is represented in the format of Eq 3 shown previously as:

$$(N_f)^{0.51} \Delta W = 4.39 \tag{Eq 8}$$

A comparison was made between Eq 8 and 6, the latter having been generated from the bulk, isothermal fatigue of a similar Sn-Ag-Cu alloy. The two equations confirm that the fatigue behavior of the Sn-Ag-Cu Pb-free solders, like their Sn-Pb counterparts, will differ between the bulk solder and solder joint, which is likely due to the effects of joint geometry.

A review was performed by Clech, which examined the state-of-the-art thermal fatigue data for Pb-free solders (Ref 45). Currently available data does not indicate a strong dependence of fatigue life on either Sn-Ag-Cu composition or on circuit board surface finish. (The sensitivity of solder joint fatigue performance to circuit board finish and, in particular, those finishes that include a protective layer [e.g., Au, Pd, etc.] that dissolves into the solder, will become more significant as the size of the joint and thus, the volume of the solder per joint, becomes smaller.)

The authors T.S. Park and S.B. Lee examined the effects of mixed stress modes on the isothermal fatigue (25 °C, or 77 °F) of 95.5Sn-3.5Ag-0.75Cu ball-grid array solder joints that

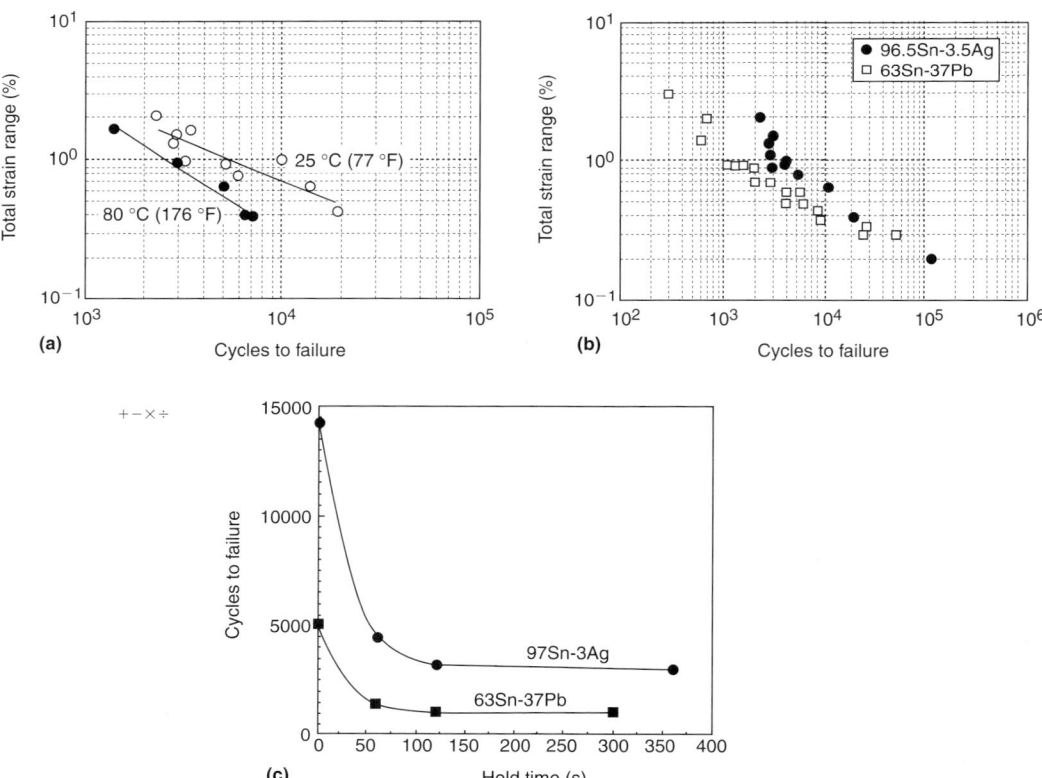

Fig. 11 Fatigue performance of bulk 96.5Sn-3.5Ag solder. The frequency was 0.5 Hz: (a) Cycles-to-failure (N_f) versus total strain range for tests performed at 20 and 80 °C (68 and 176 °F) (0 s hold times). (b) Cycles-to-failure versus total strain range at 25 °C (77 °F)—a comparison between the Sn-Ag solder and 63Sn-37Pb solder (0 s hold times). (c) Cycles-to-failure versus hold time at 25 °C (77 °F)—a comparison of Sn-Ag and 63Sn-37Pb solders. The total strain range was 0.006 and the overall frequency remained 0.5 Hz. Source: Ref 43

were made between two laminate substrates having Au-Ni plated pads (Ref 46). The test specimen allowed for the loading mode to be changed from that of a simple lap shear to uniaxial tension/compression. The strain ranges in the two limiting cases were varied from 0.01 to 0.066. The resulting data of strain range versus cycles-to-failure (at 50% load drop) clearly showed an increased loss of fatigue life when progressing from 100% shear deformation to uniaxial fatigue deformation. When the cycles-to-failure were plotted against strain energy density, ΔW, all of the data, irrespective of mode mixture, fell on the same trend line. A similar behavior has characterized the 63Sn-37Pb baseline solder. These results imply that fatigue deformation of the Sn-Ag-Cu solder, like that of the 63Sn-37Pb alloy, is well represented by the scalar strain energy density, irrespective of the particular loading mode.

Thermal mechanical fatigue studies were performed on bulk 96.5Sn-3.5Ag solder (Ref 43). The temperature range was 20 to 80 °C (68 to 176 °F) and the frequency was 10^{-2} Hz. Cycles-to-failure (50% load drop) were evaluated as a function of mechanical strain range. The data were compared to the isothermal fatigue studies (0.5 Hz) at the same limiting temperatures. (The fatigue behavior was not sensitive to the different frequencies of 0.5 versus 10^{-2} Hz.) It was observed that, similar to the Sn-Pb solders, the combination of plastic and creep deformations that occur during the temperature cycle in TMF is more detrimental to the fatigue resistance of Sn-based solders than is the degradation created under isothermal fatigue at either the low or high temperatures.

Thermal mechanical fatigue data have been obtained for Pb-free solder joints by means of accelerated aging tests performed on printed wiring assemblies. Although it is important to point out that these data are very specific to the interconnection geometries and test conditions, the findings provide valuable insight into the general performance of the solder alloys. In the review by Clech (Ref 45), TMF data were compiled from printed circuit board test vehicles assembled with the 95.5Sn-3.8Ag-0.7Cu and 95.5Sn-3.9Ag-0.6Cu alloys. There was a variety of package configurations and I/O types; the surface finishes included immersion Ag, Ni-Au, and organic solderability preservatives (OSPs). The thermal cycle ranges were 0 to 100 °C (32 to 212 °F), −40 to 125 °C (−40 to 257 °F), and −55 to 125 °C (−67 to 257 °F). The characteristic life—that is, the number of cycles required to fail 63% of the solder joints—was normalized to the individual solder joint crack areas and plotted as a function of cyclic shear strain range per a generalized Coffin-Manson relationship. A graph of the nominal trend line and 2x scatter bands appears in Fig. 12. The correlation is very good across the range of package interconnections, surface finishes, and the two Pb-free solder alloys ($R^2 = 0.96$). There was insufficient data from which to conclude that the two Pb-free compositions had significantly different fatigue behaviors. Also, Clech points out that this database is skewed towards cycle strain ranges (>0.01) that are larger than the typical conditions experienced by solder joints while in service, or when exposed to accelerated aging tests. The underlying objective used by many investigators, which is to employ large strain ranges in order to obtain fatigue data quickly, may obscure the more subtle effects caused by the solder joint size effects, surface finishes, or the small differences in Sn-Ag-Cu composition that will be critical to the performance of Pb-free interconnections under the reduced strains common to service conditions.

Clech combined the data in Fig. 12 with similar TMF data for the traditional Sn-Pb solder (Ref 45). The resulting trends are shown in Fig. 13. When the cyclic strain range is less than 0.062, the Sn-Ag-Cu solders have better TMF resistance than the Sn-Pb alloy. On the other hand, cyclic strain ranges greater than 0.062

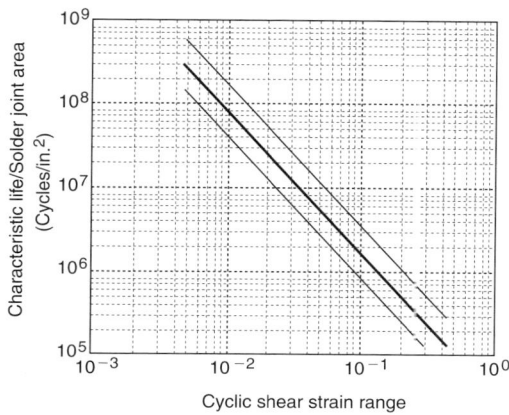

Fig. 12 Solder joint characteristic life for TMF, normalized to crack area, as a function of cyclic strain range for a variety of solder joint configurations, surface finishes and the two Pb-free solders, 95.5Sn-3.8Ag-0.7Cu and 95.5Sn-3.9Ag-0.6Cu. The "2x" band lines were drawn to show the data scatter. Source: Ref 45

cause the Sn-Pb solders to have better TMF resistance. Since the service environments of most consumer electronics expose the interconnections to cyclic strains in the range of 0.001 to 0.01, it is to be expected that the Sn-Ag-Cu solder joints will have a better reliability vis-à-vis the Sn-Pb interconnections. However, the higher cyclic strains associated with harsh service environments (e.g., automotive underhood electronics) or those used in accelerated aging tests (e.g., -55 to ≥ 125 °C, or -67 to ≥ 257 °F) that assess the reliability of high-valued military and space hardware will cause the Sn-Ag-Cu solders to have a poorer TMF performance than the Sn-Pb solders.

The underlying physical metallurgy that is responsible for the so-called "crossover" effect is not understood at this time. Because these data were generated by TMF testing, the reduced fatigue life of the Sn-Ag-Cu at $\Delta \varepsilon > 0.062$ could have been caused by (a) an intrinsic susceptibility to early failure by the alloy microstructure under large deformations; (b) a sensitivity of the solder fatigue strength to these temperature limits; or (c) a synergistic effect between these two factors. It is hypothesized that better fatigue properties of the Sn-Ag-Cu alloys at low strain ranges reflect the higher strength of these solders when compared to the Sn-Pb alloys. However, that higher strength comes at a price, which is lower ductility. At larger strain ranges, the higher strength Sn-Ag-Cu solders cannot readily accommodate the deformation and, accordingly, become more susceptible to an earlier onset of crack development than are the more ductile Sn-Pb solders.

A similar comparison was performed by Clech between the 99.3Sn-0.7Cu solder and the Sn-Pb alloy (Ref 45). The Sn-Cu solder is a leading candidate to replace Sn-Pb alloys in wave soldering applications, especially for low-cost consumer products. A crossover effect was observed; however, the crossover strain range was considerably smaller, (approximately 0.028) and moreover, the trend was *reversed*. That is, at cyclic strain ranges less than 0.028, the Sn-Cu solder has a poorer TMF performance than the Sn-Pb solders. At strain ranges that exceeded 0.028, the Sn-Cu solder had better TMF performance than the Sn-Pb. Shear strength data indicates that, in fact, the Sn-Cu solder is weaker than the Sn-Pb alloy, resulting in the reverse crossover effect according to the above strength argument (Ref 27).

Several other trends were observed in the Clech analysis of current TMF data on Pb-free solders (Ref 45). Those trends are highlighted below. The author warns that these behaviors are only of a first-order nature, owing to the limited TMF databases for the Pb-free solders and, in particular, the lack of failure data at the lower cyclic strain ranges (<0.01). Reference 45 carries a complete bibliography of the original data sources.

- The Sn-Ag-Cu alloys have a lower TMF lifetime than does Sn-Pb solder in flip chip solder joints. Although Clech cautions that this trend may be caused by the limited database, there is also the hypothesis that there is a real size scale or length-scale effect that reduces the fatigue life of Pb-free solder joints. The temperature ranges of these data were exclusively 55 to 125 °C (131 to 257 °F) or -40 to 125 °C (-40 to 257 °F).
- There appears to be no across-the-board effect of Pb contamination on the TMF of Sn-Ag-Cu solder joints. The sensitivity of Pb-free solder joints to Pb contamination from the component I/Os was package dependent. That dependence may reflect different quantities of Pb introduced into the joint or the variation in stress state caused by the I/O configuration. There was also an effect generated by the surface finishes (Ni-Au, immersion Ag, etc.) on the circuit board. The TMF tests were performed under the -55 to 125 °C (-67 to 257 °F) or -40 to 125 °C (-40 to 257 °F) temperature range.

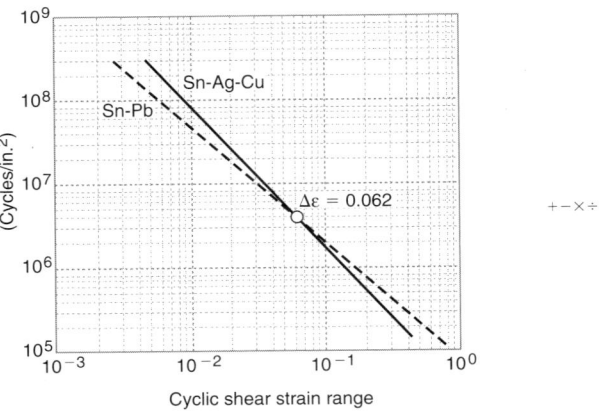

Fig. 13 Solder joint characteristic life of TMF, normalized to crack area, as a function of cyclic strain range. The data sets included those of the two Pb-free solders, 95.5Sn-3.8Ag-0.7Cu and 95.5Sn-3.9Ag-0.6Cu and data representing the traditional Sn-Pb solder. The crossover of fatigue life occurs at a strain of 0.062. Source: Ref 45

- There is no consistent trend in the TMF lifetime of area-array packages having Sn-Ag-Cu solder balls that are assembled to the circuit board with either Sn-Ag-Cu or Sn-Pb solder as long as the entire solder joint has been melted, including the SnAgCu ball. Nearly all of the supporting data were from life tests performed under the harsher −55 to 125 °C (−67 to 257 °F) or −40 to 125 °C (−40 to 257 °F) temperature ranges. These data indicated that the quantity of Sn-Pb solder and specifically, the Pb component introduced into the joint, did not significantly alter the interconnection reliability.
- The case was also evaluated for the area array packages that had Sn-Pb solder balls, which were then assembled with either Sn-Pb or Sn-Ag-Cu solder paste. The data suggest that for temperature ranges of −40 to 125 °C (−40 to 257 °F) and −55 to 124 °C (−67 to 255 °F), use of the Sn-Ag-Cu solder paste resulted in a higher reliability than was observed of solder joints comprised entirely of Sn-Pb solder. On the other hand, a milder temperature range of 0 to 100 °C (32 to 212 °F) resulted in a poorer reliability for mixed solder joints than was observed in joints made entirely of Sn-Pb solder. These trends were exactly opposite those that would be predicted, based upon the assumption that the solder joint behavior would reflect upon the introduction of the Sn-Ag-Cu solder and the aforementioned crossover effect that predicts improved fatigue performance for Sn-Ag-Cu joints at the 0 to 100 °C (32 to 212 °F) temperature range. The opposite trends signify that more subtle changes occurred to the microstructure as a consequence of the quantity of Sn-Pb and Sn-Ag-Cu solders and the specific joint geometry.

In summary, a reliability database is being compiled for the Pb-free, Sn-Ag-Cu solder alloys. Based upon the current state of the art, it can be concluded that a single, generalized comparison *cannot* be made between the fatigue reliability of Pb-free solder joints versus that of Sn-Pb solder joints. The factors that must be taken into account include: I/O and interconnection geometries, surface finishes and Pb contamination, as well as the temperature extremes and strain ranges used in accelerated aging tests.

Pb-Free Solder Microstructures

At present, the quantitative fatigue data of Pb-free solders has been accompanied by relatively little microstructural analysis. Unpublished microstructural data were compiled from low-cycle fatigue of double-lap shear samples that were made with the 95.5Sn-3.9Ag-0.6Cu alloy (Ref 47). The solder joints were tested with strain ranges, $\Delta\varepsilon$, of 0.025, 0.050, and 0.100 and at temperatures of 25, 100, and 160 °C (77, 212, and 320 °F). Separate specimens were isothermally aged for times and at temperatures that were commensurate with the isothermal fatigue conditions, in order to distinguish microstructural changes caused by annealing from those caused by the cyclic deformation. The solder microstructures exhibited an arrangement of horizontal and vertical grain boundaries as observed in Fig. 14. The grain structure did not change as a result of exposure to the simulated test conditions of elevated temperature and time. Isothermal fatigue at 25 °C (77 °F) generated cracks that propagated along existing grain boundaries (Fig. 15(a) and 15(b), 125 cycles, $\Delta\varepsilon = 0.10$). However, as the extent of deformation to the microstructure increased (Fig. 15(c) and 15(d),

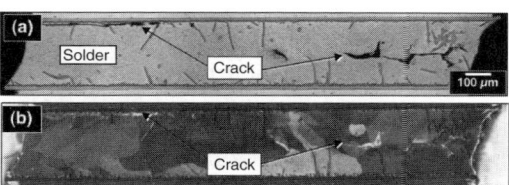

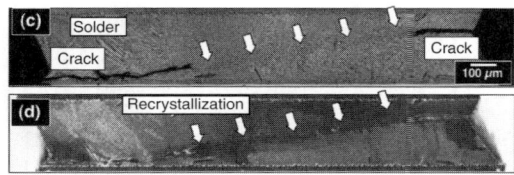

Fig. 15 Optical micrographs using (a, c) bright field and (b, d) polarized light, to reveal the grain and crack structures in the following image pairs: (a, b): 25 °C (77 °F), $\Delta\varepsilon$ = 0.10, 125 cycles; and (c, d) 25 °C (77 °F), $\Delta\varepsilon$ = 0.05, 750 cycles. In the latter case, arrows indicate a zone of recrystallization connecting the two crack tips.

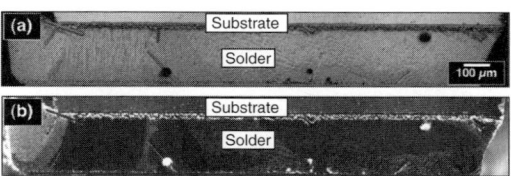

Fig. 14 Optical micrographs using (a) bright field and (b) polarized light, to document the grain structure of 95.5Sn-3.9Ag-0.6Cu solder joints between Cu subjected to low-cycle fatigue testing

750 cycles, $\Delta\varepsilon = 0.05$), a pathway of recrystallized material connected the two crack tips. The crack path continued along the newly formed grain boundaries. This phenomenon may be likened to the Pb-rich phase coarsening that is observed ahead of TMF cracks in Sn-Pb solder joints. It was also observed that cracking became relatively minimal for all tests performed at 100 °C (212 °F). Tests preformed at 160 °C (320 °F) produced fatigue cracks having a path near the solder/substrate interface.

In general, the microstructural observations that have been reported, thus far, have originated largely from TMF testing. Dunford and coworkers examined the microstructures of BGA solder joints exposed to TMF accelerated testing (-40 to 125 °C, or -40 to 257 °F temperature extremes) (Ref 48). The 95.5Sn-3.8Ag-0.7Cu and 96.5Sn-3.5Ag solders were evaluated. The authors confirmed that, like the isothermal fatigue data presented previously, the preferred path of TMF cracks is along high-angle grain boundaries. Smaller cracks were observed after prolonged thermal cycling, which appear to have formed along the boundaries of considerably smaller grains. Those latter grains may indicate grain development caused by localized recrystallization such as that observed in Fig. 15(c) and 15(d). In all cases, the fatigue cracks were not associated with the substrate IMC layers, per se.

S. Kang, et al. examined the role of Ag on the TMF performance of three Sn-Ag-Cu compositions and a Sn-Ag-Cu-Bi alloy, using the accelerated testing (0 and 100 °C, or 32 and 212 °F temperature limits) of BGA solder joints (Ref 13). The authors observed that fatigue cracks favored a propagation path near to the solder/substrate interface in the high Ag content (3.8 wt.%) alloy, which also exhibited a higher nominal strength (and lesser ductility). On the other hand, the reduced strength of the low Ag content alloys (2.1 to 2.5 wt%) resulted in crack propagation within the interior of the solder. It was also concluded that the pro-eutectic Ag_3Sn plates in the higher Ag-containing alloys had no single effect on the fatigue performance of the solder joints. Plates aligned in the direction of preferred crack propagation quicken the crack growth rate, while those oriented perpendicular to the crack path appeared to have slowed the propagation rate.

The fatigue of traditional Sn-Pb solders has often been accompanied by the observation of grain-boundary and phase-boundary sliding. Boundary sliding is a signature artifact that usually (albeit, not exclusively) indicates a significant component of creep in cyclic deformation, usually as a consequence of the elevated homologous temperature conditions. It must be understood that grain (and phase) boundary sliding is not simply the relative movement between two rigid solids, as in the case of a mechanical bearing. Rather, boundary sliding is believed to be caused by very specific, coordinated atomic movements along the interface that are likely due to vacancy motion or the activity of dislocations (Ref 49). Therefore, the propensity for grain boundary sliding to occur will depend upon (a) the specific test conditions; (b) the intrinsic mechanical properties of the solder; and (c) most importantly, the microstructure of the alloy. These contributors are briefly discussed as follows.

1. Grain boundary sliding is most likely to occur in TMF (accelerated testing) assessments because of the relatively slow temperature ramp rates and longer hold times at the temperature limits that allow creep to take place. Shown in Fig. 16(a) and (b) are examples of grain boundary sliding and phase boundary sliding that were observed in the Pb-free alloys: (a) 91.84Sn-3.33Ag-4.83Bi and (b) 58Bi-42Sn, respectively, as a result of TMF during accelerated aging (Ref 50, 51). Shown in Fig. 16(c) is an example in which extensive grain and phase boundary sliding generated microcracks would later joint together and cause failure of the interconnections.

2. The propensity for grain boundary sliding to occur depends upon the creep strength of the solder, but more so, upon the strength properties of the grain boundaries themselves. Weak boundaries are more susceptible to sliding at high temperatures. The greater the crystallographic mismatch across the boundary—high-angle boundaries as compared to low-angle boundaries that have a reduced mismatch—the weaker will be the interface strength. The absence of significant grain boundary structures being observed in metallographic cross sections of Sn-based solders implies that those boundaries are less prone to sliding during fatigue, even TMF, due to a relatively low mismatch angle across them.

3. The solder microstructure affects grain and phase boundary sliding. Grain boundary sliding can be significantly curtailed by the concentration or precipitation of alloying ele-

ments along the boundary (Ref 52). Because grain boundary sliding is based upon atomic movements, features such as precipitates, dislocation networks, etc. can alter the extent of sliding. Certainly, the presence of Ag_3Sn and Cu_6Sn_5 particle phases in Sn-Ag-Cu alloys provides a probable cause for the limited observation of grain boundary sliding in these Pb-free solders.

Creep Deformation

Introduction

Elevated temperature, time-dependent deformation, or *creep,* is a critical parameter that affects the performance of solder interconnections. This significant contribution is a result of the low solidus temperature of solder alloys. Even room temperature ($T = 25$ °C, or 77 °F) represents an elevated temperature for the Pb-free alloys, as indicated by a high homologous temperature, $T_h = T/T_s$, for these and nearly all electronic solders. Therefore, there is a significant likelihood for creep to occur in interconnections, even under modest stresses.

Phemenologically, creep is defined as strain (deformation) versus time, which results from the application of an applied stress. The general strain-time creep response is shown in Fig. 17. Three stages are identified: the *primary or transient stage* is characterized by a gradually decreasing strain rate with time; the *secondary or steady-state stage* that exhibits a time independent strain rate; and lastly, the *tertiary stage* in which the strain rate increases with time, to the point of failure or *creep rupture*. The presence, and the extent to which any one of the stages contributes to the overall creep response, de-

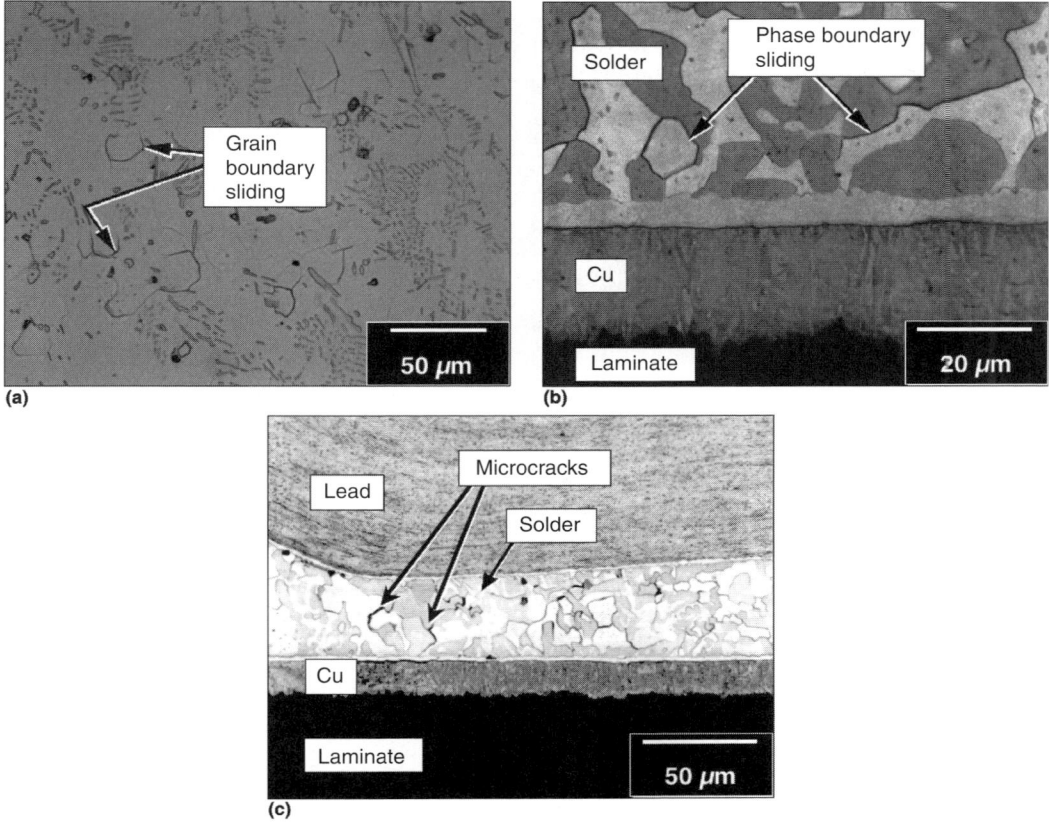

Fig. 16 Optical micrographs exemplifying the effects of grain boundary and phase boundary sliding caused by TMF during accelerated aging (0–100 °C, or 32–212 °F, 2500 cycles): (a) grain boundary sliding in a solder joint having the 91.84Sn-3.33Ag-4.83Bi alloy, (b) both grain and phase boundary sliding in the 58Bi-42Sn solder joint, and (c) grain and phase boundary sliding leading to microcracks in a 58Bi-42Sn interconnection.

pends upon the applied stress and the temperature environment. Also, the shape of the creep curve presented in Fig. 17 may not always be observed, particularly when there are simultaneous changes taking place to the material microstructure that are caused by the elevated temperature environment.

The ASTM method for creep testing is provided in Ref 53. Any one of the test methods described in Ref 4 to 7 can also be adapted for creep testing. Novel test procedures can be used to assess the creep behavior of soldered joints. Although time-dependent deformation is, in general, less sensitive to joint geometry that is the faster stress-strain test, it is not completely immune to the effects of joint dimensions, as well as possible size or length-scale effects that may occur for very small interconnections (e.g., flip chip joints).

Given the strength of most Pb-free solders and the stress levels that often prevail in an interconnection as a result of service environments or accelerated aging tests, the solder will deform primarily in the primary or transient stage. Therefore, it is important to understand the underlying deformation mechanisms that are active at this stage. A brief synopsis follows, describing several proposed theories for primary creep. A detailed account of those theories can be found in Ref 54.

Primary or transient creep is a relatively complex phenomenon because the underlying mechanisms and microstructure, and therefore, the strain rate, change with time. The defects required for deformation may be created at the moment that the stress is applied. On the other hand, an adequate density of defects may already be present in the microstructure to provide deformation in response to the applied stress.

These two bounding conditions, and conditions in-between, are represented schematically in Fig. 18 (Ref 55). When defects are absent, they must be generated to support subsequent deformation in response to the applied stress. The creep curve takes on a sigmoidal shape because there is an initial incubation time during which the defects are generated (curves "A" and "B" in Fig. 18). The length of the incubation period and subsequent stretch of accelerating strain rate diminish for a higher quantity of initial defects in the microstructure, or for a higher rate of defect creation caused by a larger applied stress, higher temperature, or a combination of these two conditions.

During the progression of creep deformation, the density of defects increases further. However, at some point, defects will begin to be annihilated at sinks, or be of such a density that they inhibit the motion of one another in the microstructure. By either course, the strain rate will then decrease with time, resulting in the traditional configuration of primary creep (curves "C" and "D" in Fig. 18 or as illustrated in Fig. 17).

The decreasing strain rate that occurs during primary creep (curves "C" and "D" in Fig. 18) has been addressed by the development of exhaustion theories. As the term implies, exhaustion theories describe the decreasing strain rate by the loss of defects that support strain, e.g., dislocations, vacancies, etc. Quantitatively, the creep strain is represented as a logarithm function of time:

$$\varepsilon = A\ ln(\nu t)^m \qquad (Eq\ 9)$$

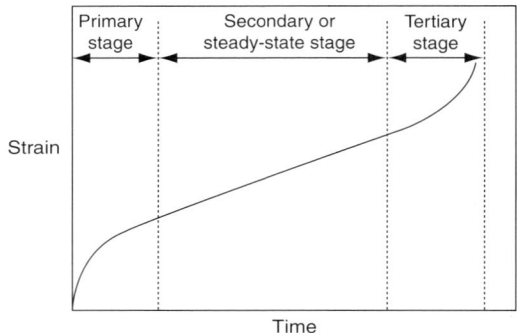

Fig. 17 Typical strain-time curve of creep deformation in a metal alloy

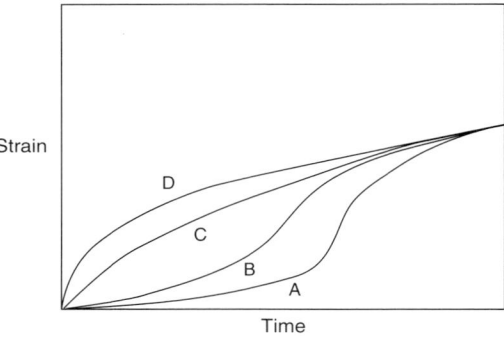

Fig. 18 Generalized creep curves illustrating the change from sigmoidal ("A" and "B") to primary ("C" and "D") creep, which is determined by the number of strain-supporting defects available at the start of testing. Either sigmoidal or primary creep can then lead into steady-state creep.

where A is a temperature-dependent constant (typically, an Arrhenius expression) and v and m are constants. The value of m is $0 < m \leq 1$.

Work-hardening theories have also been used to explain a decreasing strain rate during primary creep. They differ from exhaustion theories in that the decreasing strain rate is due to a slowing of defect (dislocation) motion rather than the loss of the defects altogether. Lattice friction and dislocation intersection processes have been proposed to explain the work hardening mechanism. These various theories have resulted in mathematical functions of strain versus time that range from a logarithmic time-dependence similar to that in Eq 9 to power law functions of time having exponents of between zero and one. Although defect annihilation is not considered in work-hardening theories, some variations include the creation of additional defects that accelerate the immobilization of defects that, in turn, causes a decrease of the strain rate.

Steady-state or secondary creep (Fig. 17) is characterized by a constant or minimum strain rate, $d\varepsilon/dt_{min}$. Two theories have typically been used to explain the constant strain rate (Ref 54): (a) There is a balance between the creation and annihilation of defects during steady-state creep. The rate processes are thermally activated and dependent upon the applied stress. The defect velocity is assumed to be a constant; and (b) There is a relatively constant supply of defects having a constant velocity that establishes the creep rate. The creep rate is determined by the stress level and temperature and is also a thermally activated process. This theory is referred to as the Nabarro-Herring creep process.

It is important to realize that, although primary and steady-state or secondary creep are often analyzed as different responses to the applied stress and temperature, they occur as a continuous transition within the material microstructure and, as such, are mechanistically linked together. The distinction between the two regimes has developed because of the difficulty of combining the two behaviors into a single mathematical expression. A number of investigators have developed theories and mathematical equations that link the two behaviors together, both mechanistically and quantitatively (Ref 56).

Because of the relatively high temperatures that are required to establish steady-state creep in metals and alloys, diffusion-controlled mechanisms typically underlie the resulting deformation. Those mechanisms include the simple diffusion of point defects—vacancies and interstitials—as well as diffusion-controlled, dislocation motion resulting from cross slip and/or jog-screw mechanisms. Diffusion-controlled process include *self-diffusion* or *bulk (lattice) diffusion*, in which atomic movements occur within the lattice structure, as well as *fast* or *short-circuit diffusion* in which atomic transport occurs along grain boundaries, internal interfaces, or other defect structures in the microstructure. Therefore, the apparent activation energy (ΔH) values for creep often reflect one of these two mechanisms. When self- or bulk-diffusion is the controlling mechanism, ΔH will have values that are typically 90 kJ/mol to 110 kJ/mol for Sn-based solders (Ref 57, 58). For example, the cross slip of dislocations and the movement of screw dislocations by jogs are based specifically on the diffusion of vacancies in the bulk lattice. Therefore, the apparent activation energy for creep, in the event that either of these two processes are rate controlling, will be similar to that of lattice self-diffusion. On the other hand, when the controlling mechanism is fast or short-circuit diffusion, the value of ΔH is typically 0.4 to 0.6 of the bulk diffusion value, or about 40 kJ/mol to 60 kJ/mol (Ref 59–61). Fast or short-circuit diffusion often controls creep deformation in the Sn-Pb solders (Ref 62). A similar condition prevails for the Pb-free solders as the discussion that follows will illustrate.

Two properties are important to the creep behavior of Pb-free solders. First, the Pb-free solders have low solidus temperatures so that, even at temperatures as low as -55 °C (-67 °F), the homologous temperature is still in excess of 0.5. Therefore, diffusion-controlled mechanisms can dominate the creep behavior. Secondly, the Sn-Ag-Cu solders can be characterized as precipitation hardened alloys having essentially a 100Sn matrix phase and two precipitate particle phases, Ag_3Sn, and Cu_6Sn_5. In general, precipitate particles will improve the creep resistance of metal and alloys, particularly when the creep mechanism is based upon dislocation glide, and to a lesser magnitude, dislocation climb. The magnitude of the effect depends upon the size, distribution, volume fraction, and coherency of the particle phases. However, it is expected that precipitate particles will have a lesser impact on creep that is diffusion-controlled, especially when the particles are incoherent, because of the contribution of thermally-activated atom and vacancy transport to deformation, whether these point defects are the sole source of mass transport or assisting dislocation climb. The short-range stress field associated with incoherent particles likely describes the effect of the relatively large Ag$_3$Sn particles on creep deformation; it is unclear whether a similar situation pertains to the Cu$_6$Sn$_5$ particles. The interfaces that form between the particles and Sn-rich matrix contribute negligibly as a short-circuit, atomic transport path because of their lack of interconnectedness. However, it is likely that particle interfaces are sources or sinks of vacancies that contribute to diffusion-controlled creep due to the material mismatch stresses/strains across them. Solid-solution alloys, such as the Bi- and In-containing Pb-free solders, present a more intriguing aspect of creep behavior. Short- and long-range stress fields can develop around coherent particles that impact vacant and atom diffusion processes.

The tertiary stage of creep is signified by a rapid increase in strain rate leading up to failure or *creep rupture* (Fig. 17). In general, the increase in creep rate is caused by a loss of load bearing area due to necking or the accumulation of microstructural damage in the material. Necking is most predominant in the tension test configuration, using relatively high applied stresses. At the homologous temperatures of 0.6 to 0.7, microstructural damage includes the nucleation and growth of cracks and cavities along grain boundaries. The time required for creep rupture to occur is dependent upon the magnitude and direction (tension, compression, or shear) of the applied stress state. A number of quantitative expressions were developed to correlate the time to creep rupture (t_r) with the steady-state strain rate. These correlations are largely empirical.

The presence or even predominance of one of the three creep stages depends upon the following factors: (a) the material properties and microstructure; (b) the temperature; and (c) the applied stress. In the case of Sn-Ag-Cu lead-free solders exposed to the accelerated aging conditions of −55 and 125 °C (−67 and 257 °F) and hold times of 15 min., time-dependent deformation will be largely primary creep. On the other hand, service conditions that expose the Sn-Ag-Cu solder to several hours at temperatures exceeding 125 °C, or 257 °F (e.g., automotive underhood applications) will potentially place (depending on the applied stress) the material into the steady-state regime. At this time, there is no data that correlate the final failure of a solder interconnection to laboratory tertiary creep data. This lack of correlation is likely due to the complex stress distribution that exists within an actual electronic interconnection.

It is apparent that quantitative analyses of creep behavior nearly always emphasize the steady-state regime. Steady-state creep represents a dynamic equilibrium condition. That equilibrium can be a balance between defect creation and annihilation rates or can result from a constant defect velocity. Irrespective of the mechanistic details, the rate kinetics of the minimum creep rate, $d\varepsilon/dt_{min}$, can provide valuable insight into the processes responsible for creep deformation, including those active in the primary creep stage.

Several mathematical expressions have been used to describe steady-state creep, specifically, the relationship between $d\varepsilon/dt_{min}$ and the applied stress and temperature. The most commonly used equation is a power law stress dependence that is represented by Eq 10:

$$d\varepsilon/dt_{min} = A\sigma^n \exp(\Delta H/RT) \qquad \text{(Eq 10)}$$

where A is a constant, n is the stress or power law exponent, ΔH is the apparent activation energy, R is the universal gas constant, and T is temperature. The power law equation has been observed to adequately represent the steady-state creep of metal and alloys that show primary creep when tested at homologous temperatures greater than 0.4. Thus, the stress levels are relatively low. Typical values of n for pure metals and alloys are as follows:

$$\text{pure metals:} \quad 4 < n < 6$$

$$\text{alloys:} \quad 2 < n < 4$$

Although Eq 10 has its roots in empirical data, dislocation theories have been proposed to explain the power law stress dependence (Ref 63). Those theories have indicated that the stress exponent should, in fact, be in the range of 3 to 5.

The concept of an internal stress, σ_i, has been used to reconcile the discrepancy between the hypothetical and experimental values of n. Computationally, the internal stress approach can be introduced into Eq 10 by replacing σ with the term, ($\sigma_{applied} - \sigma_i$). The consideration of an internal stress must be consistent with the creep mechanism. For example, an internal stress may certainly be important when dislocation glide is the dominant mechanism. In that case, the internal stresses are generated when dislocations are pinned by coherent or incoherent precipitate particles, or by the stress fields of nearby dislocation networks, low-angle grain boundaries, etc. Also, the Peierls-Nabarro stress, or lattice friction stress, can slow the glide of dislocations (Ref 54). In the case of Pb-free solders, the high homologous temperature causes creep to be supported largely by diffusion-based mechanisms and, specifically, the thermally-activated diffusion of vacancies. As such, the role of an internal stress on creep behavior is expected to have a limited effect on creep strain, even that which is generated by dislocation climb, which is a diffusion-controlled process.

A drawback of the power law equation is the so-called "power law breakdown" effect. As the applied stress increases in magnitude, the stress exponent, n, no longer remains constant and increases with stress level. Therefore, it becomes necessary to use multiple power law equations to describe the creep behavior over an extended range of applied stresses.

At higher stresses, an exponential function has been used to describe the stress dependence of the steady-state strain rate. This relationship is shown in Eq 11:

$$d\varepsilon/dt_{min} = C \exp(\beta\sigma) \exp(\Delta H/RT) \quad \text{(Eq 11)}$$

where parameter C is a constant. The constant β, which is multiplied by σ to form the *effective stress*, $\beta\sigma$, is sensitive to temperature. However, in most cases, β and therefore $\beta\sigma$, can be considered as constant over relatively small temperature ranges.

A versatile equation that relates the minimum creep rate to stress and temperature is the so-called sinh law shown in Eq 12:

$$d\varepsilon/dt_{min} = B[\sinh(\alpha\sigma)]^p \exp(\Delta H/RT) \quad \text{(Eq 12)}$$

where B is a constant; α is a temperature-dependent parameter such that $\alpha\sigma$ is an effective stress, and p is the sinh law exponent, which is also a constant (Ref 64). Although α is strictly temperature dependent, the variation is usually sufficiently small so that a constant value is presumed in the steady-state creep analysis. This is the case for solder alloys (Ref 14, 62, 65).

The sinh law equation has been derived from first principles. Underlying mechanisms are based upon thermally activated diffusion processes, ranging from theories of point defect motion (vacancies and interstitials) and dislocation cross slip at the low stresses, to dislocation glide mechanisms at higher stresses. Thus, a particular attribute of the sinh law equation is that it adequately represents creep strain rate over an extended range of stresses, thereby eliminating the need for multiple exponent values, p, as is the case with a power law representation. In fact, at low stresses, Eq 12 reverts to the power law in Eq 10 while at high stresses, the sinh law can be adequately represented by the exponential stress dependence that is shown in Eq 11.

Microstructure

The microstructural aspects of creep deformation will be discussed here. There are two microstructural scales to be considered. There is the single-crystal scale, which represents the individual grains within each phase. Generally, when the microstructure is comprised of a predominant matrix phase and one or more particle phases, deformation within the grain structure of the matrix phase has the greatest potential impact on creep behavior. Dislocation glide will occur at high stresses and low temperatures. Dislocation pile-ups and tangles will develop at grain boundaries and within the grain structure, respectively. Slip lines may develop that become evident on the sample surface. All of these structures are the precursors to crack initiation. Dislocation activity typically results in the development of low-angle grain boundaries that, in turn, form sub-grain or cell structures within the individual grains. The high stacking fault energy of Sn causes it to be more susceptible to subgrain development as opposed to stacking fault development. Twinning can also have a significant role in the creep of Sn-based phase(s).

At lower stresses and high temperatures, diffusion-controlled mechanisms will be active. Atomic-level mass transport by vacancies and interstitials can cause deformation at the macroscopic level. Vacancies can assist the motion of dislocations by cross-slip processes. As vacancy mobility and/or the density increases, the phenomenon of *grain boundary migration* will be observed.

The second microstructural scale is that of a polycrystalline structure. The most common creep deformation mode is that of *grain-boundary sliding*. Grain boundary sliding generally occurs at high-angle boundaries when subjected to an applied stress at high homologous temperatures; hence, it is often observed in the creep of solders. The apparent activation energy for grain boundary sliding in polycrystalline Sn is

79 kJ/mol, a value which lies between that of lattice or bulk diffusion and fast or grain boundary diffusion (Ref 62). However, the likelihood of grain boundary sliding occurring in Pb-free solders may be significantly curtailed by the presence of particles, Ag_3Sn and Cu_6Sn_5 particles in the case of the Sn-Ag-Cu solders, particularly when they are located along the grain boundary.

Pb-Free Solders

Studies of the creep of Pb-free solders have emphasized the reporting of steady-state creep behavior and, specifically, the rate kinetics derived from $d\varepsilon/dt_{min}$ data. Of course, these data are critical toward understanding the constitutive behavior of the material, especially when developing solder fatigue computational models. However, it is equally important to consider the entire strain-time creep curve response by Pb-free solders. Those observations can provide valuable insight into the underlying mechanisms responsible for time-dependent deformation. In particular, the primary stage often dominates the creep response of Pb-free solders under the stresses that are generated in interconnections subjected to service temperature or the thermal cycles of accelerated aging tests.

Shown in Fig. 19(a) are the strain-time curves resulting from the compression creep of 95.5Sn-3.9Ag-0.6Cu samples in the *as-cast* condition; these tests were performed at 25 °C, or 77 °F (Ref 65). All of the curves show primary creep leading into steady-state creep. The extent of creep strain is not entirely monotonically dependent upon stress level; this variability indicates a non-uniformity in the defect density and/or defect movement between different test specimens that affects the creep response. Similar behaviors were observed in tests performed at −25 and 75 °C (−13 and 167 °F).

The strain-time curves in Fig. 19(b) resulted from compression creep tests performed at 125 °C (257 °F). The typical primary creep response occurred at an applied stress of 2.7 MPa (0.39 ksi). However, as the stress was raised to 5.2 and 5.3 MPa (0.75 and 0.77 ksi), the strain-time curve transitioned from a primary stage to a sigmoidal shape. Then, at 7.3 and 7.8 MPa (1.06 and 1.13 ksi), the creep curves transitioned once again, going from a sigmoidal behavior to largely, albeit not entirely, a steady-state creep response. Similar creep behavior was observed in tests performed at 160 °C (320 °F).

The decreasing strain rate and sigmoidal creep behaviors occurred apparently in reverse order between Fig. 19(a) and 19(b). That is, it would be expected that sigmoidal creep would occur at the lower test temperatures where there would be less thermal activation energy with which to increase an initially low defect density. At higher temperatures, thermal activation would assist defect generation and/or motion, thereby eliminating the incubation period that characterizes the sigmoidal shape. Therefore, the strain-time curves shown in Fig. 19 suggest there is a metastability of the as-cast microstructure. At temperatures exceeding 75 °C (167 °F), other processes occurred simultaneously with creep deformation, such as recovery and recrystallization, or even second-order effects that resulted from Ag_3Sn or Cu_6Sn_5 particle coarsening. Separately, or in combination, these ancillary processes may have lowered the defect density, thereby causing the sigmoidal creep curve shapes at the 125 and 160 °C (257 and 320 °F) test temperatures.

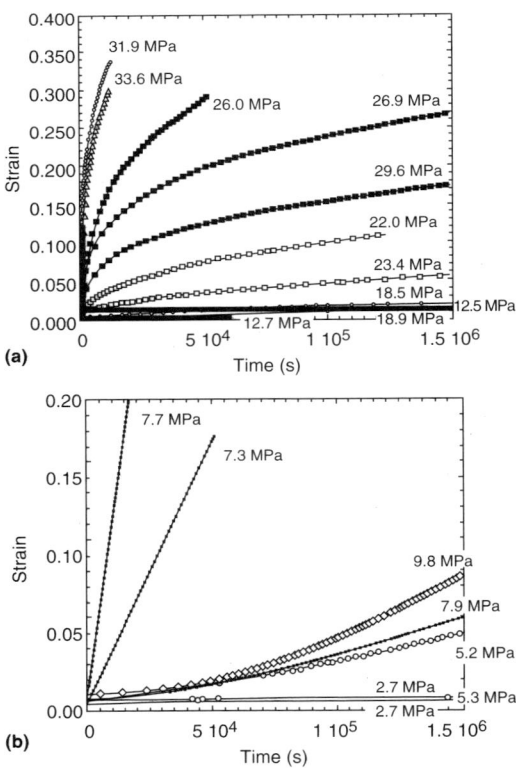

Fig. 19 Strain-time, compression creep curves of 95.5Sn-3.9Ag-0.6Cu solder in the as-cast condition. The test temperatures were (a) 25 °C (77 °F) and (b) 125 °C (257 °F). Source: Ref 65

Owing to relatively slow strain rates, time dependent deformation processes, including creep, are particularly sensitive to material microstructure. Often, samples used for creep studies are exposed to an annealing treatment prior to testing in order to stabilize the microstructure so as to improve the consistency of the data. A similar procedure was used with the 95.5Sn-3.9Ag-0.6Cu solder (Ref 14). Creep samples were exposed to annealing treatments of 125 °C (257 °F) for 24 h or 150 °C (302 °F) for 24 h. It was observed that neither annealing treatment significantly altered the shape of the strain-time curves as compared to those generated with the as-fabricated samples. This point is exemplified in Fig. 20(a), 20(b), 21(a), and 21(b) which show the strain-time curves at 25 and 125 °C (77 and 257 °F) for the annealed specimens. Primary and secondary stages were observed at test temperatures less than, or equal to, 75 °C (167 °F). The sigmoidal shape, which then transitioned into steady-state creep, was observed at test temperatures of 125 and 160 °C (257 and 320 °F). The 125 °C (257 °F) for 24 h annealing treatment did not improve the consistency of the creep curves as a function of stress (Fig. 20); however, the 150 °C (302 °F) for 24 h annealing treatment did improve the creep performance in that regard (Fig. 21). It was also observed that the annealing treatments generally did not change the magnitude of creep strain, indicating that they caused either a consistent hardening or softening of the material prior to creep testing.

Negative creep was observed in studies of the 95.5Sn-3.9Ag-0.6Cu solder. Negative creep refers to the phenomenon whereby a tensile load causes a contraction in the sample. Conversely, as in the cited work on compression creep tests, a compressive load results in an elongation of the sample. Negative creep has been observed previously in both crystalline and amorphous metallic alloys (Ref 54, 66). The phenomenon has also been predicted from theoretical grounds (Ref 67). Negative creep was observed only in

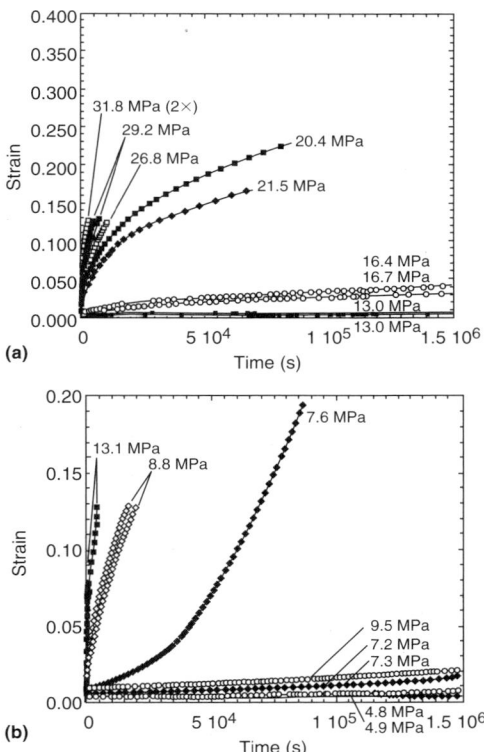

Fig. 20 Strain-time, compression creep curves of 95.5Sn-3.9Ag-0.6Cu solder after annealing at 125 °C (257 °F) for 24 h. The test temperatures were (a) 25 °C (77 °F) and (b) 125 °C (257 °F). Source: Ref 14

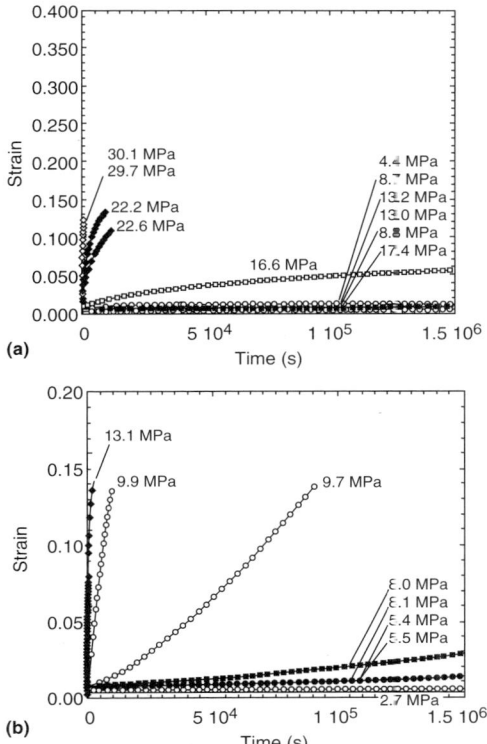

Fig. 21 Strain-time, compression creep curves of 95.5Sn-3.9Ag-0.6Cu solder after annealing at 150 °C (302 °F) for 24 h. The test temperatures were (a) 25 °C (77 °F) and (b) 125 °C (257 °F). Source: Ref 14

the 95.5Sn-3.9Ag-0.6Cu samples that had been exposed to one of the annealing treatments prior to testing; it was not observed in samples in the as-fabricated condition. Specifically, negative creep was observed with samples annealed at 125 °C (257 °F) for 24 h, but only when tested at 75 °C (167 °F). The phenomenon is illustrated in Fig. 22(a). The effect was reproducible at a given applied stress. When samples were annealed at 150 °C (302 °F) for 24 h, negative creep occurred at test temperatures of −25, 25, and 75 °C (−13, 77, and 167 °F). Negative creep was observed at −25 °C (−13 °F) with applied stresses of 8.0 and 15.8 MPa (1.16 and 2.29 ksi), and test times exceeding 10^5 s as shown in Fig. 22(b). This behavior was also observed in a companion test performed at 15.4 MPa (2.23 ksi). On the other hand, the sample tested at 16.0 MPa (2.32 ksi) showed negative creep only between 5×10^4 s and 10^5 s. For times longer than 10^5 s, positive creep was observed. Negative creep was less pronounced at test temperatures of 25 and 75 °C (77 and 167 °F). At 25 °C (77 °F), negative creep was observed at times exceeding 10^5 s for a single test performed at 4.4 MPa (0.64 ksi). Raising the stress to 8.8 MPa (1.28 ksi) resulted in only an intermittent negative creep behavior during the time interval of 70×10^4 s $< t < 1.4 \times 10^5$ s. The latter behavior was observed in the sole instance of negative creep recorded at 75 °C (167 °F), which was at a stress of 6.8 MPa (0.99 ksi).

In summary, the strain-time behavior of the 95.5Sn-3.9Ag-0.6Cu solder suggests that several microstructural processes besides deformation were active during the creep tests. The appearance of a sigmoidal strain-time plot indicates that there are dynamic changes in the density of defects that support the deformation. Negative creep provided additional evidence that extensive, ancillary microstructural changes occurred concurrently with creep.

Next, the steady-state creep behavior, based upon the minimum creep rate $d\varepsilon/dt_{min}$ parameter, will be described. A summarization of the state of the art was provided by Clech in Ref 68. In that work, Clech compiled the steady-state creep data for bulk 96.5Sn-3.5Ag solder evaluated in tension, using several sources that are cited in the reference. The $d\varepsilon/dt_{min}$ data was presented in the sinh law format and is represented by Eq 13, that follows. The error terms on p and ΔH are ± one standard deviation:

$$d\varepsilon/dt_{min} = 1044[\sinh(0.0615\sigma)]^{4.9 \pm 0.6}$$
$$\exp(-57000 \pm 9000/RT) \quad \text{(Eq 13)}$$

The units are as follows: $d\varepsilon/dt_{min}$, s^{-1}; σ, MPa; ΔH, J/mol; and T, K. Studies were performed under the auspices of the National Center for Manufacturing Sciences (NCMS) consortium, which examined the compression creep of bulk 96.5Sn-3.5Ag solder (Ref 69). There was an inadequate amount of data to be fit to the sinh law equation. However, the data were fit to the power law equation; the resulting expression appears below as Eq 14:

$$d\varepsilon/dt_{min} = 9.44 \times 10^{-5}\sigma^{6.1 \pm 0.7}$$
$$\exp(-61000 \pm 6000/RT) \quad \text{(Eq 14)}$$

As pointed out by Clech, a comparison of Eq 13 and 14 indicated that 96.5Sn-3.5Ag solder ap-

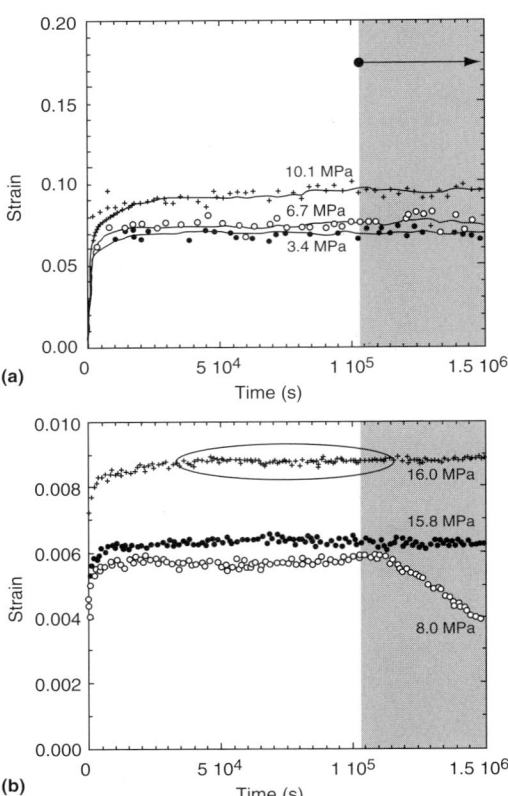

Fig. 22 Strain-time, compression creep curves of 95.5Sn-3.9Ag-0.6Cu solder showing the negative creep effect: (a) samples annealed at 125 °C (257 °F), 24 h and tested at 75 °C (167 °F); (b) samples annealed at 150 °C (302 °F), 24 h and tested at −25 °C (−13 °F). The oval indicates the region of negative creep that was subsequently followed by positive creep. Source: Ref 14

peared to have had a greater creep resistance in compression than in tension for similar stress and temperature conditions.

The sinh term and power law exponents, as well as the apparent activation energies, provided insight into the underlying creep mechanism(s). The sinh term exponent in Eq 13 and the power law exponent in Eq 14 were significantly higher than values of 2 to 4 that are expected of metal alloys. Rather, the exponent values were in the range of 4 to 6 that is characteristic of steady-state creep in pure metals. This observation suggests that the 96.5Sn-3.5Ag solder behaved as though it were simply polycrystalline Sn, which would infer that the second phase Ag_3Sn particles did not have a significant role in the creep behavior. The apparent activation energies (ΔH) were similar between Eq 13 and 14. The values are approximately 0.4 to 0.6 that of bulk diffusion in Sn (94 to 110 kJ/mol), implying that a fast-diffusion process (e.g., defect motion along grain boundaries) was the likely mechanism for creep deformation in this alloy.

Clech also compiled steady-state strain rate data in Ref 68, which originated from laboratory shear creep tests of 96.5Sn-3.5Ag solder, using lap joints or ring-and-plug specimens. The data were fit to a single sinh law in which the uniaxial, minimum strain rate and stress ($d\varepsilon/dt_{min}$ and σ, respectively) were replaced with the corresponding shear test mode parameters, $d\gamma/dt_{min}$ and τ. The resulting expression is Eq 15:

$$d\gamma/dt_{min} = 8.18 \times 10^{11}[\sinh(0.0266\tau)]^{8.7 \pm 1.1}$$
$$\exp(-77000 \pm 9000/RT) \quad (Eq\ 15)$$

where $d\gamma/dt_{min}$, s^{-1}; τ, MPa; ΔH, J/mol; and T, K are the units. There is a significant difference between the kinetics parameters (the sinh term exponent and the apparent activation energy values) of the uniaxial creep data (Eq 13) and the shear creep data (Eq 15). That discrepancy may likely be from the geometric constraints imposed by the lap shear and ring-and-plug solder joint geometries with respect to the unconstrained geometries of the tension or compression tests. Those constraints result in a complex, multiaxial stress state within the solder that includes non-homogeneous stresses across the joint thickness as well as along the lateral dimensions (i.e., parallel to the interfaces) of the joint. Non-uniform stresses can alter the apparent deformation rate kinetics. On the other hand, the fact that the apparent activation energy is similar to that of grain boundary sliding in Sn suggests that a shear load encourages the latter deformation maximum as compared to creep deformation that is generated under a uniaxial load.

The shear creep performance of the 96.5Sn-3.5Ag solder was recently studied by Morris and co-workers (Ref 70). The steady-state strain rate was measured at three temperatures (60, 95, and 130 °C, or 140, 203, and 266 °F) and stresses in the range of 3 to 15 MPa (0.44 to 2.18 ksi). The samples were aged at 160 °C (320 °F) for four h prior to testing; the solder joint thickness was relatively large at 0.160 mm (0.006 in.) so that length-scale effects were likely to have a minimal impact on creep performance. The $d\varepsilon/de_{min}$ data were described by a power law expression. The power law breakdown was observed; the stress exponent was 4.5 in the low stress regime and 10.6 in the high stress regime. The apparent activation energies were similar: 80 kJ/mol and 75 kJ/mol, respectively, for the two stress regimes. These ΔH values are higher than those indicative of a fast-diffusion process and less than those characteristic of self-diffusion as the underlying mechanism. It was possible that a combination of both mechanisms had occurred with fast-diffusion being favored at low temperatures and bulk diffusion dominating at the higher test temperatures. The apparent activation energy value is also similar to that of grain boundary sliding in Sn.

Similar experiments were performed by Kerr and Chawla (Ref 71). The 96.5Sn-3.5Ag solder joint was relatively large, having a gap of 0.500 mm (0.019 in.). The test temperatures were 25, 60, 95, and 130 °C (77, 140, 203, and 266 °F). Shear stresses ranged from 2 MPa (0.29 ksi) to approximately 35 MPa (5.08 ksi). The authors analyzed the steady-state strain rate by a power law equation. Power law breakdown was observed with stress exponents of 4 to 6 in the low stress regime and 13 to 20 in the high-stress regime. Two apparent activation energies represented the creep behavior; they were distinguished by test temperature regimes: 50 kJ/mol at low temperatures (25 to 95 °C, or 77 to 203 °F) and 120 kJ/mol at high temperatures (95 to 130 °C, or 203 to 266 °F), which indicated fast-diffusion and self-diffusion processes, respectively. The authors evaluated their data using a threshold stress mechanism. The threshold stress is the added stress required to move dislocations passed particles within the microstructure, in this case, by a climb mechanism as proposed by the authors. Transmission electron microscopy (TEM) images in Ref 72 showed the pinning of dislocations by Ag_3Sn particles in the solder. However, a point that remains to be resolved is the similarity of creep properties between 96.5Sn-3.5Ag and 100Sn, the latter not having the Ag_3Sn particles (Ref 70).

The test methodologies, whether tension/compression tests on bulk solders or shear tests on solder joints, have their merits for specific study objectives. Tension and compression tests on bulk solder provide critical input data for constitutive models. The finite element analysis within those models can account for geometric effects on creep deformation in an actual joint configuration (included spatially varying stress state). However, length-scale effects of small joints, or elemental contamination that alters the intrinsic mechanical and physical properties of the solder, require an entirely new set of constitutive equations because finite element analysis cannot account for these effects. Thus, in the latter circumstance, the most accurate creep data would be results obtained from tests on actual joints.

Creep experiments have been performed on the Sn-Ag-Cu solders. New data, as well as published studies compiled by Clech in Ref 68, will be discussed. In Clech's summary, results were described from creep tests performed on the three Pb-free alloys: 95.5Sn-3.8Ag-0.7Cu; 96.5Sn-3.0Ag-0.5Cu; and 96.2Sn-2.5Ag-0.8Cu-0.5Sb. There was very little difference between the creep properties of the three compositions at similar stress and temperature regimes. The minimum strain-rate data of these three alloys were compared to the compression creep data of the 93.6Sn-4.7Ag-1.7Cu solder. The steady-state creep performance of the latter solder was very similar to those of the three aforementioned alloys. This similarity implies that the higher Ag and Cu contents did not significantly alter the creep behavior. From the microstructure point-of-view, the insensitivity of steady-state creep to these ranges of Ag and Cu concentrations appears to substantiate the hypothesis that the diffusion-based mechanisms responsible for the minimum creep rate are not significantly impacted by the presence of Ag_3Sn and Cu_6Sn_5 particle phases. Rather, the Sn-rich matrix microstructure, per se, largely controls time-dependent deformation.

Specifically, the minimum tensile strain rate data were described as a function of stress and temperature by a sinh expression. The $d\varepsilon/dt_{min}$ results of all three Pb-free solders (95.5Sn-3.8Ag-0.7Cu; 96.5Sn-3.0Ag-0.5Cu; and 96.2Sn-2.5Ag-0.8Cu-0.5Sb) were combined together into Eq 16:

$$d\varepsilon/dt_{min} = 2631[\sinh(0.0453\sigma)]^{5.0 \pm 0.8} \exp(-59000 \pm 8000/RT) \quad \text{(Eq 16)}$$

where $d\varepsilon/dt_{min}$, s^{-1}; σ, MPa; ΔH, J/mol; and T, K are the units. The error terms are ± one standard deviation. Isotherm lines are shown in Fig. 23, which is a plot of the natural logarithm of minimum strain rate versus the natural logarithm of the stress.

It was observed that the kinetic parameters (the sinh term exponent and apparent activation energy) were similar between Eq 16 and like parameters for the 96.5Sn-3.5Ag solder as shown in Eq 13. In fact, a direct comparison of creep rates between the higher-order Pb-free solders and 96.5Sn-3.5Ag solders at similar stresses and temperature indicated that the former were only slightly more creep resistant than the binary alloy. Quantitatively, that difference stemmed from the relatively small variations in α and the A constant. Mechanistically, it indicates that the Ag_3Sn and Cu_6Sn_5 phases did not have a significant role in the creep deformation, either explicitly or even indirectly, such as by altering the Sn-rich matrix microstructure.

Clech also made a comparison of steady-state creep rates between the Sn-Ag and Sn-Ag-Cu(-Sb) Pb-free solders and the traditional Sn-Pb solder (Ref 68). It was observed that the Pb-

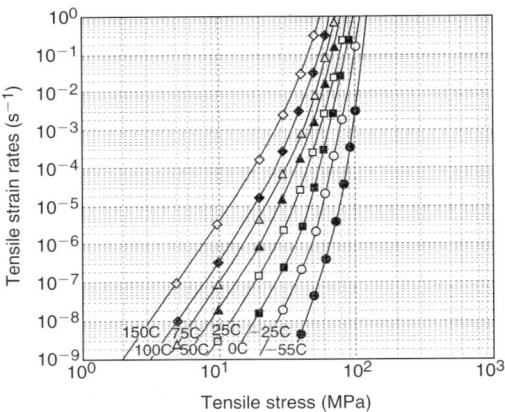

Fig. 23 Plot of steady-state strain rate versus tensile stress from bulk sample creep for the solders 95.5Sn-3.8Ag-0.7Cu; 96.5Sn-3.0Ag-0.5Cu; and 96.2Sn-2.5Ag-0.8Cu-0.5Sb per Eq 16. Source: Ref 68

free solders have better creep resistance than the Sn-Pb alloys, but only at stresses less than 20 to 30 MPa (2.90 to 4.35 ksi) for 96.5Sn-3.5Ag solder and 40 to 50 MPa (5.80 to 7.25 ksi) in the case of the higher order alloys. These demarcation points (stresses) were not particularly sensitive to temperatures in the range of −55 to 100 °C (−67 to 212 °F). At higher stress levels, which may occur in harsh service environments or in accelerated aging tests for high-reliability assemblies, the Sn-Pb solder has better creep resistance. The lower solidus temperature of the Sn-Pb solder may allow for a greater propensity of dynamic recovery and recrystallization processes at high stresses, resulting in a reduction in the density of defects available to support deformation.

The relatively high sinh term exponent value in Eq 16, 5.0 ± 0.8, indicates that the Sn-Ag-Cu(-Sb) solders behaved more like a simple metal than an alloy. Like the bulk tensile creep data of the Sn-Ag alloy, the Ag_3Sn and Cu_6Sn_5 phases did not have a significant role in the creep deformation. Rather, the Pb-free solders exhibited the creep behavior of a simple metal because the microstructure of the Sn-rich matrix was largely responsible for time-dependent deformation. Lastly, the apparent activation energy of 59 ± 8 kJ/mol for creep in the Sn-Ag-Cu(-Sb) alloys was similar to that of the binary Sn-Ag solder, indicating that a fast-diffusion mechanism was likely responsible for creep.

Bulk tension creep experiments were performed by J. Pang and co-workers on the 95.5Sn-3.8Ag-0.7Cu alloy (Ref 23). The tests temperatures were 25, 75, and 125 °C (77, 167, and 257 °F). The applied stresses were 2 to 40 MPa (0.29 to 5.80 ksi). The steady-state creep rate was expressed by a sinh law dependence on stress as shown in Eq 15:

$$d\varepsilon/dt_{min} = 501.3[\sinh(0.0316\sigma)]^{5.0} \exp(-45000/RT) \quad \text{(Eq 17)}$$

Error terms were not provided for the sinh term exponent or for the apparent activation energy. The stress exponent was identical to that in Eq 16 while the value of ΔH was less than that of the former analysis. Further investigation compared Eq 17 directly to the 95.5Sn-3.8Ag-0.7Cu data obtained by Schubert, et al. that was used to develop Eq 16 (Ref 72). The latter study of bulk tensile creep used three temperatures (20, 70, and 150 °C, or 68, 158, and 302 °F) and stresses of 2 to 30 MPa (0.29 to 4.35 ksi). The resulting sinh law exponent and apparent activation energy were 6.4 and 54 kJ/mol. The values of both parameters were considerably higher than those observed by Pang, et al. per Eq 17. A closer examination of the individual data indicated a difference of strain rates at the comparable test temperature pairs of (Schubert, Pang): (70, 75 °C or 158, 167 °F) and (150, 125 °C or 302, 257 °F). The strain rates measured by Pang, et al. were higher than those determined by Schubert, et al. In both studies, the tests that were performed at room temperature exhibited very comparable strain rates per the applied stress. The differences in the two data sets reflect variations in the solder microstructures as a consequences of testing at temperatures of approximately 70 °C (158 °F) and higher. A similar trend was observed in the compression creep data, which are discussed in the following text.

Kim and co-workers examined the effects of cooling rate on the tensile creep of the alloys 95.8Sn-3.5Ag-0.7Cu, 95.5Sn-3.0Ag-0.5Cu, and 95.5Sn-3.9Ag-0.6Cu solders (Ref 20). Applied stresses ranged from 20 to 100 MPa (2.90 to 14.50 ksi). It was observed that a faster cooling rate improved the creep resistance of the Pb-free alloys. It appears that cooling rate effects on the microstructure of Pb-free solders can alter the long-term creep performance and, therefore, the long-term reliability of these alloys, more so than is the case of the traditional Sn-Pb solders for which room temperature aging quickly nullifies the effects of cooling rate resulting from the solidification effect.

Vianco and co-workers performed compression creep experiments on the 95.5Sn-3.9Ag-0.6Cu solder (Ref 65 and 14). Those tests were conducted at temperatures of −25, 25, 75, 125, and 160 °C (−13, 77, 167, 257, and 320 °F) and stresses of 2 to 45 MPa (0.29 to 6.53 ksi). The solder material was evaluated in one of three conditions: as-cast, aged 125 °C (257 °F) for 24 h, or aged 150 °C (302 °F) for 24 h. The minimum strain rate data of the as-cast specimens were represented by the sinh law in Eq 18:

$$d\varepsilon/dt_{min} = 4.41 \times 10^5[\sinh(0.005\sigma)]^{4.2 \pm 0.6} \exp(-45000 \pm 7000/RT) \quad \text{(Eq 18)}$$

where $d\varepsilon/dt_{min}$, s^{-1}; σ, MPa; ΔH, J/mol; and T, K are the units. The error terms represent 63% confidence intervals, approximately ± one standard deviation. The low value of $\alpha = 0.005$ indicated that a power law breakdown effect would have been minimal so that a power law equation would have been adequate to describe the data. An improved regression analysis fit was realized when the data were separated into low-temperature and high-temperature regimes. The resulting sinh law expressions are:

-25 to $75\,°C$ $(-13$ to $167\,°F)$:

$$d\varepsilon/dt_{min} = 99.3[\sinh(0.005\sigma)]^{4.4\pm0.7}$$
$$\exp(-25000 \pm 7000/RT) \quad \text{(Eq 19)}$$

75 to $160\,°C$ $(167$ to $320\,°F)$:

$$d\varepsilon/dt_{min} = 5.83 \times 10^{13}[\sinh(0.005\sigma)]^{5.2\pm0.8}$$
$$\exp(-95000 \pm 14000/RT) \quad \text{(Eq 20)}$$

The sinh term exponent was statistically the same in Eq 18, 19, and 20. However, the apparent activation energy values indicated that different mechanisms were active in the two temperature regimes. A plot of Eq 19 and 20 is shown in Fig. 24; the 75 °C (167 °F) data were included in the high temperature regime. The change of ΔH is apparent in the shape of the curves. In the low temperature regime, the small ΔH value of 25 kJ/mol would imply that a very low energy barrier mechanism was active. That mechanism was thermally activated and may have even been diffusion-based, given that -25 °C (-13 °F) still represents a relatively high homologous temperature of 0.51. It was highly probable that a very fast diffusion process, perhaps aided by recovery and/or recrystallization-like processes, was responsible for low temperature creep. At high temperatures, the value of ΔH (95 kJ/mol) is nearly identical to that of self-diffusion in Sn (Ref 57, 58).

A comparison was made between Fig. 23 and 24, the former having been generated by tensile creep tests on bulk Pb-free solders. A consistent trend was not identified that distinguished the creep rate values measured by tensile tests with respect to those determined from compression tests. Any differences between the two methodologies were temperature dependent.

Compression creep experiments were repeated on samples that were aged at 125 °C (257 °F) for 24 h prior to testing (Ref 14). The aging treatment did not significantly alter the strain-time curves with the exception of negative creep that was discussed previously. The steady-state creep rate kinetics were best represented by two temperature regimes; however, the aging treatment caused a shift in low- and high-temperature

regimes to -25 to $125\,°C$ (-13 to $257\,°F$) and 125 to $160\,°C$ (257 to $320\,°F$), respectively. The minimum creep rate equations are:

-25 to $125\,°C$ $(-13$ to $257\,°F)$:

$$d\varepsilon/dt_{min} = 1.57 \times 10^{7}[\sinh(0.0050\sigma)]^{5.3\pm0.7}$$
$$\exp(-47000 \pm 6000/RT) \quad \text{(Eq 21)}$$

125 to $160\,°C$ $(257$ to $320\,°F)$:

$$d\varepsilon/dt_{min} = 1.09 \times 10^{14}[\sinh(0.0050\sigma)]^{7.3\pm1.0}$$
$$\exp(-75000 \pm 40000/RT) \quad \text{(Eq 22)}$$

where $d\varepsilon/dt_{min}$, s^{-1}; σ, MPa; ΔH, J/mol; and T, K are the units. The sinh term exponent of the low-temperature regime was very similar to those calculated for the as-cast samples. The value of ΔH was at the lower end of the range considered representative of a fast-diffusion process. Steady-state creep in the high-temperature regime was marked by a significant increase in the sinh law exponent. From the mechanistic viewpoint, the increase of the sinh law exponent suggests that the Sn-Ag-Cu solder had a creep behavior that was more like that of alloys rather than that of a single elemental metal. This trend may have been caused by changes to the Ag_3Sn precipitate structure as the latter coarsened by the annealing treatment and, to a lesser extent, subsequent creep testing. There was also a decrease of the apparent activation energy to a value that was at the upper end of the range con-

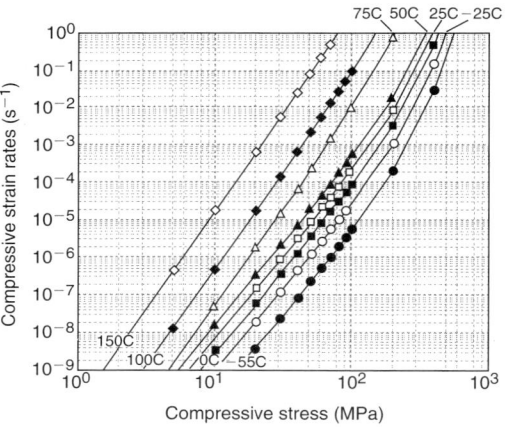

Fig. 24 Plot of steady-state strain rate versus compressive stress from bulk sample creep data for the solder 95.5Sn-3.9A-0.6Cu in the as-cast condition, per Eq 19 and 20 the 75 °C (167 °F) data were analyzed in the high-temperature regime. Source: Ref 65

sidered to indicate fast-diffusion processes and, as such, approached values commensurate with self-diffusion. Lastly, the high-temperature p and ΔH parameters were also accompanied by wider confidence intervals. The increased variability of creep behavior indicated that the annealing treatment may have actually destabilized the microstructure for tests performed at temperatures equal to, or exceeding, 125 °C (257 °F).

The 150 °C (302 °F) 24 h annealing treatment was more effective at stabilizing the solder microstructure over the entire test temperature regime. A relatively good correlation was obtained with a single sinh law equation for $d\varepsilon/dt_{min}$ (Eq 23):

$$d\varepsilon/dt_{min} = 517[\sinh(0.0938\sigma)]^{4.9 \pm 0.3}$$
$$\exp(-66000 \pm 5000/RT) \quad \text{(Eq 23)}$$

where $d\varepsilon/dt_{min}$, s^{-1}; σ, MPa; ΔH, J/mol; and T, K are the units. The sinh law exponent indicates a steady-state creep behavior like that of a pure metal (i.e., the tin-rich matrix) and the ΔH value implies the activity of a fast-diffusion process. There was a significant increase in the α coefficient of the stress, σ, to a value more commensurate with those computed for similar alloys (Ref 62, 68). A plot of Eq 23 is provided in Fig. 25. A superposition of the plots in Fig. 24 (as-cast) and 25 (annealed, 150 °C (302 °F) for 24 h) indicates that annealing caused a lower strain rate at low stress levels than was observed for the as-cast material. At higher applied stresses, the annealed sample had a faster minimum strain rate. The cut-off stress between the two behaviors was temperature dependent, but not monotonically so.

The data correlation improved only slightly when the analysis was separated into two temperature regimes:

-25 to 125 °C (-13 to 257 °F):

$$d\varepsilon/dt_{min} = 5.17[\sinh(0.0938\sigma)]^{4.5 \pm 0.4}$$
$$\exp(-54000 \pm 7000/RT) \quad \text{(Eq 24)}$$

125 to 160 °C (257 to 320 °F):

$$d\varepsilon/dt_{min} = 1.24 \times 10^8[\sinh(0.0938\sigma)]^{5.9 \pm 0.6}$$
$$\exp(-105000 \pm 10000/RT) \quad \text{(Eq 25)}$$

The low-temperature regime has an apparent activation energy that suggests a fast-diffusion mechanism. There were significantly higher values of the sinh law exponent and ΔH in the high-temperature regime. Again, the higher exponent value indicated a more "alloy-like" creep behavior. The ΔH value indicated the prevalence of a self- or bulk-diffusion mechanisms. Finally, the 150 °C (302 °F) for 24 h annealing treatment resulted in reduced error terms for the kinetics parameters in Eq 23 to 25, thereby confirming the improved stability of the microstructure.

Shear creep studies have been performed on the Sn-Ag-Cu alloy solder joints. Morris and coworkers examined the constant shear stress creep of the 96.5Sn-3.0Ag-0.5Cu solder joints having a thickness of 0.160 mm, or 0.006 in. (Ref 70). The test temperatures were 60, 95, and 130 °C (140, 203, and 266 °F); stresses ranged from approximately 3 to 15 MPa (0.44 to 2.18 ksi). The samples were aged at 160 °C (320 °F) for 4 h prior to testing. The minimum creep rate data were analyzed according to the power law equation. The expected power law breakdown phenomenon was observed with values of n equal to 6.6 and 10.7 for the low-stress and high-stress regimes, respectively. Confidence intervals were not provided with the data. The apparent activation energies were 95 kJ/mol and 75 kJ/mol, indicating a predominance of bulk diffusion at the low stresses and an emergence of a fast-diffusion mechanism or grain-boundary sliding in the high stress regime. The study cited in Ref 70 also investigated the shear creep of 100Sn, 96.5Sn-3.5Ag, and 99.3Sn-0.7Cu solder joints. The creep kinetics were nearly identical

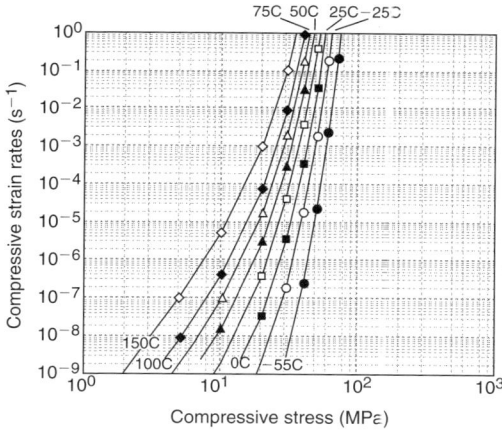

Fig. 25 Plot of steady-state strain rate versus compressive stress from bulk sample creep data for the solder 95.5Sn-3.9A-0.6Cu in the 150 °C (302 °F), 24 h annealed condition, per Eq 23. Source: Ref 14

between all four solder compositions. This observation further substantiates the hypothesis that the particle phases in the Sn-based binary and ternary solders do not have a significant, direct role in creep; rather, it is the Sn-rich matrix phase and its microstructure that controls creep performance. Of course, the particles may have a second-order effect by altering the solidification microstructure of the Sn-rich matrix phase.

Q. Zhang, et al. performed shear creep tests on 95.5Sn-3.9Ag-0.6Cu solder joints (Ref 73). The gap thickness was 0.180 mm (0.007 in.). The tests were performed at three temperatures: 25, 75, and 125 °C (77, 167, and 257 °F). The shear stresses ranged from 4 to 30 MPa (0.58 to 4.35 ksi). Initially, the authors observed the power law breakdown effect when the steady-state creep rate, shear stress and temperature were assessed with the power law function. Subsequently, the authors obtained a more consistent fit when all of the steady-state creep data were analyzed with the sinh law equation:

$$d\gamma/dt_{min} = 284.4[\sinh(0.188\tau)]^{3.8} \exp(-63000/RT) \quad \text{(Eq 26)}$$

where $d\varepsilon/dt_{min}$, s^{-1}; σ, MPa; ΔH, J/mol; and T, K are the units. Error terms were not provided in the cited report. The sinh term exponent was generally less than the same parameter determined from compression creep tests on this alloy and tension creep tests performed on the 95.5Sn-3.8Ag-.07Cu solder. The apparent activation energy value was higher than values from other creep studies performed with comparable test temperatures but, nonetheless, still in a range that is commensurate with fast-diffusion processes. It is important to note that the authors also evaluated the shear creep of the traditional 63Sn-37Pb solder. Qualitatively, the Sn-Ag-Cu solder exhibits a lower creep rate than did the Sn-Pb solder for stresses up to approximately 20 to 25 MPa (2.90 to 3.63 ksi) and test temperatures of 25 to 125 °C (77 to 257 °F). At higher stresses, the Sn-Pb solder exhibits a lower creep rate. This observation corroborates the crossover effect of TMF data compiled by Clech in Ref 45 (Fig. 13) and also substantiates the important role of creep in the cyclic deformation of the Sn-Ag-Cu alloys (Ref 68).

There is a growing number of studies that are providing creep data on actual solder joints. A study by Wiese, et al. examined the creep of 0.200 mm (0.007 in.) thick flip chip solder joints made of 95.5Sn-4.0Ag-0.5Cu (Ref 74). The joints were hour-glass in shape and relatively large so that length scale effects were likely minimal. The shear stress and strain were converted to their uniaxial counterparts and the minimum strain rate was expressed as a power law of stress. The stress exponent and apparent activation energy were 18 and 83 kJ/mol, respectively. The stress exponent was considerably larger than values reported from creep tests on bulk Sn-Ag-Cu solder and soldered joints. Certainly, the hour-glass configuration of the joints may have been a contributing factor for the unusually high exponent value.

Clech analyzed creep data from Pb-free solders (and baseline Sn-Pb solder) using the lesser known *obstacle-controlled creep model* developed by Frost and Ashby (Ref 75). That analysis examined tensile and compression test data as well as creep results derived from shear creep tests of soldered joints (Ref 76). The basic mathematical representation is:

$$d\varepsilon/dt_{min} = A\sigma^m \exp[\Delta H/RT(1 - \sigma/\sigma_o)] \quad \text{(Eq 27)}$$

where $d\varepsilon/dt_{min}$ is the strain rate, A is the scaling constant, σ is the applied stress, m is the stress exponent, ΔH is apparent activation energy, R is the universal gas constant, T is temperature; and σ_o is *athermal flow strength*. The athermal flow strength is the strength of the material at absolute zero (0K), that is, in the absence of any thermal activation that could assist the deformation mechanism. Thus, a stress-dependent activation energy is defined in Eq 27 as $\Delta H(1 - \sigma/\sigma_o)$ in which the applied stress has been normalized to the athermal flow strength parameter, σ_o. It is interesting to observe that Eq 27 can be interpreted as a combination of Eq 10, the power law expression, and Eq 11. In the latter equation, the combination of the two exponential terms—one being the effective stress, $\beta\sigma$ and the other being the Arrhenius expression—would result in a similar stress dependence for the apparent activation energy. The activity of multiple creep mechanisms can be taken into account by the superposition of multiple terms having the form of the right-hand side of Eq 27 as is represented by Eq 28:

$$d\varepsilon/dt_{min} = A_1\sigma^{m,1} \exp[\Delta H_1/RT(1 - \sigma/\sigma_{o,1})] + A_2\sigma^{m,2} \exp[\Delta H_2/RT(1 - \sigma/\sigma_{o,2})] + \ldots \quad \text{(Eq 28)}$$

It was observed that the creep data were readily fit to the two-term or two "cell" version of Eq 28. Qualitatively, two-terms for Eq 28 is suitable because the creep behavior of lead-free solders is usually a combination of bulk diffusion and fast diffusion processes as has been demonstrated by the apparent activation energy values cited previously. Values of the parameters used in Eq 28 are listed in Table 3 for a number of Pb-free solders and the Sn-Pb baseline alloy. Clech validated the resulting equations, using creep and stress-strain test results that are independent of those data used to establish the parameters in Table 3.

The parameters in Table 3 indicated that two mechanisms were potentially active during creep of the Sn-Ag, Sn-Ag-Cu, and Sn-Pb solders. The first mechanism had a low apparent activation energy. In the case of the Sn-Ag and Sn-Pb solders, the value of ΔH indicated grain boundary diffusion. An accelerated fast diffusion process—possibly a recrystallization-assisted mechanism—controlled creep of the Sn-Ag-Cu alloys as was discussed previously. The second mechanism had a higher apparent activation energy that was representative of bulk diffusion. It was interesting to note that the same parameters, with the exception of the scaling constant, applied to both Sn-Ag-Cu alloys, despite the compositional differences and the different test modes.

The Bi-Sn and Sn-Cu solders were best represented by single, but different, mechanisms. It is noted that the Sn-Cu creep data reflects a shear test configuration. Assuming a Von Mises transformation, the conversion between shear stress (τ) and tensile stress (σ), as well as that between the minimum shear strain rate ($d\gamma/dt_{min}$) and minimum tensile strain rate ($d\varepsilon/dt_{min}$), can be calculated by Eq 29 and 30, respectively.

$$\sigma = \sqrt{3}\tau \qquad \text{(Eq 29)}$$

$$d\varepsilon/dt_{min} = (1/\sqrt{3})d\gamma/dt_{min} \qquad \text{(Eq 30)}$$

The low apparent activation energy of the Bi-Sn solder indicated that grain (or phase) boundary diffusion was the predominant creep mechanism for this alloy. On the other hand, the high apparent activation energy of the Sn-Cu solder implied that creep was controlled by bulk diffusion processes.

It was more difficult to correlate the stress exponent values m_1 and m_2 with physical mechanisms in the solder. It was noted above that, during high temperature creep, values of m tend to be in the range of $2 < m < 4$ for metal alloys and in the range of $4 < m < 6$ for pure metals. In Table 3, the values of m for the Bi-Sn and Sn-Cu alloys would concur with the behavior expected of an alloy. On the other hand, the remaining alloys, including the baseline Sn-Pb solder, do not show values of m_2 (the high temperature creep regime) that collectively reflect the nature of these solders as alloys. This trend is not unexpected given similar evaluations on other metals and alloy systems. (In fact, past studies have clearly shown that the value of m is also sensitive to differences of microstructure per a given material composition).

The relative contributions of the two creep mechanisms were evaluated for the Sn-Ag-Cu alloys. In general, grain boundary diffusion processes dominated at low temperatures (and concurrently, high stresses) while bulk diffusion was predominant at high temperature (and correspondingly, low stresses). It was observed that nearly equivalent contribution occurred at approximately 75 °C (167 °F). Recall that this was also the transition temperature between the low and temperature regimes observed in the compression creep analysis of the 95.5Sn-3.9Ag-0.6Cu (Ref 65).

Pb-Free Solder Microstructures

The quantitative analyses of the minimum strain rate, $d\varepsilon/dt_{min}$, which determine the rate ki-

Table 3 Parameters for the single- and two-cell equations representing the obstacle-controlled creep model

Solder (wt%)	Test Regiment	A_1 (s^{-1})	m_1	ΔH_1 (kJ/mol)	$\sigma_{0,1}$ (MPa)	A_2 (s^{-1})	m_2	ΔH_2 (kJ/mol)	$\sigma_{0,2}$ (MPa)
58Bi-42Sn	Tension, bulk	802	3.02 ± 0.14	76.4 ± 2.9	596 ± 97	...	...	...	...
99.3Sn-0.7Cu	Shear, joint	2.23 × 10^8	2.23 ± 1.16	118.3 ± 6.7	45.2 ± 10.6	...	...	...	...
96Sn-4Ag	Tension, bulk	2.77 × 10^{-11}	8.58 ± 5.05	52.6 ± 10.9	244 ± 154	67.3	7.41 ± 0.44	105 ± 4	408 ± 82
95.5Sn-3.8Ag-0.7Cu	Tension, bulk	5.00 × 10^{-9}	5.56 ± 0.88	29.5 ± 6.9	1280 ± 4610	6800	3.02 ± 0.25	91.9 ± 7.6	181 ± 10
95.5Sn-3.9Ag-0.6Cu	Compression, bulk	0.0500	5.56 ± 0.88	29.5 ± 6.9	1300 ± 4600	0.0500	3.02 ± 0.25	91.9 ± 7.6	181 ± 10
63Sn-37Pb	Tension, bulk	17.4	3.07 ± 0.12	66.4 ± 1.5	171 ± 8	5110	5.89 ± 1.18	112 ± 11	217 ± 54

Source. Ref 76

netics of steady-state creep, provide valuable insight into the mechanism(s) responsible for the time-dependent deformation. However, it is also important to substantiate these results with the evaluations of actual solder microstructures that are present at all stages of the strain-time curve.

Kerr and Chawla observed the microstructure of 96.5Sn-3.5Ag solder after creep testing using TEM techniques (Ref 71). Dislocations were observed to extend between Ag_3Sn particles within the ternary eutectic regions. In the same study, the fracture surfaces of creep rupture samples (in shear) indicated that grain boundary sliding did not contribute significantly to creep deformation.

The microstructures were examined that originated from compression creep 95.5Sn-3.9Ag-0.6Cu samples that were tested in the as-fabricated condition (Ref 65). Metallographic cross sections were performed on samples, the tests of which were terminated at the end of the primary creep stage, or within the steady-state creep stage. There was no evidence of voiding, grain boundary sliding, or crack formation for all combinations of stress and temperature conditions in that study. The microstructure of the sample tested at $-25\ °C$ ($-13\ °F$) and a 40 MPa (5.80 ksi) stress is shown in Fig. 26 ($-25\ °C$ ($-13\ °F$), 40 MPa, or 5.80 ksi). Very fine, as well as coarse ternary eutectic structures, along with the Sn-rich dendritic regions, are observed in the micrograph. Owing to the lack of deformation in the microstructure, that micrograph would be equally representative of the as-cast condition. At test temperatures of 75 °C (167 °F) and higher, a boundary of coarsened Ag_3Sn particles, accompanied by zones depleted of Ag_3Sn particles to either side of the coarsened boundary, were formed in the larger eutectic regions between the Sn-rich dendrites. This artifact is shown in Fig. 27. At the higher test temperatures of 125 and 160 °C (257 and 320 °F), there was a general coarsening of the Ag_3Sn particles in the ternary eutectic regions as is illustrated in Fig. 28(a) and 28(b), respectively. This generalized coarsening eventually masked the Ag_3Sn boundary development shown in Fig. 27. Preliminary comparisons suggested that the applied stress slightly accelerated the coarsening of the Ag_3Sn particles. However, this observation could not be confirmed by subsequent experimentation.

A companion compression creep study of 95.5Sn-3.9Ag-0.6Cu used samples that were annealed prior to creep testing (Ref 14). The annealing conditions changed the microstructure. The 125 °C (257 °F) for 24 h annealing treatment resulted in the formation of coarsened Ag_3Sn particle boundaries within the larger, ternary eutectic regions. The 150 °C (302 °F) for 24 h aging treatment caused the formation of coarsened Ag_3Sn boundaries in the ternary eutectic regions; those boundaries had a more contiguous formation of Ag_3Sn phase. There was also an

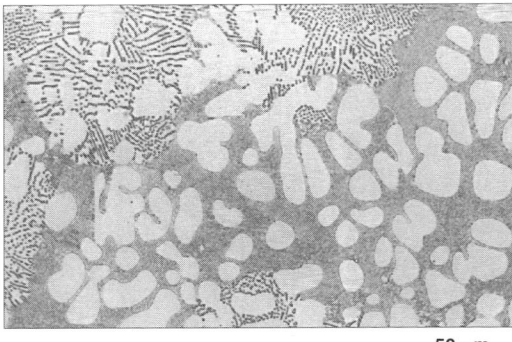

Fig. 26 Optical micrograph showing the microstructure of the 95.5Sn-3.9Ag-0.6Cu sample tested at $-25\ °C$ ($-13\ °F$) and 40.3 MPa (5.85 ksi) stress. The sample was in the as-fabricated condition. This micrograph is also representative of the untested microstructure. Source: Ref 65

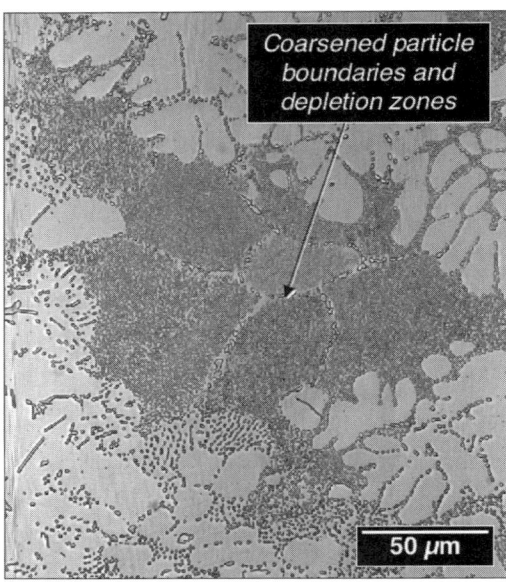

Fig. 27 Optical micrograph showing the coarsened Ag_3Sn particle boundaries that formed in the larger ternary eutectic regions between the Sn-rich dendrites of the 95.5Sn-3.9Ag-0.6Cu solder that was creep tested at 75 °C (167 °F) and a 10.9 MPa (1.58 ksi) stress. The sample was in the as-fabricated condition. Source: Ref 65

increased, general coarsening of the remaining Ag$_3$Sn particles. It was not possible to discern any further changes to these particular microstructural features that may have been accelerated or curtailed by subsequent creep testing. Also, there was no evidence of grain boundary sliding, pore formation, or cracking in the microstructures, even in those samples that were allowed to barrel-out under very extensive compressive strains. In the latter case, the microstructures did develop some evidence of shear band formation.

Lastly, as noted previously, and unlike their Sn-Pb counterparts, there was an absence of grain boundary sliding due to creep deformation of the Sn-based, Pb-free solders. It is clear from the discussion, and as evidenced in Fig. 27 and 28, that Ag can diffuse at elevated temperatures, creating the coarsened boundaries of Ag$_3$Sn particles (as well as general coarsening of the Ag$_3$Sn particles). Previous studies have shown that the precipitation of particles along interfaces can significantly reduce grain boundary sliding (Ref 52). Therefore, it appeared that the Ag$_3$Sn phase prevents significant grain boundary sliding from occurring during creep of Sn-Ag and Sn-Ag-Cu solders. This mechanism is, in fact, a very effective means of imparting high creep resistance in alloys used at elevated service (homologous) temperatures.

Summary

This review describes the fundamental aspects of fatigue and creep deformation in the Sn-based, Pb-free solders. Quantitative data and, in particular, the corroborating evidence of the "crossover" effect, confirmed that creep deformation has a significant role in the fatigue behavior of the Pb-free solders in service, or during accelerated testing conditions. Thus far, creep studies on Pb-free solders have generated an important database of minimum strain rate ($d\varepsilon/dt_{min}$) values. However, it is also important to understand the primary creep stage and its relationship to steady-state creep. Primary creep comprises a majority of the deformation that will occur during the service cycles or accelerated test regiments. Clearly, negative creep, which occurred in the annealed 95.5Sn-3.9Ag-0.6Cu alloy at low stresses, will play an important role in defining the fatigue and/or creep response of the Pb-free solder that is used in an interconnection.

An important objective of this chapter was to correlate fatigue and creep behaviors with the microstructure of lead-free solders. Few studies have critically assessed the microstructural features associated with these deformation modes in lead-free solders. For example, the Ag$_3$Sn phase particles appear to be instrumental towards preventing grain boundary sliding as a damage mode in the lead-free solders. However, those particles appear to have a lesser role in the diffusion-based mechanisms of creep deformation that precedes such damage. It will become increasingly more important to understand the solder microstructure—its effect on creep and fatigue as well as its response to these deformation modes—if size or length-scale effects are to be considered in the constitutive response of increasingly smaller interconnections.

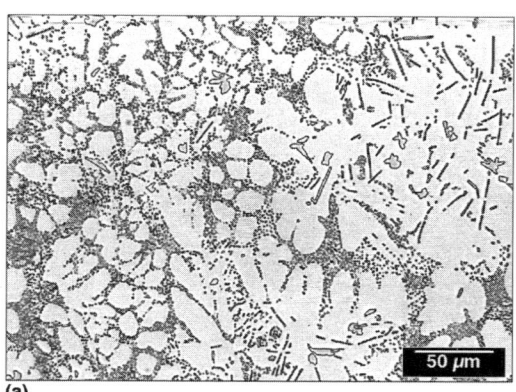
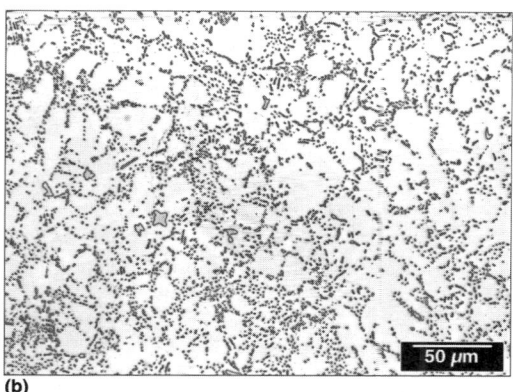

Fig. 28 Optical micrographs of the 95.5Sn-3.9Ag-0.6Cu creep sample tested at (a) 125 °C (257 °F) and 7.8 MPa (1.13 ksi) stress, and (b) 160 °C (320 °F) and 7.3 MPa (1.06 ksi) stress, showing coarsening of the Ag$_3$Sn particles in the eutectic regions between the Sn-rich dendrites. Both samples were in the as-fabricated condition.

ACKNOWLEDGMENTS

The author wishes to thank Mike Dvorack for his thorough review of this manuscript. Sandia is a multiprogram laboratory operated by Sandia Corporation, a Lockheed Martin Company, for the U.S. Dept. of Energy's National Nuclear Security Administration under contract DE-AC04-94AL85000.

REFERENCES

1. H. Solomon, Fatigue of 60/40 Solder, *IEEE Trans.*, Vol CHMT-9, 1986, p 423–433
2. E. Cutiongco, S. Vaynman, M. Fine, and D. Jeannotte, Isothermal Fatigue of 63Sn-37Pb Solder, *Trans. of the ASME*, Vol 112, 1990, p 110–114
3. R. Sandstrom, J.O. Osterberg, and M. Nylen, Deformation Behavior During Low Cycle Fatigue Testing of 60Sn-40Pb Solder, *Mater Sci Technol.* Vol 9, 1993, p 811–819
4. "Standard Test Methods for Tension Testing of Metallic Materials," *ASTM E8-03*, American Society for Testing and Materials, 2003
5. "Standard Test Methods of Compression Testing of Metallic Materials at Room Temperature," *ASTM E9-89a*, American Society for Testing and Materials, 2000
6. "Standard Test Methods for Elevated Temperature Tension Tests of Metallic Materials," *ASTM E21-03a*, American Society for Testing and Materials, 2003
7. "Standard Practice for Compression Tests of Metallic Materials at Elevated Temperatures with Conventional or Rapid Heating Rates and Strain Rates," *ASTM E209-00*, American Society for Testing and Materials, 2003
8. "Standard Test Method for Poisson's Ratio at Room Temperature," *ASTM E132-97*, American Society for Testing and Materials, 2003
9. G. Dieter, *Mechanical Metallurgy* 2nd ed., McGraw-Hill, 1976, p 350–351
10. G. Carter, *Principles of Physical and Chemical Metallurgy*, ASM International, 1979, p 167
11. C. Miller, I. Andersen, and J. Smith, A Viable Tin-Lead Solder Substitute: Sn-Ag-Cu, *J. Electron. Mater.* Vol 23, 1994, p 595–601
12. K. Moon, W. Boettinger, U. Kattner, F. Biancaniello, and C. Handwerker, Experimental and Thermodynamic Assessment of Sn-Ag-Cu Alloys, *J. Electron. Mater.*, Vol 29, 2000, p 112–1136
13. S. Kang, et al., The Microstructure, Thermal Fatigue, and Failure Analysis of Near-Ternary Eutectic Sn-Ag-Cu Solder Joints, *Mater. Trans.*, Vol 45 (3), 2004, p 695–702
14. P.T. Vianco, J.A. Rejent, and A.C. Kilgo, Creep Behavior of the Ternary 95.5Sn-3.9Ag-0.6Cu Solder: Part II—Aged Condition, in review for *J. Electron. Mater.*, Vol 33, 2004, p 1473–1484
15. P. Vianco, J. Rejent, and R. Grant, Development of Sn-Based, Low Melting Temperature Pb-Free Solder Alloys, *Trans. Jpn. Inst. Met.*, Vol 45, 2004, p 765–775
16. P. Vianco and J. Rejent, Properties of Ternary Sn-Ag-Bi Solder Alloys: Part I—Thermal Properties and Microstructural Analysis, *J. Electron. Mater.*, Vol 28, 1994, p 1131–1138
17. P. Vianco and J. Rejent, Properties of Ternary Sn-Ag-Bi Solder Alloys: Part II–Wettability and Mechanical Properties Analyses, *J. Electron. Mater.*, Vol 28, 1999, p 1139–1144
18. P. Vianco, J. Rejent, and A. Kilgo, Time-Independent Mechanical and Physical Properties of the Ternary 95.5Sn-3.9Ag-0.6Cu Solder, *J. Electron. Mater.*, Vol 32, 2003, p 142–152
19. Y. Kariya, C. Gagg, and W. Plumbridge, Tin Pest in Lead-Free Solders, *Solder. Surf. Mt. Technol.*, Vol 13, 2000, p 39–40
20. A. Kim, S. Huh, and K. Suganuma, Effect of Cooling Speed on Microstructure and Tensile Properties of Sn-Ag-Cu Alloys, *Mater. Sci. Eng.*, Vol 333 (No. 1–2) 2002, p 106–114
21. P. Vianco, Sandia National Laboratories, 2004, unpublished data
22. P. Vianco, J. Rejent, and J. Martin, The Compression Stress-Strain Behavior of Sn-Ag-Cu Solders, *J. Met.* Vol 55, (6), 2003, p 50–55
23. J. Pang, B. Xiong, and T. Low, Comprehensive Mechanics Characterization of Lead-Free 95.5Sn-3.8Ag-0.7Cu Solder, *Micromaterials and Nanomaterials* (3), 2004, p 87–93
24. F. Ochoa, J. Williams, and N. Chawla, The Effects of Cooling Rate on Microstructure and Mechanical Behavior of Sn-3.5Ag Solder, *J. Met.* Vol 55 (6), 2003, p 56–60
25. "Standard Test Method for Young's Modulus, Tangent Modulus, and Chord Modu-

25. lus," *ASTM E11-97,* American Society for Testing and Materials, 2003
26. "Standard Test Method for Dynamic Young's Modulus, Shear Modulus, and Poisson's Ratio by Impulse Excitation of Vibration," *ASTM E1876-01,* American Society for Testing and Materials, 2003
27. *Solder Alloy Data: Mechanical Properties of Solders and Soldered Joints,* International Tin Research Institute, Pub. 656, 1986
28. J. Pang, et al., Bulk Solder and Solder Joint Properties for Lead Free 95.5Sn-3.8Ag-0.7Cu Solder Alloy, *53rd Electronic Components and Technology Conf. Proc.,* 2003, p 673–679
29. G. Dieter, *Mechanics Metallurgy,* 2nd ed., McGraw-Hill, 1976, p 403–447
30. A. Argon, et al., *Mechanical Behavior of Materials,* Addison-Wesley, 1966, p 576–611
31. R. Hertzberg, *Deformation and Fracture Mechanics of Engineering Materials,* 2nd ed., John Wiley and Sons, 1983, p 519–598
32. H. Solomon, The Influence of Hold Time and Fatigue Cycle Wave Shape on the Low Cycle Fatigue of 60/40 Solder, *38th Elect. Components Conf. Proc.,* 1988, p 7–13
33. S. Majumdar and W. Jones, How Well Can We Predict the Creep-Fatigue Life of a Well-Characterized Material, *Solder Mechanics: A State of the Art Assessment,* D. Frear, et al. Ed., TMS, 1991, p 273–360
34. W. Lee, L. Nguyen, and G. Selvaduray, Solder Joint Fatigue Models: Review and Applicability to Chip Scale Packages, *Microelectronics Reliability,* Vol 40, 2000, p 231–244
35. V. Stolkarts, B. Moran, and L. Keer, Constitutive and Damage Models for Solders, *48th Electronic Components and Technology Conf. Proc.,* 1998, p 379–385
36. Y. Wei, et al., Behavior of Lead-Free Solder Under Thermomechanical Loading, *IMECE'03 Proc.,* 2003, p 1–8
37. *Fatigue: An Interdisciplinary Approach,* J. Burke and V. Weiss Ed., Syracuse University Press, 1964
38. S. Vaynman, M. Fine, and D. Jeannotte, Low-Cycle Isothermal Fatigue of Solder Materials, *Solder Mechanics: A State of the Art Assessment,* D. Frear et al., Ed., TMS, 1991, p 156–189
39. D. Frear, Thermomechanical Fatigue in Solder Materials, *Solder Mechanics: A State of the Art Assessment,* 1991, p 191–237
40. P. Vianco, D. Frear, F. Yost, and J. Roberts, Development of Alternatives to Pb-Based Solders, *Sandia Report,* SAND97-0315, 1997
41. S.M. Lee and D. Stone, Grain Boundary Sliding in As-Cast Pb-Sn Eutectic, *Scr. Metall. Mater.,* Vol 30, 1994, p 1213–1218
42. M. Gagnon, M. Suery, A. Eberhardt, and B. Baudelet, High Temperature Deformation of the Pb-Sn Eutectic, *Acta Metall.,* Vol 25, 1977, p 71–75
43. M. Fine, Physical Basis for Mechanical Properties of Solders, *Handbook of Lead-Free Solder Technology for Microelectronic Assemblies,* Marcel-Dekkar, 2003, p 211–237 Referencing: H. Maroovi, et al., *J. Electron. Mater.,* Vol 26, 1997, p 783–790
44. J. Lau, D. Shangguan, D. Lau, T. Kung, and W. Lee, Thermal-Fatigue Life Prediction Equation for Wafer-Level Chip Scale Package (WLCSP) Lead-Free Solder Joints on Lead-Free Printed Circuit Board (PCB), *54th Electronic Components and Technology Conf. Proc.,* 2004, p 1563–1569
45. J.P. Clech, Lead-Free and Mixed Assembly Solder Joint Reliability Trends, *IPC Printed Circuits Expo Proc.,* 2004, p S28-3-1–S28-3-14
46. T.S. Park and S.B. Lee, Isothermal Low Cycle Fatigue Tests of Sn/3.5Ag/0.75Cu and 63Sn-37Pb Solder Joints under Mixed Mode Loading Cases, *52nd Electronic Components and Materials Conf. Proc.,* 2002, p s23p4–s23p9
47. P. Vianco and M. Grazier, Sandia National Laboratories, Albuquerque, NM and E. Cotts and L. Lehman, SUNY, Binghamton, NY, unpublished data
48. S. Dunford, A. Primavera, and M. Meilunas, Microstructural Evolution and Damage Mechanisms in Pb-Free Solder Joints During Extended −40 °C to 125 °C Thermal Cycles, *IPC Conf. Proc.,* 2002, p S08-4-1–S08-4-13
49. L. Murr, *Interfacial Phenomena in Metals and Alloys,* Addison-Wesley, 1975, p 322–340
50. P. Vianco, I. Artaki, and A. Jackson, Reliability Studies of Surface Mount Boards Manufactured with Lead-Free Solders, *Surface Mount International Proc.,* 1994, p 437–448
51. P. Vianco and C. May, An Evaluation of Prototype Surface Mount Circuit Boars Assembled with Three Non-Lead Bearing Sol-

ders, *Surface Mount International Proc.*, 1995, p 481–496
52. *Mechanical Behavior of Materials,* F. McClintock and A. Argon Ed., Addison-Wesley, 1966, p 634–635
53. "Standard Test Method for Conducting Creep, Creep-Rupture, and Stress-Rupture Tests of Metallic Materials," *ASTM E139-00e1,* American Society for Testing and Materials, 2003
54. F. Garofalo, *Fundamentals of Creep and Creep Rupture in Metals,* MacMillian, 1965, p 156–201
55. A. Krausz and H. Eyring, *Deformation Kinetics,* McGraw-Hill, p 190–200
56. F. Garofalo, *Fundamentals of Creep and Creep Rupture in Metals,* MacMillan, 1965, p 25–45
57. J. Askin, *Tracer Diffusion Data,* Plenum, 1970
58. *Smithells Metal Reference Book,* 6th Ed., Butterworth and Co., 1983, p 13-8-13-10
59. P. Shewmon, *Transformations in Metal,* McGraw-Hill, 1969, p 63–65
60. P. Shewmon, *Diffusion in Solids,* 2nd ed., TMS, 1989, p 189–199
61. J. Christian, *The Theory of Transformations in Metals and Alloys: Part I—Equilibrium and General Kinetic Theory,* Pergamon, 1975, p 541–543
62. J. Stephens and D. Frear, Time-Dependent Deformation Behavior of Near-Eutectic 60Sn-40Pb Solder, *Metall. Mater. Trans. A,* Vol 30A, 1999, p 1301–1313
63. O. Sherby and P. Burke, Mechanical Behavior of Crystalline Solid at Elevated Temperature, *Prog. Mater. Sci.,* Vol 13, Pergamon, 1968, p 340–347
64. F. Garofalo, *Trans. Metall. Soc.* AIME, Vol 227, 1963, p 351
65. P. Vianco, J. Rejent, and A. Kilgo, Creep Behavior of the Ternary 95.5Sn-3.9Ag-0.6Cu Solder: Part I—As-Cast Condition, to be published, *J. Electron. Mater.,* Vol 33, 2004, p 1389–1400
66. P. Vianco and J. Li, Negative Creep in an Amorphous Metallic Alloy, *Mater. Sci. Eng.,* Vol 95, 1987, p 175–186
67. J. Li, Negative Creep and Mechanochemical Spinodal in Amorphous Metals, *Mater. Sci. Eng,,* Vol 98, 1988, p 465–468
68. J. Clech, "Review and Analysis of Pb-Free Solder Properties," Report to the NEMI Pb-Free Solder Project, 2003
69. "NCMS Lead Free Solder Project," Report 0401RE96, June 1998
70. J. Morris, H. Song, and F. Hua, Creep Properties of Sn-Rich Solder, *53rd Electronic Components and Technology Conf. Proc.,* 2003, p 54–57
71. M. Kerr and N. Chawla, Creep Deformation Behavior of Sn-3.5Ag Solder/Cu Couple at Small Length Scales, *Acta Mater.,* 2004, in press
72. A. Schubert, et al., Thermal Mechanical Properties and Creep Deformation of Lead-Containing and Lead-Free Solders, *2001 International Symposium on Advanced Packaging, Materials Proc.,* 2001, p 129–134
73. Q. Zhang, A. Dasgupta, and P. Haswell, Viscoplastic Constitutive Properties and Energy-Partitioning Model of Lead Free 95.5Sn3.9Ag0.6Cu Solder Alloy, *53rd Electronic Components and Technology Conf. Proc.,* 2003, p 1862–1868
74. S. Wiese, S. Jakschik, F. Feustal, E. Meusel, Fracture Behavior of Flip Chip Solder Joints, *51st Electronic Components and Technology Conf. Proc.,* 2001, p 890–902
75. H. Frost and M. Ashby, *Deformation Mechanism Maps—The Plasticity and Creep of Metals and Ceramics,* Pergamon, 1982
76. J.P. Clech, An Obstacle-Controlled Creep Model for Sn-Pb and Sn-Based Lead-Free Solders, *Surface Mount Technology Association International Conf. Proc.,* 2004, on CD

CHAPTER 4

Lead-Free Solder Joint Reliability Trends

Jean-Paul Clech, EPSI Inc.

Introduction

This chapter presents a quantitative analysis of solder joint reliability data for lead-free Sn-Ag-Cu (SAC) and mixed assembly (Sn-Pb + SAC) circuit boards based on an extensive, but non-exhaustive, collection of thermal cycling test results. The data is collected with the ultimate objective of validating life prediction models and acceleration factors for a wide range of assemblies and test conditions. The present goal is to put the data in perspective, comparing lead-free and Sn-Pb test results for a variety of components, board finishes, and test conditions. The assembled database covers life test results under multiple test conditions for a variety of components: conventional Surface Mount Technology (SMT) components such as Leadless Ceramic Chip Carriers (LCCCs) and chip resistors, Ball Grid Arrays (BGAs), Chip Scale Packages (CSPs), wafer-level CSPs, and flip-chip assemblies with and without underfill. First-order life correlations are developed for SAC assemblies under thermal cycling conditions. The results of this analysis are put in perspective with the correlation of life test results for Sn-Pb control assemblies. Fatigue life correlations show different slopes for SAC versus Sn-Pb assemblies, suggesting opposite reliability trends under low- or high-stress conditions. The analysis then looks into the effect of lead (Pb) contamination and board finish on lead-free solder joint reliability. Test data are presented to compare the life of mixed solder assemblies to that of standard Sn-Pb assemblies for a wide variety of area-array components. The trend analysis compares the life of area-array assemblies with: (a) SAC balls and SAC or Sn-Pb paste; and (b) Sn-Pb balls assembled with SAC or Sn-Pb paste. Last, we compare Sn-Pb and SAC creep data and examine differences in proposed life prediction models.

Attachment reliability is defined as the ability of solder joints to survive the planned design life of a given product. Blanket statements regarding attachment reliability, or the lack of it, of lead-free assemblies should be considered within the context of product reliability goals, and generalizations should be made with extreme caution. The acquisition of field reliability data under non-accelerated conditions is cost and time prohibitive and only late adopters of a new technology can afford to wait for historically-based field returns. Reliability is *application-specific* and is preferably established through the acquisition of failure distributions under accelerated test conditions, followed by careful extrapolation of test failure times to service conditions. To this author's knowledge, the latter step is still in the development stage. Lead-free life prediction models and acceleration factors and end-user validation studies are not readily available. However, numerous efforts to that effect are in progress (Ref 1–4, for example). The development of Sn-Pb solder joint reliability models extends over three decades, starting with the landmark paper of Norris and Landzberg (Ref 5) in 1969. It would be optimistic to expect robust lead-free reliability models to become available within a year. In the meantime, it is useful to gather accelerated test data for comparison purposes. Our experience working with Sn-Pb and lead-free assemblies has been that a wide range of test data need be considered to establish firm reliability trends and to develop reliable life prediction models.

Empirical Correlations of SAC Thermal Cycling Test Data

SAC Thermal Cycling Data. Figure 1(a) is an attempt at correlating characteristic life and average cyclic shear strain in solder joints as per the classical Coffin-Manson (Ref 6) approach for metal fatigue. The characteristic lives (cycles to 63.2% failures) are from failure distributions for 100% lead-free SAC assemblies (Ref 7 to 15) and component populations subject to accelerated thermal cycling under a range of conditions: 0 to 100 °C (32 to 212 °F), −40 to 125 °C (−40 to 257 °F), and −55 to 125 °C (−67 to 257 °F). The average cyclic shear strain is defined as follows:

$$\Delta\gamma = \frac{L\Delta\alpha\Delta T}{h}$$

where L is the maximum Distance to Neutral Point (DNP) for the outermost joints, $\Delta\alpha$ is the board-to-component mismatch in Coefficients of Thermal Expansion (CTE), h is the average component stand-off height or joint thickness, and ΔT is the temperature difference between the hot and cold sides of a given thermal profile. All components were leadless and mounted on FR-4 type boards: resistors (sizes: 0603, 1206, and 2512), 32 by 32 mm (1.25 by 1.25 in.) and 42 by 42 mm (1.65 by 1.65 in.) square ceramic ball grid arrays (CBGAs), 169 Input/Output (I/O) chip scale packages (CSP), 20 I/O leadless ceramic chip carriers (LCCC) and 1.27 mm (0.05 in.) pitch plastic ball grid arrays (PBGAs). The reader is referred to the original sources of data (Ref 7 to 15) for more detailed information on components, test vehicle parameters, other conditions, and test results. Twenty-seven data points are shown in Fig. 1(a) and (b). Nominal alloy composition varied slightly (Sn-3.8Ag-0.7Cu or Sn-3.9Ag-0.6Cu), board finishes were immersion silver (Ag), organic solderability preservative (OSP) or electroless Ni-Au. The board finish was unspecified for three of the 27 datasets.

The correlation coefficient for the power-law trendline of the data in Fig. 1(a) is $R^2 = 0.64$. Clearly, there is a discrepancy between the data and the basic Coffin-Manson approach. The data is replotted in Fig. 1(b) with the following modifications:

- Instead of characteristic lives for a population of components, cyclic lives are given on a per joint basis, as was done in the past in the development of Sn-Pb reliability models—this allows for the number of most critical joints per component (Ref 16)
- The joint characteristic life is then scaled for the solder joint crack area, as justified in previous work (Ref 16 and 17)

The scaling of characteristic lives for the solder crack area is intended to account for differences in crack propagation times through joints of varied sizes. The resulting parameter—char-

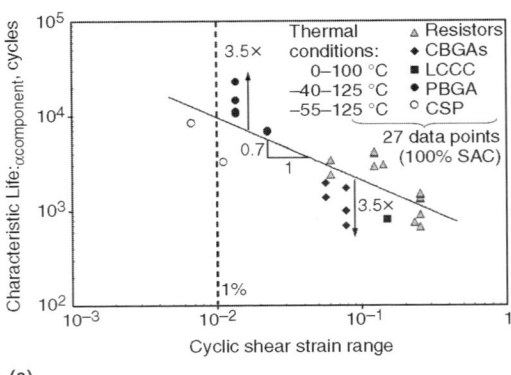

(a)

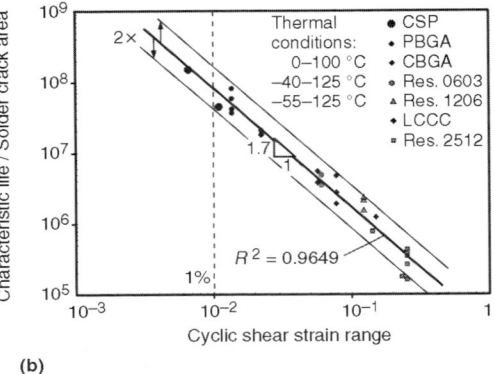

(b)

Fig. 1 Life data correlations for 100% lead-free SAC assemblies subject to accelerated thermal cycling. (a) Correlation of component characteristic life versus cyclic shear strain range. (b) Correlation of joint characteristic life scaled for solder joint crack area versus cyclic shear strain range.

acteristic life over the solder joint area—is interpreted as the inverse of an area crack propagation rate and is shown in Fig. 1(b) with units of cycles/in^2. The correlation coefficient for the power-law trend line in Fig. 1(b) is $R^2 = 0.96$. This R^2 value is remarkably high, which suggests a reasonable empirical correlation of the data within the strain range under consideration. What this means is that the data is very consistent across a range of component sizes, joint sizes, and thermal cycling conditions. However, the correlation in Fig. 1(b) is purely empirical and should not be used to extrapolate test results to field conditions, since important parameters such as dwell times or test frequency are not accounted for. The latter effects may not matter within the context of this exercise since all tests were rather rapid with similarly short dwell times, compared to long dwell periods that may be encountered in service. Nevertheless, the correlation in Fig. 1(b) is of use to investigate the effect of other parameters, such as board finish or alloy composition effects, as shown in the following section.

Interestingly, only one of the 27 data points in Fig. 1(a) and 1(b) is below the 1% shear strain level. A 1% strain is considered high, leading to a reduced cyclic life, and would definitely raise a red flag if this was to occur under service conditions. While our database of SAC test results is far from exhaustive, this suggests a lack of test results under mild conditions. It appears that many tests are run under highly accelerated conditions to gather failure data rapidly for comparison purposes. This results in a scarcity of data in low- to medium-stress conditions, slowing down efforts to validate reliability models in regions closer to the milder conditions encountered in many applications.

Alloy Composition and Board Finish Effects. The data in Fig. 1(b) is replotted in Fig. 2(a) and 2(b) where, instead of showing component types, results are shown according to alloy composition or board finish. Out of the 27 datasets of interest, five had the nominal solder composition: Sn-3.9Ag-0.6Cu (SAC 3906) while all the others had the nominal composition: Sn-3.8Ag-0.7Cu (SAC 3807). In Fig. 2(a), the SAC 3906 data points are all below the correlation center line. It might be tempting to conclude that the SAC 3807 alloy leads to a slightly longer life than SAC 3906. However, the SAC 3906 population in the analysis is small (5 data points out of a total of 27), and more data is needed to test the statistical differences between the SAC 3807 and SAC 3906 populations.

In Fig. 2(b), the same data points are grouped and labeled by board finish: immersion Ag, OSP, electroless Ni-Au, or unspecified. For a given group, the data is above or below the centerline and no definite trend is visible on the basis of board finishes. For example, looking at the four Ni-Au and OSP data in the upper left of the plot, the OSP points are above the Ni-Au points. The opposite is observed for the Ni-Au and OSP data points in the lower right region of Fig. 2(b). Over a wide range of cyclic shear strains, no board finish appears to provide consistently better results than the others.

Lead-Free to Sn-Pb Comparison

Reflowed SAC Versus Near-Eutectic Sn-Pb. The 100% lead-free SAC data in Fig. 1(b)

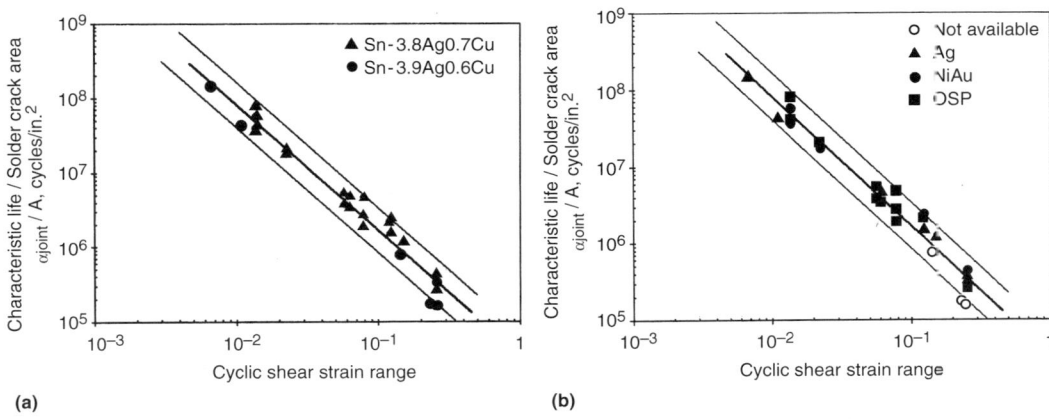

Fig. 2 Test data from Fig. 1(a) sorted out by (a) alloy composition and (b) board finish

is shown in Fig. 3, along with thermal cycling results for standard Sn-Pb assemblies from the same experiments (Ref 7 to 15). Each group shows some scatter as is expected with fatigue data. The power-law trendlines that are fitted through each dataset intersect at a shear strain level of about 6.2%. To the left of the crossover point, SAC assemblies have longer lives than Sn-Pb assemblies. For most points at cyclic shear strains less than 3%, the Sn-Pb data is clearly below the SAC data. The opposite holds for data points at shear strains above 20%.

The empirical correlations in Fig. 3 indicate that, under high-stress conditions, the life of SAC solder joints is less than that of Sn-Pb. The differences in the slopes of the SAC and Sn-Pb trendlines and the intersect of the two lines, suggest that when SAC assemblies do not perform as well as Sn-Pb assemblies under highly accelerated test conditions, SAC assemblies are likely to be more reliable than Sn-Pb assemblies under milder conditions. This trend reversal further highlights the need to extrapolate test results to field conditions in an accurate manner. Relying on accelerated test results alone may lead to the rejection of some SAC designs and assemblies that would perform as well or better than Sn-Pb assemblies under mild enough conditions. Last, assuming that the trendlines in Fig. 3 extrapolate further to the left, if SAC test results are better than Sn-Pb assemblies, the trend can possibly hold under milder service conditions. However, a word of caution is necessary here: life improvements observed in accelerated testing, with typically short dwell times at the temperature extremes, may not carry over to milder service conditions with extended dwell times. As shown in Ref 43, the life of SAC solder joints shortens significantly with longer dwell times, and is less than that of Sn-Pb joints under 0 to 100 °C (32 to 212 °F) thermal cycling conditions with dwell times of approximately 110 min.

Sn0.7Cu Versus Sn-Pb for Bare Chip Assemblies. Figure 4 shows simple correlations of characteristic lives to cyclic shear strains for bare chip assemblies using Sn-0.7Cu (5 data points in Fig. 4a), Sn-37Pb or Sn-36Pb-2Ag (9 data points in Fig. 4b) and SAC solders (8 data points in Fig. 4c) of nominal composition close to that of Sn-3.8Ag-0.7Cu. The characteristic lives were not scaled for the solder joint crack area since all assemblies had similar pad sizes, at least to a first order. For each alloy type, the test data was from multiple independent sources (Ref 2, 18–20). Components were either of the flip-chip type or wafer-level CSPs (all without underfill) with a variety of metallizations or underbump metallurgies. Thermal cycling conditions covered the following temperature extremes: −50 to 20 °C (−58 to 68 °F), 0 to 70 °C (32 to 158 °F), 50 to 120 °C (122 to 248 °F), 0 to 100 °C (32 to 212 °F), −40 to 125 °C (−40 to 257 °F) and −55 to 125 °C (−67 to 257 °F). Shear strain ranges are all above 1% because of the large CTE mismatch between silicon chips and organic substrates.

The correlations for each alloy group show scatter that is typical of fatigue data (as much as 2.5–3× above or below the trendlines). However, the Sn-0.7Cu correlation is remarkably tight. The power-law trendlines for the three groups are shown on the same graph in Fig. 4(d). The SAC and Sn-Pb trendlines intersect and have similar relative positions as in Fig. 3, thus confirming, as discussed previously, the reversal of reliability trends under low- or high-stress conditions. The two lines, which are for bare chip data alone, intersect at a shear strain level of about 6.2%. Interestingly, this is the exact same shear strain as for the intersect of the SAC and Sn-Pb trendlines for unrelated data in Fig. 3. The Sn-Cu and Sn-Pb trendlines in Fig. 4(d) intersect as well. However, their relative positions follow an opposite trend of the relative positions of the SAC and Sn-Pb trendlines. More data is needed to test the validity of this Sn-Cu trendline. If it holds, more caution will be needed when interpreting Sn-Cu test results. To the right of the intersect of the Sn-Cu and Sn-Pb lines, the trend is that Sn-Cu assemblies have longer solder joint lives than Sn-Pb assemblies. To the left of that same crossover point, Sn-Cu assemblies would have shorter lives than Sn-Pb

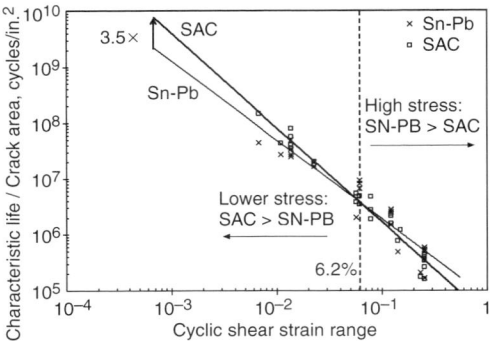

Fig. 3 Correlations of joint characteristic life scaled for solder joint crack area versus cyclic shear strain range in temperature cycling for standard Sn-Pb and for 100% lead-free SAC assemblies

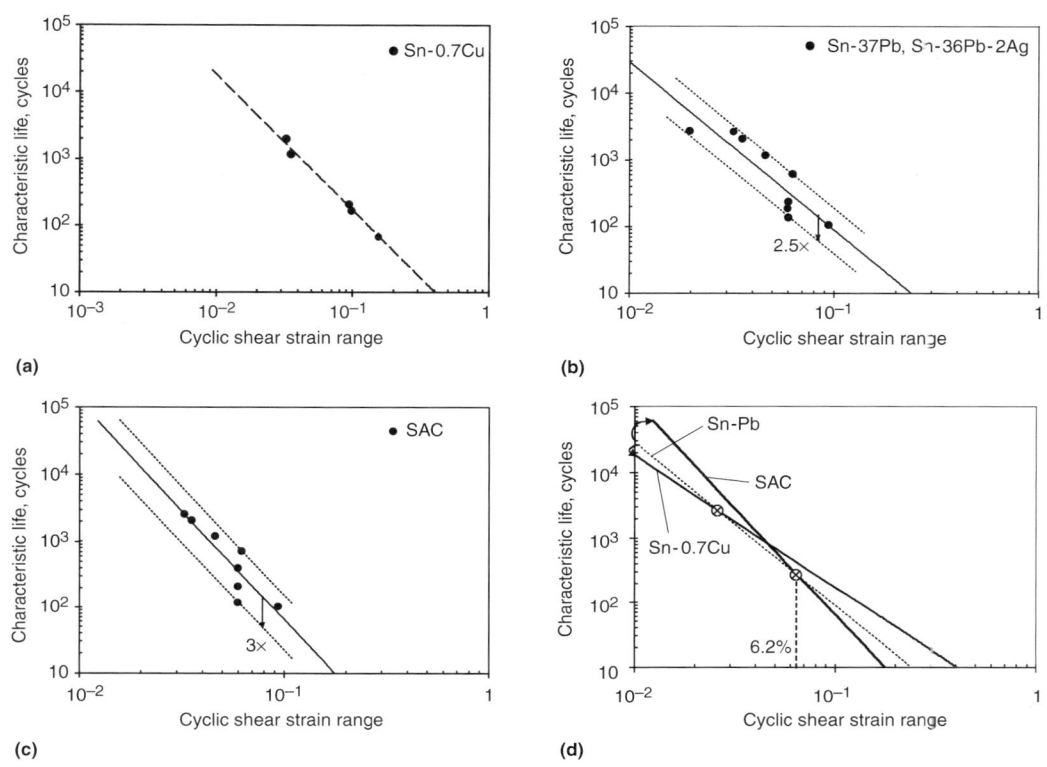

Fig. 4 Correlations of characteristic life to cyclic shear strain range for bare chip assemblies. (a) Sn-0.7Cu data. (b) Sn-Pb data. (c) SAC data. (d) Power-law trendlines for Sn-Cu, Sn-Pb and SAC assemblies.

assemblies. This lower stress region is representative of many use conditions and accurate acceleration factors are needed to assess whether Sn-0.7Cu still meets product-specific field reliability requirements.

The relative positions of the fatigue trendlines in Fig. 4(d) are supported by similar trends observed in mechanical shear fatigue tests at 20 °C (68 °F), as shown in Fig. 5. The collected data is from fatigue strength tables (Ref 21) for stress-controlled fatigue experiments on copper ring and plug joint specimens soldered with alloys of slightly different compositions (Sn-1.0Cu instead of Sn-0.7Cu, and Sn-3.5Ag instead of SAC). Figure 5 shows that Sn1Cu solder has a shorter life than that of 60Sn-40Pb solder in the lower stress area of the plot, similar to what was observed for Sn-0.7Cu in Fig. 4(d).

Critical Component Data

Life Test Data for SAC and Sn-Pb TSOP Assemblies. Figure 6 shows correlations of Alloy42 TSOP characteristic lives versus temperature swings, ΔT, in accelerated thermal cycling. The characteristic lives were read off failure distributions in two independent studies (Ref 22, 23). The paste composition (Sn-3.9Ag-0.6Cu or

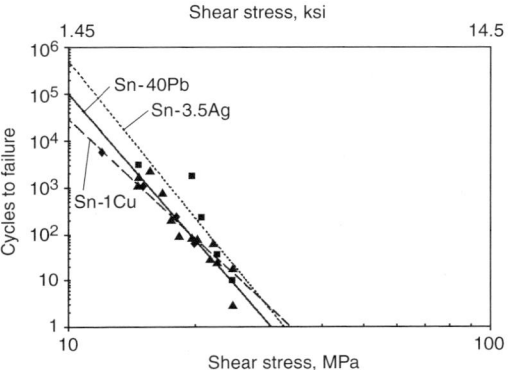

Fig. 5 Plot of mechanical fatigue strength data at 20 °C (68 °F) and crosshead speed of 0.2 mm/min. (0.008 in./min.) for copper ring and plug joints soldered with Sn-1.0Cu, Sn-40Pb or Sn-3.5Ag solder. Source: Ref 21

Sn-3.0Ag-0.7Cu) and lead finish (Sn-Pb for all except for one Sn-2Bi data point) are shown in the legend of Fig. 6. Temperature cycling conditions were: −55 to 125 °C (−67 to 257 °F), −40 to 125 °C (−40 to 257 °F), 0 to 100 °C (32 to 212 °F) (Ref 22), and −25 to 125 °C (−13 to 257 °F) (Ref 23). For components with Sn-Pb finish and for a given paste type (Sn-Pb or lead-free SAC), the data from the two sources appear to fit together, at least to a first order. No significant departure is visible for the TSOPs assembled with one type of SAC paste or the other. The data, although limited, does not show a marked effect of SAC alloy composition on life. The all-Sn-Pb data from the two studies fit together in a similar manner. Power law trendlines are shown by paste type for the merged datasets, SAC or Sn-Pb. The trendlines have similar slopes and the SAC paste/Sn-Pb finish solder joint lives are less than for the all-Sn-Pb assemblies under the stated test conditions. For example, looking at the data in the lower right region of Fig. 6, the characteristic life of the lead-free TSOP assemblies is 1.63 times lower than that of the conventional Sn-Pb TSOP assemblies. For the Sn-3.0Ag-0.7Cu paste assemblies, TSOPs with Sn-2Bi finish have a solder joint life 31% longer than in the case of TSOPs with Sn-10Pb finish.

The relative position of the two trendlines in Fig. 6 is similar to that of SAC and Sn-Pb lines in Fig. 3 and 4(d), although the slopes of the two lines are much closer. The intersect (not shown) corresponds to a crossover point at a ΔT of about 9 °C (48 °F). Assuming that the SAC and Sn-Pb lines in Fig. 6 can be extrapolated to smaller ΔT's, the life of SAC paste/Sn-Pb finish Alloy42 TSOP joints would be shorter than that of conventional Sn-Pb TSOP assemblies, even for temperature swings as low as 9 °C (48 °F). Given the small size of the TSOP datasets in the above analysis, more data is needed to validate the observed trends. However, given that Alloy42 TSOPs assembled with Sn-Pb on FR-4 boards have application limited reliability (e.g., their use is not recommended (Ref 24) in telecommunication products with a five year or longer life span), their implementation in lead-free products will require detailed board-level reliability evaluations.

Flip-Chip with Underfill. Flip-chip with underfill thermal cycling results from two sets of experiments, (Ref 25–27) and are plotted in Fig. 7 with the failure times given as mean cycles to failure for Sn-Pb and lead-free assemblies:

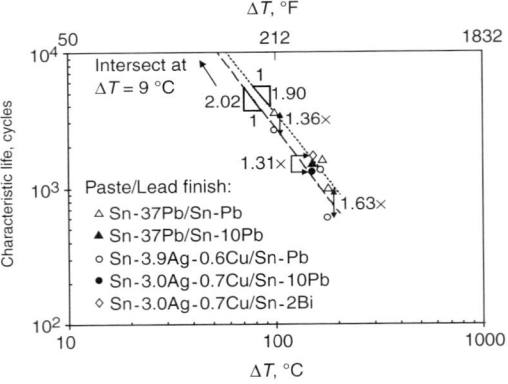

Fig. 6 Correlations of characteristic lives to temperature swings ΔT for 48 I/O Alloy42 TSOPs.

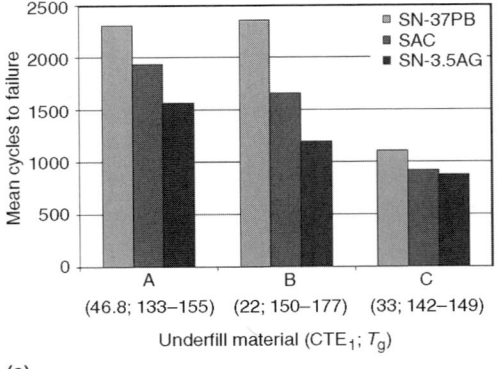

(a)

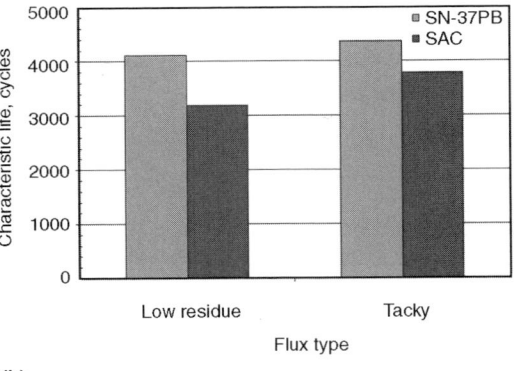

(b)

Fig. 7 Sn-Pb versus SAC thermal cycling results for underfilled flip-chip assemblies. (a) Results of experiment with three underfill materials. (b) Results of experiment with two flux types. Source: (a) Ref 25 (b) Ref 26, 27

- In the first experiment (Ref 25), failure times are for assemblies using different underfill materials with CTEs and glass transition temperatures (T_g) as shown in Fig. 7(a). Thermal cycling was between −55 and 125 °C (−67 and 257 °F) with dwell times of 10 min. at the temperature extremes, and a 30 min. cycle duration. Solder fatigue was the primary failure mode.
- In the second experiment (Ref 26, 27), Sn-Pb and SAC results are for assemblies that used low residue or tacky flux. Thermal cycling was between −40 and 125 °C (−40 and 257 °F) with dwell times of 5 min. at the temperature extremes, and a 12 min. cycle duration. The underfill material was a standard material with a CTE of 35 ppm/°C and a T_g of 130 °C (266 °F). Failure modes were mixed, including solder fatigue and underfill delamination, with more of the latter in the SAC case than in the Sn-Pb case.

Regardless of the test variable—underfill material or flux type—the SAC failure times are consistently less than failure cycles for Sn-Pb flip-chip assemblies, 20 to 30% less in Fig. 7(a) and 14 to 23% less in Fig. 7(b). Results for Sn3.5Ag underfilled assemblies in Fig. 7(a) are also worse than for SAC assemblies and 20 to 50% less than for Sn-Pb assemblies. While this is a cause for possible concern, the reliability of the SAC assemblies under service conditions may still be acceptable or even exceed that of Sn-Pb underfilled assemblies, if the stress-dependent life-trend reversal that was observed previously holds for underfilled assemblies. Again, this illustrates the heightened importance of having reliable life prediction tools to extrapolate accelerated test results to field conditions.

Impact of Pb Contaminant or Sn-Pb Alloy on Lead-Free Reliability

Reflowed Conventional Leadless SMT Assemblies with Pb Contaminant. In Fig. 8, we have plotted cycles-to-1% failures for SAC paste assemblies with various Pb contaminant, versus cycles-to-1% failures for 100% lead-free SAC assemblies under thermal cycling conditions: −40 to 125 °C (−40 to 257 °F), or −55 to 125 °C (−67 to 257 °F). When data points appear above the main diagonal, life for assemblies with Pb contamination is of a longer duration than that of 100% lead-free assemblies. The data (Ref 7, 8, 13) is for conventional leadless SMT components: 20 Input/Output LCCCs and resistors ("R") of sizes 0603, 1206, and 2512. The source of Pb contamination is the Sn-Pb HASL board finish, or the Sn-Pb component termination, in the 1206 resistors. In the 20 I/O LCCC SAC boards with Sn-Pb HASL finish, Woodrow (Ref 13) measured a 0.5% Pb contamination level. The cycles to 1% failure were calculated from Weibull parameters (characteristic life and slope of failure distributions) provided

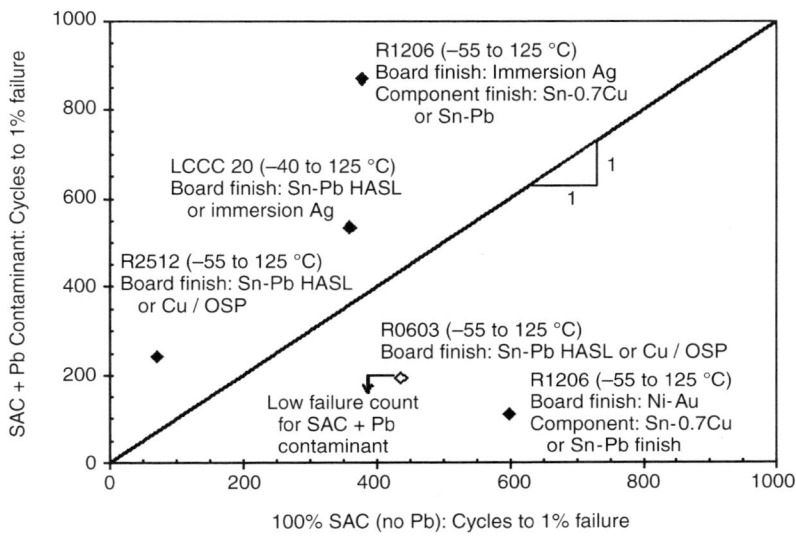

Fig. 8 Cycles to 1% failure for SAC assemblies with or without Pb contaminant. Pb is from Sn-Pb HASL boards or Sn-Pb component finish. Labels indicate: component type, temperature extremes, board and/or component finishes. Source: Ref 7, 8, 13

in the original studies (Ref 7, 8, 13). Similar trends as shown for cycles-to-1% failure would hold for characteristic lives. We chose to plot cycles-to-1% failure, since the early part of failure distributions is more relevant to product reliability, especially in the case of boards where a possible defect such as Pb contamination may reduce early life product reliability.

Since the data in Fig. 8 falls either above or below the main diagonal, the record is mixed with positive or negative effect of Pb contamination on assembly reliability under accelerated thermal cycling conditions. Of the two data points below the main diagonal, the 0603 resistor data (Ref 7) may have large confidence bands since the Weibull parameters were obtained from a population with a low failure count. The 1206/Ni-Au data point (Ref 8) did not present any such anomaly. Since the 1206/immersion Ag data point from the same study (Ref 8) is well above the main diagonal, the problem of Pb contamination is possibly confounded with board finish effects. More data will be added to Fig. 8 when available, to further investigate the sensitivity of Pb contamination effects to board metallization. Thus far, the available data (as shown in Fig. 8) suggests that Pb contamination at the 0.5% level raises a potential reliability concern for organic boards with Ni-Au finish on copper pads.

Area Array Assemblies with SAC Balls and SAC or Sn-Pb Paste. During the transition phase to 100% lead-free technology, lead-free components may be assembled with conventional, eutectic or near-eutectic Sn-Pb paste. This scenario, described as "backward compatibility," raises the issue of how the board-level reliability of lead-free components assembled with Sn-Pb paste compares to that of conventional 100% Pb assemblies.

Figure 9 shows cycles-to-failure for lead-free area array components (SAC balls) assembled with Sn-Pb paste versus cycles-to-failure for conventional Sn-Pb area array components (Sn-Pb balls) assembled with Sn-Pb paste. Board finish labels are given, along with component names. The data was gathered from several studies (Ref 9, 28–31) where test vehicles were assembled using standard Sn-Pb or SAC reflow profiles. In the case of SAC ball assemblies, the peak reflow temperature was above the melting point for typical SAC alloys (217 °C, or 422 °F). For example, for the data point (Ref 30) labeled "wbPBGA, OSP" in Fig. 9, the peak reflow temperature was 222 °C (431 °F) (soak profile) with a time above liquidus (TAL) quoted in the range of 60 to 90 s. The reader is referred to the original studies (Ref 9, 28–31) for more detailed information on the test vehicles, including component information for the various BGAs or CSPs labeled in Fig. 9. Cycles-to-failure are given as cycles to a low percentage of failures that were either read-off failure distributions at the 0.1% failures (Ref 28) level, cycles to first failure (Ref 29) (at a cumulative failure level between 1 and 5%), or calculated at the 1% fail-

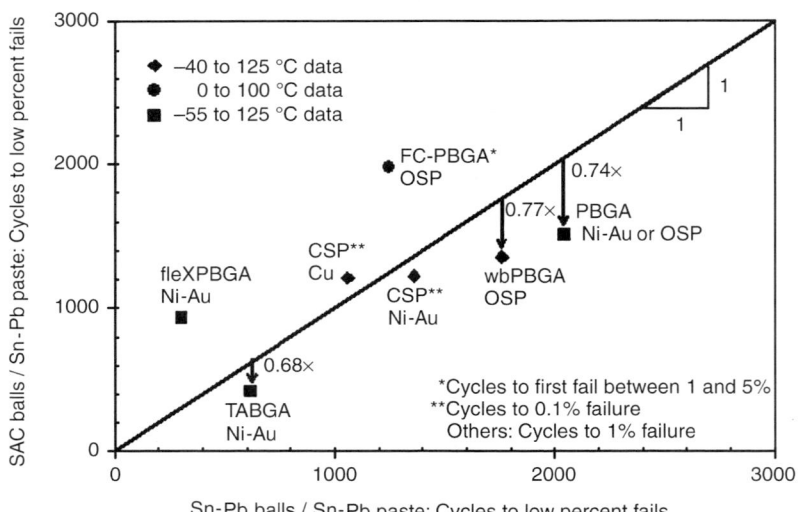

Fig. 9 Thermal cycling data for SAC or Sn-Pb ball area-array components assembled with Sn-Pb paste. For SAC ball assemblies, peak reflow temperature is above 217 °C (422 °F). Labels are for component names and board finishes.

ure level when Weibull distribution parameters were available. The data points in Fig. 9 are above or below the main diagonal, that is, thermal cycling test lives for SAC assemblies with Sn-Pb paste are either better or worse than for conventional Sn-Pb assemblies. For the four data points below the main diagonal, the life of SAC ball/Sn-Pb paste assemblies is less than that of Sn-Pb assemblies by 10 to 32%. Looking at data point labels in Fig. 9, no definite board finish effect is apparent. Similarly, Fig. 9 does not show any definite temperature profile effect. The one data point for thermal cycling between 0 and 100 °C (32 and 212 °F) is above the main diagonal, but further data under similar conditions would have to be added before any firm conclusion can be drawn with regard to the possible effect of thermal cycling profiles. Last, note that the above discussion applies to SAC ball/Sn-Pb paste assemblies for which the peak reflow temperature is above approximately 225 °C (437 °F). Due to incomplete mixing of alloys and segregated microstructures, SAC ball/Sn-Pb paste assemblies that are reflowed at peak temperatures typical of conventional Sn-Pb processes (205 to 225 °C (401 to 437 °F) peak temperature) have been observed (Ref 52) to lead to further process and reliability concerns, including little or no self-alignment of parts, poor interfacial bonding, and increased voiding.

In conclusion, the backward-compatibility record for SAC balls/Sn-Pb paste assemblies under accelerated thermal cycling conditions is mixed. A peak reflow temperature above approximately 225 °C (437 °F) is critical to attain quality and reliability levels comparable to those of conventional assemblies (Ref 52). However, to our knowledge, no life prediction model or acceleration factors are available for mixed assemblies. Thus, we are unable to extrapolate SAC balls/Sn-Pb paste assemblies test results to field conditions, and no firm conclusion can be drawn as to the field reliability of these assemblies compared to conventional Sn-Pb assemblies. Nevertheless, since the record under test conditions is mixed, close attention needs to be paid to the reliability of SAC balls/Sn-Pb paste assemblies, especially when the peak reflow temperature is below 225 °C (437 °F).

In the next phase of the transition to lead-free technology, lead-free components assembled with Sn-Pb paste will eventually be assembled with SAC paste. Relevant test data (Ref 9, 28, 29, 31) for area array components with SAC balls is shown in Fig. 10, where cycles-to-failure for SAC paste assemblies is plotted versus cycles-to-failure for Sn-Pb paste assemblies. The presentation of the data is similar to that of Fig. 9 with symbols and labels identifying temperature profiles, components, and board finishes. All seven data points in Fig. 10 are close to or above the main diagonal, suggesting that during the transition from mixed assemblies (SAC balls/Sn-Pb paste) to 100% lead-free boards (SAC balls/SAC paste), reliability concerns are likely minimized.

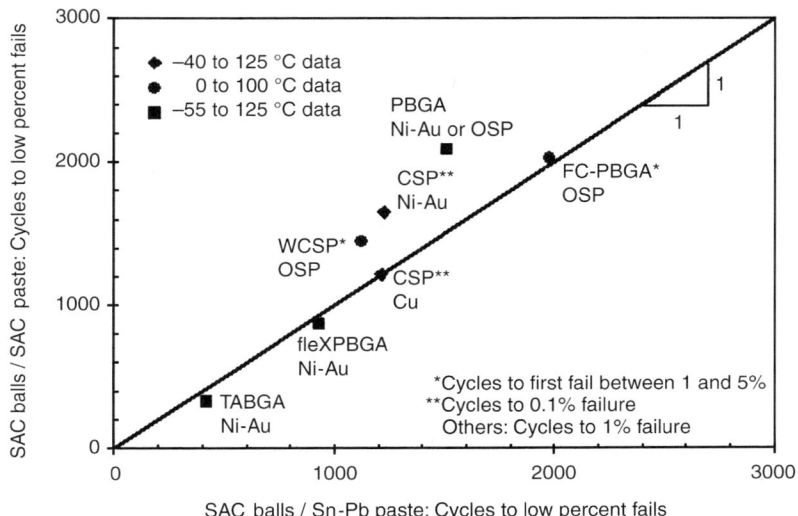

Fig. 10 Thermal cycling data for SAC ball area-array components assembled with SAC or Sn-Pb paste. For SAC ball assemblies, peak reflow temperature is above 217 °C (422 °F). Labels give component names and board finishes.

Area Array Assemblies with Sn-Pb Balls and SAC or Sn-Pb Paste. Another scenario of interest during the transition to lead-free technology is that of conventional area-array components using Sn-Pb balls assembled with SAC paste. This scenario is often described as a forward compatibility situation. In Fig. 11, we show cycles-to-1% failure for Sn-Pb ball area array components assembled with SAC paste versus cycles-to-1% failure for similar components assembled with Sn-Pb paste. The data was gathered from relevant test cells in several independent studies (Ref 3, 11, 12, 22, 29, 32, 33). Figure 11(a) shows the data for assemblies that were cycled between −40 and 125 °C (−40 and 257 °F) (6 data points). Figure 11(b) shows similar test data for thermal cycling under milder conditions: 0 to 100 °C (32 to 212 °F) (7 data points) and 15 to 95 °C (59 to 203 °F) (2 data points[3] for 144 Input/Output PBGAs assemblies with Ni-Au or Sn-Cu HASL board finish).

Figures 11(a) and 11(b) suggest a strong dependence of relative trends on thermal cycling conditions. Under harsher conditions (−40 to 125 °C, or −40 to 257 °F) data in Fig. 11a), the data for the Sn-Pb ball components using SAC paste (except for one data point labeled "PBGA 357") falls above the main diagonal. Under milder conditions, 0 to 100 °C (32 to 212 °F) and 15 to 95 °C (59 to 203 °F) in Fig. 11(b), all data points are below the main diagonal. A trendline drawn through the 0 to 100 °C (32 to 212 °F) data points in Fig. 11(b) has a slope of 0.84, which gives an average 16% life reduction under 0 to 100 °C (32 to 212 °F) thermal cycling conditions. The 15 to 95 °C (59 to 203 °F) data points are below the 0 to 100 °C (32 to 212 °F) trendline, suggesting further reliability losses under milder conditions. This raises potential reliability concerns for conventional Sn-Pb ball area-array components assembled with SAC paste.

Discussion

Creep of Sn-Pb Versus Lead-Free Solders. A cursory look at steady-state creep data (Ref 28, 34–39) for Sn-Pb and lead-free solders suggests that over a wide temperature range (−55 to 125 °C, or −67 to 257 °F) and under high enough stress, many of the common lead-free solders (except perhaps for eutectic Sn-Bi) creep at similar rates or faster than standard Sn-Pb. Figures 12 and 13 are plots of isothermal steady-state creep data in shear and tension, respectively, for near-eutectic Sn-Pb and lead-free solders of various compositions. The shear data in Fig. 12(a), 12(b), and 12(c) are for test temperatures of approximately 25, 75 and 125 °C (77, 167, and 257 °F). The tensile data in Fig. 13(a) through 13(d) are for test temperatures of approximately −55, 25, 75, and 125 °C (−67, 77, 167, and 257 °F).

From the data in Fig. 12, shear creep rates for SAC alloys of composition SAC3807 (3.8%Ag-0.7%Cu wt), or SAC3906 (3.9%Ag-0.6%Cu wt), are similar to creep rates for Sn-40Pb, starting at stress levels in the range of 10 to 20 MPa (1.45 to 2.90 ksi). For the tensile data in the temperature range 25 to 125 °C (77 to 257 °F) (Fig. 13b–d), creep rates for the lead-free solders are similar or higher than for Sn-Pb starting at a stress level of approximately 10 MPa (1.45 ksi).

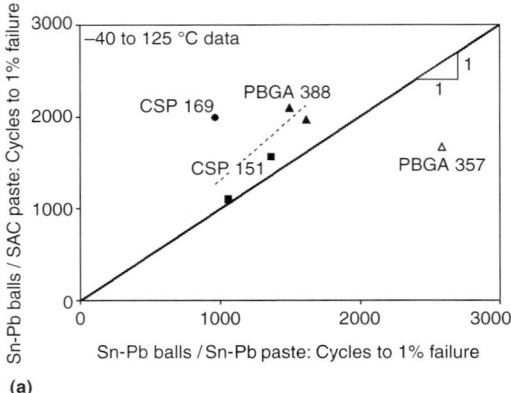

(a)

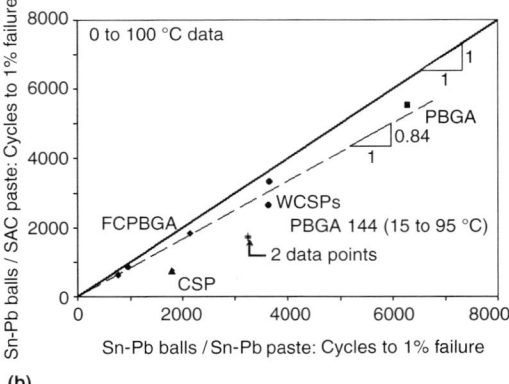

(b)

Fig. 11 Thermal cycling data for Sn-Pb ball area-array components assembled with SAC or Sn-Pb paste. (a) −40 to 125 °C (−40 to 257 °F) data. (b) 0 to 100 °C (32 to 212 °F) data (except for two 15 to 95 °C, or 59 to 203 °F PBGA 144 data points).

The transition stress is lower, at approximately 2 to 3 MPa (0.29 to 0.44 ksi), for the Sn-1.0Ag-0.75Cu alloy with reduced Ag contents. At −55 °C (−67 °F) (Fig. 13a), the creep rates for Sn-40Pb are slightly higher, but similar to creep rates for Sn-4.0Ag. The transition in relative creep rates, with similar or higher rates for common SAC alloys than for Sn-Pb under high stress conditions, can possibly explain shorter solder joint lives experienced by SAC alloys versus Sn-Pb solder under highly accelerated conditions, and large cyclic shear strains (see test life data in Fig. 3).

In Fig. 13(b) (25 °C, or 77 °F data) and 13(d) (125 °C, or 257 °F data), eutectic Sn-Bi appears to be consistently more creep resistant than eutectic Sn-Pb under high stress conditions. At lower stress, creep rates for the two eutectic alloys are very similar. The rather favorable creep resistance of eutectic Sn-Bi at all stress levels is consistent with the thermal cycling test results of the National Center for Manufacturing Sciences (NCMS) lead-free project (Ref 40) where Sn-Bi surface-mount assemblies outperformed Sn-Pb assemblies under both conditions: 0 to 100 °C and −55 to 125 °C (32 to 212 °F and −67 to 257 °F).

Life Prediction Models. The reliability test results discussed in this chapter are under accelerated conditions. For the data to be of use in gauging the reliability of actual product boards, accelerated test results need to be extrapolated to field conditions. Without acceleration factors and/or calibrated life prediction models of board level reliability under representative field conditions, product reliability claims remain unfounded. In the absence of such a bridge from test to field conditions, real products can be over-designed, at risk, or even rejected under assumptions with variable degrees of conserva-

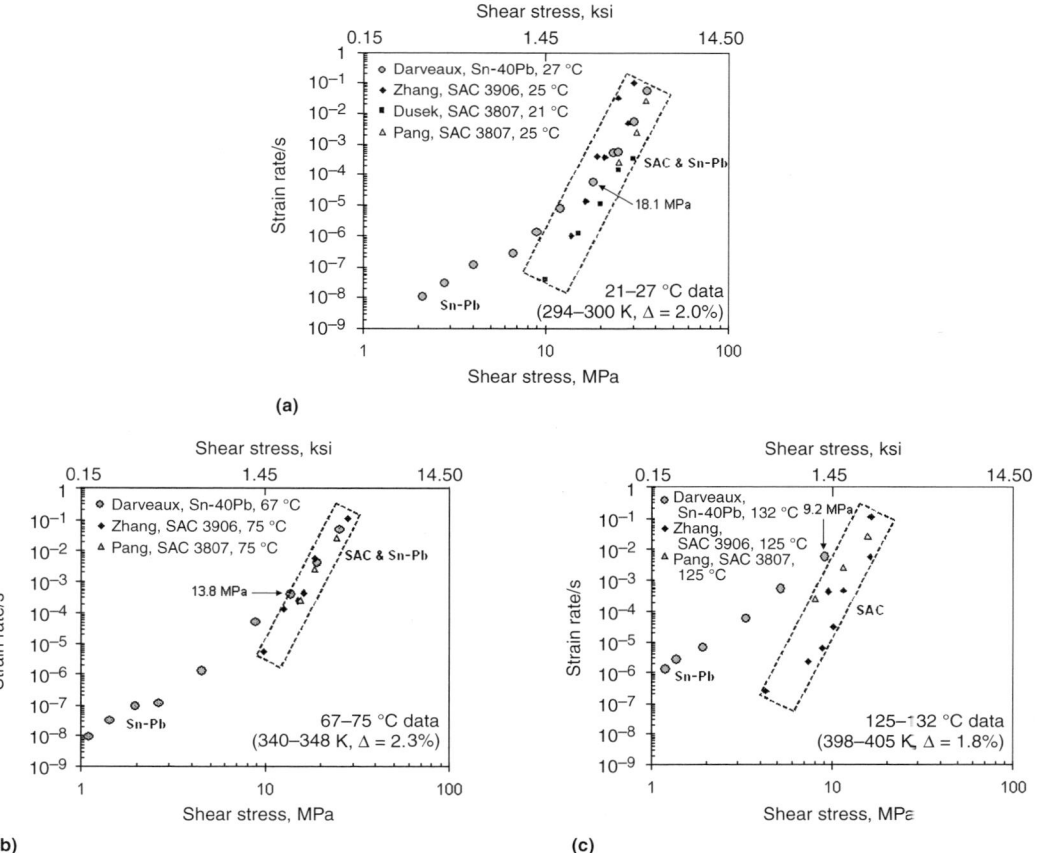

Fig. 12 Standard Sn-Pb and lead-free isothermal steady-state creep rates in shear. (a) Shear creep data at approximately 25 °C (77 °F). (b) Shear creep data at approximately 75 °C (167 °F). (c) Shear creep data at approximately 125 °C (257 °F).

tism. The latter two scenarios, high risk or rejection, are of particular concern to products in medium- to high-reliability applications.

As of this writing, several finite-element based life prediction models have been proposed (Ref 1–4, 41) for SAC type alloys without Pb contamination. Some models feature primary and secondary creep of SAC solders, others ignore primary creep. According to recent, unpublished studies by this author, tertiary creep appears to be a significant deformation stage, but has not been considered in life prediction models. Moreover, existing models have not explored the impact of modeling time-steps and element sizes on the scaling constants in life data correlations. As demonstrated in detail in Ref 42 for Sn-Pb assemblies, differences in model type (slice or three-dimensional models), time-steps and element sizes in the critical interfacial areas can lead to life predictions being off by a factor of as much as seven times. Consistency of finite-element features from one model to the next is the engineering solution to this problem.

Figures 14 and 15 show plots of finite-element based life correlations (Ref 2, 41) for SAC solders. Both use the same steady-state creep

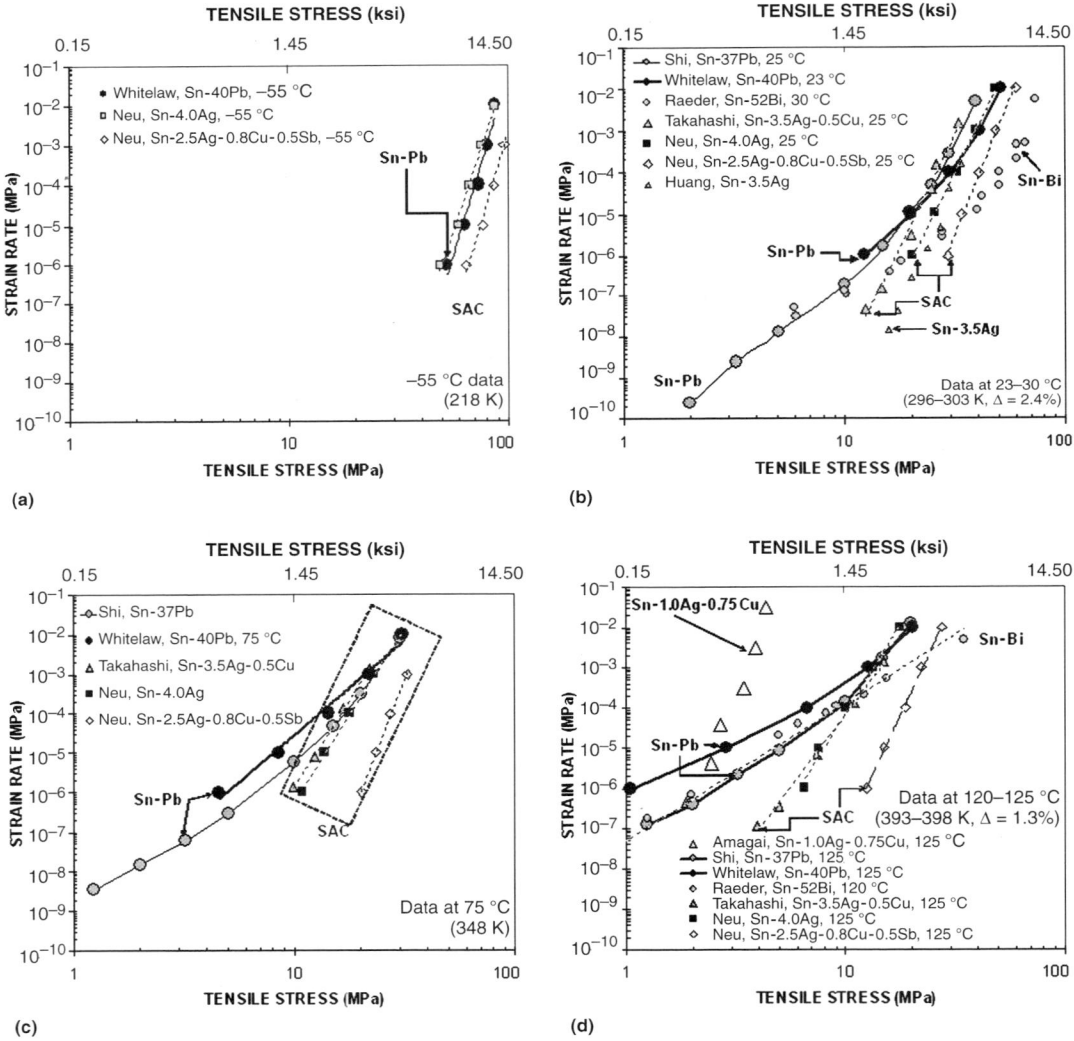

Fig. 13 Standard Sn-Pb and lead-free isothermal steady-state creep rates in tension. (a) Tensile creep data at −55 °C (−67 °F). (b) Tensile creep data at approximately 25 °C (77 °F). (c) Tensile creep data at 75 °C (167 °F). (d) Tensile creep data at approximately 125 °C (257 °F).

model, however, the model in Ref 2 accounts for primary creep as well. Reference 4 also suggests including primary creep in life prediction models for SAC solders. The life correlations in Fig. 14 and 15 are presented as cyclic life versus accumulated creep strain, ε_{acc} (Fig. 14), or accumulated creep strain energy, w_{acc} (Fig. 15). The Syed (Ref 41) life data is in terms of median life (cycles to 50% failure, $N_{50\%}$) whereas the Schubert et al. (Ref 2) failure times are given as characteristic lives (cycles to 63.2% failures, $N_{63.2\%}$). For typical values of the shape parameter β of two-parameter Weibull distributions in the range $\beta = 4$ to 12, the characteristic life is 10.3 to 4.3% above the median life, respectively. Median lives and characteristic lives are close enough to allow for a direct comparison of life correlations from the two models:

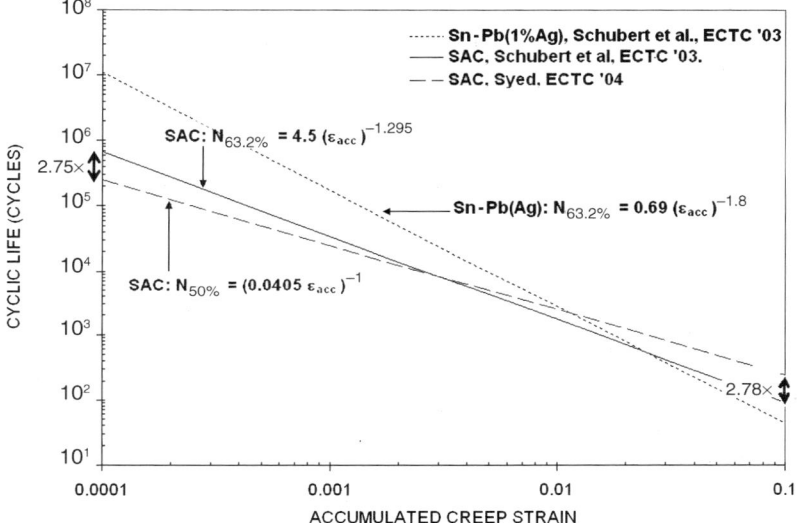

Fig. 14 Plot of SAC and Sn-Pb life versus accumulated creep strain correlations obtained by finite element analysis. Source: Ref 2, 41

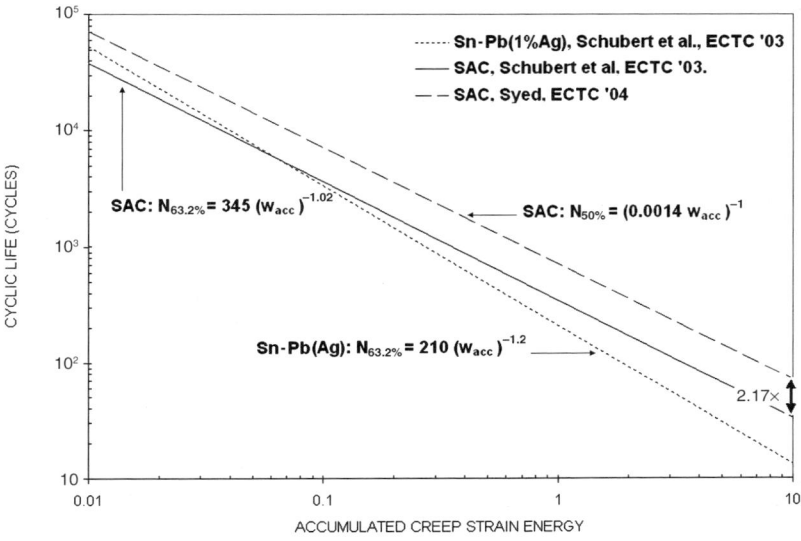

Fig. 15 Plot of SAC and Sn-Pb life versus accumulated creep strain energy correlations obtained by finite element analysis. Source: Ref 2, 41

- From Fig. 14, the SAC life correlations based on accumulated creep strains have different slopes, -1.295 versus -1. At low strains ($\varepsilon_{acc} = 0.01\%$), the Syed model is a factor 2.75 times below the Schubert, et al. model. At high strain ($\varepsilon_{acc} = 10\%$), the relative trend of the two models is opposite, and the Syed model is 2.78 times above the Schubert, et al. model. It is not clear why the slopes of the two life correlations for SAC are different but this is likely to result in different acceleration factors. Possible reasons include differences in mesh size, time-steps, as well as a possible effect of primary creep which was not included in the Syed model.
- Figure 14 also includes a life correlation proposed by Schubert, et al. for near-eutectic Sn-Pb (with 1–2% Ag). The effect of the 1–2% Ag contents on the life of standard Sn-Pb joints is small, in general. The SAC and Sn-Pb correlations by Schubert, et al. intersect at creep strains of 2 to 3% and cyclic lives of about 600 cycles. To the left of the crossover point, the Sn-Pb model lies above the SAC model. This implies that under low cyclic strain conditions, and for scenarios where dwell times are long enough to develop the maximum available strains as may occur in many applications, the life of SAC assemblies could be significantly shorter than that of Sn-Pb assemblies. This is opposite the SAC versus Sn-Pb life trends that are suggested in Fig. 3, based on a simple empirical correlation of test life results versus cyclic shear strains under thermal cycling conditions with short dwell times of 5 to 30 min. However, the trend suggested by the Schubert, et al. correlations is consistent with dwell time effects reported in Ref 43 during thermal cycling of CBGA assemblies. With dwell times of 100 to 110 min., SAC solder joint life was slightly less than Sn-Pb life under thermal cycling conditions from 0 to 100 °C (32 to 212 °F). For thermal cycling between -40 and 125 °C (-40 and 257 °F), SAC life was shorter than Sn-Pb life (by 16%) for dwell times of 7 min. and even worse for dwell times of approximately 106 min.
- The two SAC solder, strain-energy correlations in Fig. 15 have similar slopes of approximately -1. This is more encouraging than the differences discussed previously for life versus creep strain correlations. As in Sn-Pb reliability models in Ref 44 to 46, this would suggest that the cyclic, inelastic strain energy approach is more robust than the classical cyclic, inelastic strain approach. The offset between the two correlations is a factor of approximately two times. Again, this is likely due to differences in finite element mesh sizes and time-steps, whether primary creep is included or not.

A strain-energy based reliability model similar to a previously developed model for Sn-Pb assemblies (Ref 16, 44–46) is under development. In lieu of finite element analysis, the model uses strength of materials techniques to capture the deformations of substrates and components. One main item presently under investigation is the selection of an adequate constitutive model for solder in lead-free solder joints (Ref 47, 49). This is a critical part of the model development effort, given significant differences in the mechanical behavior of bulk solder and small size solder joints.

Until predictive life models become available for independent testing and validation across a wide range of conditions, uncertainties will remain as to the reliability of lead-free products. One intermediate strategy is to first estimate the time to failure of existing Sn-Pb product assemblies. When Sn-Pb life predictions are greater than the expected design life, safety factors are estimated for Sn-Pb assemblies. If those safety factors are large enough, and barring quality issues that may result in infant-mortality type of failures, chances are good that the lead-free versions of those same board assemblies will meet the product life requirements.

Accelerated Thermal Cycling Conditions. Accelerated thermal cycling of lead-free assemblies is most often conducted using temperature profiles that were developed for reliability testing of Sn-Pb assemblies. For example, a 0 to 100 °C (32 to 212 °F) cycle with 10 °C/min. (50 °F/min.) ramps and 5 to 10 min. dwells is a fairly standard test profile for commercial applications. Air-to-air or liquid-to-liquid thermal shock from -55 or -40 to 125 °C (-67 or -40 to 257 °F), with higher ramp rates and 10 to 30 min. dwells at the temperature extremes, is commonly used for harsh environment applications. While experience has shown that the just mentioned profiles are effective in providing reliable data for Sn-Pb assemblies, it is not clear whether these same conditions are optimum for SAC assemblies. Recent investigations suggest that the applicability of Sn-Pb thermal cycles to

SAC reliability testing requires careful interpretation of test results due to a stronger temperature-dependence of competing creep mechanisms for SAC solders.

The activation of creep mechanisms in SAC solders appears to be more temperature-sensitive than in Sn-Pb solders (Ref 49). Figures 16(a) and 16(b) show plots of (a) minimum creep rates versus stress (thin lines) at temperatures from -40 to 160 °C (-40 to 320 °F), and (b) creep mechanism contour lines (thick lines), for Sn-37Pb and Sn-3.8Ag-0.7Cu, respectively. The creep models that were used to plot minimum creep rates versus stress are additive, two-cell models where, for a given load and temperature, the total creep rate is the sum of creep rates for two dominant creep mechanisms: grain boundary creep (GBC) or grain boundary sliding (GBS) in the case of Sn-37Pb, and matrix creep (MC). Further details on the creep models and their validation against independent creep data are given in Ref 49. The creep mechanism contour lines in Fig. 16(a) and 16(b) give a mapping of stress and temperature conditions for which the creep rate for one of two creep mechanisms is a fixed percentage of the total creep rate. These contour lines are of use to estimate the relative importance of creep mechanisms under a given set of stress and temperature conditions.

- In Fig. 16(a), the "75% GBS, 25% MC" contour line for Sn37Pb corresponds to various combinations of stress and temperature, for which 75% of the total creep rate is due to GBS. For typical test and use conditions, with stresses less than 10 MPa (1.45 ksi) in commercial applications, the creep mechanism contour lines in Fig. 16(a) predict dominant GBS creep (more than 75% GBS and less than 25% MC) at temperatures in the range of -40 to 125 °C (-40 to 257 °F). These results are similar to those obtained from creep contour charts in a previous study of creep of Sn-36Pb-2Ag solder (Ref 50).
- The pattern of creep mechanism contours for SAC solder (Fig. 16b) is quite different from that for Sn-37Pb (Fig. 16a), a reflection of vastly different creep mechanisms. While the Sn-37Pb creep contours are somewhat orthogonal to isothermal creep rate lines (at least for stresses less than 20 MPa, or 2.90 ksi), the Sn-3.8Ag-0.7Cu contour lines form less of an angle to, or are closer to, the isothermal creep rate lines (up to approximately 20 MPa, or 2.90 ksi). This suggests that the areas of dominant creep mechanisms for SAC solders are highly temperature-dependent.
- For example, for Sn-3.8Ag-0.7Cu, the creep mechanism contour line that gives equal contribution from the two creep rates ("50% GBC, 50% MC" line) is close to the 75 °C (167 °F) creep rate line, at least up to about 25 to 30 MPa (3.63 to 4.35 ksi). This is the same temperature that Ref 51 identified as the dividing point for the analysis of steady-creep rates, fitting SAC creep data with separate models in the two temperature ranges: -25 to 75 °C (-13 to 167 °F), and 75 to

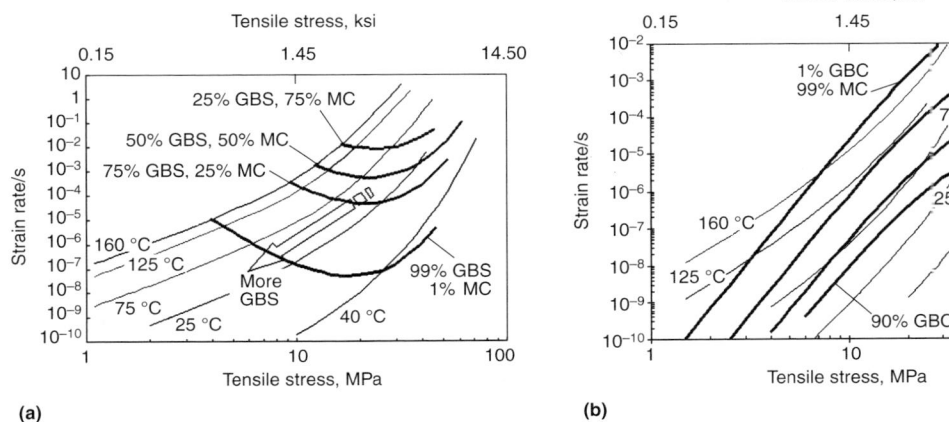

Fig. 16 Isothermal steady-state creep rates model (strain rate vs. tensile stress, thin lines) and creep mechanism contour lines (thick lines). (a) Creep rate and creep mechanism contour lines for Sn-37Pb. (b) Creep rate and creep mechanism contour lines for Sn-3.8Ag-0.7Cu. Source: Ref 49

160 °C (167 to 320 °F). For specimens that were creep-tested at or above 75 °C (167 °F), the microstructural analysis of Ref 51 revealed significant transformations within the inter-dendritic regions, namely, the formation of coarsened particle boundaries with depletion of Ag_3Sn precipitates in the vicinity of those boundaries. This mechanism was not observed in specimens tested at 25 °C (77 °F) but was accentuated in specimens tested at 160 °C (320 °F). Since this mechanism occurred in the interdentritic regions, i.e., in the dispersion-strengthened Sn-matrix, the corresponding term in our creep contour analysis is Matrix Creep (MC). In agreement with the microstructural analysis of Ref 51, the SAC creep contour lines indicate less than 10% MC at 25 °C (77 °F) (at least up to 40 MPa, or 5.80 ksi), 50% MC at 75 °C (167 °F) and more than 90% MC at 125 and 160 °C (257 and 320 °F).

- The results for creep of SAC solder suggest that the transition from one dominant creep mechanism to another occurs at a temperature within the range of common operating or accelerated testing conditions, somewhere near 75 °C (167 °F). This temperature-related transition implies that accelerated thermal cycling profiles should be designed carefully so that stress/strain cycles under test conditions accelerate the same creep mechanism(s) as in service. SAC solder joints for which the creep contours in Fig. 16(b) apply would experience one dominant creep mechanism (GBC) when thermally cycled between 25 and 75 °C (77 and 167 °F) but would experience both creep mechanisms (GBC and MC) when thermally cycled between −40 and 125 °C (−40 and 257 °F).

Brittle Fracture Under High Strain Rate Conditions. While much emphasis has been placed on thermal cycling conditions, reliability under mechanical loads (e.g., board flexure, shock, vibration) deserves as much attention. Of particular concern in the early stages of lead-free implementation is the risk of brittle fracture, or solder joint embrittlement, under high strain rate conditions. Both bulk solder and interfacial fractures have been observed during impact or drop testing, with increased risk of interfacial failures when assembled boards are subject to prolonged aging.

Figure 17 is a plot of fracture energy versus temperature for Sn-5.0Ag solder and a high lead solder, 93.5Pb-5.0Sn-1.5Ag. The data was obtained by Charpy testing of bulk solder specimens (Ref 53), whereby a swinging pendulum

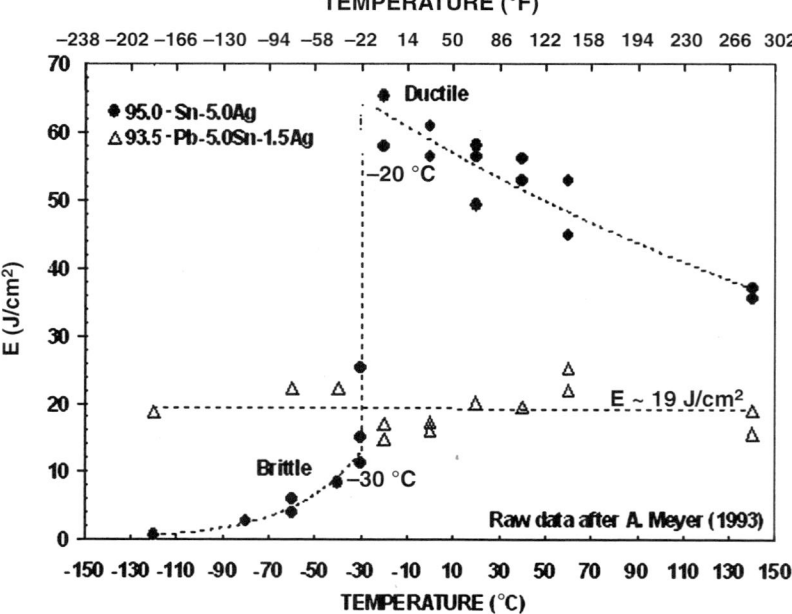

Fig. 17 Charpy test results for bulk Sn-5.0Ag and 93.5Pb-5.0Sn-1.5Ag solders: fracture energy vs. temperature data. Source: Ref 53

impacts and fractures a notched specimen under high strain rate conditions. The fracture energy data for Sn-5.0Ag displays a sharp, ductile-to-brittle transition at approximately −20 to −30 °C (−4 to −22 °F). At or above −20 °C (−4 °F), fractures were characterized as ductile, whereas at −30 °C (−22 °F) or colder temperatures, fractures were of a brittle nature. The high-lead solder did not display such a transition. Sn-3.5Ag has a ductile-to-brittle transition temperature of approximately −20 to −30 °C (−4 to −22 °F), which may be of concern for some real life applications in cold environments. As illustrated by a car manufacturer, such a transition temperature would be undesirable for door-mounted electronics, as in the case of vehicle doors being slammed closed in the harsh winters in northern regions of the world. Given that the ductile-to-brittle transition temperature of Sn-5.0Ag is −30 to −20 °C (−22 to −4 °F), that of SAC solders with 3 to 4% Ag contents is possibly near the same temperature range. To this author's knowledge, no such characterization has been conducted as of yet for SAC solders.

Other studies have reported a loss of impact toughness (Ref 54) or a degradation of drop test resistance (Ref 55) of SAC solder joints after thermal aging. Reference 55 reported an 80% degradation of drop test performance for boards with SAC balls on bare copper pads, following thermal aging of assembled boards at 125 °C (257 °F) for 10 days. Brittle interfacial failures occurred between copper pad and Cu_6Sn_5 intermetallic layers. Interfacial strength was reduced due to the brittleness of thicker intermetallic layers, but mostly to the formation of interfacial voids (possibly Kirkendall voids) during the aging treatment. While some have hypothesized impurities from copper plating baths as a potential source of pad contamination, the root cause of the problem is not well understood and the associated risks are not predictable. Further studies are warranted to elucidate the true root cause of solder joint embrittlement problems (Ref 56).

Conclusions

This chapter pulled together independent test results in an attempt to put large amounts of apparently unrelated data in perspective, and to spot lead-free reliability trends and possible areas of concern for lead-free or mixed-assembly circuit boards:

- First-order test life correlations (cycles to failure versus applied shear strains) were developed for lead-free and Sn-Pb assemblies. The empirical correlations show different slopes for SAC, Sn-0.7Cu and standard Sn-Pb assemblies. The crossover points of the various trendlines suggest that the rank-ordering of solder joint lives for the different alloys change with the applied stress or strain level. The correlations that were shown are purely empirical and should not be used to calculate acceleration factors or for life prediction purposes.
- The life test correlations show a scarcity of data for cyclic shear strains below 1%. For the most part, accelerated testing has been conducted under high stress conditions with the goal of obtaining rapid results for alloy comparison purposes. When SAC test results are inferior to Sn-Pb test results, the higher slope of SAC fatigue life correlations suggests that SAC assemblies may still outperform Sn-Pb assemblies under mild enough thermal cycling conditions. There is a definite lack of reliability test data under mild conditions, which may slow down the development and/or validation of robust life prediction tools or acceleration factors.
- The life data correlations and the analysis of mixed assembly test results did not show any clear or definite effect of common board finishes (immersion Ag, electroless Ni-Au or OSP) on solder joint reliability.
- Backward-compatibility test results over a wide range of test cycles are mixed. Area-array assemblies using SAC balls and Sn-Pb paste require a careful reliability assessment. While some studies suggest reliability levels comparable to that of conventional assemblies when the peak reflow temperature is above approximately 225 °C (437 °F), serious quality and reliability concerns arise for boards assembled with conventional Sn-Pb profiles (with typical peak reflow temperatures in the range of 205 to 225 °C, or 401 to 437 °F).
- Forward-compatibility test results are also mixed with a strong effect of thermal cycling conditions on cyclic life trends. Under conditions of 0 to 100 °C (32 to 212 °F), or even milder, area-array assemblies using Sn-Pb balls and SAC paste appear less reliable than conventional Sn-Pb area-array assemblies.

Table 1 Metrics of solder reliability studies

Metric	Sn-Pb reliability	SAC reliability	Ratio ("SAC" to "Sn-Pb" column)
Author's reference list	~2500 publications	~250 publications	10%
Years of industry experience	~50 years	~12 years	24%

- Creep data for a variety of SAC solders shows that SAC is more creep resistant than Sn-Pb under low- to medium-stress levels. Under high enough stresses, as may be encountered under accelerated test conditions, SAC creep rates are similar to Sn-Pb creep rates, and even higher for some SAC compositions.
- The comparison of finite element-based life prediction models suggest that the inelastic strain energy approach looks promising and may be more robust than the cyclic inelastic strain approach. However, the various models can be offset from each other depending on element sizes, time-steps and whether primary and tertiary creep of solders are accounted for or not.
- Last, reliability under mechanical loading deserves more attention, especially under high strain rate conditions where risks for solder joint embrittlement, in the bulk or across intermetallic layers, have been identified.

While we tried to analyze test failure cycles that cover a wide range of conditions, components, and board finishes, the analysis herein is far from exhaustive. In future work, we intend to add new data as to further exploit the correlations presented in this study and to possibly identify new parametric trends. Using two simple metrics to assess progress in SAC or lead-free reliability studies, our estimate is that the industry know-how on SAC assembly reliability is 10 to 24% up the learning curve compared to the Sn-Pb reliability knowledge base. This rough estimate is based on soft data given in Table 1:

- The number of publications in this author's reliability reference lists for standard Sn-Pb and lead-free or SAC assemblies
- The approximate number of years of industry experience: about 50 years for Sn-Pb assembly versus about 12 years for lead-free or SAC, using a starting date of 1992 for the first lead-free consortium project in North America, under the auspices of the National Center for Manufacturing Sciences (NCMS) (Ref 34).

Obviously, many lead-free reliability issues were not addressed in this chapter and much development work lies ahead for the industry's understanding of SAC reliability to come up to par with that of standard Sn-Pb solder.

REFERENCES

1. A. Schubert, R. Dudek, R. Döring, H. Walter, E. Auerswald, A. Gollhardt, B. Schuch, H. Sitzmann, and B. Michel, Lead-Free Solder Interconnects: Characterization, Testing and Reliability, 3rd *International Conference on Benefiting from Thermal and Mechanical Simulation in Microelectronics Proc.,* 2002, p 62–72
2. A. Schubert, R. Dudek, E. Auerswald, A. Gollhardt, B. Michel, and H. Reichl, Fatigue Life Models for SnAgCu and SnPb Solder Joints Evaluated by Experiments and Simulation, *IEEE 53rd Electronic Components and Technology Conference Proc.,* 2003 (CD-ROM)
3. J. Lau, W. Dauksher, J. Smetana, R. Horsley, D. Shangguan, T. Castello, I. Memis, D. Love, and B. Sullivan, HDPUG's Design for Lead-Free Solder Joint Reliability of High Density Packages *IPC SMEMA Council APEX 2003 Proc.,* Paper 5-42-2, 2003 (CD-ROM)
4. Q. Zhang, A. Dasgupta, and P. Haswell, Viscoplastic Constitutive Properties and Energy-Partitioning Model of Lead-Free Sn3.9Ag0.6Cu Solder Alloy, *IEEE 53rd Electronic Components and Technology Conference Proc.,* 2003 (CD-ROM)
5. K.C. Norris and A.H. Landzberg, Reliability of Controlled Collapse Interconnections, *IBM J. Res. Dev.,* May 1969, p 266–271
6. S.S. Manson, Behavior of Material Under Stress and Strain Cycling, *Thermal Stress and Low Cycle Fatigue,* R.E. Krieger Publishing Co., 1981, p 132–133
7. M. Dušek, J. Nottay, and C. Hunt, "Compatibility of Lead-Free Solders with PCB Materials," National Physical Laboratory,

U.K., NPL Report, MATC (A) 89, Aug 2001
8. T. Woodrow, Reliability and Leachate Testing of Lead-Free Solder Joints, *IPC Lead-Free Conference Proc.,* May 2002 (CD-ROM)
9. G. Swan, A. Woosley, N. Vo, and T. Koschmieder, Development of Lead-Free Peripheral Leaded and PBGA Components to Meet MSL3 at 260 °C Peak Reflow Profile, *IPC SMEMA Council Proc.,* APEX, Paper LF2-6, 2001
10. M. Farooq, L. Goldmann, G. Martin, C. Goldsmith, and C. Bergeron, Thermo-Mechanical Reliability of Pb-Free Ceramic Ball Grid Arrays: Experimental Data and Lifetime Prediction Modeling, *IEEE 53rd Electronic Components and Technology Conference Proc.,* 2003 (CD-ROM)
11. C. Handwerker, T. Siewert, J. Bath, E. Benedetto, E. Bradley, R. Gedney, J. Sohn, and P. Snugovsky, NEMI Lead-Free Assembly Project: Comparison Between PbSn and SnAgCu Reliability and Microstructures, *SMTA International Conference Proc.,* 2003 (CD-ROM)
12. C. Handwerker, NEMI Lead-Free Solder Projects: Progress and Results, *IPC/JEDEC 4th International Conference on Lead-Free Electronic Assemblies and Components Proc.,* 2003, p 26–39
13. T. Woodrow, The Effects of Trace Amounts of Lead on the Reliability of Six Lead-Free Solders, *IPC Lead-Free Conference Proc.,* 2003 (CD-ROM)
14. S.O. Dunford, A. Primavera, and M. Meilunas, "Microstructural Evolution and Damage Mechanisms in Pb-Free Solder Joints During Extended −40 °C to 125 °C Thermal Cycles," *IPC Conf. Proc.,* 2002, paper S08-4
15. M. Meilunas, A. Primavera, and S.O. Dunford, "Reliability and Failure Analysis of Lead-Free Solder Joints," *IPC Conf. Proc.,* 2002
16. J.-P. Clech, "Solder Reliability Solutions: a PC-Based Design-for-Reliability Tool," *Surface Mount International Conf. Proc.,* p 136–151. Republished in *Sold. Surf. Mt. Technol.,* Vol 9, No. 2, July 1997, p 45–54
17. J-P. Clech, J.C. Manock, D.M. Noctor, F.E. Bader, and J.A. Augis, A Comprehensive Surface Mount Reliability Model: Background, Validation and Applications, *Surface Mount International Proc.* Vol I, 1993, p 363–375
18. D-H. Kim, P. Elenius, and S. Barrett, Solder Joint Reliability and Characteristics of Deformation and Crack Growth of Sn-Ag-Cu Versus Eutectic Sn-Pb on a WLP in a Thermal Cycling Test, *IEEE Transactions on Electronics Packaging Manufacturing Proc.,* Vol 25 (No. 2), April 2002, p 84–90
19. L. Wetz, B. Keser, and J. White, Design and Reliability of New WL-CSP, *IMAPS Conference 2002, 35th International Symposium on Microelectronics Proc.,* 2002 (CD-ROM)
20. C. Zhang, J-K. Lin, and L. Li, Thermal Fatigue Properties of Lead-Free Solders on Cu and Ni-P Under Bump Metallurgies, *IEEE 51st Electronic Components and Technology Conf. Proc.,* 2001 (CD-ROM)
21. ITRI, *Mechanical Properties of Solders and Solder Joints,* International Tin Research Institute, publication no. 656.
22. G. Swan, A. Woosley, N. Vo, and T. Koschmieder, Development of Lead-Free Peripheral Leaded and PBGA Components to Meet MSL3 at 260 °C Peak Reflow Profile, *IPC/SMEMA Council APEX Proc.,* 2001, paper LF2-6
23. *Solder Joint Reliability Evaluation* (TSOP 48)—http://edevice.fujitsu.com/fj/CATALOG/AD81/81-00004/index_e html
24. D. Noctor, F.E. Bader, A.P. Vera, P. Boysan, S. Golwalkar, and R. Foehringer, Solder Joint Attachment Reliability Evaluation and Failure Analysis of Thin Small Outline Packages (TSOPs) with Alloy 42 Leads, *IEEE Transactions on Components, Hybrids and Manufacturing Technology,* Vol 16, No. 8, Dec 1993, p 961–971
25. A. Schubert, R. Dudek, H. Walter, E. Jung, A. Gollhart, B. Michel, and H. Reichl, Reliability Assessment of Flip-Chip Assemblies with Lead-Free Solder Joints, *52nd Electronic Components and Technology Conf. Proc.,* 2002 (CD-ROM)
26. Z. Hou, G. Tian, C. Hatcher, R.W. Johnson, E.K. Yaeger, M.M. Konarski, and L. Crane, Lead-Free Solder Flip Chip-on-Laminate Assembly and Reliability, *IEEE Transactions on Electronics Packaging Manufacturing,* Vol 24 (No. 4), Oct 2001, p 282–292
27. Z. Hou, G. Tian, C. Hatcher, R.W. Johnson, E. Yaeger, M. Konarski, and L. Crane, Assembly & Reliability of Flip Chip-on-Laminate with Lead Free Solder, *Advancing Microelectronics,* IMAPS, Vol 29 (No. 2), March/April 2002, p 7–12

28. M. Amagai, M. Watanabe, M. Omiya, K. Kishimoto, and T. Shibuya, Mechanical Characterization of Sn-Ag-Based Lead-Free Solders, *Microelectronics Reliability,* Vol 42 (Issue 6), June 2002, p 951–966
29. P. Chalco, Solder Fatigue Reliability Issues in Lead-Free BGA Packages, *SMTA Pan Pacific Microelectronics Symposium Proc.,* 2002, p 163–168
30. F. Hua, R. Aspandiar, C. Anderson, G. Clemons, C-K. Chung, and M. Faizul, Solder Joint Reliability Assessment for Sn-Ag-Cu BGA Components Attached with Eutectic Pb-Sn Solder, *SMTA International Conf. Proc.,* 2003 (CD-ROM)
31. D. Nelson, H. Pallavicini, Q. Zhang, P. Friesen, and A. Dasgupta, Manufacturing and Reliability of Pb-Free and Mixed System Assemblies (SnPb/Pb-Free) in Avionics Environments, *SMTA International Conf. Proc.,* 2003 (CD-ROM)
32. P. Chalco, and E. Blackshear, Reliability Issues of BGA Packages Attached with Lead-Free Solder, *The Pacific Rim/ASME International Electronic Packaging Technical Conf. Proc.,* 2001 (CD-ROM)
33. P. Roubaud, G. Henshall, R. Bullwith, S. Prasad, F. Carson, S. Kamath, and E. O'Keefe, Thermal Fatigue Resistance of Pb-Free Second Level Interconnect, *SMTA International Conf. Proc.* 2001, p 803–809
34. M.L. Huang, and L. Wang, Creep Behavior of Eutectic Sn-Ag Lead-Free Solder Alloy, *J. Mater. Res.,* Vol 17, No. 11, November 2002, p 2897–2903
35. R.W. Neu, D.T. Scott, and M.W. Woodmansee, Thermomechanical Behavior of 96Sn-4Ag and Castin Alloy, *ASME Transactions, J. Electron. Packaging,* Vol 123 (No.3), Sept 2001, p 238–246
36. C.H. Raeder, Thermomechanical Deformation Behavior of Eutectic Sn-Bi Solder, *Weld. J.,* American Welding Society, 1995
37. X.Q. Shi, Z.P. Wang, W. Zhou, H.L.J. Pang, and Q.J. Yang, A New Constitutive Creep Constitutive Model for Eutectic Solder Alloy, *ASME Transactions, J. Electron. Packaging,* Vol 124, June 2002, p 85–90
38. H. Takahashi, T. Kawakami, M. Mukai, I. Mori, K. Tateyama, and N. Ohno, Thermal Fatigue Life Simulation for Sn-Ag-Cu Lead-Free Solder Joints, *International Conf. on Electronics Packaging ICEP Proc.,* Tokyo, Japan, 2003, p 215–220
39. R.S. Whitelaw, R.W. Neu, and D.T. Scott, Deformation Behavior of Two Lead-Free Solders: Indalloy 227 and Castin Alloy, *ASME Transactions, J. Electron. Packaging,* Vol 121 (No. 2), June 1999, p 99–107
40. NCMS, "Lead-Free Solder Project Final Report" Report 0401RE96, National Center for Manufacturing Sciences, Aug 1997 (CD-ROM)
41. A. Syed, Accumulated Creep Strain and Energy Density-Based Thermal Fatigue Life Prediction Models for SnAgCu Solder Joints, *54th Electronic Components and Technology Conf. Proc.,* June 2004, p 734–746
42. R. Darveaux, Effect of Simulation Methodology on Solder Joint Crack Growth Correlation and Fatigue Life Predictions, *ASME Transactions, J. Electron. Packaging,* Vol 124 (No. 3), Sept 2002, p 147–154
43. J. Bartelo, S.R. Cain, D. Caletka, K. Darbha, T. Gosselin, D.W. Henderson, D. King, K. Knadle, A. Sarkhel, G. Thiel, and C. Woychik, Thermomechanical Fatigue Behavior of Selected Lead-Free Solders, *IPC SMEMA Council Proc. APEX,* paper LF2-2, 2001
44. J.-P. Clech, Solder Reliability Solutions: From LCCCs to Area Array Assemblies, *Nepcon West Proc.,* 1996, p 1665–1680
45. J.-P. Clech, Flip-Chip/CSP Assembly Reliability and Solder Volume Effects, *Surface Mount International Conf. Proc.,* 1998, p 315–324
46. J.-P. Clech, Solder Joint Reliability of CSP Versus BGA Assemblies, *SMT ESS & Hybrids Conf. Proc.,* 2000, p 19–28
47. J.-P. Clech, Comparative Analysis of Creep Data for Sn-Ag-Cu Solder Joints in Shear, *Micromaterials and Nanomaterials* (Issue 3) IZM/MMCB Fraunhofer Institute, Berlin, Germany), 2004, p 144–155
48. J.-P. Clech, Review and Analysis of Lead-Free Solder Material Properties, to appear in IEEE/Wiley NEMI's Lead-Free Project, 2004. On-line version can be viewed at: http://www.metallurgy.nist.gov/solder/clech/
49. J.-P. Clech, An Obstacle-Controlled Creep Model for Sn-Pb and Sn-Based Lead-Free Solders, *SMTAI Conf. Proc.,* Chicago, IL, 2004
50. G. Grossmann, G. Nicoletti, and U. Soler, Results of Comparative Reliability Tests on Lead-Free Solder Alloys, *52nd Electronic*

Components and Technology Conf. Proc., 2002

51. P. Vianco, Creep Behavior of the Ternary 95.5Sn-3.9Ag-0.6Cu Solder: Part I—As-Cast Condition, *J. Electron. Mater.,* 2004

52. D. Shangguan, A Holistic Approach to Lead-Free Transition and Environmental Compliance, *SMTAI Conf. Proc.,* Chicago, IL, 2004

53. A. Meyer, Investigations Physico-Chimiques Préliminaires sur les Alliages de Brasage sans Plomb, *Conférence BRASAGE 93,* Lannion, France, 1993, p 337–359

54. M. Date, T. Shoji, M. Fujiyoshi, K. Sato, and K.N. Tu, Impact Reliability of Solder Joints, *54th Electronic Components and Technology Conf. Proc.,* 2004, p 668–674

55. T-C. Chiu, K. Zeng, R. Stierman, D. Edwards, and K. Ano, Effect of Thermal Aging on Board Level Drop Reliability for Pb-Free BGA Packages, *54th Electronic Components and Technology Conf. Proc.,* 2004, p 1256–1262

56. P. Borgesen, "Sn-Ag-Cu Solder Joint Fragility," web presentation, Universal Instruments Corporation, Sept 2004

CHAPTER 5

Chemical Interactions and Reliability Testing for Lead-Free Solder Interconnects

Laura J. Turbini, Centre for Microelectronics Assembly and Packaging, Canada

Introduction

The chemical and material interactions related to soldering and assembly processes are many and varied. At preheat and soldering temperatures, the solder flux or paste reacts with the metallization on the printed wiring board (PWB) and component leads to remove oxide and surface contamination, in order to prepare the surface for good metallurgical bonding with the molten solder. Residues created by the flux contain metal salts, as well as organic and inorganic materials. The soldering temperature will affect the nature of these residues.

Molten solder is above the glass transition temperature of the PWB, such that the polymer component of the composite will soften, allowing diffusion of some flux residues into the board. In the early 1980s, it was demonstrated that water-soluble fluxes containing polyethylene glycol had a significant negative effect on the electrical properties (Ref 1). The presence of this chemical in the board was proven by extracting it with acetonitrile (Ref 2). This dissolution of the polyethylene oxide in the epoxy backbone was explained as an example of "like dissolves like" (Ref 1).

In the 1990s, ionic residues were extracted from processed PWBs by soaking them in a solution of 75 vol% isopropanol and 25 vol% deionized water at 80 °C (176 °F) for 1 h (Ref 3). Ion chromatography (IC) was used to characterize the residues and demonstrated that hot air solder leveling fluxes containing HBr in their formulations left high levels (30 µg/cm^2) Br$^-$ in the epoxy-glass board, whereas PWBs that were not processed with HASL fluid had less than 1 µg/cm^2 Br$^-$.

The thermal and chemical interactions described above are enhanced at the higher soldering temperatures required by lead-free solder alloys. The rate of diffusion of contaminants into the epoxy matrix during the soldering process is increased. These contaminants can lead to field failures for electronics. Thus, it is important to provide a review of soldering fluxes, electrochemical migration, and the resulting failure modes that can occur and might be exacerbated by the higher processing temperatures. In addition, solder flux/paste residues can degrade high speed circuits in the RF and microwave frequencies, and an understanding of these issues is important. These effects are also covered in this chapter.

Background on Solder Flux Chemistry

Soldering (Ref 4) is defined as the process of joining metallic surfaces with solder without the melting of the basis metal. In order for this joining to take place, the metal surfaces must be clean of contamination and oxidation. This cleaning action is performed by the flux (Ref 2), a chemically active compound which, when heated, removes minor surface oxidation, minimizes oxidation of the basis metal, and promotes the formation of an intermetallic layer between solder and basis metal. Soldering fluxes contain several components:

- The *vehicle* is a solid or non-volatile liquid that coats the surface to be soldered, dissolves the metal salts formed in the reaction of the activators with the surface metal oxides, and ideally provides a heat transfer medium between the solder and the components or PWB substrate.
- The *solvent* serves to dissolve the vehicle, activators, and other additives. It is normally an alcohol, glycol or glycol ether, and may also be water.
- *Activators* are present in the flux formulation to enhance the removal of metal oxide from the surfaces to be soldered. They are reactive even at room temperature, but their activity is enhanced as the temperature is raised during the preheat step of the soldering process.
- Soldering fluxes often contain small amounts of other constituents that serve a specialized function. For example, a surfactant may be added to enhance the wetting properties. Solder paste formulations require the presence of additives to provide good viscosity or flow characteristics, low slump during the preheat step, and good tack characteristics for holding the component in place until reflow occurs. Finally, cored-wire flux used for hand soldering contains a plasticizer to harden the flux ingredients that are present in the core of the wire.

It is important to note that lead-free soldering requires higher processing temperatures. Thus, it is anticipated that flux chemistries are changing to deal with this fact.

Rosin/Resin Flux. Traditionally, rosin has been used as the vehicle in soldering fluxes for electronic applications. Rosin, a naturally occurring resin from the sap of the pine tree, contains a mixture of resin acids, predominantly abietic and pumeric (Fig. 1). Rosin dissolves the metal salts produced by the reaction of the flux activators in removing surface oxides, and then solidifies when cooled, entrapping, and for the most part immobilizing, the contaminants. Replacing rosin with a synthetic resin, pentaerythritol tetrabenzoate, whose composition would be fixed, was proposed in 1979 (Ref 5). This would act as the flux vehicle. Both rosin and resin fluxes leave insulating residues that are usually benign from a corrosion standpoint. However, the residue may impede bed-of-nails electrical testing. Thus, boards processed with these fluxes are normally cleaned.

Water Soluble Flux. Water-soluble fluxes have been called "organic acid" fluxes. The name is misleading since all fluxes used for electronic soldering contain organic ingredients and many contain organic acid activators. Activators for these fluxes include organic acids, halide-containing salts, and amines. Polyglycols, polyglycol surfactants, and glycerin are among the major chemicals used as the flux vehicle. Fluxes in the water soluble category are soluble in water and their residues should also be soluble in water. These fluxes are much more active than rosin fluxes and have higher soldering yield, a broader processing window, and create less defects. However, water-soluble fluxes contain corrosive residues that, if not properly removed, will cause corrosion of electronics in the field and long-term reliability problems. A proper aqueous cleaning process must follow the soldering step.

Low Solids Flux. Until the mid-1980s, liquid soldering fluxes were formulated in 25 to 35 wt% solids or non-volatile liquid. Then flux chemistries changed and new formulations,

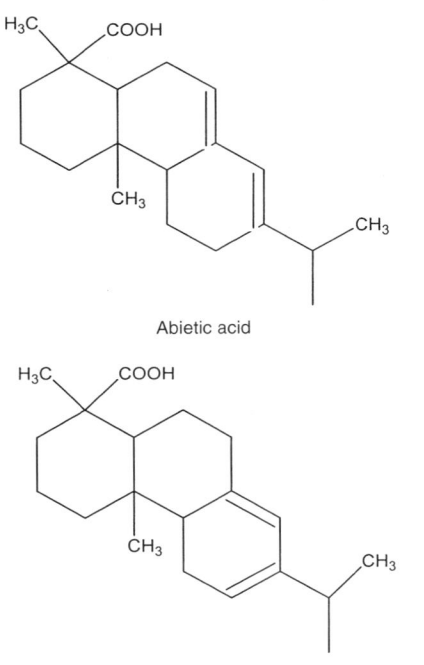

Fig. 1 Molecular structures of abietic and pimaric acid, primary constituents in rosin

which were lower in total solids content, came on the scene. These fluxes are composed principally of weak organic acids and frequently contain a small amount of resin or rosin. Early formulations had 5 to 8% solids, but today's low solid fluxes are 1 to 2% solids composition. Nomenclature of these fluxes varies from "low solids flux" to "low residue flux" to "no clean flux." The solvent used for these fluxes may be alcohol or water.

Electrochemical Migration

A basic electronic assembly consists of a series of circuit elements connected by conductor traces. The substrate material is chosen to serve as an insulator, preventing current flow between adjacent traces. This resistance to current flow is termed "insulation resistance."

Ohm's law describes the linear relationship between voltage, current, and resistance:

$$V = IR$$

where V = the voltage drop; I = the current in amps, R = the resistance in ohms.

Resistance is not an intrinsic property of a material. Rather it depends on geometric factors, as well as the intrinsic resistivity of the substrate. For surface resistance,

$$R = \rho \left(\frac{d}{A}\right)$$

where ρ = resistivity of the material; d = the separation between the conductor traces; A = the cross-sectional area.

Conductivity (σ) is the inverse of resistivity.

$$\sigma = \frac{1}{\rho}$$

Both resistivity and conductivity can be affected by processing chemicals.

Electrochemical migration (Fig. 2) is the movement of an ionic species under the influence of a DC voltage. It can lead to electrical failure if (a) a short occurs due to surface dendrites or (b) an open circuit condition occurs due to depletion of the anode.

Electrochemical corrosion increases as circuit spacing decreases. This is due to the increase in the electric field, which is inversely proportional to the spacing between the conductors:

$$E = \frac{V}{d}$$

where E is the electric field; V is the voltage; d is the spacing between the conductors.

Moisture is essential to electrochemical migration. In the presence of moisture, metal ions may form at the anode and migrate toward the cathode where they plate out forming dendrites. When a dendrite bridges the gap, a short occurs, which may burn out quickly due to the high current within the fragile dendrite.

Ionic contaminants that dissolve in the moisture on the surface of the board, increase the conductance (lower the resistance) of the insulating layer between the anode and cathode. Their presence will enhance electrochemical migration, where the amount of enhancement will

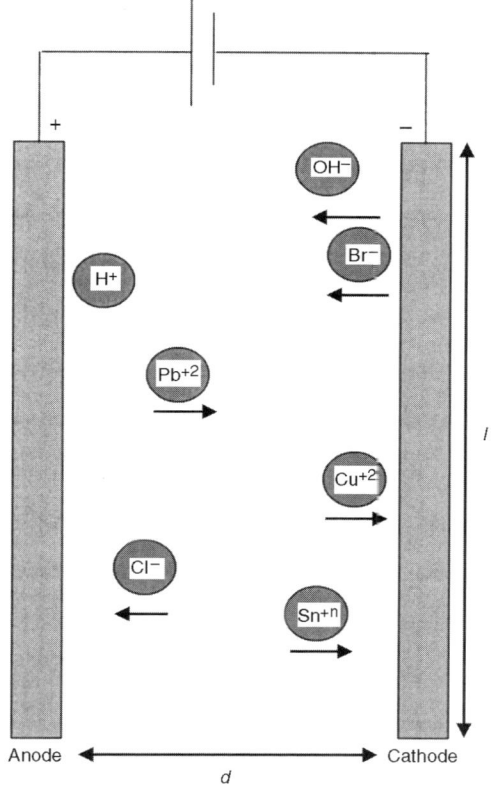

Fig. 2 Electrochemical migration occurs under a bias voltage. Anions migrate toward the anode (positive electrode) while cations migrate toward the cathode (negative electrode) where they are reduced, forming metal dendrites.

depend on a number of factors: (a) the solubility of the ion, (b) the mobility of the ion, (c) the effect of pH on solubility, (d) the reactivity of the ion, (e) temperature, and (f) relative humidity.

The solubility of the ion and its mobility in water are important because conductivity depends upon both of these factors:

$$\sigma = \Sigma N_i q_i \mu_i$$

where N_i = the number of charge carriers of a particular type (e.g., Na^+ cations, or Cl^- anions); q_i = the charge of the carrier; μ_i = the mobility of the charge carrier.

The mobility of a given ion is dependent upon temperature and its diffusion coefficient per the following equations:

$$\mu_{ion} = \left(\frac{q}{kT}\right) D_{ion}$$

where k = Boltzmann's constant; T = degrees Kelvin; D_{ion} = diffusion constant for a particular ion.

The actual diffusion rate for an ion is temperature dependent per:

$$D_{ion} = D_0 \exp \frac{-E_a}{kT}$$

Tables of ionic conductance in aqueous solution, as well as diffusion constants, can be found in the *Handbook of Chemistry and Physics* (Ref 6).

Most contaminants from processing do not exist in isolation from one another. Rather, there may be several ionic species present, some of which will interact creating new ionic species. A simple example of this is the reaction of moisture with copper. Normally the formation of the Cu^{2+} ion is favored. However, in the presence of chloride, Cu^+ ions are favored and complexes such as $(CuCl_2)^-$ are formed (Ref 7).

Solubility of the contaminant salts will also depend on the pH. Water ionizes under a bias voltage creating an acidic medium at the anode and a basic medium at the cathode, as shown in the following reactions:

Anode: $H_2O = \frac{1}{2}O_2 + 2H^+ + 2e^-$

Cathode: $2H_2O + 2e^- = 2OH^- + H_2$ or
$O_2 + H_2O + 4e^- = 4(OH)^-$

Temperature is another important factor to consider. An increase in temperature increases ion solubility, mobility, and reactivity.

In summary, there are a number of factors that affect electrochemical migration. These include the nature of substrate and metallization, the presence of contaminants, the voltage gradient, and the presence of sufficient moisture. This last element is key, for without sufficient monolayers of water on the surface, ion mobility is impossible.

In electrochemical migration, failure can occur due to dendrite growth, open circuit short, or conductive anodic filament (CAF) formation. Dendrites can form on the surface due to contamination left by the solder flux (paste) or other residues. Under a bias voltage, the metal at the anode goes into solution, migrates toward, and plates-out at the cathode. The nature of the dendrite will depend upon the surface metallization. The following oxidation reactions can occur at the anode:

$$Cu \Rightarrow Cu^{n+} + ne^-$$

$$Pb \Rightarrow Pb^{2+} + 2e^-$$

$$Sn \Rightarrow Sn^{n+} + ne^-$$

Examples of dendrites include (a) copper dendrite (Fig. 3) and (b) lead needle-like dendrites with faceted tin oxide precipitate (Fig. 4). An open circuit short can result when a segment of the anode has completely eroded as shown in Fig. 5. Green and/or blue corrosion products indicate copper corrosion.

Surface Insulation Resistance (SIR)

Insulation resistance measurements are important test data used to characterize printed wir-

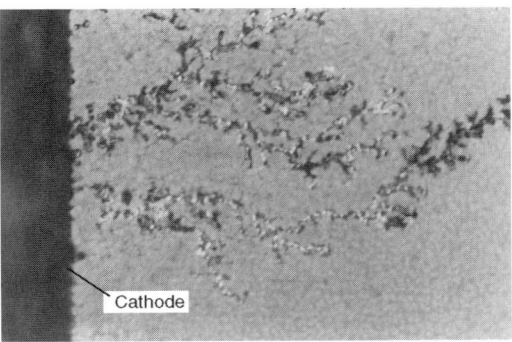

Fig. 3 Copper dendrite grows from the cathode (left).

ing board laminates, soldering fluxes, solder masks, and conformal coating. Data of this type have been collected to study the effect of aging a sample at accelerated conditions (temperature, humidity, and/or bias voltage) to determine if there are any detrimental effects to the insulation resistance.

Insulation resistance is the integrated effect of both volume and surface resistivity according to the ASTM definition (Ref 8). The measured bulk resistance will depend upon the nature of the laminate, solder mask or conformal coating material under investigation, and the degree of cure. It will also be affected by soldering flux/paste residues if they dissolve into the polymeric material during the soldering process, depending upon the nature of any surface contamination and the amount of moisture present during the measurement.

Although the SIR readings are a combination of both bulk and surface resistance, for the case of FR4 epoxy glass laminate, 99.9% of the current leakage will occur on the surface of the laminate (the ratio of the surface resistivity to the volume resistivity is $1/10^3$ (Ref 9) Test patterns for measuring volume resistance will either use electrodes on the top and bottom of the substrate (for Z axis measurements) or hole-to-hole patterns for measuring the XY resistance. Test patterns for surface resistance are typically interdigitated comb patterns such as the IPC-B-24 coupon (Fig. 6).

SIR Test Procedures

The most commonly used test vehicle for measuring SIR is an interdigitated comb pattern. It can exist in a variety of configurations with spacing between conductors ranging from 0.15 to 1.25 mm (0.006 to 0.05 in.). SIR tests performed with a bias voltage applied during the duration of the test are also electrochemical migration tests. Bias voltages range from 10 V to 100 V. Periodically, a test voltage is applied and the insulation resistance is measured. This test voltage is typically 100 V with a reverse polarity

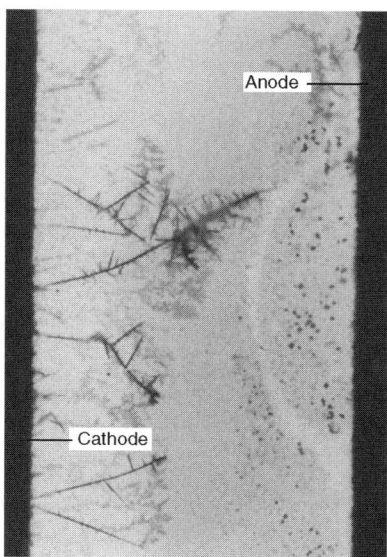

Fig. 4 Needle-like lead (Pb) dendrites grow from the cathode (left) while faceted tin oxide crystals have precipitated near the anode (right). Source: Ref 31

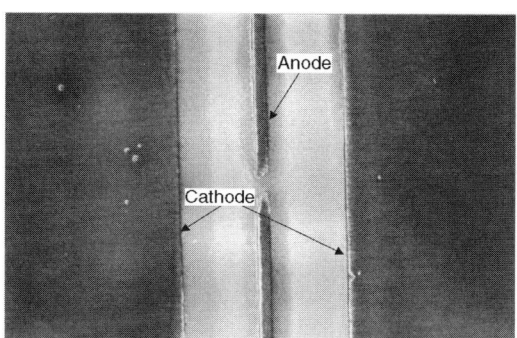

Fig. 5 The anode (center trace) has been completely eroded, leaving an open circuit condition.

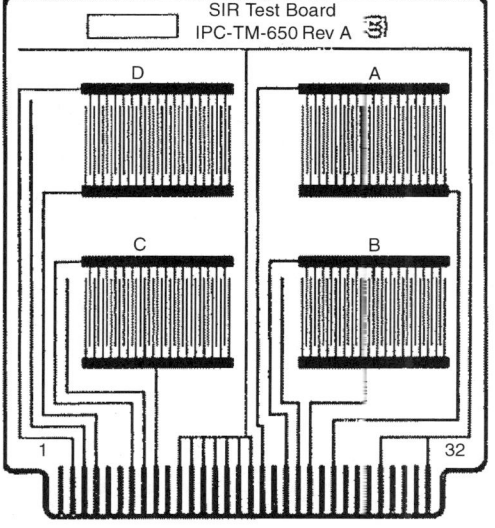

Fig. 6 IPC-B-24 test pattern. Each comb has 0.4 mm (0.016 in.) lines and 0.5 mm (0.019 in.) spacing.

relative to the bias voltage. In the case of electrochemical migration tests required by Telcordia (Ref 10) the bias voltage and the test voltage have the same polarity. IPC lists two test methods for SIR evaluation in the *Test Methods Manual* (IPC-TM-650): TM 2.6.3.3 (Surface Insulation Resistance, Fluxes) and TM 2.6.3.6 (Surface Insulation Resistance—Fluxes—Telecommunications), both of which use different test coupons.

SIR tests are normally performed at accelerated conditions of temperature and humidity. These range from 35 °C (95 °F) and 95% RH to 85 °C (185 °F) and 85% RH with a test duration varying between 4 and 21 days. Pass criteria also vary from 10^8 Ω to 10^{12} Ω depending upon the test pattern and the temperature/humidity conditions of the test.

The electrochemical migration test requires that any degradation in insulation resistance value be less than a decade. The Telcordia Electrochemical Migration test method has also become an IPC test method (TM 2.6.14.1). Table 1 compares the variation in SIR and electrochemical migration test conditions for telecommunications test versus the non-telecommunication applications.

As indicated previously, the insulation resistance is an extrinsic property of the material under investigation. This property will be affected by the test pattern chosen, temperature, humidity, bias voltage, and duration of test, as well as the contamination associated with previous processing steps. This contamination may result in electrochemical corrosion.

Geometry. The SIR reading is dependent upon a number of factors. The geometry of the test pattern is the first factor to consider. When placed under a bias voltage, an interdigitated comb pattern experiences a distributed resistance across the number of parallel traces over which the measurement is taken. The length of the interacting conductors divided by the separation between conductors is defined as the number of squares. In comparing data from two different comb patterns, the readings are sometimes reported as ohms/square.

Relative Humidity. When a monolayer of water is absorbed onto the surface of an epoxy-glass PWB, the water molecules' hydrogen bond to the epoxy making them essentially immobile. As subsequent water layers are added, thicker films are formed allowing the dissolution of contaminants and formation of hydrated ions that are mobile under the influence of an electric field (Ref 11). Conductivity measurements made on aluminum oxide (Ref 12) revealed that for films of thickness less than 3 monolayers, the surface conductivity is two orders of magnitude below that of bulk water. The surface conductivity increased asymptotically with the increase in the number of monolayers, with equilibrium being reached above 20 monolayers. The presence of contamination on the surface will increase the moisture adsorption. Evidence indicates that there is a *critical relative humidity* at which a compound exhibits a surge in moisture adsorption (Ref 13, 14). It was demonstrated that the critical relative humidity for epoxy coatings is 70% and that the epoxy degrades over time when exposed to humid environments (Ref 15).

Contamination. The critical relative humidity can be lowered by the presence of contaminants. The nature of these contaminants will determine how much moisture is adsorbed at a given relative humidity. If these contaminants are ionic in nature, they can enhance electrochemical reactions that occur in the presence of a bias voltage.

Voltage. The bias voltage applied across the insulator will set up a response in the dipolar

Table 1 SIR test parameters. A comparison of IPC and Telcordia test criteria

Parameters	IPC-TM-650 Solder flux	Telcordia SIR TA-NWT-000078	Telcordia electromigration
Test voltage, V	100	100	45–100
Bias voltage, V	50	50	10
Polarity	Reverse	Reverse	Same
Environment:			
Temperature, °C (°F)	85 (185)	35 (95)	85 (185)
Relative humidity, %	85	85–93	85–93
Duration	7 days	4 days	500 h
Lines/spacing, mm (in.)	0.4/0.5 (0.016/0.019)	0.64/1.27 (0.025/0.05) 0.32/0.32 (0.01/0.01)	0.32/0.32 (0.01/0.01)
Number of squares	~1000	~100 ~500	~500
Failure criteria	100 MΩ	10^5 MΩ 2×10^4 MΩ	SIR less than 1 decade decline

polymer substrate. In performing SIR testing, it is important that the bias voltage chosen be realistic. Typical test methods require 45 to 50 V bias because it represents a moderate accelerating condition relative to the ±15 V circuits common in today's electronics. Excessively high voltage tests for routine circuits can lead to an error in interpretation.

SIR tests are accelerated aging tests. Accelerated life tests (ALT) are used to determine the expected lifetime in the use environment. To obtain this data, a statistical number of test coupons are exposed to accelerated conditions and tested to failure. Several temperature conditions are needed to determine the activation energy for the process, as indicated in the Arrhenius equation below:

$$k_T = k_0 \exp\left(\frac{-E_a}{k_b T}\right)$$

where k_T = the reaction rate at temperature T (Kelvin); k_0 = a constant; E_a = the activation energy for the reaction; k_b = Boltzmann's constant.

The effect of relative humidity must also be evaluated by collecting data at several different humidity conditions and determining the constants in the equation that follows:

$$k_H = k_0^1 \exp C(RH)^b$$

where k_H = the reaction rate at a given relative humidity; k_0^1 = a constant; C = a constant (related to a given temperature of the reaction); b = an exponent observed in the range between 1 and 2.

Once these values are obtained, the acceleration factor can be determined by using the following equation

$$\text{Acceleration Factor} = \exp\left[\frac{E_a}{k}\left(\frac{1}{T_{\text{life}}} - \frac{1}{T_{\text{test}}}\right)\right.$$
$$\left. + C(RH_{\text{test}}^b - RH_{\text{life}}^b)\right]$$

where *life* refers to normal operating conditions; *test* refers to the accelerated test condition.

Corrosion Test Method

Standard SIR tests are used to measure the effect of soldering fluxes and their residues on the dielectric properties of PWBs. However, SIR measurements define the laminate insulation and provide no information on the effect of the soldering flux residues on the metallic circuits of the PWB.

The development of a semi-quantitative electrochemical corrosion test for soldering flux residues was proposed in 1972 (Ref 16). Copper wires of varying diameters were soldered between contacts on a PWB having additional conductors, which were used to apply a potential across the flux residues. The test coupon was placed under accelerating temperature and humidity conditions and an electrical continuity test was used to determine when the wires had broken.

In 1989, this concept was extended and a test was proposed for assessing the corrosivity of soldering flux residues by measuring the rate of corrosion of a fine copper wire printed on a PWB (Fig. 7) (Ref 17). The thin copper laminate (7.5 µm thick) was etched to 75 µm wide traces, which became the anode in an electrochemical cell where the cathode consisted of broad traces separated from the anode by 2 mm (0.08 in.) spacing. The copper traces were in a 12 V circuit with a current limiting resistor. A standard resistor equivalent to the trace resistance was also provided to monitor the stability of the power supply. The cathode was connected to a 250 V power source with a limiting resistor in that circuit also. The coupons were placed in a dessicator at room temperature to create a humidity of 95 to 100%. The subsequent corrosion of the wire was measured by monitoring the changes in the resistance of the copper trace over time. A corrosion factor (CF) was defined as the percent change in the resistance of the track relative to its initial resistance.

$$CF = \frac{(R_t - R_0)}{R_0} \times 100$$

The Turbini-modified (Ref 18) corrosion coupon is shown in Fig. 8. This represents a shrinking of the space between anode and cathode from 2 to 0.5 mm (0.08 to 0.02 in.) to coincide with the IPC-B-24 comb dimensions. The current limiting resistors in Bono's design have been placed on a resistor block, which is outside the temperature/humidity chamber in which the boards are placed. Electrical modifications in the test method included a lowering of the cathode voltage to 50 V and a change in the current lim-

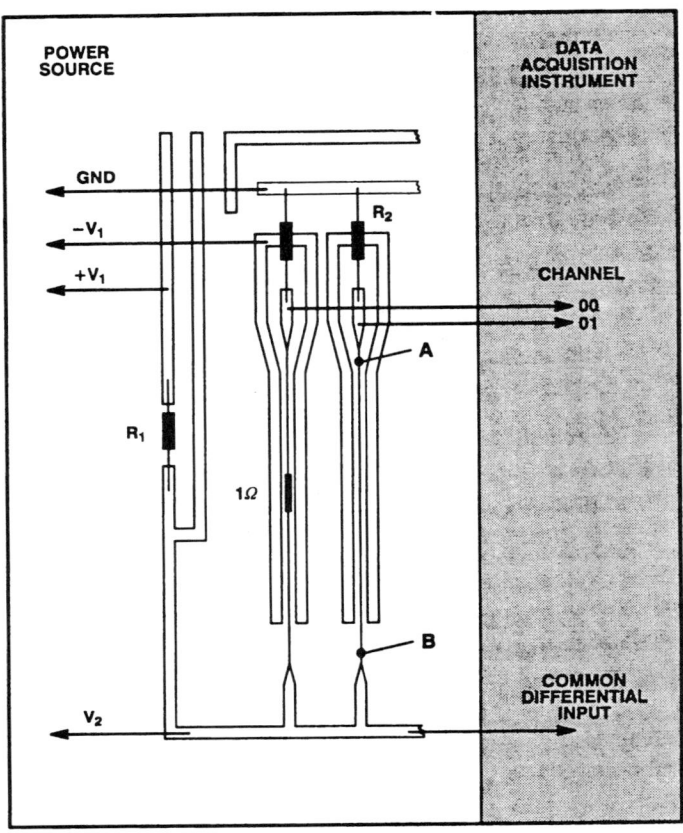

Fig. 7 Original Bono corrosion test coupon. Source: Ref 18

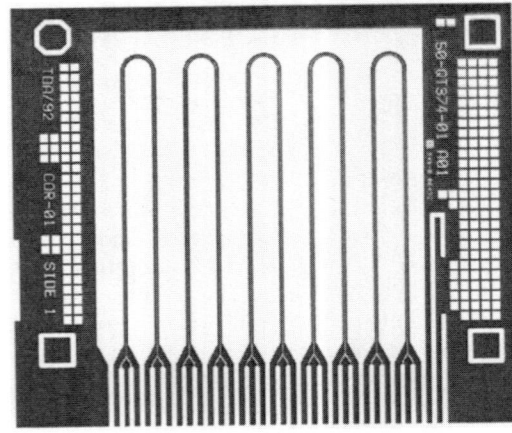

Fig. 8 Turbini corrosion test coupon. Source: Ref 19

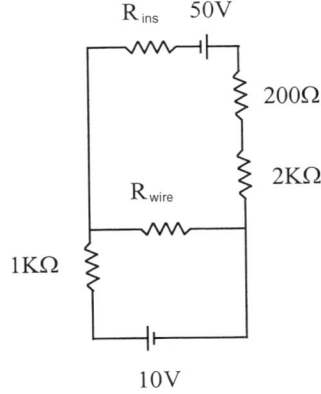

Fig. 9 Schematic electrical drawing associated with Fig. 8

iting resistor for the anode to allow a 1 mA current to flow through the trace (Fig. 9). Table 2 shows the difference between the two tests. In order to accelerate the testing time required, the test has been performed at 85 °C (185 °F) and 85% RH. By monitoring the resistance of the standard resistor and measuring the power supply voltage at each time measurement, the test setup presently shows a drift for the standard resistor measurement of less than ±0.1% corrosion.

The corrosion test is designed to measure the rate of corrosion of a very thin copper wire. It is critical that the copper wire be thin because the test depends upon the difference in resistance of the wire over time. Therefore, no solder is used in preparing the test coupons—only flux, solder paste flux, or cored-wire flux should be applied to the copper anode and heated to soldering temperatures. This test provides complementary information to the SIR test that measures the insulating characteristics of the laminate. SIR readings, however, reflect the combined effect of (a) residues that are corrosive to the conductor tracks and (b) residues that interact with the laminate.

Many of the factors discussed previously for the SIR test are equally applicable to the corrosion test. The spacing between the anode and cathode will affect the corrosion rate because it affects the electric field driving the electrochemical migration. Temperature and humidity both provide accelerated conditions as described previously. Contamination will be more strongly coupled to electrochemical migration, which results in the increase in corrosion factor over time. This test is so sensitive that corrosion can be measured electrically before corrosive residues are visible microscopically. This results because the corrosion process being monitored involves a uniform thinning of the entire conductor.

The Corrosion Test is being developed to serve as a quantitative test for flux residues. Figure 10 shows the corrosion data for a low solids flux-treated coupon. Based on the data collected for a number of different fluxes, the following corrosion factors are suggested for a 7-day test at 85 °C (185 °F) and 85% RH with the following criteria for flux classification:

Corrosion Factor for $L = 0 \leq CF \leq 0.5\%$
Corrosion Factor for $M = 0.5\% < CF \leq 4.0\%$
Corrosion Factor for $H = CF > 4.0\%$

These proposed corrosion factors apply to flux residues that remain uncleaned on the test coupon (Ref 19).

Conductive Anodic Filament Formation

Description

In the mid-1970s researchers at Bell Labs (Ref 20) and at Raytheon (Ref 21) discovered a new failure mode characterized by a sudden and unpredictable loss in insulation resistance between conductors. This failure mode, conductive anodic filament (CAF) formation, involves the subsurface growth of a copper-containing filament from the anode, toward the cathode (Fig. 11). This growth occurs under a bias voltage and under humid conditions. It was initially thought to have little impact upon service lifetimes. In 1976, a prediction was proposed (Ref 20), "that for operating voltages below 100 V, median lifetime of circuits in typical air-conditioned equip-

Table 2 Differences between Bono and Turbini corrosion tests

Parameter	Bono Original	Turbini 1991	Turbini 1991	Turbini 1999
Copper thickness, oz	¼	½		¼
Power source, V	250	200		50
	12	10		10
Flux application	Dip	Spray		Spray
Heat transfer	Wave solder, face-up	Float		Reflow, float
Test conditions:				
Temperature, °C (°F)	20 to 22 (68–71)	85 (185)	85 (185)	40 (104)
Relative humidity, %	95–100	85	85	93
Duration, days	>160	7	7	21
Resistors:				
Current limiting, KΩ	50	1		500
Verifying, Ω	1	3		4.7
Circuit limiting, KΩ	5	5		2
Sensing, Ω	...	...		200

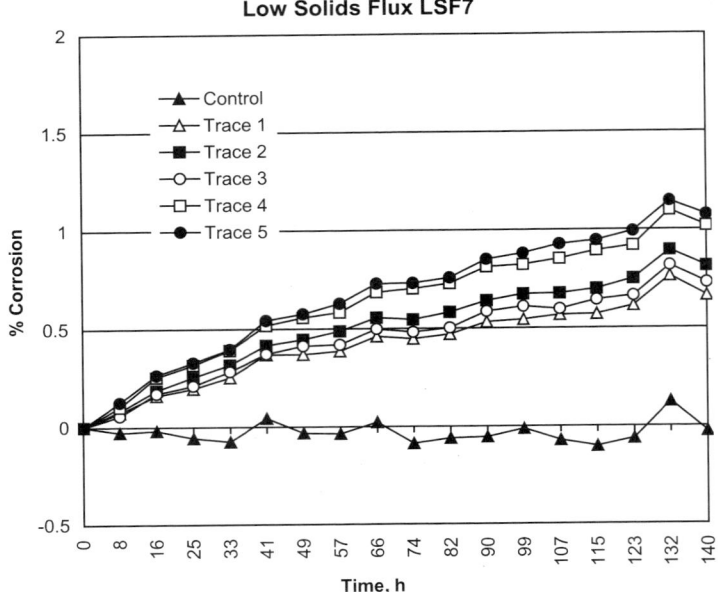

Fig. 10 Example of a corrosion test plot. Source: Ref 19

ment space (25 to 35 °C, or 77 to 95 °F and 40 to 60% RH) will be conservatively measured in tens of years." This prediction was based on extensive accelerated life testing with failure rates extrapolated to operating and use conditions. The increased circuit density of today's electronics has led to increase incidence of CAF.

A model developed by Bell Labs researchers (Ref 22) in the late 1970s (Ref 23) details the mechanism by which CAF formation and growth occurs. The first step is a physical degradation of the glass/epoxy bond. Moisture absorption then creates an aqueous medium along the separated glass/epoxy interface that provides an electrochemical pathway and facilitates the transport of corrosion products.

CAF can be observed microscopically on double-sided PWBs using back lighting (Fig. 12). Subsequent scanning electron microscopy with energy dispersive x-ray spectroscopy (SEM/EDS) analysis shows that it contains copper and chloride (Fig. 13). A CAF sample was extracted and transmission electron microscopy (TEM) analysis was used to obtain electron diffraction data. Analysis of these results identify CAF as atacamite, $Cu_2(OH)_3Cl$ (Ref 24). The Pourbaix diagram (Ref 25) (Fig. 14) of the copper-chlorine-water system shows that copper-hydroxy-chloride is insoluble in an acidic medium (pH below 4), and thus it precipitates and grows from the anode, which is acidic. CAF formation can lead to catastrophic field failures (Fig. 15) because the $Cu_2(OH)_3Cl$ has semiconductor properties.

Factors That Affect CAF Formation

Substrate Material Choice. FR4 was compared with several substrates: G-10 (a non-fire retardant epoxy/woven glass material), polyimide/woven glass (PI), triazine/woven glass,

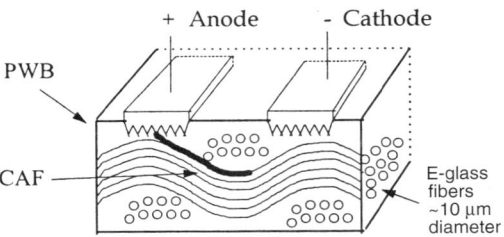

Fig. 11 Conductive anodic filament (CAF) grows subsurface along the epoxy-glass interface from anode (+) to cathode (−). Source: Ref 27

epoxy/woven kevlar, and finally, polyester/woven and chopped glass (Ref 4). An extensive experimental comparison among the substrates: bismaleimide triazine (BT), cyanate esters (CE) and FR4 was performed (Ref 26). In addition, the CAF susceptibilities of FR4 with CEM-3 (a substrate similar to G-10 except with chopped glass) and MC-2 (a blended polyester and epoxy matrix with woven glass face sheets, and a chopped glass core) were compared (Ref 27). Of all materials tested by these investigators, the BT material proved to be most resistant to CAF formation (Ref 4) (due to its low moisture absorption characteristics). Conversely, the MC-2 substrate proved to have the least resistance to CAF formation. The susceptibility of the materials follows the trend:

MC-2 ≫ Epoxy/Kevlar > FR-4
     ≈ PI > G-10 > CEM-3 > CE > BT

To ensure immunity to CAF, the laminate of preference is BT. However, there is a cost penalty to consider. Newer CAF resistant materials have appeared in the past few years. A number of these materials have been examined, as well as the manufacturing variations that affect CAF formation (Ref 28).

Conductor configuration, (Fig. 16) is a critical variable in susceptibility to CAF formation. The hole-to-hole configuration was the most susceptible to CAF as shown in Ref 4. This is due to the direct contact of the plated through-hole (PTH) barrel with the e-glass fibers (Ref 4). The track-to-track configuration was the least susceptible (Ref 4). The susceptibility of other configurations is between these two extremes. The data (Ref 8) confirmed these findings. An

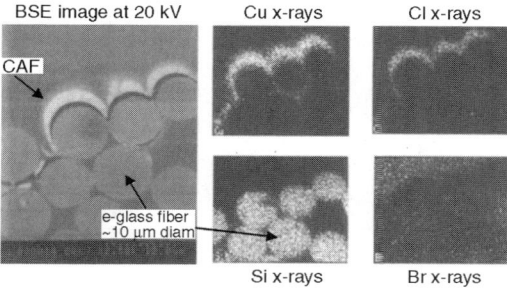

Fig. 13 CAF revealed in SEM micrograph (above left) as traveling along the separated fiber/epoxy interface. EDS elemental map (above right) of CAF shows that it is copper and chlorine containing. Source: Ref 27

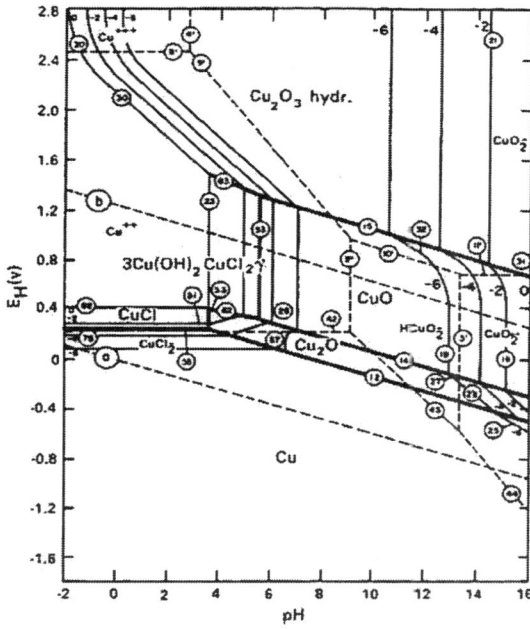

Fig. 12 Tip of IPC-B-24 comb finger. Shadows to the right of the anode tip are subsurface CAF viewed with transmitted light. Source: Ref 27

Fig. 14 Pourbaix plot of the copper chloride system. Note, copper hydroxy chloride salt insoluble below pH 4. Source: Ref 25

off-weave hole-to-hole pattern that shows reduced CAF was proposed (Ref 27) (Fig. 17).

Voltage Gradient Effects. Another critical factor used in determining CAF susceptibility is the voltage gradient. A study on the effect of voltage and spacing on CAF susceptibility determined the following relationship for mean time to failure (MTTF):

$$MTTF = C \exp \frac{E_a}{k_b T} + d \frac{L^4}{V^2}$$

where C and d are constants, E_a is the activation energy, k_b is Boltzmann's constant, T is temperature in Kelvin units, L is the spacing between conductors, and V is the voltage. Earlier works by Welsher (Ref 29) had suggested the relationship was L^2/V. Later, Ref 30 stated that the voltage dependence was closer to V^2.

Solder Flux/HASL Fluid Composition. It has been shown that polyglycols (Ref 1) diffuse into the epoxy during soldering (Ref 31). This absorption occurs when the PWB is above its glass transition temperature. The absorption has been shown to reduce performance by increasing moisture uptake by the substrate (Ref 2). The use of polyglycols in soldering fluxes and fusing fluids to increased susceptibility to CAF formation were first linked in Ref 32. Furthermore, Ref 33 detailed a field failure that occurred on only certain production lots. It showed that this failure resulted from the use of a polyglycol containing HASL fluid during production. This fluid also contained hydrobromic acid that diffused into the brominated epoxy substrate, resulting in an increased bromide concentration in the board. Both of these constituents increased the assembly's susceptibility to CAF formation by enhancing moisture absorption and providing an appropriate anion for the electrochemical reaction. Therefore, the use of hydrobromic acid-containing fluxes and fusing fluids should be avoided to reduce the likelihood of CAF formation.

Thermal Excursions. Diffusion of polyglycols into the PWB substrate occurs during soldering. Since the diffusion rate is temperature dependent, the length of time the board is above the glass transition temperature will have an ef-

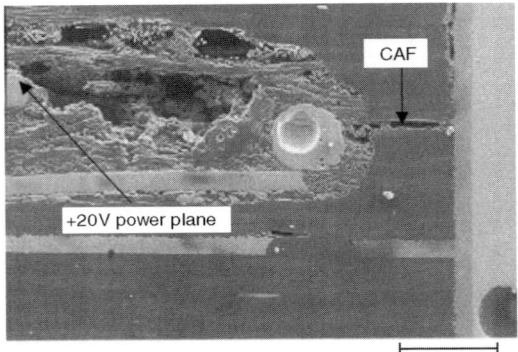

Fig. 15 Catastrophic field failure due to CAF shorting between grounding pin (right) and 20 + V power plane. Source: Ref 24

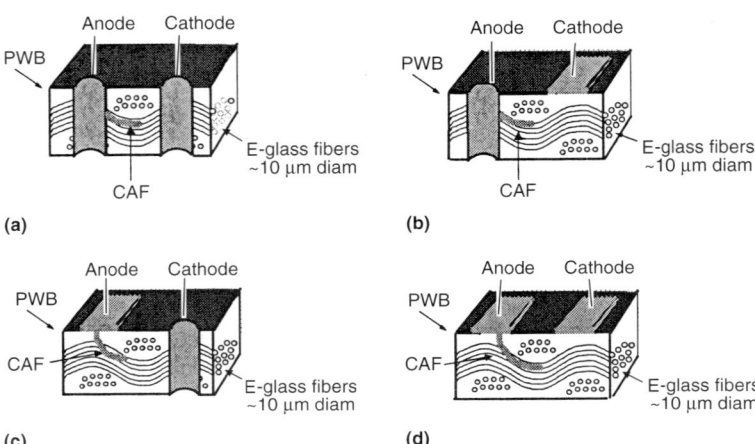

Fig. 16 CAF configurations. (a) Hole to hole. (b) Hole to pad. (c) Pad to hole. (d) Pad to pad. Source: Ref 24

fect on the amount of polyglycol absorbed into the epoxy and that will, in turn, affect its electrical properties. Water-soluble flux-treated test coupons that were prepared using two different thermal profiles was reported in Ref 14. Those which experienced the higher thermal profile exhibited an SIR level that was an order of magnitude lower than those processed under less aggressive thermal conditions. A significant increase in CAF formation at the higher soldering temperatures required for lead-free soldering have been shown (Ref 34). They processed IPC-B-24 test coupons with a series of water soluble flux formulations at either 201 or 241 °C (393 or 465 °F), cleaned them, and then exposed them to 100 V bias for 28 days with SIR measurements taken every 24 h. Comb patterns were then examined with backlighting, and the number of CAF from 2 coupons (8 comb patterns) were counted. Table 3 shows the SIR readings, as well as the total number of CAF for each flux chemistry tested at the two different temperatures. The increased CAF formation at higher processing temperatures is due to the weakening of the bond between the epoxy and glass fibers caused by the different coefficient of thermal expansion characteristics of these two materials. Lead-free soldering conditions are expected to result in an increased incidence of CAF failures.

PWB Storage and Use: Ambient Humidity Effects. There is a humidity threshold below which CAF formation will not occur (Ref 35). This relative humidity threshold depends upon operating voltage and temperature. It is important to remember that humidity need not be present in the operating environment. Moisture absorption can occur during any part of the assembly's lifetime. This is particularly critical during transportation or storage, when the assembly may experience harsh environmental conditions.

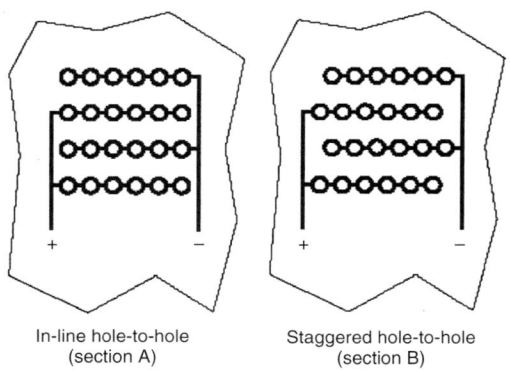

Fig. 17 Staffed hole-to-hole pattern designed by Sauter and used in the new IPC CAF test method. Source: Ref 37

Testing

Test methods for CAF detection and quantification have been proposed (Ref 27, 36). In Ref 36 the focus is on research; the authors showed

Table 3 Comparison of SIR levels and number of CAF associated with two different reflow temperatures

	SIR, Ω		CAF	
Flux	201 °C (393 °F) reflow	241 °C (465 °F) reflow	201 °C (393 °F) reflow	241 °C (465 °F) reflow
Polyethylene glycol-600(PEG)	$<10^6$	$<10^6$	90	55
PEG/HCl	$<10^6$	High 10^8	None	None
PEG/HBr	$<10^6$	High 10^8	None	None
Polypropylene glycol 1200 (PPG)	$>10^{10}$	$>10^{10}$	None	455
PPG/HCl	$>10^{10}$	$>10^{10}$	None	379
PPG/HBr	$>10^{10}$	$>10^{10}$	1	423
Polyethylene propylene glycol 1800 (PEPG 18)	High 10^9	High 10^9	1	406
PEPG 18/HCl	High 10^9	High 10^9	10	135
PEPG 18/HBr	10^{10}	High 10^9	9	279
Polyethylene propylene glycol 2600 (PEPG 26)	High 10^9	High 10^9	None	91
PEPG 26/HCl	High 10^9	High 10^9	6	218
PEPG 26/HBr	10^{10}	High 10^9	None	51
Glycerine (GLY)	$>10^{10}$	High 10^9	None	56
GLY/HCl	$>10^{10}$	High 10^9	None	583
GLY/HBr	$>10^{10}$	High 10^9	3	104
Ocyl phenol ethoxylate (OPE)	Low 10^9	Low 10^9	None	83
OPE/HCl	Low 10^9	Low 10^9	14	62
OPE/HBr	$>10^{10}$	High 10^9	2	599
Linear aliphatic polyether (LAP)	Low 10^9	Not tested	None	Not tested
LAP/HCl	Low 10^9	Low 10^9	15	203
LAP/HBr	Low 10^9	Low 10^9	None	272

that SIR measurements taken daily, or even hourly, failed to reveal the presence of CAF. Thus, they developed a linear circuit with an operational amplifier to monitor the resistance of the test pattern and to remove the accelerating voltage when resistance dropped below a certain level. The test method in Ref 27 was focused on evaluating different laminate materials, PWB design, and manufacturing processes. This test method has been developed into IPC-TM-2.6.25 (Ref 37).

Flux Residues and RF Signal Integrity

Wireless devices such as pagers and cellular phones are becoming common consumer items. These products require low-loss RF signal propagation, which is affected by material choices and processing conditions. The effect of solder flux residues on RF signal propagation has been a concern, as high frequency signals are easily affected by residues and skin effects associated with solder paste residues. Early work in Ref 38 reported stray capacitance due to solder flux residues on RF circuit boards. A series of no-clean solder paste residues at 900 MHz and 2 GHz were examined (Ref 39). The test circuit included a GaAs antenna switch, and signal transmission and leakage measurements were used to distinguish those solder pastes that degraded the signal from those that did not. They also observed the effect of processing temperatures on the results because the chemical degradation products from the solder paste will differ as the soldering temperature changes. Thus, a solder paste that has good RF signal properties when processed at Sn-Pb soldering temperatures may not, when processed at lead-free soldering temperatures.

A test method for examining the effect of flux residues at frequencies as high as 10 GHz was developed (Ref 40). A bidirectional coupler was used (Fig. 18) and a lumped element model to simulate the coupler was developed. This model for describing the effect of flux residue on a coupler circuit was useful up to 7 GHz.

A two-port T-resonator was used (Ref 41) (Fig. 19) to characterize the effect of solder flux residue on the transmission of RF signals up to 10 GHz. Bare copper conductors were used on an FR4 substrate. The change in effective dielectric constant ($\Delta\varepsilon_{eff}$) between bare and processed conditions was the parameter used to compare different solder pastes on the basis of residue-attributable RF signal transmission loss.

A T-resonator was used in a study (Ref 42) to detect qualitative differences in insertion loss frequency response caused by several surface mount materials. The study highlights the necessity of finite element modeling to analyze the effect of these materials on RF signal transmission. Recent work (Ref 43) used the Ansoft HFSS software to establish the circuit dimensions for a T-resonator test vehicle that produced resonant frequencies in the bands of interest near 5 (5.1 to 5.8 GHz) and 10 (10.0 to 10.7) GHz. These frequency bands are of interest because they have been allocated internationally for use by commercial broadband wireless systems and have greater bandwidth than allocated frequencies below 5 GHz. A finite element model was designed for this circuit data was obtained for several no-clean solder pastes, and the experimental data and the model were used to show that the dielectric constant of the flux residues, and the shape of the residues were important factors in determining the amount of RF degradation. Figure 20 shows the sideview of the T-resonator with and without flux residues (control). Figure 21 shows the difference in insertion loss with and without flux residues. Reference 43 showed that for some solder pastes, the drift in RF signal changed after aging for 15 days at 85 °C (185 °F) and 85% RH. Using the finite element model and the experimental data, the dielectric constant of the flux residues were able to be determined.

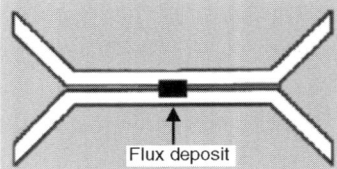

Fig. 18 Bi-directional coupler topology. Source: Ref 43

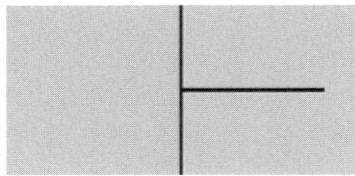

Fig. 19 T-resonator circuit. Source: Ref 43

A circuit test vehicle with both active and passive elements was designed (Ref 44). The passive circuits consisted of band pass filters (6 to 7 GHz range) and T-resonators (5, 10, and 15 GHz), while the active circuits included a mixer device (4 to 6.5 GHz), a VCO module and a wide-band amplifier circuit. The assemblies were tested after soldering and reflow, and again after flux/paste residues had been removed. Variations in signal transmission for two different solder pastes were observed on the passive circuit elements. For the mixer device, variation between signal transmissions for the measurements with/without paste residues was approximately 1%.

Summary

This chapter has provided a number of test methods whose application to new lead-free soldering materials will be important in assessing the chemical reliability questions. The chapter began by discussing soldering fluxes and their constituents. Surface insulation resistance and electrochemical migration were described in detail, along with the factors that affect the results. A new corrosion test method and several RF test methods are described. In addition, failure modes in PWBs including CAF failure are explained.

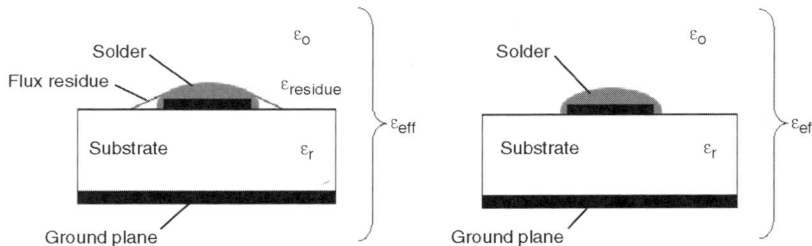

Fig. 20 Side view of T-resonator with flux residues visible on the left, and flux residues removed on the right. Source: Ref 43

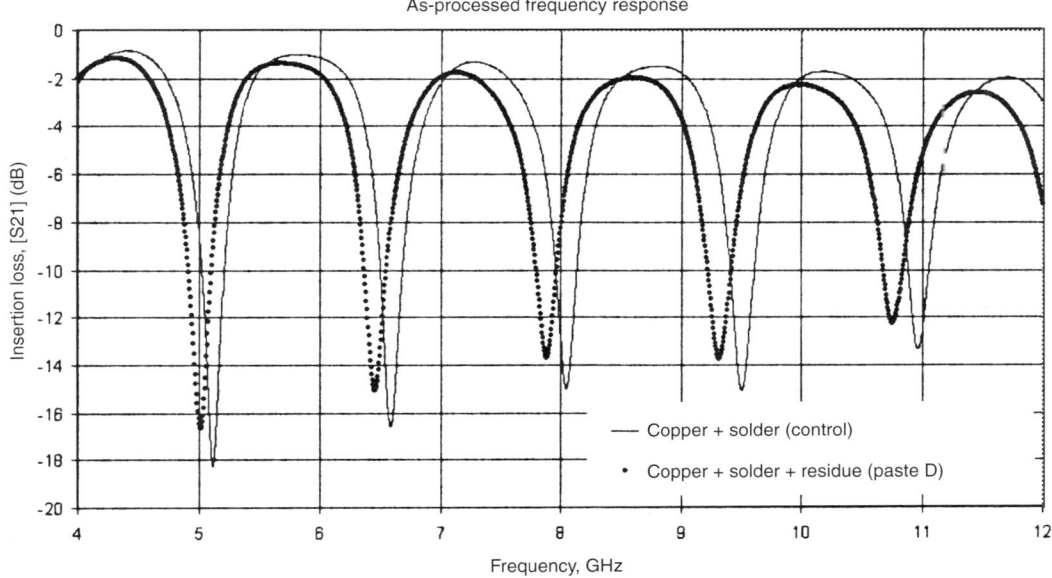

Fig. 21 Difference in insertion loss frequency response of T-resonator circuit with residues removed (black dots) and with residues from paste D remaining (gray dots). Source: Ref 43

REFERENCES

1. F.M. Zado, Effects of Non-ionic Water Soluble Flux Residues, *Western Electric Engineer*, Issue 1, 1983, p 41–48
2. J. Brous, Water Soluble Flux and its Effect on PC Board Insulation Resistance, *Electronic Packaging and Production*, Vol 21 (No. 7), 1981, p 80–85
3. T. Munson, unpublished private communication
4. Terms and Definitions for Interconnecting and Packaging Electronic Circuits, (ANSI/IPC-T-50) published by the IPC—Association Connecting Electronic Industries, 3000 Lakeside Drive, Suite 3095, Bannockburn, IL 60015
5. W. Rubin, Xersin Flux, *Electronic Production*, June 1979, p 43–52
6. D.R. Linde, Ed., *Handbook of Chemistry and Physics*, 80th ed. CRC Press, New York, 1999
7. H.H. Uhlig, *Corrosion and Corrosion Control*, 2nd ed., John Wiley and Sons, Inc., 1971, p 232
8. "D-C Resistance or Conductance of Insulating Materials," D 257-78, *Annual Book of ASTM Standards*, ASTM
9. E.J. Gorondy, Surface Insulation Resistance—Part 1: The Development of an Automated SIR Measurement Technique, "IPC Technical Paper IPC-TP-518," IPC, 1984
10. Telcordia, Generic Physical Design Requirements for Telecommunications Products and Equipment, *TR-NWT-000078*, Section 13.1, Dec 1991, p 157–168
11. G.B. Cvijanovich and A.R. Baily, "Technical Report on the Conductivity and Electrolytic Properties of Adsorbed Layers of Water," Advanced Product and Manufacturing Department, AMP Inc., Harrisburg, PA, 1980
12. B.D. Yan et al., Water Adsorption and Surface Conductivity Measurements on Aluminum Substrates, *36th Electronic Components Conf.*, IEEE, 1986
13. P.A. Sweitzer, *Atmospheric Degradation and Corrosion Control*, Marcel Dekker, Inc., New York, 1999
14. C.W. Jennings, Effect of High Moisture Environments on Printed Wiring Board Insulation, *Printed Circuit World Convention II*, Munich, Germany, 1981
15. D.R. Lefebvre, K.M. Takahashi, A.J. Muller, and V.R. Raju, *J. Adhes. Sci. Technol.*, Vol 5 (No. 3), 1991, p 201–227
16. W. Rubin and B.M. Allen, The Chemistry and Behavior of Fluxes, *Trans. Inst. Metal Finishing*, Vol 50, 1972, p 133–137
17. D. Bono, The Assessment of the Corrosivity of Soldering Flux Residues Using Printed Copper Board Tracks, *Solder. Surf. Mt. Technol.*, No. 2, June 1989, p 22–29
18. L.J. Turbini, J. Schodorf, J. Jachim, L. Lach, R. Mellitz, and F. Sledd, Characterizing the Corrosion Properties of Flux Residues: Part 2: Test Method Modification, *Solder. Surf. Mt. Technol.*, No. 12, Oct 1992, p 50–54
19. L.J. Turbini, M.S. Ramanachalam, R. Saied, B. Smith, and V. Yelander, A Corrosion Test for Characterizing Solder Flux Residues, *Proceedings of Pan Pacific Microelectronics Symposium*, 1996, p 201–203
20. P.J. Boddy, R.H. Delaney, J.N. Lahti, and E.F. Landry, Accelerated Life Testing in Flexible Printed Circuits, *14th Annual Proceedings of Reliability Physics*, 1976, p 108–117
21. A. DeMarderosian, Raw Material Evaluation through Moisture Resistance Testing, *Proceedings of the IPC Fall Meeting*, IPC-TPP-125, 1976
22. D.J. Lando, J.P. Mitchell, and T.R. Welsher, Conductive Anodic Filaments in Reinforced Polymeric Dielectrics: Formation and Prevention, *17th Annual Proceedings of Reliability Physics*, 1979, p 51–63
23. J.N. Lahti, R.H. Delaney, and J.N. Hines, The Characteristic Wearout Process in Epoxy-Glass Printed Circuits for High-Density Electronic Packaging, *17th Annual Proceedings of Reliability Physics*, 1979, p 39–43
24. W.J. Ready and L.J. Turbini, The Effect of Flux Chemistry, Applied Voltage, Conductor Spacing, and Temperature on Conductive Anodic Filament Formation, *J. Electron. Mater.*, Vol 31 (No. 11), 2002, p 1208–1224
25. M.J.N. Pourbaix, *Atlas of Electrochemical Equilibria in Aqueous Solutions*, Pergamon Press, 1966, p 384–392
26. B. Rudra, M. Pecht, and D. Jennings, Assessing Time-to-Failure Due to Conductive Filament Formation in Multi-Layer Organic Laminates, *IEEE Transactions on Components, Packaging, and Manufacturing Techniques—Part B*, Vol 17 (No. 3), 1994, p 269–276
27. W.J. Ready, "Factors Which Enhance Con-

ductive Anodic Filament (CAF) Formation," Master's thesis, Georgia Institute of Technology, 1997
28. K. Sauter, Electrochemical Migration Testing Results: Evaluating PWB Design, Manufacturing Process and Laminate Material Impacts on CAF Resistance, *CircuiTree,* July 2002
29. T.L. Welsher, J.P. Mitchell, and D.J. Lando, CAF in Composite Printed Circuit Substrates: Characterization, Modeling and a Resistant Material, *18th Annual Proceedings of Reliability Physics,* IEEE, 1980, p 235–237
30. M.S. Gandhi, J. McHardy, R.E. Robbins, and K.S. Hill, Measles and CAF in Printed Wiring Assemblies, *Circuit World,* Vol 18 (No. 23), 1992, p 23–25
31. J. Brous, Electrochemical Migration and Flux Residues—Causes and Detection, *Proceedings of NEPCON West,* 1992, p 386–393
32. J.A. Jachim, G.B. Freeman, and L.J. Turbini, Use of Surface Insulation Resistance and Contact Angle Measurements to Characterize the Interactions of Three Water Soluble Fluxes with FR4 Substrates, *IEEE Transactions on Components, Packaging, and Manufacturing Technology—Part B,* Vol 20 (No. 4), 1997, p 443–451
33. W.J. Ready, B.A. Smith, L.J. Turbini, and S.R. Stock, Analysis of Catastrophic Failures Due to Conductive Anodic Filament (CAF) Formation, *Electronic Packaging Materials Science X,* Vol 515, D.J. Belton, M. Gaynes, E.G. Jacobs, R. Pearson, and T. Wu, Ed., Materials Research Society, 1998
34. L.J. Turbini, W.R. Bent, and W.J. Ready, Impact of Higher Melting Lead-Free Solders on the Reliability of Printed Wiring Assemblies, *J. Surf. Mt. Technol.,* Vol 13 (No. 4), 2000, p 10–14
35. J.A. Augis, D.G. DeNure, M.J. LuValle, J.P. Mitchell, M.R. Pinnel, and T.L. Welsher, A Humidity Threshold for Conductive Anodic Filaments in Epoxy Glass Printed Wiring Board, *Proceedings of the 3rd International SAMPE Electronics Conference,* 1989, p 1023–1030
36. W.J. Ready, L.J. Turbini, R. Nickel, and J. Fischer, A Novel Test Circuit for Detecting Conductive Anodic Filament Formation, *J. Electron. Mater.,* Nov 1999, p 1158–1163
37. Conductive Anodic Filament (CAF) Resistance Test: S-Y Axis *IPC-TM-650 Test Method 2.6.26,* IPC
38. M.S. Heutmaker, LM. Fletcher, and J.E. Sohn, Measurement of Stray Capacitance Due to Solder Flux Residue at Radio Frequency Circuit Boards, *IEEE MTT-S Symposium,* Technologies for Wireless Applications, Feb 20–22, 1995
39. L.J. Turbini, B.A. Smith, J. Brokaw, J. Williams, and J. Gamalski, The Effect of Solder Paste Residues on RF Signal Integrity, *J. Electron. Mater.,* Vol 29 (No. 10), 2000, p 1164–1169
40. J. Csonka-Peeren, Quantifying Parasitic Induced by No-Clean Solder Paste Residues at RF Frequencies, *Proceedings of APEX,* 2002
41. M. Duffy et al., RF Characterization of No-Clean Solder Flux Residues, *IMAPS International Symposium on Microelectronics,* 2001
42. M.J. Liberatore, RF Characterization of No-Clean Solder Fluxes and Other SMT Materials, *Proceedings of APEX,* 2002
43. D. Shirley, G. Bendzsak, L.J. Turbini, and G. Munie, Evaluating the Effect of Solder Paste Residues on RF Signals Between 5 and 10 GHz, *J. Surf. Mt. Technol.,* Vol 17 (No. 1), 2004
44. D. Geiger and D. Shangguan, Investigation of Solder/Flux Residue Effect on RF Signal Integrity Using Real Circuits, *Proceeding of SMTA International,* Sept 2003

CHAPTER 6

Tin Whisker Growth on Lead-Free Solder Finishes

K.N. Tu, J.O. Suh, and Albert T. Wu, UCLA

Introduction

The leadframes used on surface mount technology are finished with a layer of solder for surface passivation and for enhancing wetting during the joining of the legs of the leadframe to printed circuit boards. When the eutectic tin-lead (Sn-Pb) solder finish is replaced by Pb-free solders, especially the eutectic tin-copper (Sn-Cu) finish, Sn whiskers have been found to grow spontaneously on the Pb-free finish. Some of the whiskers can grow to several hundred microns, which are long enough to become shorts between neighboring legs of a leadframe. The whisker growth has become a reliability issue in the application of pure Sn and the eutectic Sn-Cu solder. For this reason, there is a renewed interest in studying the growth mechanism of Sn whiskers, with the goal of prevention. How to suppress Sn whisker growth and perform accelerated testing of Sn whisker growth are challenging tasks in the electronic packaging industry.

Figure 1 is a schematic diagram of the cross section of a leadframe leg bonded to a substrate. The Pb-free solder finish is typically eutectic Sn-Cu or pure Sn. On the Sn-Cu finish, as shown in Fig. 2, many long Sn whiskers have been found to grow at room temperature, and one of them is long enough to short two neighboring legs of the leadframe. It is also possible that when there is a high electrical field across the narrow gap between the tip of a whisker and the point of contact on the other leg, a spark may

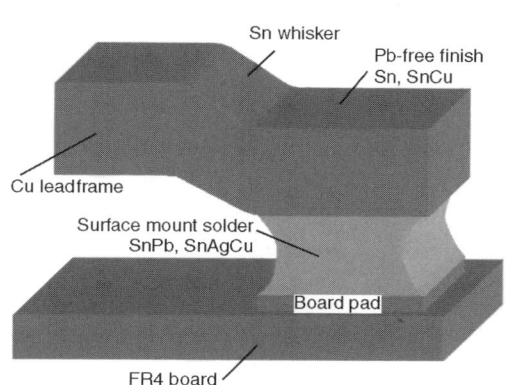

Fig. 1 Schematic diagram of the cross-section of a leadframe leg mounted on a board

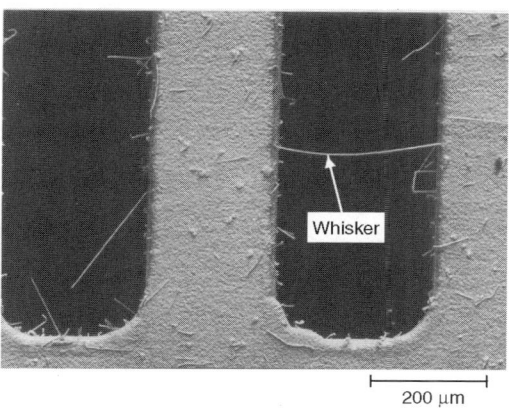

Fig. 2 SEM image of Sn whiskers grown on eutectic Sn-Cu finish on a leadframe, showing that a whisker shorts two of the leadframe legs.

ignite fire. The fire may result in failure of the device or instrument. Actually, there is a ban on the use of Sn-based solder in military devices. Due to environmental concerns, the electronic industry is removing Sn-Pb solder from consumer products and replacing eutectic Sn-Pb with Pb-free solders which are mostly Sn-rich. The probability of the occurrence of Sn whisker on Pb-free solder is high, especially when the leadframe is finished by eutectic Sn-Cu. Since Sn whisker growth is spontaneous, we need to understand its morphology, driving force, kinetic process, mechanism of growth, and how they are interrelated.

Tin whisker growth is a well-known surface relief phenomenon (Ref 1–9). The subject of Sn whisker growth has often been reviewed. One of the latest reviews (Ref 10), covers most of the early work and the proposed mechanisms of Sn whisker growth and will not be repeated here. In this chapter, we limit our review based on our own findings of spontaneous Sn whisker growth at room temperature on leadframes finished by eutectic Sn-Cu and pure Sn (Ref 11–14). A three-dimensional nonlinear stress analysis on the initiation of Sn whisker has been reported (Ref 15). Many on-line resources can be found at the National Electronics Manufacturing Initiative website (Ref 16). The topic of tin pest (Ref 17,18) and tin cry (Ref 19) will not be covered.

Spontaneous whisker growth is a kinetic process in which both stress generation and stress relaxation occur simultaneously at room temperature. We find that fast room temperature reaction between Sn and Cu, compressive stress, and stable surface oxide on Sn are three necessary and sufficient conditions for whisker growth. We have used cross-sectional scanning and transmission electron microscopy to examine Sn whiskers, with samples prepared by focused ion beam. We have also used x-ray microdiffraction in synchrotron radiation to study the structure and stress distribution in eutectic Sn-Cu. A simple kinetic model of whisker growth, in agreement with the measured growth rate, is presented. Suppression of Sn whisker growth, based on reduction of driving force and kinetics, will be discussed. An accelerated test of whisker growth by electromigration is suggested.

Morphology of Sn Whisker on Pb-Free Solder Finishes

In Fig. 3(a), an enlarged scanning electron microscopic (SEM) image of a long whisker on the eutectic Sn-Cu finish is shown. The whisker in Fig. 3(a) is straight and its surface is fluted. Many other whiskers are bent at a sharp angle, as shown in Fig. 3(b).

On the pure Sn finish surface, short whiskers were observed (Fig. 4). The surface of the whisker in Fig. 4 is faceted. Besides the difference in morphology, the rate of whisker growth

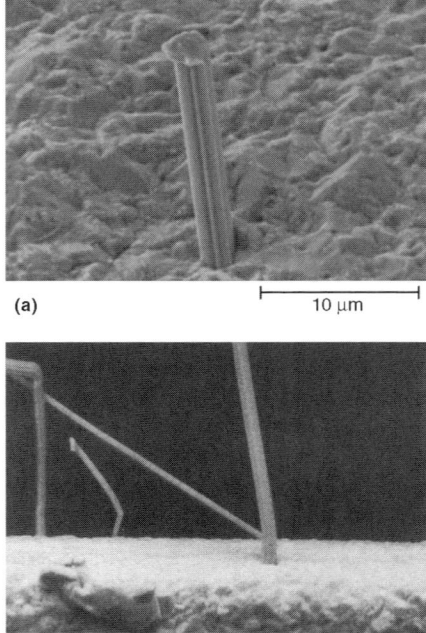

Fig. 3 Tin whisker morphologies. (a) A fluted whisker on the Sn-Cu finish surface. (b) Many other whiskers are bent at a sharp angle.

Fig. 4 SEM image of a short whisker on a pure Sn finish.

on the pure Sn finish is much slower than that on the Sn-Cu finish.

Comparing the whiskers formed on Sn-Cu and pure Sn, it seems that the Cu in eutectic Sn-Cu enhances Sn whisker growth. Although the composition of eutectic Sn-Cu consists of 98.7 at.% of Sn and 1.3 at.% of Cu, the smaller amount of Cu seems to have had a very large effect on whisker growth on the eutectic Sn-Cu finish.

In Fig. 5(a), a cross-sectional SEM image of a leadframe leg with Sn-Cu finish is shown. The rectangular core of Cu is surrounded by an approximate 15 μm thick Sn-Cu finish. A higher magnification image of the interface between the Sn-Cu and Cu layers, prepared by focused ion beam, is shown in Fig. 5(b). An irregular layer of Cu_6Sn_5 compound can be seen between the Cu and Sn-Cu. No Cu_3Sn was detected. The grain size in the Sn-Cu finish is approximately several microns. More importantly, there are Cu_6Sn_5 grains in the grain boundaries of Sn-Cu. The grain boundary precipitation of Cu_6Sn_5 is the source of stress generation in the Cu-Sn finish. It provides the driving force of spontaneous Sn whisker growth (Ref 10–13). We shall address this critical issue later.

In Fig. 5(c), a cross-sectional SEM image, prepared by focused ion beam, of pure Sn finish on Cu leadframe is shown. While the layer of Cu_6Sn_5 compound can be seen between the Cu and Sn, there is almost no Cu_6Sn_5 precipitates in the grain boundaries of Sn. The grain size in the Sn finish is also approximately several microns. The lack of grain boundary Cu_6Sn_5 precipitates is the most important difference between the eutectic Sn-Cu and the pure Sn finish.

The Stress Generation (Driving Force) in Sn Whisker Growth

The growth of whiskers (or surface hillocks) relieves the compressive stress in the matrix on which they grow. The Sn whisker is known to grow from the bottom, not from the top. This was deduced from the fact that the morphology of the tip does not change during the growth. Also, during the growth of a bent whisker, the part of the whisker below the bent grows longer, but not the part above the bent. So a whisker is being pushed out by compression.

The origin of the compressive stress can be mechanical, thermal, and chemical. But the mechanical and thermal stresses are finite in magnitude, so they cannot sustain a spontaneous growth of whiskers for a long duration. The chemical force is essential for spontaneous Sn whisker growth, but not obvious. The origin of the chemical force is due to the room temperature reaction between Sn and Cu to form the intermetallic compound (IMC) of Cu_6Sn_5. The chemical reaction provides a sustained driving force for spontaneous growth of whiskers.

Since whisker growth is a kinetic process in which stress generation and stress relaxation oc-

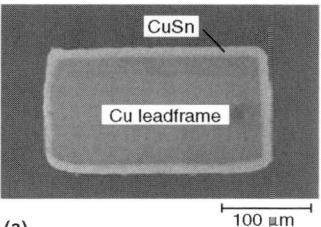

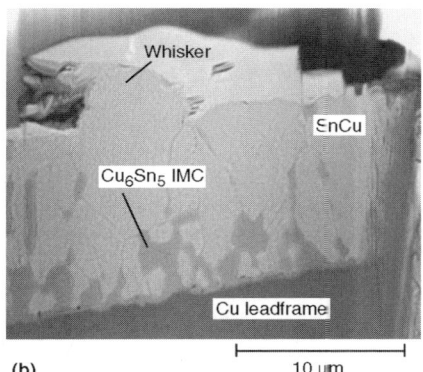

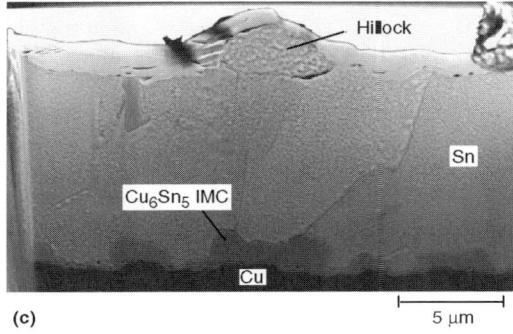

Fig. 5 Whiskers formed on SnCu and Cu leadframes. (a) Cross-sectional SEM images of a leg of the Sn-Cu finished leadframe. The rectangular core of the Cu is surrounded by an Sn-Cu finish approximately 15 μm thick. (b) An enlarged magnification of the interface between the Sn-Cu finish and the Cu leadframe. A scallop-type Cu_6Sn_5 compound can be seen. (c) A cross-sectional SEM image, prepared by focused ion beam, of pure Sn finish on Cu leadframe.

curs simultaneously, we need to consider both stress generation and stress relaxation.

Stress is generated by interstitial diffusion of Cu into Sn and the formation of an IMC in the Sn; it generates compressive stress in the Sn (Ref 20–23). When the Cu atoms from the leadrame diffuse into the finish to grow the grain boundary IMC (Fig. 5), the volume increase due to the IMC growth will exert a compressive stress to the grains on both sides of the grain boundary. In Fig. 6, if we consider a fixed volume "V" in the Sn finish that contains an IMC precipitate, the growth of the IMC due to the diffusion of a Cu atom into this volume to react with Sn will produce a stress (Ref 13):

$$\sigma = -B\frac{\Omega}{V} \quad \text{(Eq 1)}$$

where σ is the stress produced, B is bulk modulus, and Ω is the partial molecular volume of a Cu atom in Cu_6Sn_5 (we ignore the molar volume change of Sn atoms in the reaction for simplicity). The negative sign indicates that the stress is compressive. In other words, we are adding an atomic volume into the fixed volume. If the fixed volume cannot expand, a compressive stress will occur. When more and more Cu atoms, (n Cu atoms), diffuse into the volume, V, to form Cu_6Sn_5, the stress in the above equation increases by changing Ω to $n\Omega$.

In diffusional processes, such as the classic Kirkendall effect of interdiffusion in a bulk diffusion couple of A and B, the atomic flux of A is not equal to the opposite flux of B. If we assume that A diffuses into B faster than B diffuses into A, we might expect that there will be a compressive stress in B, since there are more A atoms diffusing into it than B atoms diffusing out of it. However, in Darken's analysis of interdiffusion, there is no stress generated in either A or B. But Darken has made a key assumption that vacancy concentration is in equilibrium everywhere in the sample. To achieve vacancy equilibrium, we must assume that lattice sites can be created and/or annihilated in both A and B, as needed. Hence, provided that the lattice sites in B can be added to accommodate the incoming A atoms, there is no stress. The addition of a large number of lattice sites implies an increase in lattice planes if we assume that the mechanism of vacancy creation and/or annihilation is by dislocation climb mechanism. It further implies that lattice planes can migrate, which in turn implies marker motion if markers are embedded in the sample. Hence, we have the marker motion equation in Darken's analysis. However, we must recall that in some cases of interdiffusion in bulk diffusion couples, vacancy may not be in equilibrium everywhere in the sample. Very often, Kirkendall void formation has been found, due to the existence of excess vacancies.

For the fixed volume of V in the finish (Fig. 6) (Eq 1) to absorb the added atomic volume due to the indiffusion of Cu, we must be able to add lattice sites in the fixed volume. If we assume Darken's mechanism and allow the volume to expand, we should have no stress. If not, stress will be generated. In a later section, when we discuss the oxide on Sn surfaces, we shall explain why the volume cannot expand and why stress is generated.

Sometimes it is puzzling to find that Sn whiskers seem to grow on a tensile region of a Sn finish. For example, when a Cu leadframe surface was plated with Sn-Cu, the initial stress state of the Sn-Cu layer immediately following plating was tensile, yet whisker growth was observed. If we consider the cross-section of a Cu leadframe leg coated with a layer of Sn (Fig. 5), the leadframe experienced a heat-treatment of reflow from room temperature to 250 °C (482 °F) and back to room temperature. Since Sn has a higher thermal expansion coefficient than Cu, the Sn should be under tension at room temperature after the reflow cycle. Yet with time, Sn whisker grows, so it seems that Sn whisker grows under tension. Furthermore, if a leg is bent, one side of it will be in tension and the other side in compression. It is surprising to find that whiskers grow on both sides, whether the side is under compression or tension. These phe-

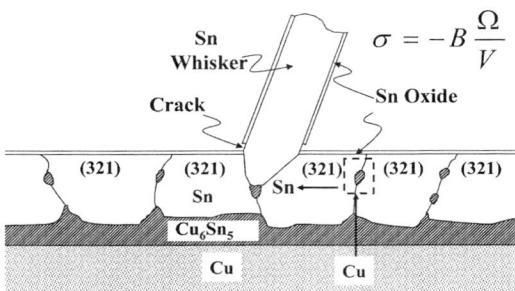

Fig. 6 Schematic diagram of cross-section of a whisker on a leadframe surface. The diffusion of Cu into the volume, V, will expand the volume and generate a compressive stress in it.

nomena are hard to understand until we recognize that the thermal stress or the mechanical stress, whether it is tensile or compressive, is finite. It can be relaxed or overcome quickly by atomic diffusion at room temperature. After that, the continuing chemical reaction will develop the compressive stress needed to grow whiskers (Fig. 5). So, the chemical force becomes dominant and it persists. When we consider the driving force of spontaneous whisker growth on Sn or Sn-Cu solder finish on Cu, we must consider the compressive stress induced by chemical reaction at room temperature.

Room Temperature Reaction Between Sn and Cu to Form Cu_6Sn_5

Room temperature reaction between Sn and Cu was studied by using thin film samples (Ref 20, 21). A bilayer of Cu-Sn thin films was prepared by deposition of Cu, first by e-beam evaporation, and then Sn by resistance heating, without breaking the vacuum at 10^{-7} torr. The bilayer was deposited at room temperature on quartz disks of 26 mm (1 in.) in diam and 1.53 mm (1/16 in.) in thickness. The quartz substrate is so thick that no bending or mechanical stress can be applied to the bilayer film. The thickness of Cu and Sn varied from 0.2 to 0.6 micron, and from 0.4 to 2.5 micron, respectively. The samples were kept at room temperature for up to one year, so there was no thermal stress, but whiskers were observed on the Sn surface. Using a glancing incidence x-ray diffractometer, we detected the growth of intermetallic compound Cu_6Sn_5 in the samples. Figure 7 shows the x-ray spectrum of a sample after one year at room temperature. In the spectra, the diffraction peaks of Cu, Sn, and Cu_6Sn_5 were identified. No reflection of Cu_3Sn was detected, even when the samples were annealed at 60 °C (140 °F). Yet when the samples were annealed at 80 °C (176 °F), both Cu_6Sn_5 and Cu_3Sn were detected. It is interesting to note that the growth rate of Cu_6Sn_5, as measured by the x-ray intensity change, does not fit the parabolic law of a diffusion-controlled growth. No morphology of Cu_6Sn_5 in the thin film sample was studied by cross-sectional electron microscopy. During the growth of Cu_6Sn_5, the dominant diffusing species was determined to be Cu by using an ultra-thin film of W as a diffusion marker (Ref 21). For comparison, a single layer of Sn film was deposited and kept at room temperature on a thick-fused quartz, but no whisker growth was observed.

Whether the Sn whisker has a preferred orientation of growth was not reported. The density of whiskers was approximately 10^3 to 10^4 whiskers/cm^2. The spacing between them was approximately several hundred microns. The average growth rate of whiskers was approximately 0.2×10^{-8} cm/sec, and the typical length of a whisker was approximately a few

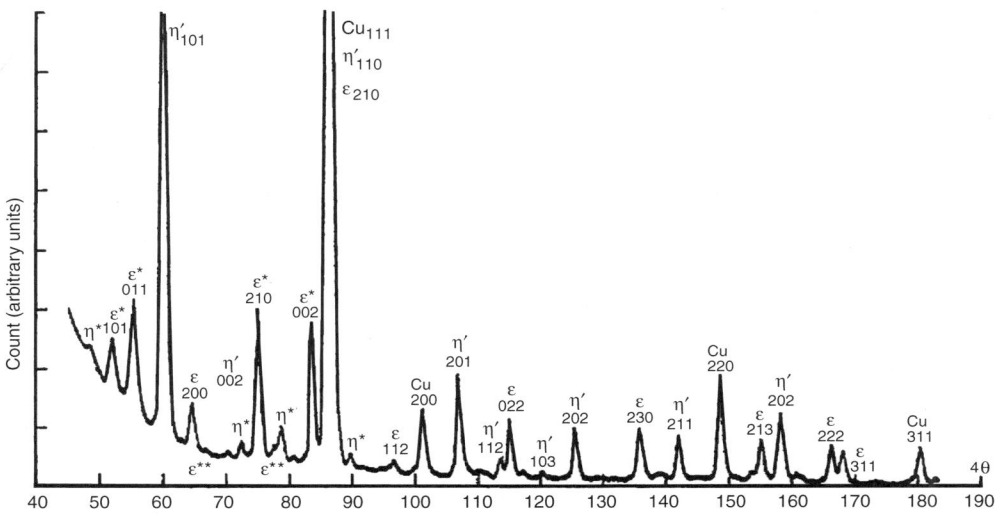

Fig. 7 X-ray spectrum of a sample after 1 yr at room temperature. In the spectra, the diffraction peaks of Cu, Sn, and Cu_6Sn_5 were identified. No reflection of Cu_3Sn was detected.

tenths of a millimeter after one year of growth at room temperature. Using the glancing incidence x-ray diffraction, the lattice parameters of Sn were measured, and the Sn film was found to be under compression. The strain was 0.16%.

The Stress Relaxation (Kinetic Process) in Sn Whisker Growth

Since the reaction of Sn and Cu occurs at room temperature, the reaction continues as long as there are free or unreacted Sn and Cu. The stress in the Sn will build up with the reaction. Yet it cannot build up forever, and the stress must be relaxed. Hence, either the added lattice planes in the volume, V, (Fig. 6) can be migrated out of the volume, or a number of Sn atoms will have to diffuse out from the volume, V, to a stress-free region.

Since room temperature is a relatively high homologous temperature for Sn, which melts at 232 °C (449 °F), the self-diffusion of Sn along Sn grain boundaries is fast at room temperature. Therefore, the compressive stress in the Sn induced by the chemical reaction at room temperature can also be relaxed at room temperature by atomic rearrangement via self grain boundary diffusion. The relaxation occurs by the removal of atomic layers of Sn normal to the stress, and these Sn atoms can diffuse along grain boundaries to the root of a stress-free whisker and push it up. Hence, whisker growth is driven by stress, and at the same time, a whisker is also a stress relief center. The room temperature reaction and the growth of the IMC will maintain a compressive stress to maintain spontaneous whisker growth.

However, compressive stress alone has been found to be a necessary but insufficient condition for whisker growth. In addition to stress generation, we must also consider the process of stress relaxation, which seems to depend upon the condition of the Sn surface oxide.

In an ultra-high vacuum, no surface hillocks were found to grow on aluminum (Al) surfaces under compression (Ref 24). Hillocks grow on Al surfaces only when the Al surface is oxidized, and Al surface oxide is known to be protective. Without surface oxide, the free surface of Al is a good source and sink of vacancies, so a compressive stress can be relieved uniformly on the entire surface of the Al, based on the Nabarro-Herring model of lattice creep or the Coble model of grain boundary creep. In these models (Fig. 8), the relaxation can occur in each of the grains by diffusion to the free surface of each grain. The free surfaces are a good and effective source and sink of vacancies. Therefore, the relaxation is uniform over the entire Al film surface; all the grains just thicken slightly. Consequently, no localized growth of hillocks or whiskers will take place.

A whisker or hillock is a localized growth on a surface. To have a localized growth, the surface cannot be free of oxide, and the oxide must be a protective oxide so that it effectively blocks all the vacancy sources and sinks on the surface. Furthermore, a protective oxide also means that it pins down the lattice planes in the matrix of Al (or Sn), so that no lattice plane migration can occur to relax the stress in the volume, V, considered previously. In the classical model of interdiffusion in a bulk couple, no stress is assumed in the model. It is because the lattice planes are allowed to migrate and to relax stress during interdiffusion; local equilibrium of vacancies is assumed. Only those metals which grow protective oxides, such as Al and Sn, are known to have hillock or whisker growth. When they are in thin film or thin layer form, the surface oxide can pin down the lattice planes easily. On the other hand, it is obvious that if the surface oxide is very thick, it will physically block the growth of any hillock and whisker. No hillocks or whiskers can penetrate a very thick oxide. Thus, the second necessary condition of whisker growth is that the protective surface oxide must not be too thick so that it can be broken at certain weak spots on the surface, and from these spots whiskers grow to relieve the stress.

Since whisker growth is localized and occurs on certain spots on the Sn surface, we ask why these spots are unique for whisker growth. Why

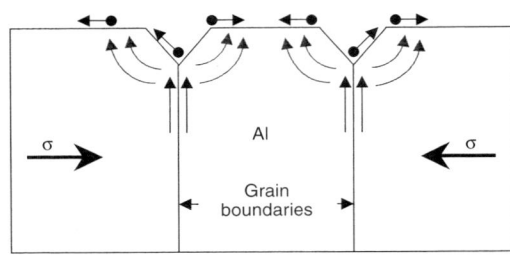

Fig. 8 Nabarro-Herring model of stress relaxation; it can occur in each of the grains by diffusion to the free surface of each grain, which is a good and effective source and sink of vacancies.

can the surface oxide at these spots be broken so easily that a whisker can nucleate and grow? Intuitively, it seems that the Sn grains or the microstructure of Sn at these spots should be different from their surrounding grains. In other words, there is a certain kind of structural discontinuity at these spots so that the surface oxide on these spots can be broken easily. For example, if we assume that the grains on a surface of Sn have an (001) texture, any grain that has a different orientation from the surrounding textured grains is a discontinuity. This grain, under compressive stress, may become the seed of a whisker. Its surface oxide is under tension, and the tension may be able to break the surface oxide along the grain boundaries between this grain and its surrounding grains. We shall discuss the use of x-ray microdiffraction in synchrotron radiation to verify this hypothesis in a later section.

Measurement of the Parameters Which Affect Sn Whisker Growth

From the point of view of device reliability, how to conduct an accelerated test of Sn whisker growth, and how to suppress Sn whisker growth are the two most relevant tasks. To achieve these tasks, we need to measure the parameters which affect whisker growth. Concerning the microstructure parameters, we need to know the intermetallic compound formation between Pb-free solder and Cu, the surface oxide on Pb-free solder, and the ambient effect on oxide growth. The grain size and texture in Pb-free solder, the whisker growth direction, and its relationship to neighboring grains are also important. The effect of impurities (Cu seems to enhance whisker growth) and Bi retards the growth, should be studied. Concerning the driving force, we need to know the effect of pressure (compressive stress drives the growth) and temperature (optimal temperature of growth is approximately 60 °C, or 140 °F) on the growth rate. Finally, we need to know the atomic mechanism of whisker growth.

In the following, we discuss cross-sectional scanning and transmission electron microscopy study of Sn whisker, the effect of surface oxide on whisker growth, synchrotron radiation microdiffraction study of Sn whisker, and kinetic model of Sn whisker growth.

Cross-Sectional Scanning and Transmission Electron Microscopy Study of Sn Whiskers. The crystal structure of Sn is a body-centered tetragonal with the lattice constant "a" = 0.58311 nm and "c" = 0.31817 nm. The growth direction, or the axis along the length of the whisker, has been found mostly to be the "c" axis, but growth along other axis such as [100] and [311] has also been found.

Fig. 9(a) and (b) show transmission electron microscopic images and the corresponding electron diffraction pattern of the cross-section of an Sn whisker (Ref 12). The cross section is normal to the growth direction of the whisker. The diffraction pattern has been indexed to be a single crystal of 001 orientation. In the image shown in Fig. 9(a), there are several point images. Whether they are the images of screw dislocations or not is unclear. The cross section of these whiskers is not rounded, rather it is faceted and fluted.

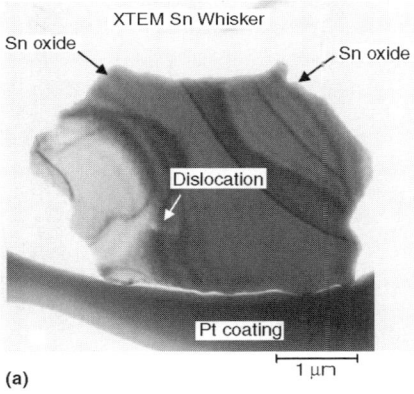

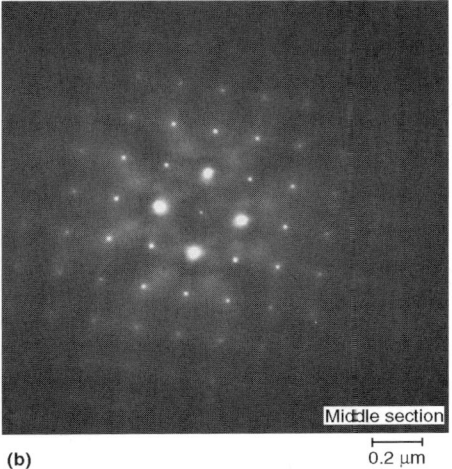

Fig. 9 Cross-section of a tin whisker. (a) Transmission electron microscopic image. (b) Corresponding electron diffraction pattern.

In Fig. 10, an SEM image of an area of the surface of an Sn-Cu finish with two rectangular holes, etched into the Sn-Cu by using a focused ion beam (FIB) is shown (Ref 12). A thin piece of Sn-Cu remained between the two holes, and this thin piece was prepared for transmission electron microscopy (TEM). This is the typical way of using FIB to prepare a thin sample at a specific location for cross-sectional TEM study. In the vicinity of the two holes, several Sn whiskers can be seen. Actually, there was a whisker at the location between the two holes before the FIB etching. The thin sample prepared by FIB enables us to study the image of cross-sections parallel to the growth direction of a whisker. Such an image also allows us to examine the grains surrounding and below the root of a whisker.

Fig. 11(a) and (b) show respectively the corresponding images of FIB and TEM of a thin slice similar to the one presented in Fig. 10. In the cross-sectional FIB image in Fig. 11(a), we can see a crack in the upper left part of the image. The same crack can also be seen in the upper left part in the corresponding cross-sectional TEM image in Fig. 11(b). The top layer above the crack is the protective layer needed to prevent the etching of the thin slice by FIB in preparing the sample. The image of a whisker which is below the top layer is indicated by an arrow in Fig. 11(b). While it is a single crystal, it nevertheless contains observable defects. Most likely, these defects were introduced during the ion beam preparation of the thin sample. All the other grains below the whisker are Sn grains. However, the most interesting finding is that there are small precipitates of Cu_6Sn_5 in the Sn grain boundaries. They appear bright and the composition of these precipitates has been confirmed by EDX analysis to be Cu_6Sn_5. The finding of these Cu_6Sn_5 precipitates enables us to explain both the compressive stress needed to drive the whisker growth and why eutectic Sn-Cu finish enhances the whisker growth as compared to pure Sn finish.

Most of these grain boundary Cu_6Sn_5 particles precipitated out during the electroplating of the eutectic Sn-Cu alloy. If the Sn-Cu finish has experienced a reflow, most of these grain boundary Cu_6Sn_5 particles should have precipitated out during the solidification. In the molten state, the 0.7 wt% (or 1.3 at.%) Cu is in solution with Sn. In solidification, the Cu becomes supersaturated and must come out (most of them precip-

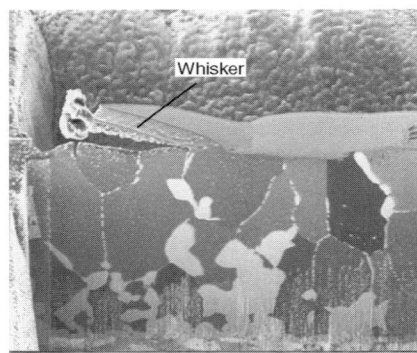

(a)

(b)

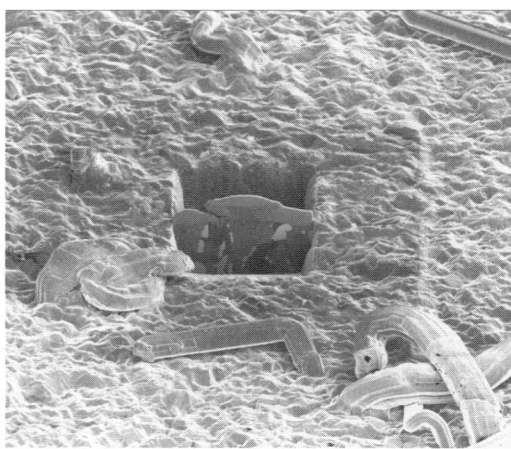

Fig. 10 SEM image of an area of the surface of Sn-Cu finish with two rectangular holes etched into the Sn-Cu by using focused ion beam.

Fig. 11 A thin slice similar to the one presented in Fig. 10. (a) Focused ion beam image. (b) Transmission electron microscopy image.

itate out during the cooling to room temperature). Since the grain size in the Sn matrix (Fig. 11a) is approximately several microns, it takes less than a second for Cu in the interior of a grain to diffuse to the surrounding grain boundaries to form Cu_6Sn_5 there. This is because the interstitial diffusivity of Cu in solid Sn is extremely fast; it is faster than 10^{-10} cm^2/s at room temperature. During room temperature aging, more Cu will diffuse from the Cu leadframe to the surface finish to grow the grain boundary precipitates; the growth increases the volume of Cu_6Sn_5 and produces the compressive stress in the finish.

We expect a stress concentration gradient around each particle, and the stress distribution will be random, because the distribution of grain boundary precipitates is random. In a region where there is a high density of grain boundary precipitates, the chance of forming a whisker is higher, too. The existence of many Cu_6Sn_5 particles in the Sn-Cu finish compared to much fewer Cu_6Sn_5 particles in pure Sn finish (Fig. 5b and c), is the main reason why a faster whisker growth is observed on the Sn-Cu finish.

The grain boundaries between the whisker and neighboring grains are of keen interest, since on the whisker side, it should have no strain, or it should have much less strain than the other side. During the growth of the whisker, the grain boundary does not move; instead, the whisker is being pushed out. If we regard the whisker as a grain, it is like a grain growth without a moving grain boundary. Hence, high resolution TEM images of the grain boundary structure are of interest and they may enable us to study the atomic growth mechanism of whiskers.

Effect of Surface Oxide on Whisker Growth. In an ambient environment, we assume that the surface of the finish is covered with oxide. The oxide at certain spots of the Sn surface must be broken, in order to grow a hillock or whisker (Fig. 6) (Ref 13). In Fig. 12(a), an FIB image of a group of whiskers on the Sn-Cu finish is shown. Most of the whiskers in Fig. 12(a) are straight, but some are bent at sharp angles. In Fig. 12(b), the oxide on a rectangular area of the surface of the finish was sputtered away by using a glancing incidence ion beam to expose the microstructure beneath the oxide. The thickness of the oxide was not measured. The sputtered area appears darker than the surrounding unsputtered area. The contrast depends on the degree of channeling of the ions; it appears dark when there is a high degree of channeling and it appears bright when there are more backscattered ions. Many bright images of particles of Cu_6Sn_5 are observed.

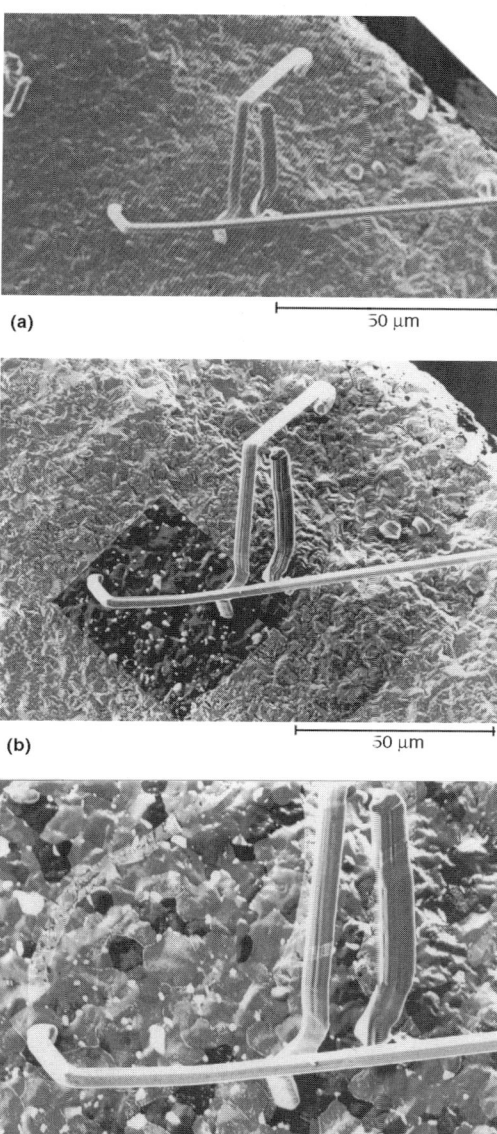

Fig. 12 Effect of surface oxide on whisker growth. (a) FIB image of a set of long whiskers on the surface of eutectic Sn-Cu. (b) FIB image of the same area shown in (a) with a rectangular area sputtered clean of surface oxide. The sputtered area appears darker than the surrounding area. Many bright particles of Cu_6Sn_5 can be seen. (c) An enlarged FIB image of the microstructure in the sputtered area. The grain structure in the Sn matrix is clear and the Cu_6Sn_5 particles are located in the grain boundaries.

In Fig. 12(c), a higher magnification image of the sputtered area is shown, in which the microstructure of Sn grains and grain boundary precipitates of Cu_6Sn_5 are clear. Due to the channeling effect, some of the Sn grains appear darker than others. The Cu_6Sn_5 particles distribute mainly along grain boundaries in the Sn matrix, and they are brighter than the Sn grains, due to less channeling. The diameter of the whiskers is approximately a few microns. It is the same as or comparable to the grain size in the Sn-Cu finish. By combining Fig. 12(a) and (b), we conclude that some of the Cu_6Sn_5 precipitates have reached the interface between the Sn-Cu and its surface oxide.

The growth of a hillock or whisker is an eruption from the oxidized surface. Stress relief takes place at these spots, and hillocks or whiskers continue to grow from them. The surface of the whisker is oxidized, too. If we imagine that the entire finish and the whiskers are all covered with oxide, the oxide is a skin. In order to grow a whisker, the growth has to stretch the surface oxide and break it. It seems that the easiest place to break the oxide is at the base of the whisker. Then to maintain the growth, the break must remain oxide-free so that it behaves like a free surface, and vacancies can diffuse into the Sn layer to sustain the long range diffusion of the Sn atoms needed to grow the whisker.

In Fig. 6, we depict that the surface of the whisker is oxidized, except for the base. The surface oxide of the whisker serves the very important purpose of confinement, so that the whisker growth is essentially a one-dimensional growth. The surface oxide of the whisker prevents it from growing in lateral direction, thus it grows with a constant cross-section and has the shape of a pencil. Also, the oxidized surface may explain why the diameter of an Sn whisker is just a few microns (Ref 12). This is because the gain in strain energy reduction in whisker growth is balanced by the formation of surface of the whisker. By balancing the strain energy against the surface energy in a unit length of the whisker, $\pi R^2 \varepsilon = 2\pi R \gamma$, we find:

$$R = \frac{2\gamma}{\varepsilon}. \qquad \text{(Eq 2)}$$

where R is radius of the whisker, γ is surface energy per unit area, and ε is strain energy per unit volume. Since strain energy per atom is approximately four to five orders of magnitude smaller than the chemical bond energy or surface energy per atom of the oxide, the radius or diameter of a whisker is found to be several microns, which is approximately four orders of magnitude larger than the atomic diameter of Sn.

The oxide surface may provide a lower bound of the diameter of a whisker; hence, we do not expect to find nano-diameter Sn whiskers. The growth rate or stress relief rate may provide an upper bound of the diameter; hence, we do not expect to find Sn whiskers as large as our hair with a diameter of approximately 100 μm. A whisker with a larger diameter may gain in volume to surface ratio, but it will grow slower due to a slower rate of vacancy supply from the circumference at the root of the whisker, and in turn, a slower strain relief rate. For this reason, the ambient condition, in which the partial pressure of oxygen and the content of moisture that affect the oxidation rate, may also affect the whisker growth rate and whisker diameter.

Synchrotron Radiation Micro-Diffraction Study of Sn Whiskers. The micro-diffraction apparatus in Advanced Light Source (ALS), Lawrence Berkeley National Laboratory, was used to study Sn whiskers grown on Sn-Cu finish on Cu leadframe at room temperature (Ref 13). The white radiation beam was 0.8 to 1 μm in diameter and the beam step-scanned over an area of 100 micron by 100 micron at steps of 1 micron. We scanned several areas of the Sn-Cu finish. We had chosen areas in each of them where there was a whisker, especially the areas that contained the root of a whisker. During the scan, the whisker, and each grain in the scanned area, can be treated as a single crystal to the beam. This is because the grain size is larger than the beam diameter. At each step of the scan, a Laue pattern of a single crystal is obtained. The crystal orientation and the lattice parameters of the Sn whisker and the grains of SnCu matrix surrounding the root of the whisker were measured by the Laue patterns. The software in ALS is capable of determining the orientation of each of the grains and displaying the distribution of the major axis of these grains. Using the lattice parameters of the whisker as stress-free internal reference, the strain or stress in the grains in the Sn-Cu matrix can be determined and displayed. Figure 13 shows a low magnification picture of an area of finish wherein a whisker is circled and scanned. Figure 14 is a micro-beam diffraction scan of a whisker and its matrix. The axis along the length of the whisker was determined to be (001) of Sn, and the grains in the matrix were found to show a texture of (211). The pole figures of the grains are shown in Fig. 15, in which the pole of the axis of (100), (110) and (321), and (211), are shown in Fig. 15(a) to (d), respectively. A high concentration of (321) and (211) at the center of the pole can be seen in Fig. 15(c) and (d). Therefore, the surface of the finish has a texture of (321) and (211) grains.

The orientation distribution on the solder finish around the root of whisker is shown in Fig. 16. Starting from Fig. 16, if we remove the x-ray data of the whisker to allow the stress around the whisker root to be analyzed, we obtain Fig. 17, which has a slightly larger grid than those in Fig. 16. Physically, it means that we have polished the whisker away, yet the root of the whisker is left in the finish. We note that the coordinates of the root of the whisker are (x, y) = (−0.8415, −0.5475) in both Fig. 16 and 17. Figure 16 shows the grain orientation distribution, but Fig. 17 shows the corresponding average biaxial compressive stress distribution. In the latter, a light-colored cross can be seen at the root of the whisker, where (x, y) = (−0.8415, −0.5475).

Our x-ray micro-diffraction study shows that at a local area of 100 µm by 100 µm, the stress is highly inhomogeneous with local variations from grain to grain. The finish is therefore under a biaxial stress only on the average. This is because each whisker has relaxed the stress in the region surrounding it. But the stress gradient around the root of a whisker does not have a radial symmetry. What Fig. 17 shows is a plot of $-\sigma'_{zz}$, which is the deviatoric component of the stress along the surface normal. Since, $\sigma'_{xx} + \sigma'_{yy} + \sigma'_{zz} = 0$, by definition, $-\sigma'_{zz}$ is a measure of the in-plane stress (note that for a blanket film, with free or passivated surface, on the average the total normal stress $\sigma_{zz} = 0$), from that, σ_b (biaxial stress) = $(\sigma_{xx} + \sigma_{yy})/2 = (\sigma'_{xx} + \sigma'_{yy})/2 - \sigma'_{zz} = -3\sigma'_{zz}/2$. This relation is always true, on average. A positive value of $-\sigma'_{zz}$ indicates an overall tensile stress, whereas a negative value indicates an overall compressive stress. The absolute value of stress

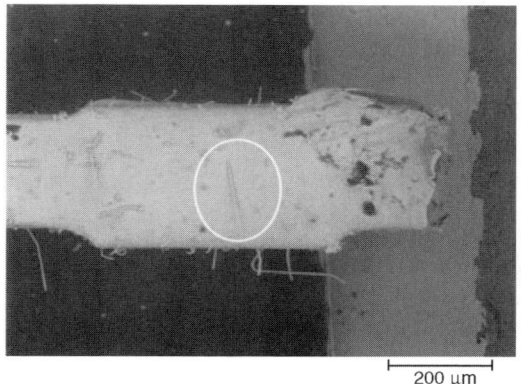

Fig. 13 A low magnification picture of an area of finish wherein a whisker is circled and scanned.

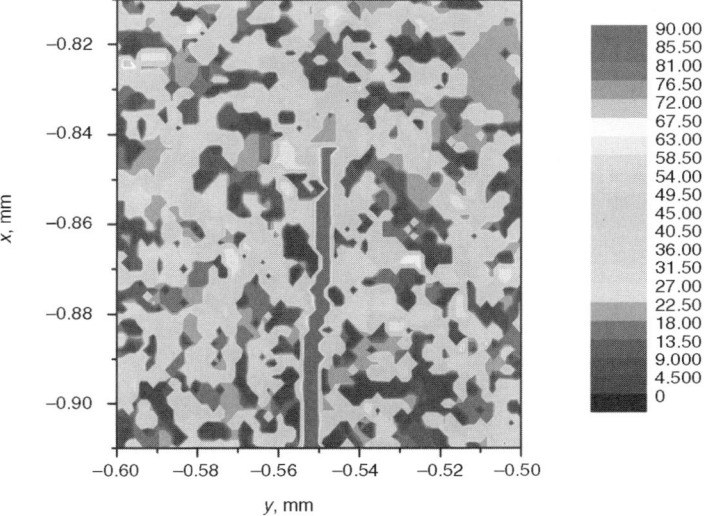

Fig. 14 A micro-beam diffraction scan of a whisker and its matrix. The axis along the length of the whisker was determined to be (001) of Sn, and grains in the matrix were found to show a texture of (211).

in the whisker root (the light-colored cross) is higher than that in the surrounding grains. If we assume the whisker to be stress-free, the surface of Sn-Cu finish is under compressive stress. However, the stress values, corresponding to a strain of less than 0.01%, are only slightly larger

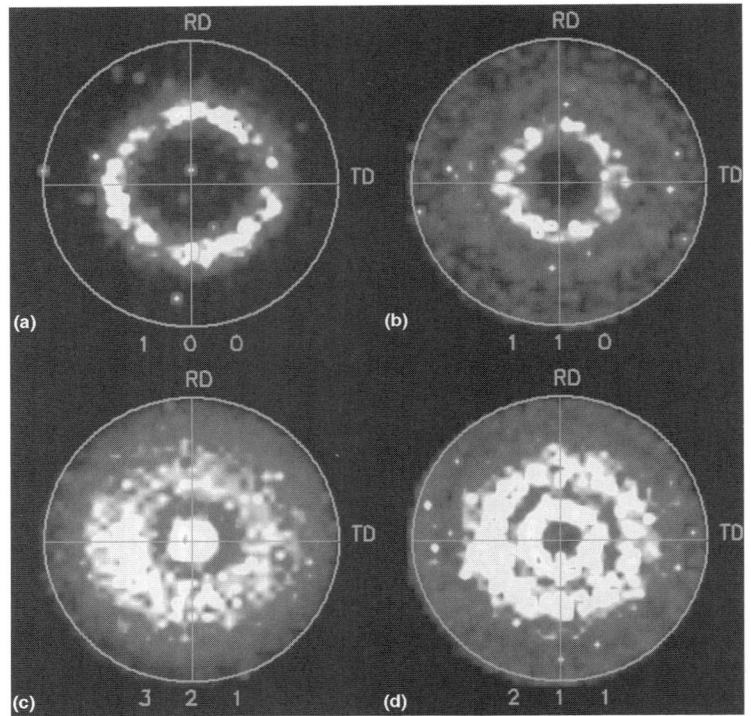

Fig. 15 The pole figures of grains in the Cu-Sn finish. (a) (100) pole, (b) (110) pole, (c) (321) pole, and (d) (211) pole.

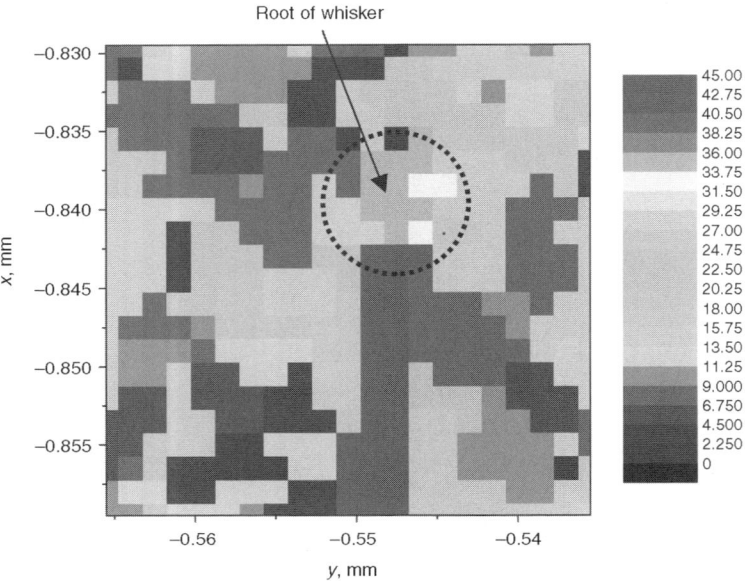

Fig. 16 The orientation distribution on the solder finish around the root of a whisker.

than the strain/stress sensitivity of the white beam Laue technique (sensitivity of the technique is 0.005% strain). Very long-range stress gradient has not been observed around the root of a whisker, indicating that the growth of a whisker has only relaxed most of the local compressive stress in the distance of several surrounding grains.

The numerical value and the distribution of stress are shown in Fig. 17, where the root of the whisker is again at the same "$x = -0.8415$" and "$y = -0.5475$" coordinates, as in Fig. 16 and 17. Overall, the compressive stress is quite low, on the order of several MPa; however, we can still see the slight stress gradient going from the whisker root area to the surroundings. It means that the stress level just below the whisker is slightly less compressive than the surrounding area. This is because the stress near the whisker has been relaxed by whisker growth. In Fig. 18, the long and dark-colored arrows indicate the directions of stress gradient. In the figure, some blocks or grids next to each other show the similar stress level, which might mean that they belong to the same grain.

Kinetic Model of Whisker Growth. In Fig. 19, we assume that the whisker has a constant diameter of 2a, and a separation of 2b, and it has a steady-state growth in a diffusional field which can be described by a 2-dimensional continuity equation in cylindrical coordinates (Ref 22). We recall that stress can be regarded as an energy density, and a density function obeys the continuity equation.

$$\nabla^2 \sigma = \frac{\partial^2 \sigma}{\partial r^2} + \frac{1}{r}\frac{\partial \sigma}{\partial r} = 0 \qquad (Eq\ 3)$$

The boundary conditions are:

$$\sigma = \sigma_0 \text{ at } r = b, \text{ and } \sigma = 0 \text{ at } r = a.$$

The solution is $\sigma = B\sigma_0 \ln(r/a)$, where $B = [\ln(b/a)]^{-1}$ and σ_0 is the stress in the Sn film. Knowing the stress distribution, we can evaluate the driving force:

$$X_r = -\frac{\partial \sigma \Omega}{\partial r} \qquad (Eq\ 4)$$

Then the flux to grow the whisker is calculated at $r = a$,

$$J = C\frac{D}{kT}X_r = \frac{B\sigma_0 D}{kTa} \qquad (Eq\ 5)$$

We note that in a pure metal, $C = 1/\Omega$. The volume of materials transported to the root of the whisker in a period dt is:

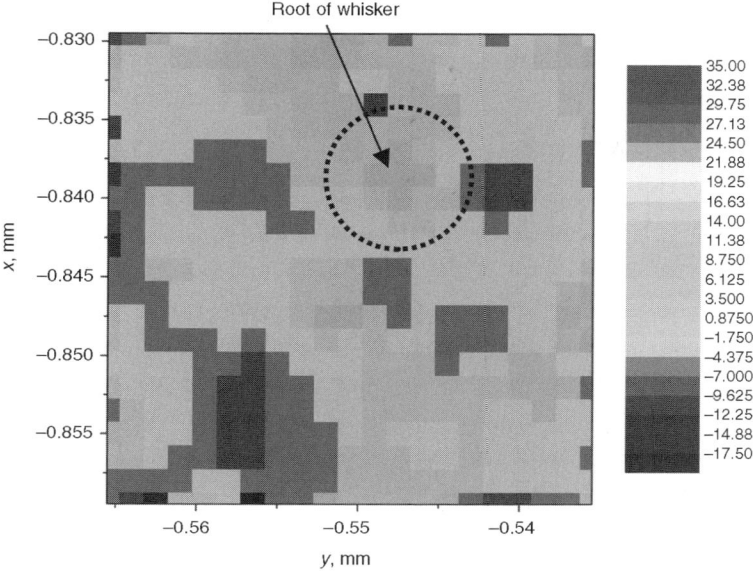

Fig. 17 Analysis of the stress around the whisker root. Deviatoric stress along the Z-direction (equivalent to the biaxial stress) around the whisker root. In the analysis, the diffraction contribution of the whisker has been removed.

(Unit: MPa)

	-0.5400	-0.5415	-0.5430	-0.5445	-0.5460	-0.5475	-0.5490	-0.5505	-0.5520	-0.5535	-0.5550
-0.8340	-2.82	-3.21	-2.26	0.93	0.93	-0.23	-8.17	2.22	1.49	1.6	-0.03
-0.8355	-2.26	-2.64	-2.64	-1.04	1.37	1.37	-1.31	0.87	0.87	0.87	-0.7
-0.8370	-2.53	-3.21	-3.21	-2.64	-1.04	3.61	0.75	0.87	0.7	0.7	-0.19
-0.8385	-7.37	-9.62	-6.57	-2.64	3.61	4.52	3.61	0.29	-1.31	0	-4.79
-0.8400	-7.37	-8.22	-6.57	-1.18	0.75	4.23	0.75	-2.25	-2.27	-2.91	-6.91
-0.8415	-4.17	-4.84	-4.17	-1.81	-0.67	-1.96	-1.96	-1.96	-3.74	-5.08	5.08
-0.8430	-4.17	-4.17	-3.63	-1.81	-1.81	-2.29	-2.29	-1.96	-1.96	-3.27	-3.27
-0.8445	-4.14	-4.17	-3.86	-3.63	2.79	-4.64	-4.78	-0.84	-1.4	-1.49	-3.27
-0.8460	-3.14	-3.63	-3.86	-3.63	-3.13	-4.78	-4.78	0.04	0.04	-1.41	-2.33
-0.8475	-4.14	-4.49	-4.49	-4.64	-3.86	-6.84	-1.72	3.55	3.55	-0.41	-2.33
-0.8490	-3.33	-5.67	-6.28	-6.29	-2.66	-2.08	-1.72	-1.79	0	-1.79	-3.73

Fig. 18 Deviatoric stress distribution around the whisker

$$JAdt\Omega = \pi a^2 dh \quad (Eq\ 6)$$

Where $A = 2\pi as$ is the peripheral area of the growth step at the root, s is the step height, and dh is the increment of height of the whisker in dt. Therefore, the growth rate of the whisker is:

$$\frac{dh}{dt} = \frac{2}{\ln(b/a)} \frac{\sigma_0 \Omega s D}{kTa^2} \quad (Eq\ 7)$$

To evaluate the whisker growth rate, we take $a = 3$ μm, $b = 0.1$ mm, $\sigma_0\Omega = 0.01$ eV (at $\sigma_0 = 0.7 \times 10^9$ dyne/cm^2), $kT = 0.025$ eV at room temperature, $s = 0.3$ nm, and $D = 10^{-8}$ cm^2/s (the self-grain-boundary diffusivity of Sn at room temperature), and we obtain a growth rate of 0.1×10^{-8} cm/s. At this rate, we expect a whisker of 0.3 mm (0.011 in.) after one year, which agrees with the observed result. The growth rate at room temperature has been measured to be close to 1 mm (0.04 in.) per year. This rate is fast enough to grow a whisker to bridge two neighboring legs of a leadframe in a few months. Since we assume grain boundary diffusion, we note that there are only several grain boundaries connecting the base of a whisker to the rest of the Sn matrix. Hence, in taking the total atomic flux which supplies the growth of a whisker to be $JAdt\Omega$, where $A = 2\pi as$, we have assumed that the flux goes to the entire peripheral of the whisker $2\pi a$ but only for a step height of s for its growth.

Volume Fraction of Cu$_6$Sn$_5$ in the Pb-Free Finish Layer. The effect of volume fraction of Cu$_6$Sn$_5$ in the Pb-free finish layer on whisker growth can be significant if the thickness of the finish is extremely thin. If the Sn layer is less than half a micron, for instance, the formation of Cu$_6$Sn$_5$ during reflow will consume most of the Sn, resulting in a large volume fraction of Cu$_6$Sn$_5$ in the finish. It is unlikely that Sn whisker will grow in this case, since there is not much Sn left. On the other hand, if we have a very thick Sn finish layer, say more than 100 μm, the lattice expansion due to the in-diffusion of Cu can be absorbed effectively by source and

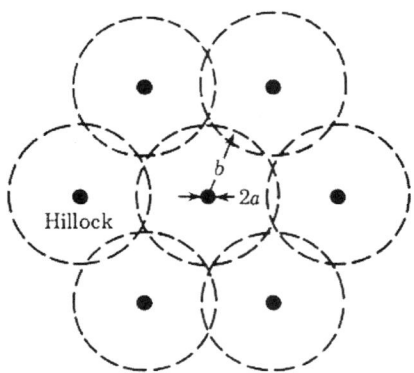

Fig. 19 A schematic diagram of the growth of whiskers of diameter of 2a and a separation of 2b on a planar surface.

sink of vacancies in the bulk of the finish. Since surface oxide is approximately 100 μm away, its effect on interaction with vacancies and pinning of lattice plane migration will be insignificant. The grain size in a very thick layer will also be larger. In the extreme case, if we prepare a bulk diffusion couple of Cu and Sn at a thickness of 1 cm (0.40 in.) and age it at room temperature, it is unlikely that we will find Sn whiskers, even though Cu_6Sn_5 still forms at the interface.

Typically, if the Sn finish layer has a thickness from 2 to 20 microns, it will be prone to Sn whisker growth. A small vol%, approximately 1% of Cu_6Sn_5 in the finish, will be sufficient to produce the compressive stress needed to grow whiskers. As we have observed in the eutectic Sn-Cu finish, the eutectic composition of only 0.7 at.% of Cu has a very large effect on Sn whisker growth. Besides, the elastic limit of a metal is 0.2%. For a pure Sn finish, the thicker the finish, the less the volume fraction of Cu_6Sn_5 with a given time of aging. The effect of surface oxide is lesser in a thicker finish, as mentioned. Since whisker growth is a localized phenomenon, the distribution of volume fraction of Cu_6Sn_5 in the finish will have an effect, too.

Suppression of Sn Whisker Growth

In order to prevent Sn whisker growth, we can either remove the driving force or reduce the kinetics, or do both. The common approach to removing the driving force is to add a diffusion barrier between Sn and Cu to stop the diffusion of Cu into Sn so that no Cu_6Sn_5 will form. It can be achieved by electroplating a thin layer of Ni before the electroplating of Sn. It helps since the room temperature reaction between Sn and Ni is much slower than that between Sn and Cu, and the Ni and Ni_3Sn_4 compound layers will serve as the barrier to Cu diffusion. The diffusivity of Cu in Ni_3Sn_4 compound has not been measured, yet is expected to be slower than that in Cu_6Sn_5 compound because of the higher melting point than the Cu-Sn compounds. Another approach is to use Cu-Sn compounds themselves as the diffusion barrier. A heat-treatment of the leadframe at 100 to 150 °C (212 to 302 °F) for 10 to 30 min. forms both Cu_6Sn_5 and Cu_3Sn at the interface between Sn and Cu. Since the Cu_3Sn does not form between Sn and Cu near room temperature, it is a good diffusion barrier to room temperature diffusion of Cu into Sn.

To reduce the kinetics, it is possible to add a certain amount of a third element such as Bi into the solder to slow down the growth of the Cu_6Sn_5 precipitates or to weaken the protective nature of Sn oxide. Another approach is to modify the electroplating process to control the grain size and texture in the Sn finish. For example, we can have a grain size either much larger or much smaller than the diameter of a whisker, i.e., a few microns. However, it is worth mentioning that there is a reflow step in joining the leadframe to surface mount. If the finish melts during the reflow, it is the microstructure after solidification, not before, that is important. Even though it is possible to change the grain size by annealing, it takes time to do so.

Accelerated Testing of Sn Whisker Growth

Accelerated testing of whisker growth may help predict the lifetime of whisker growth on Pb-free solder finish. It must be performed above room temperature. It is possible to conduct the tests up to 60 °C (140 °F), but the rate of whisker growth is still quite slow due to slow atomic diffusion. When the temperature is approaching 100 °C (212 °F), the diffusion will be faster, yet the stress will be relaxed due to fast atomic diffusion. Hence, we run into a competition between driving force and kinetics. Although we can add Cu, as in eutectic Sn-Cu solder, to enable a faster whisker growth, the rate is not fast enough. Besides, we need to isolate the effect of Cu on whisker growth.

We consider here the use of electromigration to conduct the accelerated testing of whisker growth. The advantage of using electromigration to study whisker growth is that not only can we vary the applied current density (driving force), but we can also conduct the test at higher temperatures (kinetics). Hence, we can control both the driving force, as well as the kinetics. Figure 20 shows the growth of an Sn whisker at the anode of test sample of pure Sn under electromigration (Ref 25). Measuring the growth rate and the diameter of the whisker, we obtain the volume change per unit time of the whisker, $V = JAdt\Omega$, where J is the electromigration flux, A is the cross-section of the whisker, dt is unit time, and Ω is atomic volume. Then we can calculate the stress which drives the growth, based on Eq 7. In addition, knowing J, we have:

$$J = C\frac{D}{kT}\left(\frac{d\sigma\Omega}{dx} + Z^*ej\rho\right) \quad \text{(Eq 8)}$$

where $C = 1/\Omega$ in pure Sn, D is diffusivity, kT is thermal energy, σ is stress at the anode, and we may assume the stress at the cathode is zero, $d\sigma/dx$ is the stress gradient along the short strip of Sn of length dx, Z^* is the effective charge number of the diffusing Sn atoms in electromigration, e is electron charge, j is current density, and ρ is resistivity of Sn at the test temperature. We can determine σ from Eq 7, and then check Z^* by using Eq 8, since we know all the other parameters.

To control the growth of the diameter of a whisker at the anode, we can sputter a thin coating of quartz over the entire Sn strip with an etched hole of a given diameter at the anode. With an applied current density, a whisker can be pushed out from the hole at the anode. We can measure the growth rate and the volume of the whisker as a function of current density, temperature, and time. However, the accelerated test may not be meaningful until we can confirm that the whisker driven by electromigration has the same growth mechanism as the whisker grown spontaneously on the Pb-free finish.

Conclusions

From the point of view of growth mechanism of spontaneous Sn whiskers, it requires both stress generation and stress relaxation to occur simultaneously. The three necessary and sufficient conditions are fast atomic diffusion at room temperature in Sn, compressive stress, which comes from the room temperature reaction between Cu and Sn from the Cu_6Sn_5 compounds in the grain boundaries of the Sn, and the native oxide, which is protective.

From the technology point of view, the two most interesting questions about Sn whiskers are how to conduct an accelerated test and how to prevent it. We can use electromigration to provide the driving force and perform the accelerated tests at high temperatures. To prevent Sn whisker growth, the common approach is to remove the driving force of chemical reaction by adding a diffusion barrier between Sn and Cu, e.g., electroplating a thin layer of Ni before the electroplating of Sn. Another approach is to use Cu-Sn compounds themselves as the diffusion barrier. A heat-treatment of the leadframe at 100 to 150 °C (212 to 302 °F) for 10 to 30 min. forms both Cu_6Sn_5 and Cu_3Sn at the interface between Sn and Cu. The Cu_3Sn is a good room temperature diffusion barrier of Cu into Sn.

ACKNOWLEDGMENT

The authors would like to thank the SRC Contract No. NJ-853, supported by the Semiconductor Research Corporation and National Semiconductor Corporation.

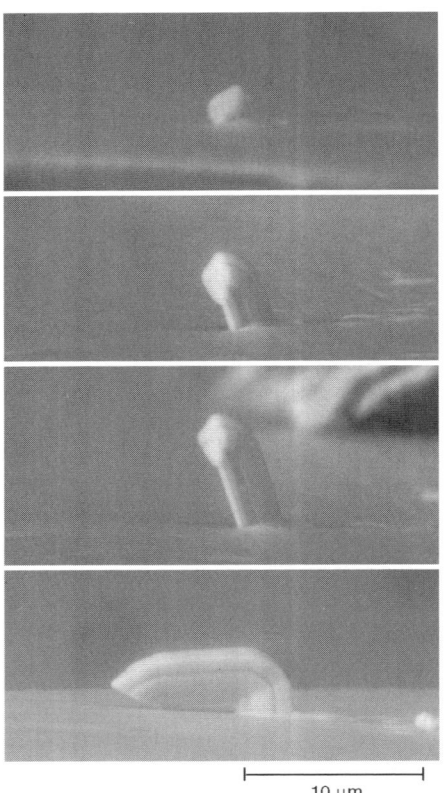

Fig. 20 Growth of a tin whisker at the anode of a test sample of pure Sn under electromigration.

REFERENCES

1. C. Herring and J.K. Galt, "Elastic and Plastic Properties of Very Small Metal Specimens," *Phys. Rev.* Vol. 85, p 1060, 1952
2. G.W. Sears, "A Mechanism of Whisker Growth," *Acta Metall.* Vol. 3, p 367 1955
3. A.P. Levitt, in *Whisker Technology,* Wiley-Interscience, New York, 1970
4. T. Nagai, K. Natori, and T. Furusawa, "Rate

of Short-Circuit Caused by Whisker Growth on Zn Electroplated Steels in Electronic Appliance," *J. Japan Inst. Met.* 53, 303 1989

5. I.A. Blech, P.M. Petroff, K.L. Tai, and V. Kumar, "Whisker Growth in Al Thin-Films," J. Cryst. Growth Vol. 32, No. 2, 1975, p 161–169

6. N. Furuta and K. Hamamura, "Growth Mechanism of Proper Tin Whisker," *Japan J. Appl. Phys.* Vol. 8, No. 12, 1969, p 1404–1410

7. W.C. Ellis, D.F. Gibbons, and R.C. Treuting, *Growth and Perfection of Crystals,* R.H. Doremus, B.W. Roberts, and D. Turnbull Ed., John Wiley, New York, 1958, p 102

8. R. Kawanaka, K. Fujiwara, S. Nango, and T. Hasegawa, Influence of Impurities on the Growth of Tin Whiskers, *Japan J. Appl. Phys.* Part I, Vol. 22, No. 6, 1983, p 917–922

9. U. Lindborg, A Model for the Spontaneous Growth of Zinc, Cadmium and Tin Whiskers, *Acta Metall.* Vol. 24, No. 2, 1976, p 181–186

10. W.J. Choi, G. Galyon, K.N. Tu, and T.Y. Lee, The Structure and Kinetics of Sn Whisker Formation and Growth on High Sn Content Finishes, in *Handbook of Lead-Free Solder Technology for Microelectronic Assemblies,* K.L. Puttlitz and K.A. Stalter Ed., Marcel Dekker, Inc., New York, 2004, p 851–914

11. W.J. Choi, T.Y. Lee, K.N. Tu, N. Tamura, R.S. Celestre, A.A. MacDowell, Y.Y. Bong, L. Nguyen, and G.T.T. Sheng, Structure and Kinetics of Sn Whisker Growth on Pb-Free Solder Finish, 52nd Electronic Component & Technology Conf. Proc. (IEEE Catalog number 02CH3734-5), San Diego, CA, 2002, p 628–633

12. G.T.T. Sheng, C.F. Hu, W.J. Choi, K.N. Tu, Y.Y. Bong, and L. Nguyen, Tin Whiskers Studied by Focused Ion Beam Imaging and Transmission Electron Microscopy, *J. Appl. Phys.*, 92, 2002, p 64–69

13. W.J. Choi, T.Y. Lee, K.N. Tu, N. Tamura, R.S. Celestre, A.A. MacDowell, Y.Y. Bong, and L. Nguyen, Tin Whiskers Studied by Synchrotron Radiation Micro-Diffraction, *Acta Mat.,* Vol. 51, 2003, p 6253–6261

14. J.B. LeBret and M.G. Norton, Electron Microscopy Study of Tin Whisker Growth, *J. Mater. Res.,* 18, 2003, p 585–593

15. J.H. Lau and S.H. Pan, 3D Nonlinear Stress Analysis of Tin Whisker Initiation On Lead-Free Components, *J. Electron. Packaging,* Vol. 125, 2003, p 621–624

16. *Tin Whisker Accelerated Test Project,* International Electronics Manufacturing Initiative (iNEMI), Herndon, VA, http://www.nemi.org/projects/ese/tin_whisker.html (accessed March 2005)

17. Y. Kariya, C. Gagg, and W.J. Plumbridge, Tin Pest in Lead-Free Solder, *Surf. Mt. Technol.,* Vol. 13, 2001, p 39–40

18. Y.J. Joo and T. Takemoto, Transformation of Sn-Cu Alloy from White Tin to Gray Tin, Materials Letters, 56, 2002, p 793–796

19. K.N. Tu and D. Turnbull, Direct Observation of Twinning in Tin Lamellae, *Acta Met.* Vol. 18, 1970 p 915

20. K.N. Tu, Interdiffusion and Reaction in Bimetallic Cu-Sn Thin Films, *Acta Met.,* Vol. 21, 1973, p 347

21. K.N. Tu and R.D. Thompson, Kinetics of Interfacial Reaction in Bimetallic Cu-Sn Thin Films, *Acta Met.,* Vol. 30, 1982, p 947

22. K.N. Tu, Irreversible Processes of Spontaneous Whisker Growth in Bimetallic Cu-Sn Thin Film Reactions *Phys. Rev.* B49, 1994, p 2030–2034

23. B.-Z. Lee and D.N. Lee, Spontaneous Growth Mechanism of Tin Whiskers, *Acta Mater.* Vol. 46, No. 10, 1998, p 3701–3714

24. C.Y. Chang and R.W. Vook, "The Effect of Surface Aluminum Oxide Film on Thermally Induced Hillock Formation," Thin Solid Films, 228, 1993, p 205–209

25. C.Y. Liu, C. Chen, and K.N. Tu, Electromigration of Thin Stripes of Sn-Pb Solder as a Function of Composition, *J. Appl. Phys.,* 88, 2000, p 5703–5709

CHAPTER 7

Accelerated Testing Methodology for Lead-Free Solder Interconnects

G. Grossmann, Swiss Federal Laboratories for Materials Testing and Research, Switzerland

Introduction

The purpose of accelerated testing is to provoke the same failure in a product during the test as it occurs in the real world but in shorter time than in the application. This can be achieved either by increasing the stress levels in a test compared to the application, or by applying the same stress as experienced in reality but more frequently in a given time frame than in the application in the case of cyclic stress. In both cases, the test time is shortened as a function of the parameters applied. Of course, both test strategies can also be combined to achieve a maximum acceleration of the test. However, there are limits. Highly accelerated tests have a high potential to produce misleading results because degradation mechanisms are activated that never occur in reality, or the tests do not activate the mechanisms that play the leading role in the application. Thus, to design an accelerated test for solder joints, it is essential to consider the material properties of the solder alloy tested, to understand the limits of the conclusions drawn from an accelerated test, and to recognize that the test is always an image of the reality but never the reality itself.

Metallurgical Background

Creep. If a tin-based solder is subjected to a constant load, one can observe a constant deformation. This means that deformation and load are not directly connected, but to each load a corresponding rate of deformation can be observed:

$$F = C \frac{\partial l}{\partial t} \quad \text{(Eq 1)}$$

where F is load, C is creep resistance, ∂l is deformation, ∂t is time difference.

Creep is always present, but it becomes an important part of the deformation when a material comes close to its liquidus temperature. How close a material is to its liquidus is expressed with the homologous temperature:

$$T_H = \frac{T_{Use}}{T_{Liquidus}} \quad \text{(Eq 2)}$$

where T_H is homologous temperature, T_{Use} is temperature at use (K), $T_{Liquidus}$ is liquidus temperature (K).

If a material is used at a homologous temperature above 0.6, creep deformation is typically significant. Although usage of a material with a $T_H > 0.6$ is usually avoided, the electronic industry uses Sn-based solder at $T_H \sim 0.9$ (Fig. 1). Thus, the main deformation regime in soft solder joints is creep.

Mechanisms of Creep Deformation. Three creep deformation mechanisms occur:

- Diffusion creep
- Grain boundary sliding (GBS)
- Dislocation climb/dislocation glide (DC)

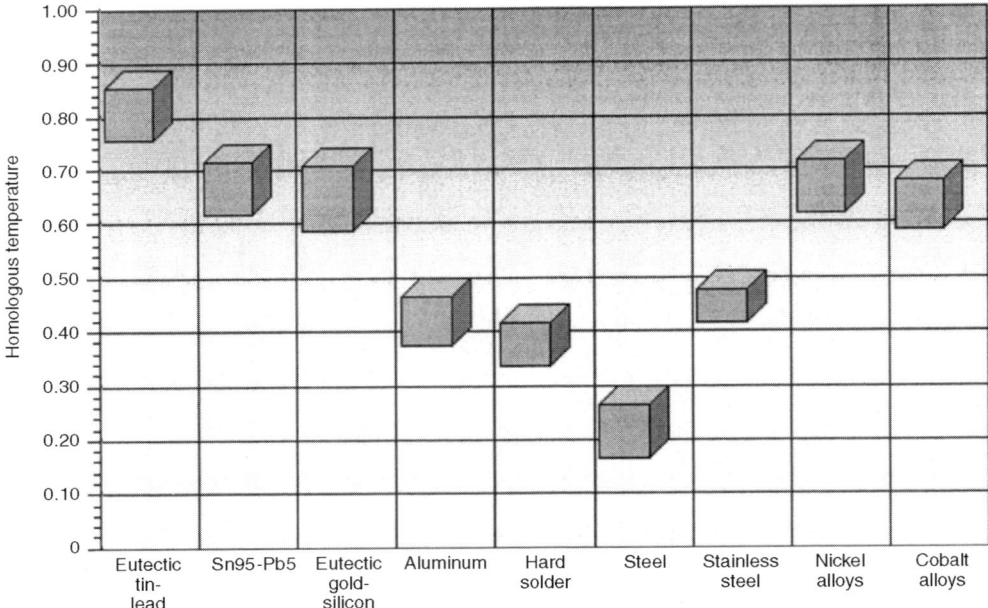

Fig. 1 Homologous temperature of selected materials

The creep rate occurring depends on the stress and the temperature.

In diffusion creep, atoms diffuse along the grain boundaries from zones with compression stress to zones with tensile stress. This creep mechanism is very slow and is activated with low stress levels and high temperatures. The creep rate also depends on the diffusion length the atoms have to travel and thus on the grain size. Diffusion creep does not play an important role in solder joints of electronic assemblies because of its slowness.

In grain boundary sliding, the crystals are moving above each other. However, in a solidified metal the crystals fit in only one constellation. Thereby, the crystals move on complex 3-dimensional curves rotating around various axes. During this movement, the crystals deform either by diffusion similar to the diffusion creep or by dislocation glide due to local stress concentration.

In dislocation climb/dislocation glide, the stress applied to the crystals activates dislocations to run through the crystal, thus moving the crystal layers upon each other. This deformation occurs at high stress levels and is very fast (Fig. 2).

One can see these three deformation mechanisms if the results of creep tests on Sn-Pb are displayed in a double logarithmic chart. At high-stress levels DC is dominant, at medium-stress levels GBS is the main player, and at low-stress levels diffusion creep is the main deformation mechanism. However, at low levels, the data show a wide spread because of the inaccuracy of long-term measurements and because of the influence of the diffusion path (Fig. 3).

Modeling Creep. One possibility, to model creep deformation among others (Ref 1, 2, 3), is

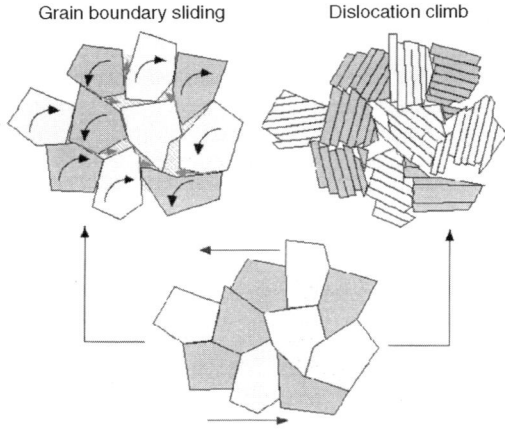

Fig. 2 Deformation mechanisms in creep

the Norton law where GBS and DC can be defined, together with an Arrhenius extension to account for the influence of the temperature (Eq 3). Since the two deformation mechanisms can be observed in measurements, this approach seems quite sensible (Fig. 3) (Ref 4):

$$\gamma' = A\left(\frac{\tau}{G}\right)^n \exp\left(-\frac{Q_{GBS}}{RT}\right) + B\left(\frac{\tau}{G}\right)^m \exp\left(\frac{Q_{DC}}{RT}\right) \quad \text{(Eq 3)}$$

where γ' is strain rate, t is shear stress, A is constant GBS, B is constant DC, n is stress exponent GBS, m is stress exponent DC, G is dynamic shear modulus, T is temperature (K), Q_{GBS} is activation energy GBS, and Q_{DC} is activation energy DC, R is gas constant.

Degradation. The degradation of a material in creep is caused by the accumulation of microvoids. These voids form in GBS as well as in DC.

As can be seen in Fig. 2, the grains of a crystalline material fit in only one configuration. If these grains are forced to move above each other as in GBS, voids will form at the grain boundaries. These voids are filled up by diffusion along the grain boundaries. In cyclic deformation however, depending on the rate of the formation of voids during deformation and the time where no additional deformation occurs, at the temperature, the microvoids will partially remain at the end of a cycle and each cycle generates new microvoids. During the test, the microvoids accumulate and form macroscopic voids and cracks that grow.

Dislocation climb/dislocation glide also forms microvoids. Each dislocation that arrives at a grain boundary causes a disturbance. Accumulation of such disturbances leads to microvoids and again, these microvoids grow to cracks, due to the accumulation at each cycle. This means that in a material where strain of varying magnitudes throughout the volume occurs, a damage field will form. The larger the strain, the larger the damage. Thus, the crack formed by the accumulation of the microvoids follows the path of the largest strain (Fig. 4).

Influence of Deformation Mechanism on Degradation. Each deformation regime has its own degradation behavior as is stated in the Miner-Palmgren law (Eq 4), which says that one has to divide the number of cycles with deformation mechanism 1 (n_1) by the number of cycles to failure for deformation mechanism 1 (N_{f1}). The same has to be done for all occurring deformation mechanisms $i = 1 \ldots k$. If the sum of all the ratios n_i/N_{fi} reaches 1, the specimen will fail.

$$\sum_{i=1}^{k} \frac{n_i}{N_{F_i}} = 1 \quad \text{at failure} \quad \text{(Eq 4)}$$

where n_i is the number of cycles with deformation mechanism i and N_{fi} is the number of cycles to failure for deformation mechanism i.

With some mathematical transformation where n_i is replaced by the fraction F of a cycle that is occupied by a deformation mechanism (e.g., 0.3) and by setting GBS as deformation mechnism 1 and DC as deformation mechanism 2, Miner-Palmgren becomes Eq 5, which has been published by Shine and Fox (Ref 5):

$$\frac{1}{N_f} = \frac{F_{GBS}}{N_{fGBS}} + \frac{F_{DC}}{N_{fDC}} \quad \text{(Eq 5)}$$

where N_f is the number of cycles to failure, F_{GBS} is the fraction of cycle occupied by GBS, F_{DC} is the fraction of cycle occupied by DC, N_{fGBS} is the number of cycles to failure GBS, and N_{fDC} is the number of cycles to failure DC.

Modeling Degradation. There have been several approaches to model the degradation of

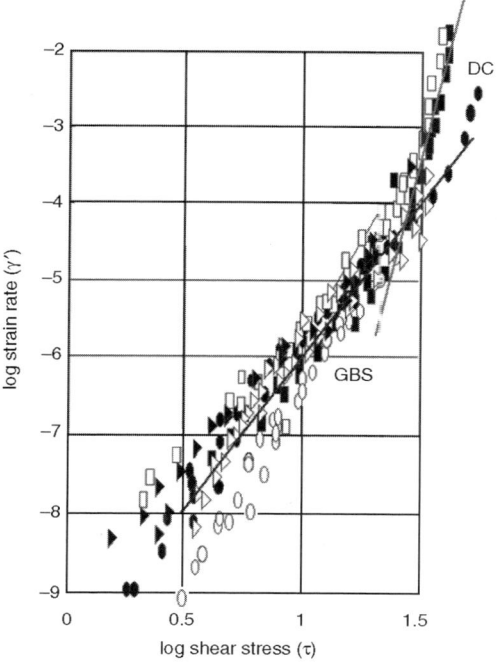

Fig. 3 Measurement results of the creep behavior of Sn-Pb36-Ag2

low cycle fatigue (Ref 6–10). The best known and easiest to use is the Coffin-Manson approach which states that the product of the number of cycles to failure and the plastic deformation per cycle raised to a power is a constant (Eq 6):

$$N_f \cdot \varepsilon_{pl}^{\alpha} = C \qquad (\text{Eq 6})$$

where N_f is the number of cycles to failure, ε_{pl} is the plastic deformation per cycle, and α is the fatigue exponent.

For shear load, as it usually occurs in fatigue of solder joints, the maximum plastic deformation is determined by:

$$\varepsilon_{pl} = \gamma = \frac{\Delta CTE \cdot l_0 \cdot \Delta\vartheta}{d} \qquad (\text{Eq 7})$$

Table 1 Parameters for Eq 3 for fully recrystallized Sn-Pb36-Ag2 as stated in Ref 1

Parameter	Value
Constant grain boundary sliding (GBS)	3×10^{15}
Constant dislocation climb/dislocation glide (DC)	3.5×10^{25}
Stress exponent	
GBS	3.3
DC	7
Dynamic shear modulus, MPa	$15290 - 23T$
Activation energy, kJ	
GBS	48
DC	52
Gas constant, J/mol · K	8.314

where γ is shear strain, ΔCTE is mismatch of the coefficient of thermal expansion, l_0 is the strain controlling length, $\Delta\vartheta$ is temperature change, and d is the thickness of solder deformed.

This means that the degradation in low-cycle fatigue is strain driven and it is important how much strain a specimen experiences during each cycle. The fatigue exponent α for Sn-Pb has been determined to be 1.6 to 2.

Deformation of Tin-Based Solder Alloys

Creep of Sn-Pb. The creep behavior of tin-lead has been well investigated. One set of data to describe the deformation behavior of fully recrystallized Sn-Pb36-Ag2 (Eq 3) is given in Table 1 (Ref 1).

As seen in Ref 1, the deformation behavior of solder is strongly dependent on the structure of the solder. Sn-Pb, as soldered, consists of large domains that recrystallize due to plastic deformation. This leads to a refinement of the tin phases and to a coarsening of the Pb phases. Once the Sn has recrystallized, enabling GBS, the material is much easier to deform, and creep rates are higher.

Creep of Lead-Free Alloys. Parameters for the deformation of lead-free solder have been published (Ref 11, 12) but are not necessarily consistent with each other. A comprehensive set

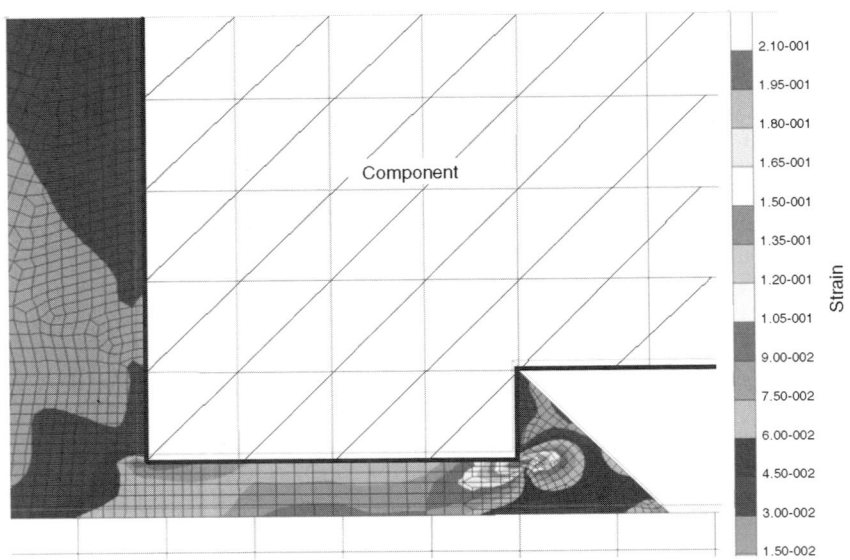

Fig. 4 Simulation of the strain distribution in the solder joint of a ceramic chip capacitor in a thermal cycle. Courtesy of C. Affolter, EMPA

of creep data has been published in Ref 13. In this work the data suggest that lead-free solder alloys creep considerably slower than Sn-Pb (Table 2) under the loading conditions included in the tests.

Another work (Ref 14) shows different values of creep rates for lead-free solder and Sn-Pb solder than Ref 13 but similar ratios of creep rates for lead-free solder and Sn-Pb solder (Table 3).

Further measurements (Ref 15) with Sn-Ag3.8-Cu0.7 gave a comparison of the creep rates with Sn-Pb-Ag (Table 4).

In Ref 15, a set of parameter for Eq 3 for Sn-Ag3.8-Cu0.7 has been published (Table 4). As seen in Ref 15, the grain size and the distribution of the phases also play an important role in the deformation of lead-free solder. As anisotropic deformation due to local resrystallization occurs, and intermetallic precipitations hinder the creep deformation (Fig. 7) (Ref 15).

Sn-Ag-Cu as soldered consists of large monocrystalline dendrites that recrystallize if plastic strain occurs (Ref 16) (Fig. 5, 6). Thus the same principles as the SnPb apply.

All of these influences are not fully understood to date. However, it is also due to this strong influence of the microstructure on the creep behavior that the data available to date differ so strongly.

Acceleration

Possibilities to Accelerate a Test. The possibilities to accelerate a test are limited by the deformation behavior outlined previously. When shortening the test cycle, the degradation mechanism relevant to the application (GBS/DC) must be activated, and the material should be allowed to creep as closely as possible to the extent occurring in reality.

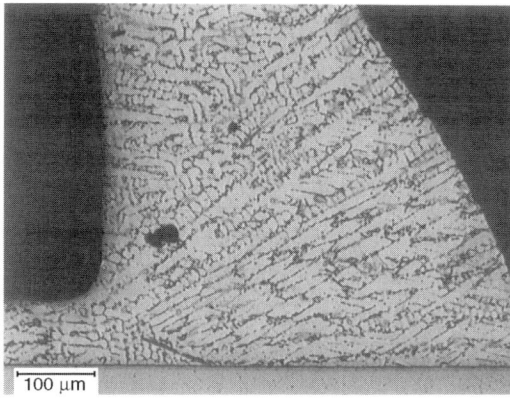

Fig. 5 SnAg3.8Cu0.7 as soldered

Fig. 6 SnAg3.8Cu0.7 after thermal cycling

Table 2   Creep rate of Sn-Pb and Sn-Ag-Cu as found in Ref 13

	Creep rate, s^{-1}			
	At 20 °C (68 °F) and 20 MPa	At 75 °C (167 °F) and indicated stress level		
Alloy		5 MPa	10 MPa	20 MPa
Sn-Pb-Ag	2×10^{-6}	2×10^{-6}	10^{-5}	9×10^{-5}
Sn-Ag-Cu	4×10^{-7}	...	...	2.5×10^{-6}

Table 3   Creep rate of Sn-Pb and Sn-Ag-Cu as shown in Ref 14

	Creep rate, rad/s, at 21°C (70 °F) and indicated stress level	
Alloy	10 MPa	20 MPa
Sn-Pb36-Ag2	5×10^{-7}	10^{-5}
Sn-Ag3.8-Cu0.7	7×10^{-8}	7×10^{-6}

Table 4   Parameters for Eq 3 for Sn-Ag3.8-Cu0.7 as stated in Ref 15

Parameter	Value
Constant grain boundary sliding (GBS)	1×10^5
Constant dislocation climb/dislocation glide (DC)	4.2×10^{39}
Stress exponent	
GBS	2
DC	9
Dynamic shear modulus, MPa	$22267 - 19.5T$
Activation energy, kJ	
GBS	48
DC	105
Gas constant, J/mol · K	8.314

In a thermal cycle, this means that, whether it is done in a temperature cycling chamber or by powering a product, one cannot run unlimited steep temperature ramps, keep the dwell time as short as desired, or extend the temperature swing beyond reasonable limits.

Accelerating mechanical tests is difficult. There are test standards (Ref 17) that state a correlation between test levels and time to failure, but the approaches to do so are highly empirical.

Passive Temperature Cycling Tests. The thermal cycling test is the most popular approach to estimate the expected lifetime or the expected failure rate of solder joints occurring in a given environment. In a thermal cycling test, the specimen is subjected to a temperature-time profile, usually with two temperature extremes. At each temperature extreme, the specimen is held at a certain dwell time and the temperature transfer is accomplished by a controlled temperature gradient.

The dwell time. Reducing the dwell time is an easy way to save test time. Assuming that a device has a duty cycle of 8 h on, 16 h off, applying dwell times of 30 min. already gives an acceleration factor of 24. During heat-up/cool-down, the PCB and the components are elastically deformed due to their mismatch in the coefficients of thermal expansion (CTE). During the dwell, the elastic stress is relaxed by the creep of the solder. However, as shown previously, the creep rate depends on the stress and the temperature. This results in an asymptotic relaxation where, theoretically, the difference of the thermal expansion will never be fully equalized, especially at temperatures below 0 °C (32 °F) (Fig. 8).

Practically, if relaxation results in the stress dropping below a certain level, the creep becomes so slow that the difference between the strain in the test and the one in the application becomes negligible. The time necessary for this relaxation however, can be rather long, depending on the dwell temperature. Further, the dwell time necessary to achieve a given relaxation due to creep also depends on the temperature gradient applied in the test. For example, if a slow

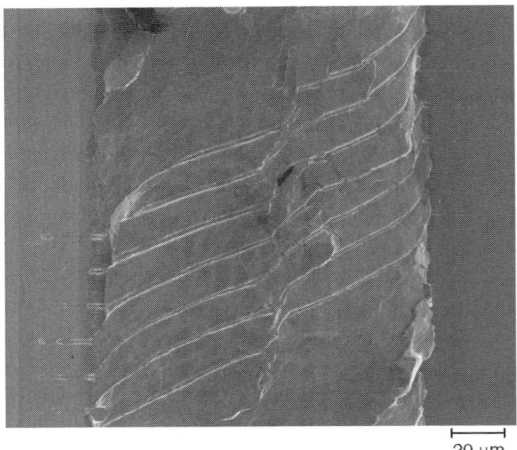

Fig. 7 Anisotropic deformation of Sn-Ag3.8-Cu0.7 due to shear strain. Straight lines have been etched across the solder joint of a single lap shear test specimen with FIB and have been deformed during the test. Courtesy of P. Jud, EMPA

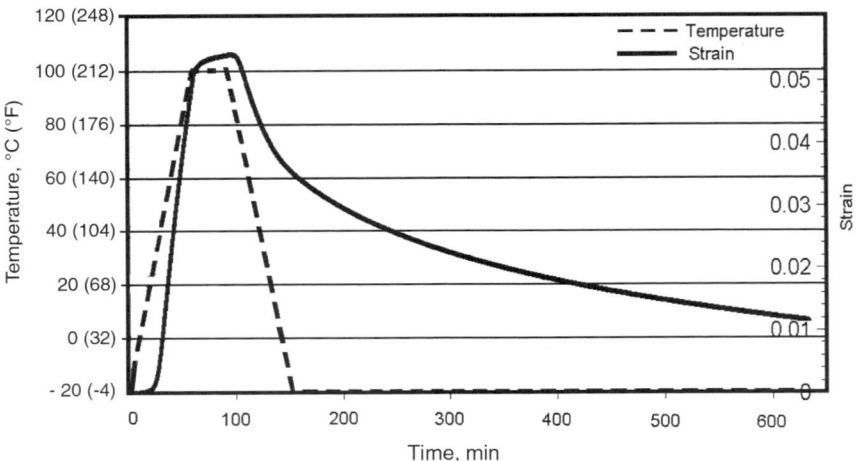

Fig. 8 Calculated relaxation of Sn-Pb and a ceramic chip capacitor 2220 on FR4 in a temperature cycle −20 to 100 °C (−4 to 212 °F), 2 °C/min. (35 °F/min.)

test with 2 °C/min. (35 °F/min.) is used, a large fraction of the relaxation already takes place during the ramp. For tests with steep temperature ramps, such as 40 °C/min. (104 °F/min.), this is not the case. During the transfer from one temperature extreme to the other, a considerable amount of the strain is stored as elastic deformation and has to relax during the dwell. Consequently, the dwell has to be longer to come to the point where the main fraction of plastic strain due to creep is induced into the solder joint. This leads to a complex relationship between the temperature dependency of creep, the stress built up during the ramp, and the time applied during the dwell. Based on the available models, the relaxation due to creep for SnAg3.8Cu0.7 in the dwell can be estimated (Ref 18) (Table 5, 6).

As can be seen in Tables 5 and 6, creep of Sn-Ag3.8-Cu0.7 is slow at temperatures below 0 °C (32 °F). This means that the dwell times to achieve total relaxation have to be very long. What is also clear from Tables 5 and 6 is that the dwell time at the upper temperature can be considerably shorter than the one at the lower temperature.

- At 100 or 120 °C (212 or 248 °F), 10 to 30 min. are sufficient to induce most of the maximum strain possible into a solder joint. The additional strain occurring with a longer dwell is negligible compared to the total strain of 5.6%.
- At 20 °C (68 °F), 30 to 60 min. dwell is long enough since the gain in strain from longer dwell times is quite marginal compared to the total strain.
- At 0 °C (32 °F) and below, it does not make much sense to wait any longer than 90 min. The additional strain brought into the solder joint between 90 and 120 min. is very small.
- Below 0 °C (32 °F), it might even be questioned whether it makes any sense at all to apply a dwell since a considerable relaxation of the solder needs several hours of dwell time to develop.

The temperature gradient. Increasing test acceleration by running steeper temperature ramps is not advised if the deformation mechanisms activated do not occur in the actual application. Temperature gradients larger than 2 °C/min. (35 °F/min.) are rare in actual applications. This means that GBS is the dominating deformation mechanism experienced by the solder joints of electronic equipment. It is therefore dangerous to apply temperature ramps in a test that activate large fractions of DC since the results of such a test are misleading when used to predict the performance of the solder joints in the real world.

Table 5 Calculated strain remaining in Sn-Ag3.8-Cu0.7 for various dwell times in a 30 μm solder gap of a ceramic capacitor 2220 on FR4 for different dwell temperatures in a thermal cycling test with a temperature swing of 120°C (248 °F) and a temperature gradient of 2 °C/min. (35 °F/min.). Maximum strain possible is 5.6%

Temperature		Calculated strain, %, at indicated dwell time							
°C	°F	5 min	10 min	30 min	60 min	90 min	120 min	8 h	16 h
120	250	0.34	0.17	0.11	0.07	0.05	0	0	0
100	210	0.34	0.29	0.2	0.14	0.13	0.1	0	0
20	70	1.9	1.8	1.5	1.3	1.1	1	0.3	0.1
0	30	2.6	2.5	2.3	2	1.8	1.6	0.7	0.3
−20	−5	3.4	3.3	3.1	2.8	2.6	2.4	1.2	0.5
−40	−40	4.2	4.2	4	3.7	3.5	3.3	1.8	0.9

Table 6 Calculated strain remaining in Sn-Ag3.8-Cu0.7 for various dwell times in a 30 μm solder gap of a ceramic capacitor 2220 on FR4 for different dwell temperatures in a thermal cycling test with a temperature swing of 120 °C (248 °F) and a temperature gradient of 10 °C/min. (50 °F/min.). Maximum strain possible is 5.6%

Temperature		Calculated strain, %, at indicated dwell time							
°C	°F	5 min	10 min	30 min	60 min	90 min	120 min	8 h	16 h
120	250	0.21	0.17	0.12	0.07	0.05	0.0	0.0	0.0
100	210	0.35	0.29	0.2	0.14	0.1	0.1	0.0	0.0
20	70	2.3	2.1	1.6	1.3	1.2	1	0.3	0.1
0	30	3.2	3	2.6	2.2	1.9	1.7	0.7	0.3
−20	−5	4.2	4	3.7	3.2	2.9	2.7	1.2	0.6
−40	−40	4.9	4.9	4.6	4.2	3.9	3.7	2.0	1.0

At low temperatures (−40 °C, or −40 °F), it is even possible to induce a brittle fracture in a solder joint if steep temperature ramps are applied. The main fraction of creep deformation in Sn-Pb solder takes place in the tin phase. Given the absence of Pb-free specific data, it is necessary to use the data available for Sn-Pb solder for the design of tests for lead-free solder. If a Norton law is used to model the deformation behavior of solder (Eq 3), it is possible to evaluate how much DC and GBS are activated for a given strain rate at a certain temperature (Fig. 9).

For stiff components, the strain rate in the solder gap can be estimated by Eq 8. (A more comprehensive method is discussed in Ref 18.)

$$\gamma' = \frac{\Delta CTE \cdot l_0 \cdot \Delta\vartheta'}{d} \quad (\text{Eq } 8)$$

where γ' is strain rate, ΔCTE is mismatch of the coefficient of thermal expansion, l_0 is the strain controlling length, $\Delta\vartheta'$ is the rate of temperature change, and d is the thickness of solder deformed.

If very steep temperature ramps are applied to a specimen, the assembly will warp due to the large stress build-up in the solder joints. This leads to a mixture of shear stress and normal stress in various axes in the solder joints that will not occur in actual applications. Results from such an experiment are of limited value if the assembly to be tested is not subjected to a similar stress in the field.

In addition, the time saved in the ramp may be more than offset during the dwell at low temperatures to achieve a significant amount of strain.

The dwell temperatures. As outlined previously, the plastic strain a material experiences in low-cycle fatigue has a large influence on the degradation behavior. With a damage exponent of approximately 2, doubling the plastic strain induced during a temperature cycle leads to one quarter of the number of cycles to failure. Thus, applying a temperature swing as large as possible during a thermal cycling test is preferred.

Given that the Coffin-Mansion approach holds, the acceleration factor of the test can be calculated by:

$$\frac{\Delta T_1}{\Delta T_2} = \left(\frac{N_{f2}}{N_{f1}}\right)^n \quad (\text{Eq } 9)$$

where ΔT_1 is the temperature swing 1 (e.g., application), ΔT_2 is the temperature swing 2 (e.g., test), N_{f1} is the number of cycles to failure condition 1, N_{f2} is the number of cycles to failure condition 2, and n is the damage exponent ≈ 2.

Note that this simple relationship only holds for total relaxation during the dwell, which means that sufficient dwell time must be given. If this is not possible, the strain brought into the solder joints during testing must be estimated. If the number of cycles to failure for various temperature swings is drawn in a double logarithmic diagram, a straight line through the measurement points allows an extrapolation to the real world (Fig. 10). However, one has to be aware of the dangers of extrapolating over a long range.

The upper temperature of the cycle is limited by the melting temperature of the solder, the destruction of components, or by the glass transition temperature (Tg) of the laminate material of the printed circuit board (PCB), which is between 120 and 150 °C (248 and 302 °F) for FR4. If the test temperature is higher than the softening point of a component's material, the mechanical properties of that material change dramatically or warpage occurs, consequently leading to undefined strain in the solder joint, which makes any use of the data found for reliability estimations questionable. Surpassing the Tg of the PCB material has a similar effect. When epoxy resin is heated above its Tg, its

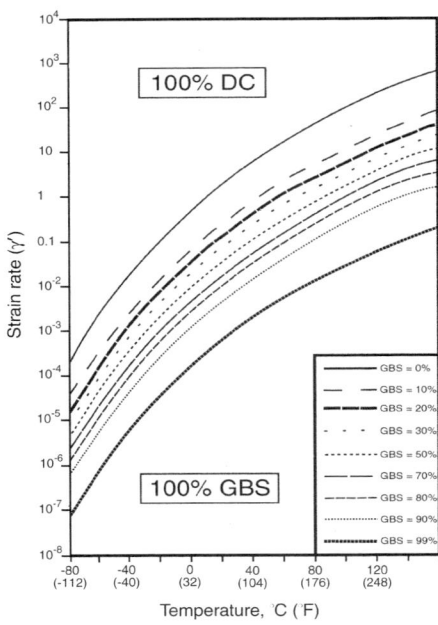

Fig. 9 Fractions of GBS and DC activated in Sn-Pb36-Ag2 at various temperatures and strain rates

properties change to a large extent (Fig. 11). This effect is not so dramatic in the X-Y plane of the PCB, at least not in the vicinity of the Tg

since, as the material softens, the glass fibers of the laminate will take a governing role in the deformation behavior of the PCB. However, exceeding the Tg in a test is not recommended, since a variable with an unpredictable influence is brought into the test.

The lower temperature is limited by the creep behavior of the solder as outlined previously. At −40 °C (−40 °F) for example, creep is so slow that the dwell time necessary to induce a strain, which is comparable to the real world of most applications, is so long that there is no such thing as an acceleration in the test. Additionally, cycling at low temperature causes large stresses that have to be borne by the components, and consequently, the destruction of components due to the fatigue test of the solder joints and brittle fracture of the solder becomes a threat.

Active Temperature Cycling. The functional test, where the specimen is subjected to an electrical load, thus producing heat, is the test which is closest to reality. In passive thermal cycling, only the strain due to the CTE mismatch is accounted for. In active temperature cycling, the strain due to the temperature difference between the components and the PCB is also part of the test. Basically, the same principles as outlined previously apply to the design of a test. The dwell time is the parameter that allows for most of the acceleration, but always under the restrictions outlined in this chapter. There are limitations in the temperature extremes, since

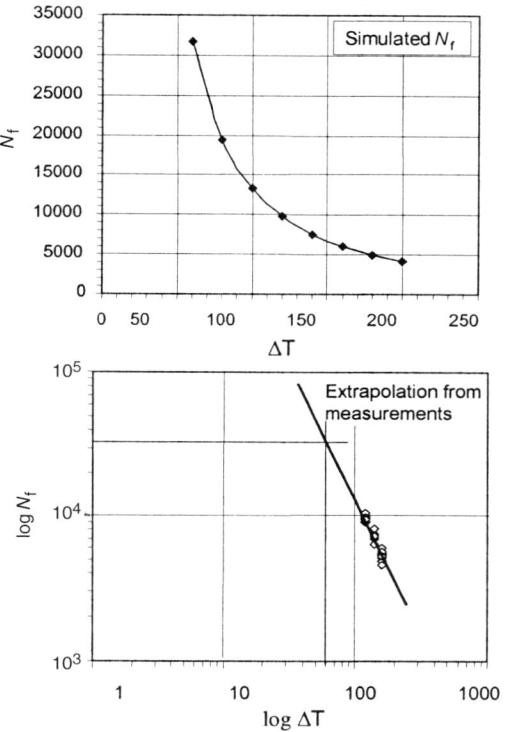

Fig. 10 Determination of N_f from data of accelerated tests

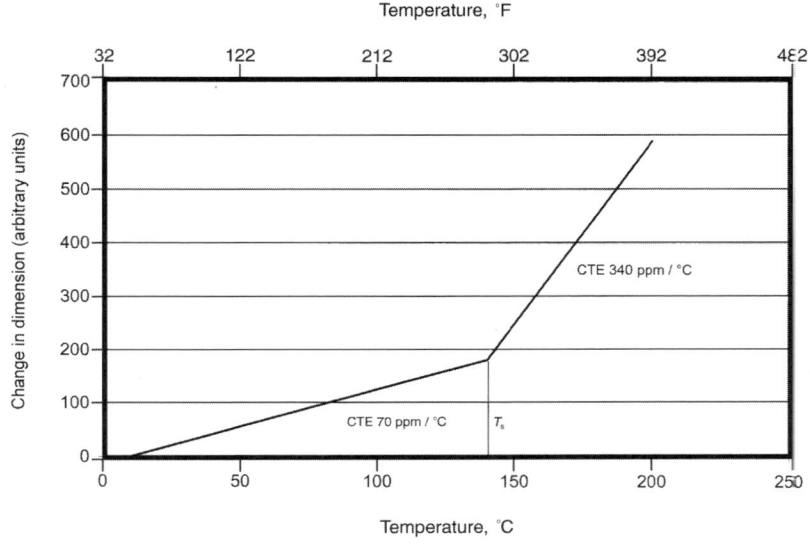

Fig. 11 Thermal expansion of an FR4 in the Z-axis

the electrical load that can be applied is limited and thus the temperature swing is restricted by the equilibrium of heat dissipation and transformation of electrical energy to heat. Changes in the overall setup of the assembly, such as turning off the cooling or insulating the housing are not recommended since this alters the thermal management considerably and thus eliminates the advantage of the active temperature cycling. Also, the application of electrical overload is dangerous since it can provoke the destruction of components. This means that an extension of the temperature swing resulting in a significant acceleration can usually not be achieved. One method to expand the plastic strain per cycle is by mounting the components on a substrate with a high ΔCTE to the components.

Mechanical Tests Include Mechanical Cycling and Vibration.

Mechanical cycling. In mechanical cycling, the same degradation mechanisms must be activated in a solder joint as in the application. It allows for the same strain to be applied to different solders independent of the temperature restrictions given by the materials of the PCB and the components. The solder joints of single components can be subjected to shear load by lateral loading, either by extending and relaxing the substrate or by moving the component in the PCB plane back and forth on the substrate (Fig. 12). Alternatively, the PCB of an assembly can be subjected to bending where the specimen rests loose on two supports and is deflected in the center (Fig. 13).

In both cases, the principles outlined in the passive temperature cycling apply. However, the test can be conducted at elevated temperatures where the creep behavior of the material allows for short dwell times and high creep rates.

In the case of lateral loading, the strain brought into a solder joint is well known and can be used for calculations of N_f. Additionally, in lateral loading, the degradation of the solder can be monitored by the decrease of the applied force necessary to maintain a given strain rate. In bending, the strain induced into the solder joints is different at each point of the PCB. First of all, the bending curve of a beam is not a radius but a curve:

$$w(x) = w_{max}\left(3\frac{x}{l} - 4\frac{x^3}{l^3}\right) \quad \text{(Eq 10)}$$

where w_{max} is deflection of the PCB in the center, l is the length of the specimen to bend, and $x \leq 1/2$ is the position where $w(x)$ is to be determined.

Secondly, due to the varying stiffness of the PCB caused by the components fixed on the board, the bending curve varies across the PCB. This means that the stress situation in the solder joints is not defined. Normal stress combined with shear stress occurs that make the extrapolation of the results found to the real world rather difficult. Also, in bending, the creep behavior of the solder has to be taken into account. Bending frequencies above 5Hz usually lead to failure in the leads of components or the components themselves rather than failure in the solder joints.

Vibration. Many electronic assemblies are subject to vibration stress. Two different kinds of vibration can be distinguished: sinusoidal vibration and random vibration.

Especially in an environment with rotating masses, sinusoidal vibrations, often with a constant frequency, are common. A sinusoidal excitation can be characterized by its frequency and its acceleration amplitude. An important test of sinusoidal vibration is the sine sweep test. In a given frequency band, the frequency changes at a given rate and the specimen is monitored to

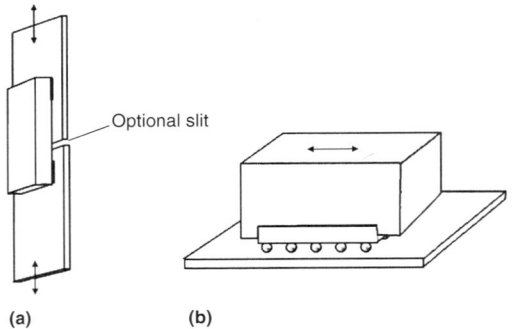

Fig. 12 Test set-up for mechanical cycling of components

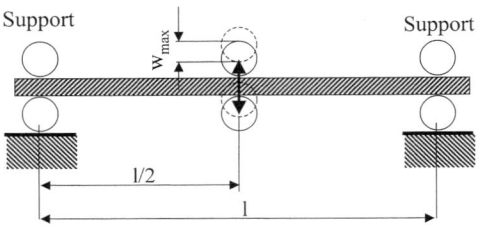

Fig. 13 Test set-up for bending tests

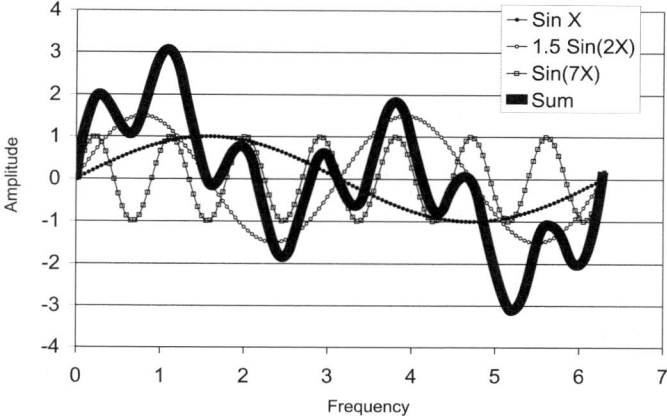

Fig. 14 Synthesis of a random spectrum from sinusoidal vibration

find the resonant frequencies that might lead to the destruction of solder joints or components, due to bending of the PCB. Principally, in sinusoidal vibration, one can vary the maximum acceleration, thus inflating the local deflection. However, there are limits. The local deflection of a PCB due to resonant excitation is highly non-linear and, if the specimen is mounted on a mounting plate, the PCB will bounce on the supporting plate causing mechanical shocks with several thousand grams of acceleration.

The main kind of vibration an electronic assembly is subjected to is random vibration where all frequencies are present within a certain frequency band. Therefore, a random spectrum can be set together by a synthesis of many sinusoidal vibrations (Fig. 14).

To characterize random vibration, a Fourier transformation of the spectrum is performed. The vibration is transformed from the time domain to the frequency domain, where each curve of Fig. 14 becomes a single line (Fig. 15). In the frequency domain, a broadband vibration might look like the one shown in Fig. 16. Using g_{rms} as a measure for the load applied by the test would give varying levels of g_{rms} depending on the width of the frequency band used to determine the g_{rms}. If the g_{rms} is squared and divided by the bandwidth, a constant measure called the Power Spectral Density (PSD) is found that has the unit g^2/Hz. To characterize the intensity of a random vibration, the area below a PSD frequency plot can be used having the unit g^2 or:

$$\sqrt{g^2} = g_{rms} \qquad (\text{Eq 11})$$

can be calculated (Ref 19).

Acceleration of failure due to vibration is difficult. Very often it is not the stress due to the acceleration of the mass of the component that drives the degradation of the solder joint. Rather, it is the deformation of the PCB, the bending, and the local build-up of hills at the resonance frequencies that destroy the solder joints.

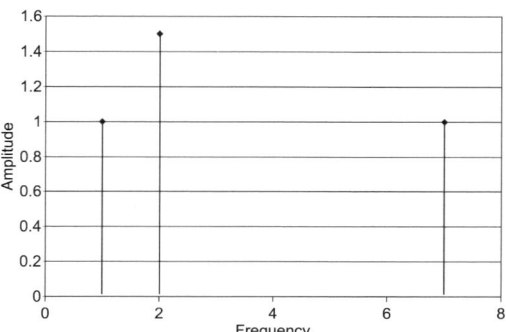

Fig. 15 Vibrations of Fig. 14 in the frequency domain

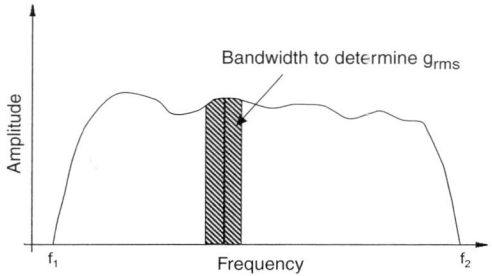

Fig. 16 Broadband vibration in the frequency domain

In random vibration, Ref 20 proposes a fatigue relationship for relating vibration levels to test time:

$$\frac{W_0}{W_1} = \left(\frac{T_1}{T_0}\right)^{1/M} \quad \text{(Eq 12)}$$

where W_0 is the functional level (PSD), W_1 is the endurance level (PSD), T_0 is time at functional level, T_1 is time at endurance level, and M is the material constant (slope of line in the log-log W/T plot, to be empirically determined).

This relationship leads to a straight line if the PSDs are plotted against the time to failure in a log-log graph. However, it has to be emphasized that the test-spectrum must have a relationship to the vibrations occurring in reality. Extrapolation from empirical data is always of questionable practice if extended far beyond the tested range. Since M determines the slope in the log PSD-log time graph, any extrapolation is very sensitive to variations of M.

One application of Eq 12 is defined in MIL STD 810F method 514.5, where the test duration is adapted to the PSD level the test is run at (Fig. 17).

Designing a Test

From the Application to the Test. The first step in designing a test is to know the stress occurring in the application. Depending on the test to be designed, the determination of the environmental load a specimen experiences during service might need considerable measurement and analysis. It is common that the preparation of a test and its design takes a multiple of the time the actual test needs.

The second step is to determine how much (how many cycles, how long) the specimen to be tested has to endure in the field. This task involves many people including the customer, and can be conducted by direct negotiations, a questionnaire, or by observation of the market. It is also important to agree on the definition of failure. This can be an electric open, a given crack length, a drop in shear strength, etc.

The third step is the design of the test, or the decision about a standard test procedure to be used. If a test is to be designed, the test strategy within the limits discussed earlier has to be determined, as the acceleration can be achieved either by increasing the stress level or by applying the stress more frequently than in the actual application, or a combination of both. Increased stress levels can be achieved by extending the temperature swing compared to the application, using materials with large differences in the CTE, inducing large plastic strain in mechanical cycling, using substantial deflection in bending, increasing the peak acceleration in sinusoidal vibration, and/or enhancing the PSD level in random vibration. The stress can be applied more frequently than in the application by shortening the dwell, running steeper temperature ramps, running on-off cycles as fast as possible, apply-

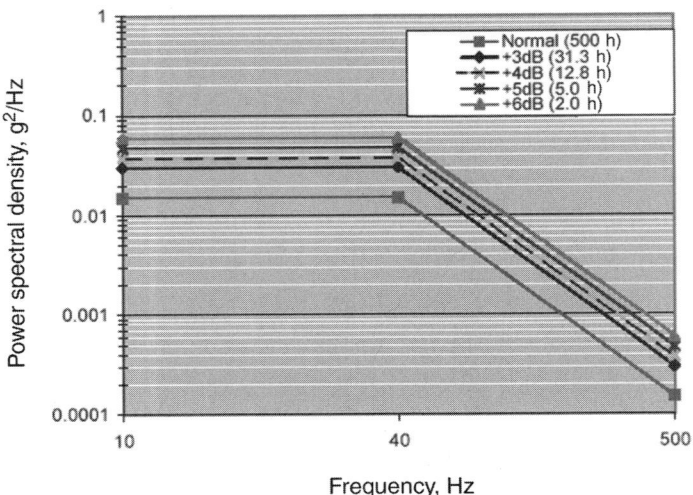

Fig. 17 Test levels for accelerated random testing according to MIL-STD-810F, Method 514.5

ing rapid movement in the mechanical test, and/or running the bending test at high frequencies.

Passive Thermal Cycling. Electronic assemblies are used in a wide range of climates. Many of these have been listed in Ref 20. However, it is advisable to measure the heating, due to the operation of the electronic devices which have considerable variation between different locations in equipment. This temperature swing, plus possible temperature excursions (e.g., going from an air-conditioned room to a hot, outdoor environment) defines the temperature swing. Additionally the temperature gradient in the application must be determined, especially in industrial, military, and aerospace applications where steep temperature ramps are possible.

Once the thermal stress has been worked out, the range of the temperature extremes has to be fixed. If the assembly is used above 0 °C (32 °F), the lower temperature of the test should not be below 0 °C (32 °F). The upper limit is given by the materials of the assembly, usually 100 to 120 °C (212 to 248 °F). If the assembly is used in an environment well below 0 °C (32 °F), it is advisable to extend the test temperatures to this limit, but accelerating the test becomes difficult due to the slow creep of the solder.

The dwell time is determined by the creep behavior of the solder. At and above 100 °C (212 °F), 15 min. of dwell should be allowed, and between 80 to 100 °C (176 to 212 °F), 20 min. is advisable. Unfortunately, it is rare that enough time can be given to allow for creep at low temperatures. Tables 5 and 6 can be used as guidelines once the specimen has reached the dwell temperature. Even if the numbers are only valid for the 2220 chip capacitors used for this estimate, the decrease in strain compared to the maximum strain possible can serve as a general guideline.

The temperature gradient does not have to be constant during a test. However, it is important to apply strain rates that activate mainly GBS, if this is the case in the real application (Fig. 9). If steep temperature ramps occur in the application, it is recommended to apply the same ramps in the test. However, no acceleration in the ramp is possible in this case.

Active thermal cycling. As already mentioned, only the dwell time can be varied if an electronic assembly is to be tested in an accelerated test. If the performance of a single component is to be evaluated, one can enlarge the strain per cycle by mounting the component on a substrate with a larger ΔCTE to the component, but the temperature ramp has to be limited in order to avoid the destruction of the component.

A measurement of the temperature-time behavior of the specimen gives clues about the equilibrium temperature and the time until the equilibrium is achieved in the "on" cycle.

The maximum temperature range is given by the maximum electric load allowed. The dwell times should not be shorter than the ones mentioned previously. In active cycling, the time until the specimen has reached its temperature equilibrium has to be considered in the definition of the thermal cycle. The temperature gradient is given by the thermal management of the assembly. In the case of using a stiff substrate to achieve a large CTE difference, it is advisable to limit the temperature gradient below the 100% GBS line (Fig. 9). This avoids the activation of degradation mechanisms that are irrelevant in the application, unless the application requires high temperature gradients.

Mechanical cycling. Mechanical cycling is a test that serves mainly for comparative purposes. A correlation to the application is not possible.

In cyclic testing, the specimen to be tested should be attached to a support in order to avoid bending of the substrate. The mechanical loading equipment must be very stiff and the mechanical attachment of the specimen to the apparatus must be very firm. Pulling a specimen in the form of Fig. 8(a) apart does not cause problems. Pushing them together often leads to a deformation sideways, which leads to uncontrolled strain making any use of the data questionable. The test should be run as strain-rate controlled with a defined strain at each cycle. However, the deformation of the measurement apparatus has to be taken into account. Because the lead-free solder is quite creep resistant, the force needed in the test will lead to a considerable deformation of the test equipment. Also, the specimen to be tested is a solder layer which is approximately 30 μm thick. This means that 50% plastic shear strain results in 15 μm deformation. However, 15 μm elongation of the testing equipment (causing 100% failure in the measurement) is not much in a testing apparatus. Further, the test must be done in a temperature-controlled environment because of the strong influence of the temperature on the deformation behavior of the solder and because temperature variations lead to a measurable deformation of the testing equipment.

In bending testing, the component should be fixed in such a way that allows the specimen to move freely (Fig. 13). Otherwise the deformation of the PCB becomes very complex. The strain in the solder joints can be estimated with Eq 10. Note that the components in the center of the specimen experience more strain per cycle than the ones at the edges. High strain rates causing high stress might lead to the destruction of components.

Sinusoidal Vibration. The main parameters of a sinusoidal vibration test to be run as outlined in IEC 68-2-6. The assembly to be tested must be fixed in the same way as in the application. The acceleration measuring point must be as close as possible to the specimen to be tested. The resonances of the mounting equipment must be determined prior to the test and might have to be excluded from the test. To find the resonances, it is often not sufficient to monitor the response of the system given by the accelerometer fixed at the vibration table as the control; a second measuring point on the specimen is advisable. On the other hand, the mass of an accelerometer fixed on the specimen to determine resonances influences the vibration

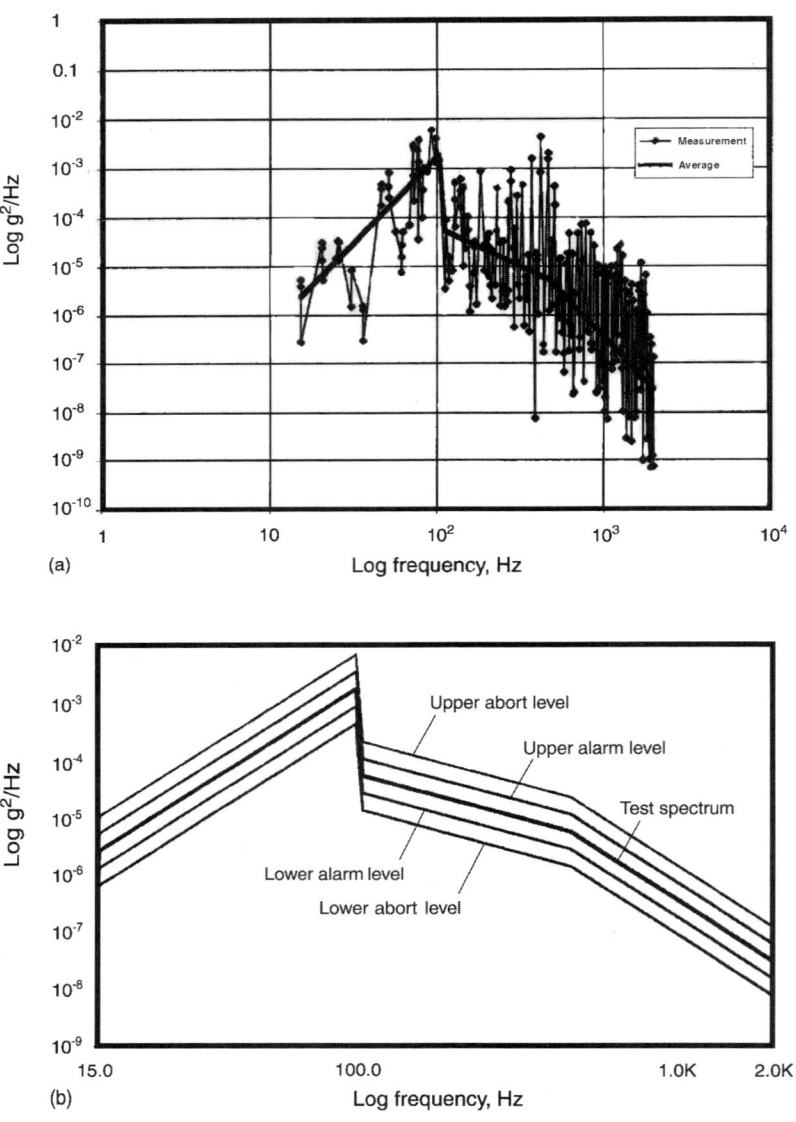

Fig. 18 Measurements of a random vibration and final test spectrum

behavior of the specimen. Whenever possible, contact-free measuring methods such as laser interferometry should be applied.

Random Vibration. If no standard test can be found that suits the needs of the application, a random vibration test spectrum has to be generated. First, the vibration of the application in the PSD frequency domain should be measured. In the measured vibration, a spectrum through the average of the acceleration excursions is defined (Fig. 18). At least three tests at varying levels should be run. The time to failure for each test level is drawn in a double logarithmic graph against the test levels, thus finding M in Eq 11. Of course, more than one specimen can be tested at each level to account for the statistical spread of the results. Finally, with Eq 11 the time to failure in the application can be estimated.

ACKNOWLEDGMENTS

The author wants to express his thanks to the Swiss Federal Laboratories for Material Testing and Research (EMPA), Ascom Switzerland, Enics Switzerland, Oerlikon-Contraves, Schindler Electronics, Siemens Building Technologies and Siemens Switzerland, as well as to the Swiss Commission for Technology and Innovation (KTI) for their support of the research in lead-free soldering.

REFERENCES

1. D. Grivas, K.L. Murty, and J.W. Morris, Deformation of Superplastic Alloys at Relatively High Strain Rates, *Acta Metall.*, Vol 27, Issue 5, 1979, p 731
2. F. Garofalo, An Empirical Relation Defining the Stress Dependance of Minimum Creep Rate in Metals, *Trans. Mat. Soc. AIME* 227, 1963, p 351
3. E.W. Hart, A Theory for Flow in Polycrystals, *Acta Metall.*, Vol 15, Issue 9, 1967, p 1545
4. L. Weber, "Creep and Fatigue Behavior of Eutectic SnPb36Ag2 Solder," thesis at ETH Zürich, 1997
5. M.C. Shine and L.R. Fox, "Fatigue of Solder Joints in Surface Mount Devices" ASTM STP 942, 1988, p 588
6. J.D. Morrow, "Symposium on Internal Friction, Damping and Cyclic Plasticity" ASTM STP 1964, p 378
7. L.F. Coffin Jr., "Fatigue at Elevated Temperatures," ASTM STP 520, 1973, p 112
8. W. Engelmaier, "Reliable Surface Mount Solder Attachments Through Design and Quality Manufacturing," International Workshop on SMT Reliability and Manufacturing Issues, 1992
9. Z. Guo, A.F. Sprecher, and H. Conrad, Plastic Deformation Kinetics of Eutectic Pb-Sn Solder Joints in Monotonic Loading and Low-Cycle Fatigue, *J. Electron. Packaging* 114, 1992, p 112
10. H.D. Solomon, The Solder Fatigue Life Acceleration Factor, *J. Electron. Packaging*, 1991, p 186
11. H. Mavoori, J. Chin, S. Vayman, B. Moran, L. Keer, and M. Fine, Creep, Stress Relaxation and Plastic Deformation in SnAg and SnZn Eutectic Solders, *J. Electron. Mater.* 26, 1997, p 783
12. Y.H. Pao, S. Badgley, R. Govila, and E. Jih, An Experimental and Modelling Study of Thermal Cyclic Behavior of SnCu and SnPb Solder Joints, Proceedings Mat. Res. Sec. Symposium, Vol 323, 1994, p 128
13. "Lead Free Solder Project," National Center for Manufacturing Sciences, CD, 1999
14. Lead-Free Soldering, National Physical Laboratory, United Kingdom, http://www.npl.co.uk/ei/research/leadfree.html (accessed Feb 2005)
15. P. Jud, G. Grossmann, U. Sennhauser, and P.J. Uggowitzer, Local Creep in SnAg3.8Cu0.7 Lead-Free Solder, *J. Electron. Mater.*, Vol 34, Issue 9, 2005
16. G. Grossmann, G. Nicoletti, and U. Solèr, Results of Comparative Reliability Tests on Lead-Free Solder, 52nd ECTC Proc., 2002, p 1232
17. "Environmental Engineering Considerations and Laboratory Tests: Vibration," MIL-STD-810F, Method 514.5, U.S. Department of Defense
18. G. Grossmann and L. Weber, Lifetime Assessment of Soft Solder Joints on the Base of the Metallurgical Behavior of Sn62Pb36Ag2, 21st IEMT Symposium Proc., 1997, p 256
19. Spectral Dynamics, Training Documentation Vibration Testing
20. "Guidelines for Accelerated Reliability Testing of Surface Mount Solder Attachments," IPC-SM-785, Institute for Interconnecting and Packaging Electronic Circuits, 1992

CHAPTER 8

Thermomechanical Reliability Prediction of Lead-Free Solder Interconnects

Suresh K. Sitaraman and Karan Kacker, Georgia Institute of Technology

Introduction

Microelectronic packaging assemblies typically consist of stacked layers of materials with dissimilar material properties. When such packaging structures are fabricated, assembled, and used, they are subjected to a wide range of thermal excursions. Under such thermal excursions, thermally-induced stresses develop in various parts of the microelectronic packaging assembly due to the coefficient of thermal expansion (CTE) mismatch among the dissimilar materials, and the microelectronic packaging assembly could prematurely fail, if not carefully designed.

Thermomechanically reliable design of microelectronic packages is a time-consuming and costly task. Current industrial practice involves building and assembling prototype systems and subjecting the prototype systems to extensive qualification tests. Although such a build-and-test approach may be appropriate for "legacy" systems, when it comes to new materials and new designs, there is no past or legacy experience to guide, and therefore, the initial few prototypes could lead to misleading results, if not carefully designed and assembled. As the experimental prototyping and testing process takes several weeks to complete, every time a wrong design choice is made, time and money are lost in the qualification process. Therefore, prior to the build-and-test approach, physics-based thermomechanically reliable designs need to be developed to be able to obtain "first-pass success" designs. As the electronic packaging industry has very limited or no legacy-based experience with the relatively new lead-free solder alloys, it is essential to develop thermomechanically reliable designs prior to embarking on expensive and time-consuming experimental prototyping and testing.

Among the various lead-free solder alloys, the National Electronics Manufacturing Initiative (NEMI) recommends Sn-3.9Ag-0.6Cu ($\pm$0.2%) for reflow soldering (Ref 1), while JEITA (Japan Electronics and Information Technology Industries Association) recommends Sn-3.0Ag-0.5Cu. Under most typical use conditions, Sn-Ag-Cu (SAC) alloy in general exhibits a greater resistance to creep as compared to Pb-Sn alloy and hence creeps 10 to 100 times slower (Ref 2). The microstructure, the plastic and creep behavior, and the failure mechanism in Sn-Ag-Cu solder are vastly different compared to Pb-Sn solder, and therefore, it is important to develop appropriate thermomechanical predictive models for Sn-Ag-Cu solder.

Constitutive Models for Lead-Free Solder Alloy

The material constitutive model plays an important role in the development of thermomechanical models for microelectronic packaging assembly. Under thermomechanical loading, solder alloy undergoes elastic and inelastic deformation. Elastic deformation is recoverable, while inelastic deformation, consisting of time-independent plastic deformation and time-dependent creep deformation, is not recoverable. The constitutive behavior can be represented by a combination of elastic, plastic (isotropic or kinematic hardening), viscoelastic, viscoplastic/creep models.

The development of thermomechanical models typically involves (a) the development of constitutive models for the solder alloy and other materials in the packaging assembly, (b) the development of geometry models to represent the solder interconnects and the packaging assembly, and (c) the development of failure predictive models for the various material systems in the assembly. The thermomechanical models, when subjected to thermal excursions seen during accelerated qualifications or during field-use, will identify the location of failure, mode of failure, and time to failure in various parts of the packaging assembly. Prior to usage, the models should be validated with experimental test data.

Thermoelastic Behavior

Isotropic elastic properties, E, the Young's modulus, and υ, the Poisson's ratio can be used to represent the elastic behavior of the solder alloy. The temperature-dependent elastic behavior of two lead-free solder alloys: Sn-Ag-Cu and Sn-Ag are presented in this section. For example, E is given by:

$$E = E_0 + E_1 T \quad \text{(Eq 1)}$$

where E is the Young's modulus, T is the absolute temperature, and E_0 and E_1 are constants. Table 1 presents the elastic constants for Sn-Ag-Cu and Sn-Ag solder alloys. Similar to Eq 1, the shear modulus of the material can also be represented as a linear function of the absolute temperature.

Inelastic Deformation Behavior

The permanent deformation after the removal of the applied load is called the inelastic deformation and is divided into an instantaneous plastic deformation component and a time-dependent creep deformation component.

Time-Independent Plastic Deformation. The time-independent plastic deformation of a solder alloy can be represented as:

$$\varepsilon_{pl} = (\sigma/A)^{1/n} \quad \text{(Eq 2)}$$

where ε_{pl} is the time-independent plastic strain, σ is the shear stress, A is commonly referred to as the monotonic strength coefficient, and n is the strength hardening exponent (Ref 6). The plasticity model can be of different types—bilinear kinematic hardening, multilinear kinematic hardening, multilinear isotropic hardening, etc. Bilinear kinematic hardening model represents both the elastic region as well as the plastic region as linear. This model requires the Young's modulus for the elastic region, the yield strength to designate the start of the plastic region, and the tangent modulus for the plastic region. The tangent modulus represents the slope of the stress-strain curve past the yield strength, in the region of plastic deformation. Multilinear kinematic hardening is a more accurate representation of the hardening behavior. Unlike bilinear hardening where the hardening curve is represented by a single linear segment, in the multilinear kinematic hardening, the plastic hardening curve is approximated by a number of linear segments. In kinematic hardening, the yield surface remains constant in size but moves around in the stress space such that, after a tensile plastic deformation, the material will yield in compression before the monotonic compressive yield stress is reached. This behavior is known as the Bauschinger effect. In general, solder alloys are represented using kinematic-hardening models. In isotropic hardening, on the other hand, the yield surface expands and the yield stress increases. This means that no further yield occurs until the new yield stress level is

Table 1 Elastic constants for lead-free solders

	Solder	Reference No.	E_0, GPa	E_1, GPa/K	Specimen type
1	95.5Sn3.9Ag0.6Cu	3	53 GPa	−0.08	Bulk specimen
2	95.5Sn3.9Ag0.6Cu	4	18.6 GPa	−0.0206	0.18 mm thick solder joints deformed in nominal shear
3	96.5Sn3.5Ag	5	19.30 GPa	0.06895	Soldered assemblies employing both shear and tensile loading

reached, and therefore, the Bauschinger effect is not taken into consideration.

As discussed, the elastic and plastic behavior of Sn-3.5Ag and Sn-3.8Ag-0.7Cu can be represented through a multilinear model. The constants for the elastic-multilinear hardening model have been obtained by a reversed shear loading on flip-chip solder interconnects (Ref 7). Table 2 provides the constants for Sn-Ag-Cu.

As seen in Table 2 and the associated figure, the elastic-plastic behavior is represented through three straight lines. The first straight line from (0, 0) to (ε_1,σ_1) represents the elastic behavior, the second straight line from (ε_1,σ_1) to (ε_2,σ_2) represents the initial plasticity, and the third straight line from (ε_2,σ_2) to (ε_3,σ_3) represents the saturated plasticity.

Creep—Time-Dependent Inelastic Deformation. In general, the creep behavior of materials consists of three different stages: primary creep, secondary creep, and tertiary creep. In the primary creep regime, the material undergoes strain hardening, resulting in a decreasing strain rate with time. In the secondary stage, also known as steady-state creep regime, the creep strain rate is essentially constant, showing a very slow decrease. In the tertiary creep regime, the strain rate increases with time, ultimately resulting in failure of the material. The different stages of creep behavior are illustrated in Fig. 1.

The solder alloys are often subjected to steady-state creep regime under typical thermomechanical loading conditions. Steady-state creep deformation can generally be expressed by the following relationship (Ref 8) for low- and medium-stress values:

$$\frac{d\gamma_s}{dt} = \frac{CGb}{kT}(b/d)^p(\tau/G)^n D_0 \exp(-Q/kT) \quad \text{(Eq 3)}$$

where $d\gamma_s/dt$ is the steady state strain rate, G is the shear modulus, b is the Burgers vector, k is the Boltzmann's constant, T is the absolute temperature, d is the grain size, τ is the applied stress, D_0 is the frequency factor, Q is the activation energy, n is the stress exponent, and p is

Table 2 Sn-Ag-Cu elastic-plastic behavior

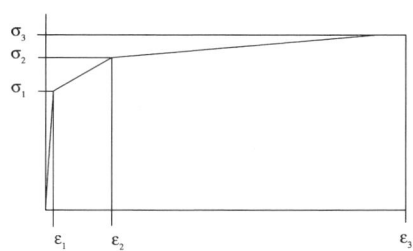

Sn-Ag-Cu T [K]	ε_1	ε_2	σ_1 [MPa]	σ_2 [MPa]	σ_3 [MPa]
As soldered					
278	1.4×10^{-3}	4×10^{-3}	57.4	80	2500
323	1.4×10^{-3}	4×10^{-3}	53.2	72	1900
Aged at T = 150 °C (302 °F) for 1500 h					
278	0.6×10^{-4}	1.5×10^{-3}	24.6	36	2100
323	0.6×10^{-4}	1.5×10^{-3}	22.8	32	1500

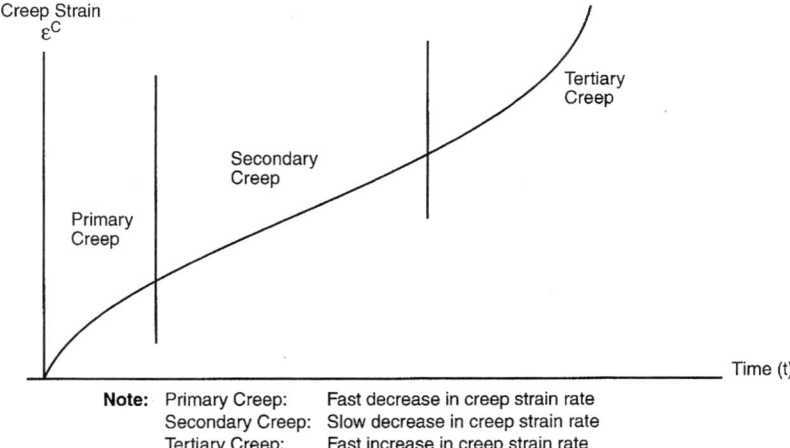

Note: Primary Creep: Fast decrease in creep strain rate
Secondary Creep: Slow decrease in creep strain rate
Tertiary Creep: Fast increase in creep strain rate

Fig. 1 Typical creep strain rate curve

the grain size. By combining certain constants, Eq 3 can be rewritten as:

$$\frac{d\gamma_s}{dt} = C_1 \tau^n \exp(-Q/kT) \quad \text{(Eq 4)}$$

where C_1 is another constant. This relationship does not hold for high values of stress where the creep strain rate is an exponential function of stress rather than a power-law function. This is known as creep power-law breakdown. Creep behavior in this high-stress region is expressed as:

$$\frac{d\gamma_s}{dt} = C_2 \exp(C_3 \sigma) \quad \text{(Eq 5)}$$

where C_2 and C_3 are constants. In order to obtain a closed form solution which captures both the power-law behavior, as well as the exponential behavior, the following form is utilized (Ref 9):

$$\frac{d\gamma_s}{dt} = C_4 \frac{G}{T} \left(\sinh\left(\alpha \frac{\tau}{G}\right)\right)^n \exp(-Q/kT) \quad \text{(Eq 6)}$$

where α represents the stress level at which the power law dependence no longer holds and C_4 is a constant. Equation 6 can be re-cast as:

$$\frac{d\gamma_s}{dt} = C_1 (\sinh(C_2 \tau))^{C_3} \exp(-Q/kT) \quad \text{(Eq 7)}$$

The creep behavior of solder can also be expressed by an equation of the following form (Ref 7):

$$\frac{d\gamma_s}{dt} = A_{II} \frac{\sigma^{n_{II}}}{d^p} \exp\left(-\frac{Q_{II}}{kT}\right) + A_{III} \sigma^{n_{III}} \exp\left(-\frac{Q_{III}}{kT}\right) \quad \text{(Eq 8)}$$

where d is the grain size, p is the grain size exponent and A_{II} and A_{III} are pre-factors. This form of the creep equation is sometimes referred to as the double-power-law creep equation. The first term in the equation is representative of grain boundary sliding and the second term is representative of dislocation climb. A summary of the constants for different creep models for Sn-Ag and Sn-Ag-Cu solder is provided in Table 3.

For lead-free solder, Dutta (Ref 16) has developed a creep model which takes into consideration strain-enhanced coarsening of intermetallic particles under thermomechanical cycling. This model captures material degradation over a period of time in terms of intermetallic particle size and consequently results in a stress-strain relationship which continues to evolve and change as the solder alloy is subjected to thermomechanical loading. For the sake of brevity, Dutta's model is not discussed in detail in this chapter.

Unified Plasticity Model The time-independent plastic deformation and the time-dependent creep deformation arise from the same fundamental mechanism of dislocation motion. Hence, a constitutive model which captures both of these deformation mechanisms is desirable. Such a constitutive model is referred to as a unified plasticity model. A commonly-used unified plasticity model is the Anand's model. This is a rate-dependent phenomenological model (Ref 17 and 18). There are two basic characteristics of the Anand's model. First, no explicit yield criterion is specified, and second, a single internal state variable (ISV) s, the deformation resistance, represents the isotropic resistance to inelastic strain hardening. Anand's model can represent the strain rate and temperature sensitivity, strain rate history effects, strain hardening, and the restoration process of dynamic recovery. Equation 9 shows the functional form of the flow equation that accommodates the strain rate dependence on the stress:

$$\dot{\varepsilon}_p = A e^{(-Q/RT)} \left[\sinh\left(\xi \frac{\sigma}{s}\right)\right]^{1/m} \quad \text{(Eq 9)}$$

where, $\dot{\varepsilon}_p$ is the inelastic strain rate, A is the pre-exponential factor, Q is the activation energy, m is the strain rate sensitivity, ξ is the multiplier of stress, R is the gas constant, and T is the absolute temperature. The ISV enters the flow equation only as a ratio with the equivalent stress. The temperature dependency is incorporated through a classical Arrhenius term. The stress and strain rate dependency, however, is of the Garafalo form. The evolution of s is given by Eq 10 and 11:

$$\dot{s} = \left\{ h_o \left|1 - \frac{s}{s^*}\right|^a \cdot \text{sign}\left(1 - \frac{s}{s^*}\right) \right\} \cdot \dot{\varepsilon}_p a > 1 \quad \text{(Eq 10)}$$

$$s^* = \hat{s} \left[\frac{\dot{\varepsilon}_p}{A} e^{(Q/RT)}\right]^n \quad \text{(Eq 11)}$$

where h_o is the hardening/softening constant and a is the strain rate sensitivity of hardening/softening. The quantity s^* represents a saturation value of s associated with a set of given temperature and strain rate as shown in Eq 9, $\hat{s}$ is a coefficient of saturation, and n is the strain rate sensitivity for the saturation value of deformation resistance.

As can be seen from the model, we need to determine nine material constants to be able to model the material behavior adequately. They are: A, Q, ξ, m, h_o, $\hat{s}$, n, a, plus the initial value of the deformation resistance, s_o. The constants for a similar model for the case of Sn-3.5Ag-

Table 3 Creep model constants for lead-free solder

Model	Equation	Reference No.	Composition	Constant	Value	Specimen type
Norton creep model	$\dot{\varepsilon} = C_1[(C_3\sigma)]^{C_2} \exp\left(\dfrac{-C_4}{T(K)}\right)$	10	SnAg4.0Cu0.5	C_1, s^{-1} C_2 C_3, MPa^{-1} C_4, K	8×10^{-11} 12 1 8,996.57	Flip chip assembly under reversed shear
			SnAg3.5	C_1, s^{-1} C_2 C_3, MPa^{-1} C_4, K	5×10^{-6} 11 1 9,994.8	As-cast
		5, 11	SnAg3.5	C_1, s^{-1} C_2 C_3, MPa^{-1} C_4, K	9.64×10^{-6} 6.5 1 6,557	Soldered assemblies employing both shear and tensile loading
	$\dot{\varepsilon} = A(\sigma)^n e^{-(Q/R_gT)}$	12	SnAg3.5Cu0.75	Q/R, K n A, s^{-1}	13,180 15.795 4.77×10^{12}	Bulk specimen
Double power law	$\dfrac{d\varepsilon}{dt} = A\left(\dfrac{\sigma}{\sigma_n}\right)^{n_1} \exp\left(-\dfrac{Q_1}{RT}\right) + A_2\left(\dfrac{\sigma}{\sigma_n}\right)^{n_1} \exp\left(-\dfrac{Q_2}{RT}\right)$	7	SnAg4.0Cu0.5	A_1, s^{-1} n_1 Q_1/R, K A_2, s^{-1} n_2 Q_2/R, K	10^{-6} 3 34.6 10^{-12} 12 61.1	Bulk specimen
			SnAg3.5	A_1, s^{-1} n_1 Q_1/R, K A_2, s^{-1} n_2 Q_2/R, K	7×10^{-4} 3 46.8 2×10^{-4} 11 93.1	Bulk specimen
Garofalo creep	$\dfrac{d\gamma_s}{dt} = C_1[\sinh(C_2\sigma)]^{C_3} \exp\left(\dfrac{-C_4}{T}\right)$	5	SnAg3.5	C_1, s^{-1} C_2, MPa^{-1} C_3 C_4, K	$(568.4 + T)/T$ $1/(13.789 + 0.0492T)$ 5.5 5,802	Soldered assemblies employing both shear and tensile loading
		13	SnAg3.5	C_1, s^{-1} C_2, MPa^{-1} C_3 C_4, K	98,437 0.103 6.65 9,562	...
		3	Sn-Ag-Cu	C_1, s^{-1} C_2, MPa^{-1} C_3 C_4, K	441,000 5×10^{-3} 4.2 5,412	Bulk specimen
		14	Sn3.8Ag0.7Cu	C_1, s^{-1} C_2, MPa^{-1} C_3 C_4, K	32,000 37×10^{-9} 5.1 6524.7	Creep specimen loaded in tension and measurements made using a video extensometer
		4	Sn3.9Ag0.6Cu	C_1, s^{-1} C_2, MPa^{-1} C_3 C_4, K	248.4 0.188 3.789 7,567	0.18 mm thick solder joints deformed in nominal shear
		15	Sn-Ag-Cu	C_1, s^{-1} C_2, MPa^{-1} C_3 C_4, K	277,984 24.47×10^{-9} 6.41 6,500	...

0.75 have been obtained (Ref 19). The constants for Anand's model for the case of Sn-3.5Ag solder, that are given in Table 4, were obtained from bulk solder specimens subjected to tensile loading on a screw-driven tensile machine at different temperatures (233K, 253K, 313K, 353K, and 398K) and at different strain rates (0.005s^{-1}, 0.01s^{-1}, 0.02s^{-1}, and 0.1s^{-1}). Also, a temperature-dependent value of the initial deformation resistance s_0 was obtained, rather than keeping s_0 constant (Ref 20).

Geometric Modeling

Three types of geometry models, commonly used to represent microelectronic packaging assemblies, will be discussed in this section: (a) 2D model, (b) 2.5D or strip model, and (c) 3D model. The choice of these three models depends on the computational accuracy needed, the computing resources available, and the focus of the modeling study.

For illustrative purposes, models of a Flip Chip on Board (FCOB) and a Chip Scale Package (CSP) on an organic board with underfill are described as follows, based on the work of Hanna (Ref 21) and Hanna et al. (Ref 22).

Two-Dimensional Model

Two-dimensional plane models can be used to represent a cross-section of a packaging assembly as shown, for example, in Fig. 2 and 3. Figure 2 shows a 2D model of a FCOB with underfill, while Fig. 3 shows a 2D model of a CSP assembly on an organic board. The plane elements have two degrees of freedom per node—displacements along the orthogonal axes in the plane of cross-section. The 2D cross-section can be taken along a transverse plane aligned with a centerline or a diagonal of the packaging assembly. By making use of geometry, material, and loading symmetry, only half the section needs to be modeled, as shown in Fig. 2 and 3. In such models, symmetry boundary conditions are applied: for example, in Fig. 2, y displacement is constrained along the left edge of the model. In addition, one node, typically at the left bottom corner of the model, is constrained in the vertical direction as well, to prevent rigid body motion. Plane-strain or plane-stress assumption can be used for thermomechanical modeling purposes and to bound the results.

Plane models provide a good understanding of the qualitative behavior of a package and the effect of variation of different material, geometry, and processing parameters. However, plane models do not capture the shape of the solder ball accurately, as they represent only a particular section. Also, they cannot capture the out-of-plane variation of material properties. Consequently, plane models do not provide quantitatively accurate results. The advantage of a 2D model is that it is computationally less intensive, as compared to a 2.5D and a 3D model.

Generalized Plane Deformation (GPD) or 2.5D Model

GPD models represent a compromise between 2D and 3D models. In a GPD model, a predetermined width of the package is modeled utilizing solid elements. Typically the width is chosen such that, at least one row of the solder balls is represented in the model. Also, only half of the package is modeled as a result of symmetry.

Table 4 Anand's Model Constants for Sn3.5Ag

Constant	Value
A, s^{-1}	177,016
Q, J/mol	85,459
ξ	7
m	0.207
h_0, MPa	27,782
ŝ, MPa	52.4
n	0.0177
a	1.6
S_0	$-0.0673T + 28.6$

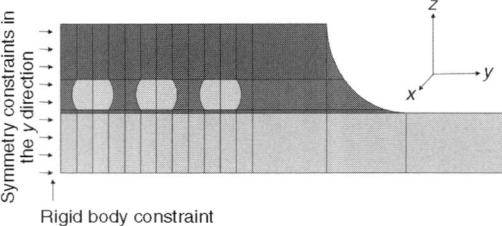

Fig. 2 2D model of a flip chip on an organic board

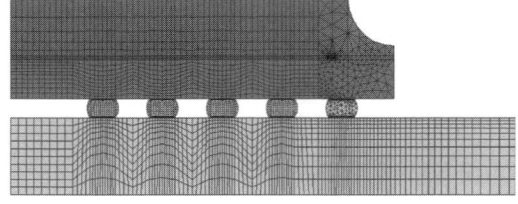

Fig. 3 Meshed 2D model of a chip scale package on an organic board

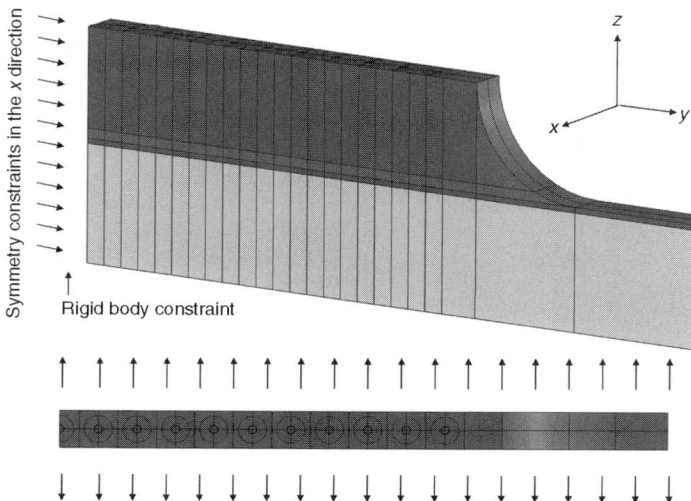

Fig. 4 Generalized plane deformation model with constraints for a flip chip on an organic board

Figure 4 illustrates the GPD model for FCOB and Fig. 5 illustrates the GPD model of a CSP on a board. In Fig. 4 the nodes are constrained in the y direction along the left plane of the package to satisfy the symmetry conditions. Also, the nodes along the front face and the nodes along the back face are separately coupled to deform together in the x direction such that the deformed cut faces remain parallel to the initial configuration. Also, to prevent rigid body motion, one of the nodes is fully constrained.

The advantage of utilizing a GPD model is that it is less computationally intensive compared to a full 3D model and, at the same time, is more accurate than a plane model.

Three-Dimensional Model

In a 3D model, no geometric assumptions are made and the complete geometry of the package is represented. A quarter or one-eighth symmetry model with appropriate symmetry boundary conditions can be used. Figure 6 illustrates an example of a quarter-symmetry 3D model with solid elements for a flip chip on an organic substrate with an underfill. As shown in Fig. 6 and 7, the nodes on the symmetry planes are constrained in a direction perpendicular to the symmetry plane. One node at the corner bottom of

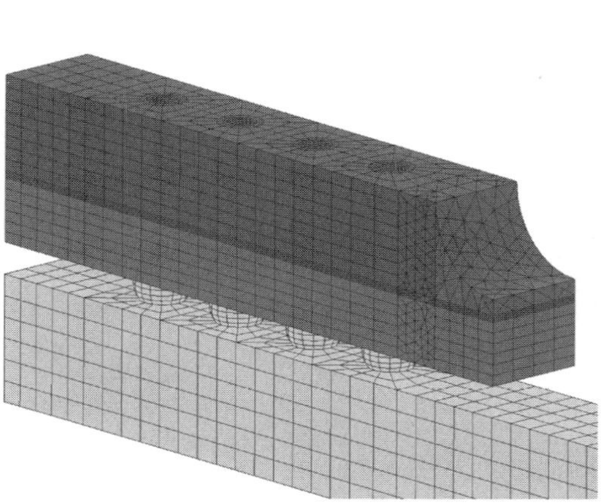

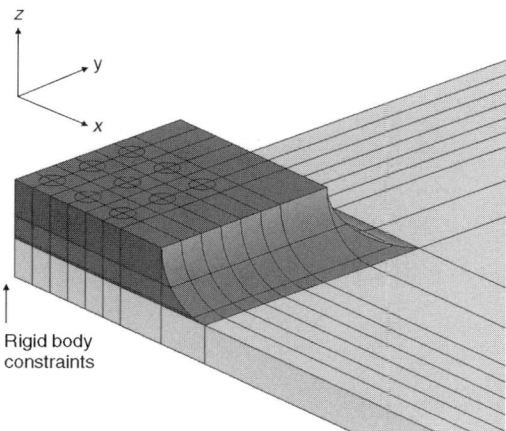

Fig. 5 Meshed generalized plane deformation model for a chip scale package on an organic board

Fig. 6 Quarter symmetry 3D model of a flip chip on an organic board

the model is fully constrained in all directions to prevent rigid body motion.

The 3D model represents the geometry, as well as the material behavior in the three orthogonal directions as accurately as possible. However, the 3D models are computationally more expensive, compared to the 2D and 2.5D models.

Other Considerations

In addition to the previously discussed elements, equivalent beam elements can also be used to represent solder column or solder ball interconnects. Also, shell elements can be used to represent the substrate and the package layered structures (Ref 23). As beam elements cannot be used to obtain a detailed distribution of stress-strain contours in solder balls/columns, a combination of beam and solid elements can be used to reduce computational time, as well as to obtain the necessary stress-strain contours. In such a modeling scenario, 3D solid elements are often used for critical solder balls/columns which are likely to fail first, while equivalent beam elements are used for the remaining solder balls/columns. Such an approach will be computationally more efficient, compared to a model which uses 3D solid elements for all of its solder balls/columns. Furthermore, viscoplastic constitutive models can be used for the critical solder balls/columns, while time-independent strain-hardening models can be used for the remaining solder balls/columns. This approach is likely to reduce the computational time without losing accuracy, as all of the solder balls/columns are not modeled with computationally-expensive viscoplastic constitutive behavior. In addition to the previous techniques, the models can be meshed so that a finer finite element mesh is used for the critical solder balls/columns, while a coarser mesh is used for the other solder balls/columns (e.g., Ref 24, 25).

In addition to the aforementioned models, axisymmetric models can also be used (Ref 26, 27), although their usage is somewhat limited to axisymmetric structures such as plated-through holes, microvias, etc. Global-local models can also be used where the displacement results from a global model are applied to the cut boundaries of a local model to analyze fine features.

Loading Conditions and Thermomechanical Stresses

Process Profile

Once the geometry and the material models are created and the boundary conditions are applied, the thermal loading conditions are then applied on the model to determine the thermally-induced stress-strain distribution in the geometry. The thermal profiles associated with the fabrication and assembly processes are considered first, and the resulting residual stresses due to such processing can be determined. It is extremely important to account for such process-induced stresses prior to assessing the reliability of the packaging assemblies. A discussion on the complete process mechanics modeling steps is beyond the scope of this chapter, and therefore, only specific highlights are discussed here. For example, it is possible to determine the warpage and stresses in various layers due to the sequential thin film processing of multilayered high-density interconnect substrate structures (Ref 28). Similarly, on the package side, the stresses induced due to die attachment and encapsulation and the chances for delamination propagation can be determined through process mechanics models (Ref 29). In addition to substrate and package models, the stresses induced due to the assembly of packages and flip chips on boards can also be determined (Ref 30).

For flip chips on organic substrates with an underfill, a two-part process profile is often used in which the thermal profile associated with solder reflow and attach is modeled in the first part, and the assembly is then simulated to be cooled down to room temperature. In this case, the entire assembly is assumed to be stress-free at the

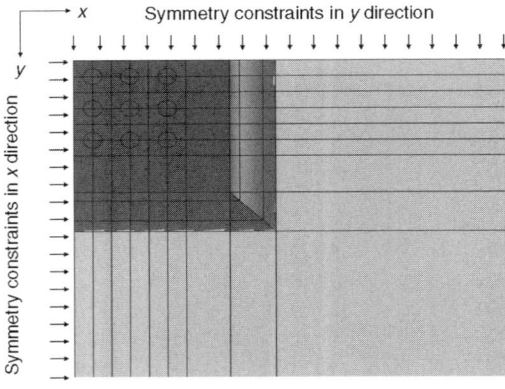

Fig. 7 Boundary conditions for a flip chip on organic board quarter symmetry 3D model

solder melting temperature. The underfill material elements are then activated to simulate underfill dispensing, and the entire assembly is then cooled down to room temperature from the underfill cure temperature. When a no-flow underfill material is used where the solder reflow and the underfill cure are performed in a single process step, the entire assembly model (flip chip with solder bumps on a substrate) is assumed to be stress-free at the solder melting temperature. The assembly is then cooled to underfill cure temperature where the underfill elements are activated. Then the entire assembly with the underfill is cooled to room temperature to determine the assembly-induced stresses. Whether it is a two-step process (reflow first and dispense and cure an underfill next) or a single-step process (simultaneous reflow and cure using a no-flow underfill), it is seen that the stress-strain contours in the solder and the die are almost identical when the material properties and process peak temperatures are the same. Therefore, a single-step process can be used even for a traditional dispense-and-cure underfill.

Accelerated Thermal Cycling

Accelerated Thermal Cycles (ATCs) are utilized for the purposes of qualification and life prediction of electronic packages. An ATC attempts to accelerate the mechanisms that drive failures and therefore can be used to assess packaging reliability in a shorter amount of time, compared to the intended lifetime of the packaging assembly. An ATC can be defined in terms of high and low temperatures, dwell times at high and low temperatures, and ramp rates between high and low temperatures. A number of recommended ATC profiles are available in literature as well as in standards. The choice of an ATC profile to be utilized for qualification is governed by the intended application and the field-use conditions of the packaging assembly.

Although the experimental accelerated thermal cycling is one effective way to qualify the packaging assemblies, it still takes time and money, due to the build-and-test approach and the lack of "legacy" knowledge when it comes to lead-free solder. Therefore, prior to performing the accelerated thermal cycling, it is essential to simulate the ATC conditions on a packaging assembly model to assess its reliability and make appropriate material, geometry, and process modifications to enhance the overall reliability.

Figure 8 illustrates an example of thermal profile for simulation purposes. As seen, the simulation will start from the stress-free temperature of solder solidification. The underfill, if present, will then be activated at the underfill cure temperature. The entire assembly is then cooled to room temperature, and then cycled through the ATC profile. In the profile shown in Fig. 8, the packaging assembly is cycled between -55 to 125 °C (-67 to 257 °F) for a cycle time of 120 min. In thermal cycling simulations, it is usually assumed that the entire packaging assembly is at a uniform temperature without thermal gradients and is at equilibrium with the thermal chamber temperature through ramps and dwells. Although such an assumption is often not correct, especially when the thermal mass of the assembly is high, for the sake of simplicity the uniform temperature assumption is often used in the simulations, and the resulting stress-strain distribution in the solder interconnects due to the thermal loading can be obtained. It is also possible to obtain the thermal contours in the packaging assembly by performing thermal simulations of accelerated thermal cycling, and use such thermal contours to determine the thermally-induced stress-strain distribution in the assembly. Also, in selected instances, temperatures in various parts of the packaging assembly can be experimentally measured in an actual thermal cycle and then used for thermomechanical simulations.

Field Use Simulation

In addition to ATC simulations, field-use simulations can also be performed to assess the long-term reliability of microelectronic pack-

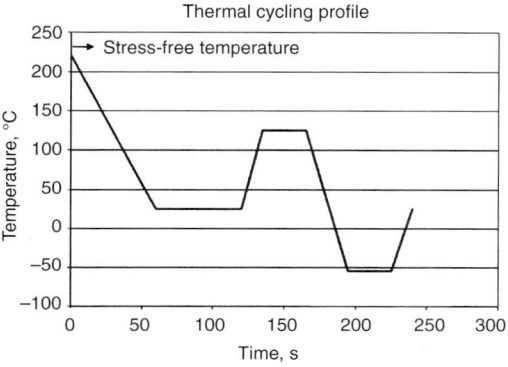

Fig. 8 Typical accelerated thermal cycle profile

aging assemblies. The available information on field-use simulations is limited. There are field-use conditions including random loading due to power cycling, thermal excursions, vibration and shock. During extended filed-use conditions, the behavior of solder, especially lead-free solder, interms of grain growth and intermetallic compound growth is not clearly understood. Also, limited studies are available on failure modes and failure prediction models. Among the available publications, for example, (Ref 11, 30, 31) are ones that have mapped the solder creep and plastic behavior over extended field-use conditions with accelerated thermal cycling conditions, and have developed ATC guidelines to mimic field-use solder behavior.

Table 5 shows thermal excursion ranges for various applications. As can be seen, different applications require different temperature ranges. Although not shown in Table 5, different applications require different lengths of field usage time.

Deformation Mechanisms

Unlike Pb-Sn eutectic solder where both Sn and Pb have comparable microstructure distribution, in Sn-Ag-Cu and Sn-Ag alloys, more than 95% by weight is Sn, and therefore, the IMCs (Ag_3Sn and Cu_6Sn_5) play an important role in the lead-free solder reliability. Under most typical loading conditions, SAC alloy in general exhibits a greater resistance to creep as compared to Pb-Sn alloy and hence, creeps 10 to 100 times slower (Ref 2). The bimaterial Sn and Ag or Sn and Cu solidify in a complex manner resulting in the formation of two significant phases—Ag_3Sn or Cu_6Sn_5. The increase in creep resistance for Sn-Ag and Sn-Ag-Cu alloys compared to Pb-Sn eutectic solder is attributed to the presence of finely dispersed IMC (Ag_3Sn and Cu_6Sn_5) particulates in the β Sn matrix (Ref 33). It has been observed that under low stress, the creep rate of lead-free solders is lower than that of lead-tin alloys. However, under conditions of high stress, the creep rate of lead-free solders is higher.

The mechanical behavior of SAC alloy also depends on the cooling rate during solder reflow and the Ag content in the solder. For example, in Sn-3.8Ag-0.7Cu alloy, when the cooling rate is slow (approximately 0.2 °C/s, or 0.36 °F/s), the volume fraction of Ag_3Sn plates and Cu_6Sn_5 rods is high (Ref 34), and when the cooling rate is high (approximately 3.0 °C/s, or 5.4 °F/s) such Ag_3Sn plate formation can be kinetically suppressed. Once large Ag_3Sn plates are formed, there may not be enough finely-distributed Ag_3Sn particulates to strengthen the large β-Sn matrix of the solder joint. Some studies recommend lower Ag content to prevent Ag_3Sn plate formation, regardless of the cooling rate (Ref 35). However, some other studies report that higher Ag content (3% and above) will nevertheless result in enough dispersion or precipitation strengthening Ag_3Sn particulates and therefore, will enhance the fatigue resistance of SAC solder (Ref 36). It should be noted that selected other studies have reported that Ag_3Sn plates could help by blocking the crack propagation in solder joint, and this is often more by chance than by design.

Depending on the nature of the loading conditions, different deformation mechanisms are activated in the solder. If higher temperatures are employed, then creep-induced deformation will be significant. A large amount of time spent at a higher temperature would also increase the contribution of creep. On the other hand, time-independent plastic deformation would be higher if the temperature range employed is higher. These concepts are well illustrated in Ref 11, where a comparison is made between field use conditions and a conventional ATC profile, in terms of time-independent plastic and time-independent creep deformation. Such a separation of time-dependent inelastic deformation serves as one of the basis for mapping field use conditions to ATC conditions.

The double power-law creep equation (Eq 8) separates the creep component into climb-controlled and combined climb-glide-controlled. During the field use conditions where the thermal excursions are milder and stresses are lower compared to typical accelerated thermal cycling

Table 5   Thermal excursion ranges for selected field use conditions Ref 32

Applications	Thermal excursion, °C (°F)
Consumer electronics	0 to 60 (32 to 140)
Computers	15 to 60 (59 to 140)
Telecommunications	−40 to 85 (−40 to 185)
Commercial aircraft	−55 to 95 (−67 to 203)
Military aircraft	−55 to 125 (−67 to 257)
Space	−40 to 85 (−40 to 185)
Automotive—passenger	−55 to 65 (−67 to 149)
Automotive—under the hood	−55 to 150 (−67 to 302)
Implantable medical electronics	37 to 40 (98 to 104)

profile, the lead-free solder is expected to have minimal dislocation glide/climb. The presence of IMC particulates would obstruct dislocation glide, requiring higher stresses and temperatures for the dislocation to climb, and therefore, the creep resistance will be higher in such Sn-Ag-Cu solder, strengthened by IMC particulates. On the other hand, compared to typical field use thermal profile, ATCs have harsher thermal excursions, and therefore, stresses are higher, and thus the lead-free solder would have significant dislocation glide/climb. Hence, separating creep-induced deformation in terms of climb-controlled and combined climb-glide-controlled can also serve as a basis for mapping field-use conditions to ATC conditions.

Thermomechanical Reliability in Packages

Fatigue Behavior of Solder

Under field-use conditions or under accelerated thermal cycling qualification conditions, the solder joints experience cyclic thermomechanical loads due to CTE mismatch or thermal gradients among various parts of a packaging assembly and failure due to such thermomechanical loads. In addition to thermomechanical fatigue loads, the solder joints experience stresses due to mechanical loads such as vibration, shock, etc. However, such mechanically-induced failures and other chemically or electrically-induced failures are not the focus of this chapter.

In general, two approaches can be adopted to characterize the number of cycles to failure due to fatigue. A stress-based approach is used to characterize the fatigue behavior when the stress is in the elastic regime and does not exceed the yield point. This is known as high-cycle fatigue. On the other hand, a strain-based approach is adopted to characterize fatigue behavior where the material experiences plastic deformation. This is known as low-cycle fatigue. Thermomechanical solder fatigue behavior is generally characterized by low-cycle fatigue, due to the fact that it undergoes a considerable amount of plastic deformation.

Before discussing the fatigue failure prediction models, a brief overview of the microstructural effects is presented here. The microstructure of Sn-3.8Ag-0.7Cu is shown in Fig. 9(a) (Ref 37). The dark-colored regions represent Sn dendrite arms and the lighter network-like regions represent dense array of IMC particulates located at the dendrite arm impingement zones. As seen, the β-Sn grain structure, as soldered (Fig. 9b) does not have any preferred orientation (Ref 37). However, when subjected to shear loading at a homologous temperature of $0.81T_m$ (125 °C, or 257 °F), the Sn-rich phase, as well as Ag_3Sn intermetallic, have coarsened, and the larger Sn-rich structure is oriented along the shear direction (Fig. 9c). Further plastic deformation at high temperatures can result in recrystallization in the β-Sn matrix which could lead to intergranular micro cracks leading to solder joint fracture.

A study (Ref 38) compared fatigue failure modes of lead-free solders to lead-tin solder joints, for a BGA package mounted on an organic board. In the study, the assemblies were subjected to thermal cycling until failure, and a comparison of the failure modes for lead-free solder and lead-tin solder was made. For the case of Sn-Pb solder, the crack path lies in the bulk of the solder material near the component body (Fig. 10). For lead-free solders, however, two distinct fracture paths are observed. The first fracture path is similar to that of lead-tin solder passing through the bulk of the solder near the component side. This is shown for SAC solder (Fig. 11) and is also observed for Sn-Ag solder. The second fracture path is characteristic only of the Sn-Ag and SAC solders. This fracture path consists of very fine cracks with multiple fronts, has a shattered appearance, and is seen near both the component and the board sides (Fig. 12).

Life-Prediction Models for Lead-Free Solder

Coffin-Manson type equations are commonly used to predict the low-cycle fatigue life of solder. The general form of the Coffin-Manson equation is as follows:

$$N_f = A(\Delta\varepsilon_{in})^m \quad \text{(Eq 12)}$$

where N_f is the number of cycles to failure, $\Delta\varepsilon_{in}$ is the inelastic strain range per cycle, and m and A are constants. In this form of the Coffin-Manson equation, the inelastic strain range per cycle is used as the damage metric—an indication of the amount of damage or degradation in the solder per cycle. Other commonly used damage metrics are accumulated inelastic strain per cycle, accumulated creep strain per cycle, strain energy density per cycle, creep strain energy per cycle, etc. For the energy-based damage metric, the fatigue life prediction equations can be written as:

$$N_f = B(\Delta W)^n \quad \text{(Eq 13)}$$

where N_f is the number of cycles to failure, ΔW is the strain energy density per cycle; and n and B are constants. The relationship between the number of cycles to failure and the damage metric is generally obtained through a regression analysis of experimentally obtained failure data.

A number of fatigue life models are available for the lead-tin solders and have been summarized in Ref 39. For the case of lead-free solders, however, only recently have such models been developed (see Table 6). Among them, for one particular model, fatigue tests were performed at two temperatures and frequency conditions (25 °C at 1 Hz and 125 °C at 10^{-3} Hz) on bulk SAC solder specimens (Ref 40). For each of the test

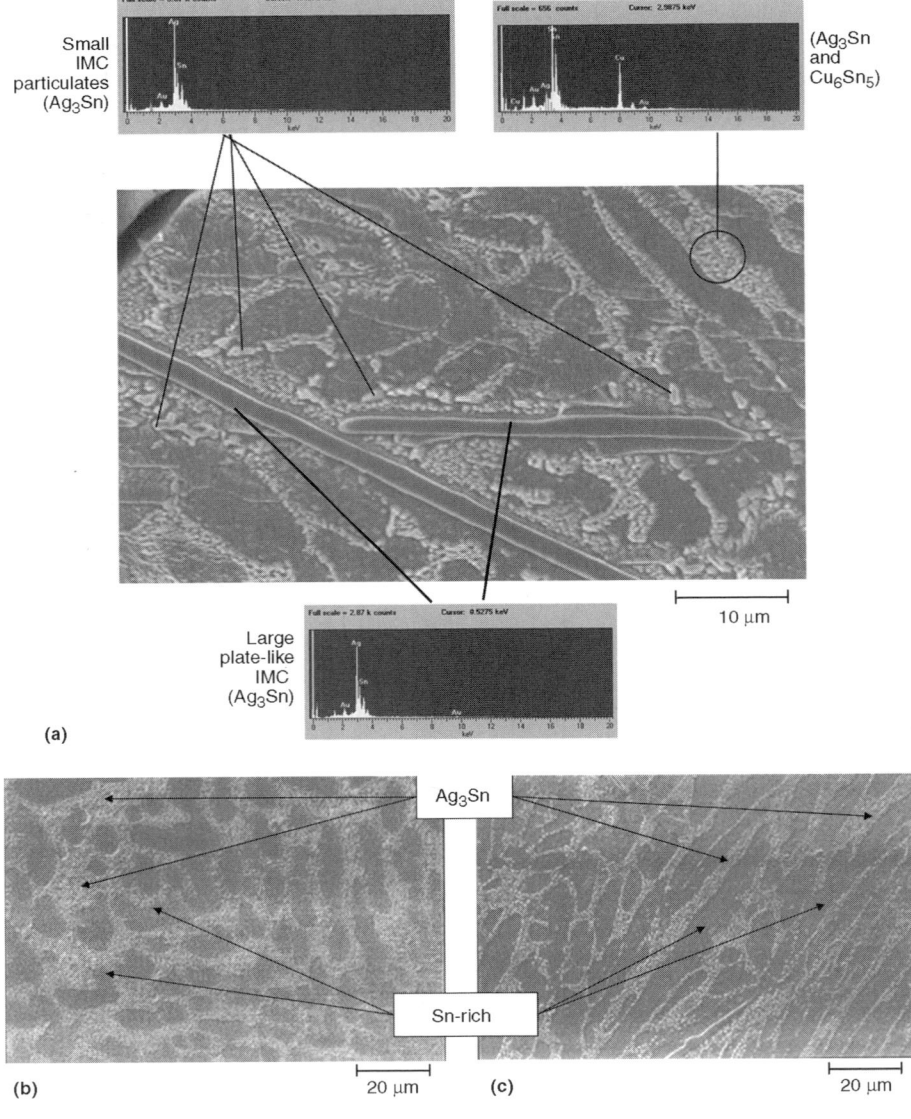

Fig. 9 Microstructure of Sn3.8Ag0.7Cu. (a) Solder joint microstructure (b) as soldered (c) after shear loading at 125 °C (257 °F) for 3 h (Ref 37)

conditions, four total strain ranges were utilized (2.0, 3.5, 5.0, and 7.5%). The criterion for failure was taken as 50% reduction of the maximum tensile load. A strain-based fatigue model was then obtained with the inelastic strain range as the damage metric, and an energy-based model was also obtained with the inelastic strain energy density as the damage metric. Equations 12 and 13 with cycling frequency v can be written (Ref 40):

$$[N_f v^{(0.95-1)}]^{0.93} \Delta\varepsilon_p = 12.9 \quad \text{(Eq 14)}$$

$$[N_f v^{(0.82-1)}]^{0.877} W_p = 1487.3 \quad \text{(Eq 15)}$$

where $\Delta\varepsilon_p$ is the inelastic strain range per cycle, and W_p is the inelastic strain energy density per cycle. Constants in Equation 12 and 13 have also been obtained using ATC test data and finite element models of three packages—Flip Chip On Board (FCOB) without underfill, FCOB with underfill, and a PBGA package (Ref 41). In a different paper, the constants for equation 13 are obtained by subjecting a wafer-level chip-scale package (WLCSP) to a force-controlled isothermal fatigue test (Ref 42). An energy partitioning model has also been developed (Ref 4) to predict the number of cycles to failure for SAC solder. The number of cycles to failure is given by:

$$1/N_f = 1/N_{fc} + 1/N_{fp} \quad \text{(Eq 16)}$$

where N_f is the number of cycles to failure, and N_{fc} and N_{fp} are the cycles to failure due to creep and plastic damage respectively. N_{fc} and N_{fp}, in turn, are given by the following equations:

$$W_c = W_{co} N_{fc}^d$$
$$W_p = W_{po} N_{fp}^c \quad \text{(Eq 17)}$$

where W_c and W_p are the deviatoric energy densities for creep and plasticity respectively. W_{po},

Fig. 10 Fracture path for lead-tin solder. Source: Ref 38

Fig. 11 First type of fracture path for SAC solder. Source: Ref 38

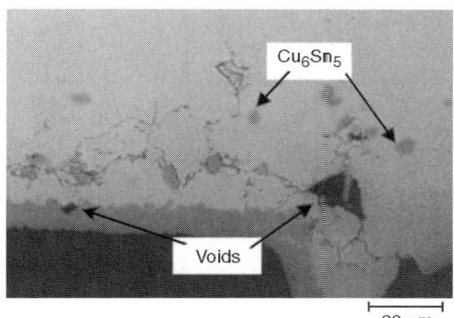

Fig. 12 Second type of fracture path for Sn-Ag solder. Source: Ref 38

Table 6 Fatigue-life models for lead-free solders

Strain-based $N_f = A(\Delta\varepsilon_{in})^m$		Energy-based $N_f = B(\Delta W)^n$			
A	m	B	n	Reference No.	Comments
26.3	−0.913	14865	−1.103	25 °C (77 °F), 1 Hz	Fatigue failure life defined at 50% reduction of maximum tensile load for a bulk specimen subjected to isothermal fatigue loading.
9.2	−0.853	311.7	−0.897	125 °C (257 °F), 10⁻³ Hz 40	
4.5	−1.295	345	−1.02	41	Fatigue life corresponds to different accelerated tests performed on PBGA packages and FCOB with and without underfill. Strain and energy refer only to creep components, and are evaluated using FEA.
...	...	18.15	−1.96	42	Force controlled isothermal fatigue tests performed on a WLSCP. Failure criterion is the doubling of cyclic displacements.

W_{co}, c and d are constants with values of 106.45 N/mm^2, 30.025 N/mm^2, -0.51 and -0.44 respectively. A life prediction model has been obtained (Ref 43) for SAC solder which is based on the partitioning of the creep deformation mechanism into climb controlled (dominating at low-stress levels) and combined glide/climb behavior (dominating at high-stress levels) based on Wiese's double power-law constitutive model for SAC solder. The number of cycles to failure is given as a function of the accumulated creep strain (Eq 18) partitioned into the two separate deformation mechanisms described previously:

$$N_f = (.013\varepsilon_{acc}^I + .036\varepsilon_{acc}^{II})^{-1} \qquad (Eq\ 18)$$

The constants for the various life prediction models are given in Table 6.

Damage Mechanics-Based Approach

A damage mechanics approach incorporates the experimentally-observed degradation of material over time, unlike previously-described constitutive models which do not take this factor into consideration. In other words, the solder constitutive models described earlier do not change with material degradation, and remain unaltered from the time a package is assembled to the time the solder interconnects fail due to fatigue failure. A typical damage mechanics approach, on the other hand, involves supplementing a viscoplastic constitutive model with damage rate equations. A suitable parameter, which can either be a scalar or vector, is utilized to evaluate the material degradation and failure. The damage parameter is utilized to influence the stress-strain relationship. Failure occurs when the damage metric, which evolves over time, attains a certain predetermined threshold value. Thus, the damage mechanics approach is able to capture material degradation over time. The damage mechanics-based approach has been utilized to characterize the behavior of lead-tin solders (e.g., Ref 44 to 47).

Model Validation

The thermomechanical models can be validated through various approaches, and in this section, only the most popular approaches are discussed. One approach is to design, fabricate, and assemble a test vehicle, and to subject the test vehicle to accelerated thermal cycling. The test vehicle typically consists of a dummy die or no die (for example, for second-level solder interconnect reliability assessment in some BGAs) and the solder interconnects form several loops of daisy chains. The electrical resistance of these daisy chain loops is either monitored in situ during thermal cycling or measured at room temperature outside the thermal chambers after a set of prescribed number of thermal cycles. An assembly is considered to have failed when the percentage change in electrical resistance in any of the daisy chain loops goes above a certain value. For statistical purposes, twenty or more samples are tested for a given thermal cycling profile, and the failure data is plotted to obtain the mean number of cycles to failure. Such a mean number of cycles is used to validate the predictions from the models.

Another approach to model validation is through Moiré interferometry. For example, laser Moiré interferometry is an optical method which provides whole field contour maps of in-plane displacements with a nm-scale sensitivity. Laser Moiré interferometry is performed on a cross-section of a sample. In Moiré interferometry, there exists a reference temperature at which no fringes or very few fringes are produced in the u-field (corresponding to deformation along the x-axis) and in the v-field (corresponding to deformation along the y-axis). On changing the temperature of the thermal chamber, the sample deforms, resulting in fringes being produced in the u-field and the v-field which can be related to deformation 'x' and 'y' directions. Similar loading conditions can be applied on the model developed to represent the behavior of the package, and corresponding deformation contours can be obtained. Such contours obtained from the simulation can be compared with those obtained through the experiment, and a good match validates the model. As an example, for a PBGA package, fringes were obtained from laser Moiré interferometry and compared against deformation contours predicted by the FEA model. Figure 13a shows the u-field contours with laser Moiré fringes in the top image and FEA deformation contours in the bottom image (Ref 11). Similarly Figure 13b shows the v-field contours.

Conclusions

Up-front thermomechanical modeling and physics-based reliability prediction of lead-free solder joints could save significant amount of cost and time in microsystem product development and qualification. Inadequate understanding of the lead-free solder material constitutive behavior, especially at the length scales associated with packaging interconnects, inadequate development of failure predictive models associated with various failure mechanisms, inadequate knowledge of lead-free solder microstructure evolution over extended field-use conditions, lack of suitable computational algorithms to model effectively hundreds of solder interconnects without compromising the accuracy of results, and computational and modeling difficulties with the heterogeneous mix of several materials combined with die/package/sol-

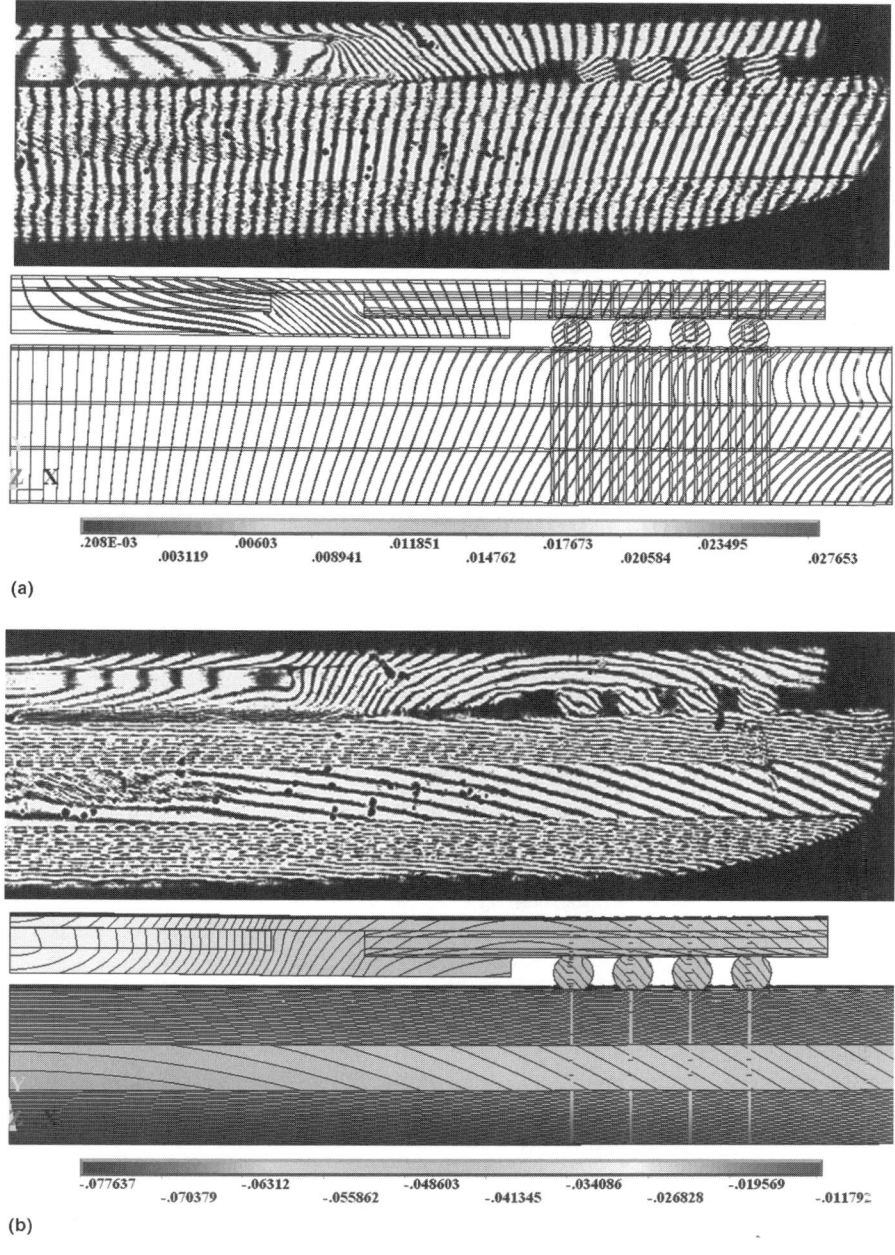

Fig. 13 (a) *u*-field and (b) *v*-field

der/board dimensions that are different by orders of magnitude, are some of the challenges associated with the thermomechanical modeling. This chapter has discussed some of these challenges, providing details on important developments over the past decade and the current status, particularly on the use of thermomechanical modeling and reliability prediction in microsystems package development.

REFERENCES

1. R. Pfahl, Pb-Free Electronics: Drivers, R&D and Transition, *Supply Network Conference,* Sept 18, 2003, (San Jose, CA), NEMI
2. G. Grossmann, G. Nicoletti, U. Solèr, Results of Comparative Reliability Tests on Lead-Free Solder Alloys, *52nd Electronic Components and Technology Conference,* 2002, p 1232–1237
3. J. Lau, W. Dauksher, and P. Vianco, Acceleration Models, Constitutive Equations, and Reliability of Lead-Free Solders and Joints, *53rd Electronic Components and Technology Conference,* 2003, p 229–236
4. Q. Zhang, A. Dasgupta, and P. Haswell, Viscoplastic Constitutive Properties and Energy Partitioning Model of Lead-Free Sn3.9Ag0.6Cu Solder Alloy, *53rd Electronic Components and Technology Conference,* 2003, p 1862–1868
5. R. Darveaux, and K. Banerji, Constitutive Relations for Tin-Based Solder Joints, *42nd Electronic Components and Technology Conference,* 1992, p 538–551
6. S. Suresh, Cyclic Deformation in Polycrystalline Ductile Solids, *Fatigue of Materials,* Cambridge University Press, 2001, p 86–131
7. S. Wiese, E. Meusel, K.J. Wolter, Microstructural Dependence of Constitutive Properties of SnAg and SnAgCu Solders, *53rd Electronic Components and Technology Conference,* 2003, p 197–206
8. T.G. Langdon, Deformation at High Temperatures, Strength of Metals and Alloys, *Proceedings of the 6th International Conference,* Aug 1982, p 1105–1120
9. H.J. Frost and M.F. Ashby, *Deformation Mechanism Maps,* Pergamon Press, 1982, Chapter 2
10. S. Wiese and E. Muesel, Characterization of Lead-Free Solders in Flip Chip Joints, *J. Electron. Packaging,* Dec 2003, Vol 125, p 531–538
11. K. Tunga, K. Kacker, R.V. Pucha, and S.K. Sitaraman, Accelerated Thermal Cycling: Is It Different for Lead-free Solder?, *54th Electronic Components and Technology Conference,* June 2004, p 1579–1585
12. M. Amagai, Characterization of Chip Scale Packaging Materials, *Microelectronics Reliability,* 1999, p 1365–1377
13. B.Z. Hong, Thermal Fatigue Analysis of a CBGA Package with Lead-Free Solder Fillets, *InterSociety Conference on Thermal Phenomena,* 1999, p 205–211
14. J.H.L. Pang, B.S. Xiong, T.H. Low, Creep and Fatigue Characterization of Lead-Free 95.5Sn-3.8Ag-0.7Cu, *53rd Electronic Components and Technology Conference,* 2004, p 1333–1337
15. A. Schubert, R. Dudek, E. Auerswald, A. Gollhardt, B. Michel, H. Reichl, Fatigue Life Models for SnAgCu and SnPb Solder Joints Evaluated by Experiments and Simulation, *Micromaterials and Nanomaterials,* Issue 3, 2004, p 30–41
16. I. Dutta, A Constitutive Model for Creep of Lead-Free Solders Undergoing Strain-Enhanced Microstructural Coarsening: A First Report, *J. Electron. Mater.,* 2002
17. L. Anand, Constitutive Equations for Hot Working of Metals, *Int. J. Plasticity,* Vol 1 (No. 2), 1985, p 213–231
18. S.B. Brown, K.H. Kim, and L. Anand, An Internal Variable Constitutive Model for Hot Working of Metals, *Int. J. Plasticity,* Vol 5 (No. 2), 1989, p 95–130
19. M. Amagai, M. Watanabe, M. Omiya, K. Kishimoto, T. Shibuya, Mechanical Characterization of Sn-Ag based Lead-Free Solders, *Microelectronics Reliability,* 2002, p 951–966
20. C. Xu, C. Gang, S. Masao, Modified Anand Constitutive Model for Lead-Free Solder Sn3.5Ag, *Proc ITHERM 2004,* p 447–452
21. C. Hanna, "Study of the Thermomechanical Reliability of Area-Array Packages," M.S. thesis, Georgia Institute of Technology, June 1999
22. C.E. Hanna, R. Raghunathan, and S.K. Sitaraman, Qualification Guidelines for Implantable Medical Devices, *ASME International Mechanical Engineering Congress and Exposition,* Nov 1999, 99-IMECE/EEP-27
23. J.S. Corbin, Finite Element Analysis for

Solder Ball Connect (SBC) Structural Design Optimization, *IBM Journal of Research and Development,* Vol 37, Sept 1993, p 585–595
24. A. Perkins, and S.K. Sitaraman, Vibration-Induced Solder Joint Failure of a Ceramic Column Grid Array (CCGA) Package, *54th Electronic Components and Technology Conference,* June 2004, p 1271–1278
25. A. Perkins, and S.K. Sitaraman, Thermomechanical Failure Comparison and Evaluation of CCGA and CBGA Electronic Packages, *53rd Electronic Components and Technology Conference,* May 2003, p 422–430
26. G. Ramakrishna, F. Liu, and S.K. Sitaraman, Experimental and Numerical Investigation of Microvia Reliability, *The Eighth Intersociety Conference on Thermal and Thermomechanical Phenomena in Electronic Systems,* ITHERM'2002, p 932–939
27. D.B. Barker, and A. Dasgupta, Thermal Stress Issues in Plated-Through-Hole Reliability, *Thermal Stress and Strain in Microelectronics Packaging,* J.H. Lau, Ed., Van Nostrand Reinhold, New York, 1993, p 648–683
28. R.C. Dunne, and S.K. Sitaraman, An Integrated Process Modeling Methodology and Module for Sequential Multilayered Substrate Fabrication Using a Coupled Cure—Thermal-Stress Analysis Approach, *IEEE Transactions–Electronics Packaging Manufacturing,* Vol 25 (No. 4), 2002, p 326–334
29. R.J. Harries and S.K. Sitaraman, Numerical Modeling of Interfacial Delamination Propagation in a Novel Peripheral Array Package, *IEEE Transactions on Components and Packaging Technologies,* Vol 24 (No. 2), June 2001, p 256–264
30. S.K. Sitaraman, R. Raghunathan, and C.E. Hanna, Development of Virtual Reliability Methodology for Area-Array Devices used in Implantable and Automotive Applications, *IEEE Transactions on Components and Packaging Technologies,* Vol 23 (No. 3), Sept 2000, p 452–461
31. R.V. Pucha, J. Pyland, and S.K. Sitaraman, Damage Metric-Based Mapping Approaches for Developing Accelerated Thermal Cycling Guidelines for Electronic Packages, *International Journal of Damage Mechanics,* July 2001, Vol 10, p 214–234
32. Internal communication with the industry collaborators
33. Q. Xiao, L. Nguyen, W.W. Armstrong, Aging and Creep Behavior of Sn3.9Ag0.6Cu Solder Alloy, *54th Electronic Components and Technology Conference,* 2004, p 1325–1332
34. D.W. Henderson, *TMS Annual Meeting,* Charlotte, NC, March 2004
35. S.K. Kang, P. Lauro, D.Y. Shih, D. Henderson, T. Gosselin, J. Bartelo, S. Cain, C. Goldsmith, K.J. Puttlitz, and T.K. Hwang, Evaluation of Thermal Fatigue Life and Failure Mechanisms of Sn-Ag-Cu Solder Joints with Reduced Ag Contents, *54th Electronic Components and Technology Conference* 2004, p 661–667
36. S. Terashima, Y. Kariya, T. Hosoi, and M. Tanaka, Effect of Silver Content on Thermal Fatigue Life of Sn-xAg-0.5Cu Flip Chip Interconnects, *J. Electron. Mater.,* Vol 32 (No. 12), 2003, p 1527–1533
37. J.H.L. Pang, B.S. Xiong, and T.H. Low, Creep and Fatigue Characterization of Lead-Free 95.5Sn-3.8Ag-0.7Cu, *54th Electronic Components and Technology Conference,* 2004, p 1333–1337
38. M. Meilunas, A. Primavera, and S.O. Dunford, Reliability and Failure Analysis of Lead-Free Solder Joints, *Proc. IPC Annual Conference,* New Orleans, 2001
39. K. Tunga, J. Pyland, R.V. Pucha, and S.K. Sitaraman, Study on the Choice of Constitutive and Fatigue Models in Solder Joint Life Prediction, *Proceedings of IMECE 2002*
40. J.H.L. Pang, B.S. Xiong, and T.H. Low, Comprehensive Mechanics Characterization of Lead-Free 95.5Sn-3.8Ag-0.7Cu, *Micromaterials and Nanomaterials,* Issue 3, 2004, p 86–93
41. A. Schubert, R. Dudek, E. Auerswald, A. Gollhardt, B. Michel, and H. Reichl, Fatigue Life Models for SnAgCu and SnPb Solder Joints Evaluated by Experiments and Simulation, *Micromaterials and Nanomaterials,* Issue 3, 2004, p 30–41
42. J.H. Lau, D. Shangguan, D.C.Y. Lau, T.T.W. Kung, and S.W.R. Lee, Thermal-Fatigue Life Prediction Equation for Wafer-Level Chip Scale Package (WLSCP) Lead-Free Solder Joints on Lead-Free Printed Circuit Board, *54th Electronic Components and Technology Conference,* June 2004, p 1563–1569

43. A. Syed, Accumulated Creep Strain and Energy Density-Based Thermal Fatigue Life Prediction Models for SnAgCu Joints, *54th Electronic Components and Technology Conference,* 2004, p 737–746
44. R.V. Pucha, J. Pyland, and S.K. Sitaraman, Damage Metric-Based Mapping Approaches for Developing Accelerated Thermal Cycling Guidelines for Electronic Packages, *International Journal of Damage Mechanics,* July 2001, Vol 10, p 214–234
45. S.-H. Ju, B. Sandor, M. Plesha, Creep Rupture Investigation of 63Sn-37Pb Solder by Experiments and Damage Mechanics, *Journal of Testing and Evaluation,* JTEVA, Vol 24 (No. 6), Nov 1996, p 411–418
46. S.H. Ju, B. Sandor, and M.E. Plesha, Life Prediction of Solder Joints by Damage and Fracture Mechanics, *J. Electron. Packaging,* Dec 1996, Vol 118, p 193–200
47. C. Basaran, and C. Yan, Thermodynamic Framework for Damage Mechanics of Solder Joints, *J. Electron. Packaging,* Vol 120 (No. 4), 1998, p 379–384

CHAPTER 9

Design for Reliability—Finite Element Modeling of Lead-Free Solder Interconnects

Walter Dauksher, Agilent Technologies

Introduction

Solder is a fundamental material used in the assembly of electronic devices to printed circuit boards (PCBs). Applications range from the connection material for leadframe packages and discrete devices to the complete interconnect in constructions such as the ball grid array (BGA) and the column grid array. In all cases, solder acts as parts of the electrical circuit, facilitates the dissipation of thermal energy, and provides the structural connection between the package and the PCB. The ball grid array (Ref 1) is the prevalent interconnect structure between the package and the PCB and will be the focus of this chapter.

Often there is a mismatch between the coefficients of thermal expansion of the package and the PCB. Consequently, when the package/PCB assembly is subject to changes in temperature, thermally driven forces develop within the assembly. Examples of temperature changes include those due to power on/power off and the mini-cycles due to periods of varying computational activity. These temperature changes subject the solder joints to loads which may eventually lead to fatigue failures. The failure of a single solder joint could compromise the functionality of a device.

Both integrated circuit vendors and their customers, therefore, are interested in the solder joint fatigue life of electronic packages when subjected to thermal loads. The character and quality of the solder joints is often evaluated with accelerated thermal cycling tests (Ref 2). In performing these tests, packages are assembled on a PCB, and the assembly is subject to a cyclic thermal environment. Typical thermal environments are 0 to 100 °C (32 to 212 °F) and −40 to 125 °C (−40 to 257 °F), with ramps between extremes and dwells at the extremes of 10 or more min. During a test, the integrity of the package, with an eye toward the second-level interconnect, is monitored through continuous or interval resistance measurements. When the resistance exceeds a specified threshold, the part is considered failed. Failure analysis is then used to identify the failure location and mode. For those failures that occurred through the solder interconnect, a statistical distribution is developed which characterizes the solder joint reliability under the accelerated test conditions.

These second-level solder joint reliability tests provide a number of benefits. First, weaknesses such as insufficient solder volume may be experimentally identified and corrected in the design stages. Second, the information may provide a customer with a quantitative measure of the performance of various vendors' components. Finally, the accelerated test results form the basis for predictions of field-use solder joint reliability.

While these accelerated test methods are generally standardized, there are significant difficulties associated with solder joint reliability testing, such as cost. Second-level tests often use specially designed daisy chain substrates, die, and PCBs. Considering the cost of masks and tooling, a complete set of test articles may easily cost tens of thousands of dollars. Accelerated reliability tests also take months to run (approximately 1000 cycles may be accumulated in a month). Therefore, the generation of data (worse, the discovery of a design issue) can adversely impact time-to-market. Finally, regardless of the specific configuration tested, a customer will be interested in the solder joint reliability of a package with a slightly different die size, temperature profile, board thickness, etc.

An efficient and accurate predictive methodology for the solder joint reliability under accelerated test conditions can be an essential design tool. Specifically, the prediction of the solder joint fatigue life that incorporates geometric and material characteristics can significantly reduce the number of design-test cycles, speed time-to-market, and satisfy customer requirements that differ from already tested configurations. In fact, IPC9701 (Ref 2) permits the use of modeling to predict the solder joint fatigue life for new packages that are similar to already tested packages. While these capabilities can be significant cost savers, the predictive tool is especially topical in the transition to lead-free solders. Here, analysis can bring focus to limited research budgets, fill in the gaps between test data and provide competitive design insight.

The design for reliability procedure (Ref 3) focuses on the evaluation of design alternatives in the context of specific reliability goals. Typical design variants include those with differing geometries or material systems. In the case of electronic packaging design, specifically solder joint reliability, the design for reliability process relies heavily on numerical modeling. Accurate numerical tools, encompassing geometric models, material constitutive relations and failure theories, are required. In the early stages of a new technology, such as lead-free soldering, the design for reliability process may be involved in the design of test articles and certainly may include multiple cycles of design, test, and failure analysis. As confidence increases in the predictive model's accuracy, the design for reliability tools may be used to produce "final" designs. The greatest benefits will be attained when the design for reliability process is implemented early in the design cycle.

This chapter is focused on the numerical tools needed for lead-free solder joint fatigue predictions in accelerated thermal cycling. In particular, the finite element method has become the standard tool for making such evaluations. This chapter reviews the general class of finite element model configurations, required material properties, and prevalent failure theories. Review is made in the context of the model's computational costs, as well as the correlation between the finite element-based reliability predictions using various failure theories and published lead-free reliability test data. The chapter includes an example of the lead-free solder joint design for reliability procedure using the tools developed herein and a discussion of future research needs. While specific emphasis is placed on ball grid array packages, the philosophy, input parameters, and approaches presented should be equally pertinent to other second-level interconnect configurations. The finite element models are created in ANSYS (Ref 4), which is one of the prevalent codes in the electronics industry.

The lead-free solder whose composition is nominally 95.5Sn-3.9Ag-0.6Cu will be the focus of the analysis. The material property information presented should be appropriate for slight deviations from this composition. For simplicity, the alloy will be referred to as Sn-Ag-Cu. Select comparisons will be made with standard eutectic lead-tin solder with 2%Ag (62Sn-36Pb-2Ag). For convenience, the lead-tin eutectic material with 2%Ag will be designated as Sn-Pb.

Modeling

There are at least three distinct components to the finite element modeling activity and these include the definition of the geometric and discretization characteristics of the model, the definition of material properties, and the examination of finite element results with a failure theory. In each step, simplifications, idealizations, and approximations are made. Unfortunately, in the case of solder fatigue predictions there is no approach that is generally acknowledged to be the most accurate and certainly no approach that produces correct predictions in all cases. In the sections that follow, a brief review will be made of prevalent modeling strategies, as well as material property data and failure theories.

Modeled Geometry

For the purposes of this discussion, the package is presumed to have octant symmetry. That is, in plan view the package is square and has reflection symmetry, both through the opposite sides and along the diagonals that connect opposite corners. Most packages follow this construction. For those that do not, some of the geometric simplifications discussed may not be envoked. Figure 1 presents a plan view of a typical package along with a notation of the geometry modeled for both slice and octant models. Three-dimensional (3D) slice models are presented throughout the later sections of this chapter.

2D Slice. The two-dimensional (2D) slice is the simplest finite element model. In this approach, a slice of the package, usually along the diagonal to a corner, captures the through-the-thickness construction with planar (2D) elements. In the unmodeled direction, a plane stress or plane strain boundary condition is usually applied. As a result, the 2D slice approximates the package as being infinite in the unmodeled direction. While this may be acceptable for the PCB and the package, the discrete solder balls are also modeled as having infinite extent in the thickness direction. As such, the 2D slice will improperly (under) predict the stresses and strains in the solder interconnects. Significant advantages are that this model is extremely easy to construct, is not prone to the user errors common in more complex models, and is very computationally efficient. There has been a long track record of successful analyses being performed with 2D slice models.

3D Slice. As with the 2D model, only a slice is used in order to save on computation time and model complexity. Again, the slice captures the through-the-thickness construction but also has a depth dimension that allows full geometric definition of the discrete solder balls. Three-dimensional, preferably hexahedral, elements are used in the 3D models. The model typically runs along a diagonal from the geometric center of the package to a corner. As presented in Fig. 2, the modeled thickness includes the package, solder balls, and PCB material that are contained within the slice. The model imposes symmetry boundary conditions on one surface of the slice. This may correspond with the center line of the solder balls. On the other cut surface, a state of general plane strain is imposed. As a result, the 3D slide model actually simulates a package that is of infinite extent in the direction perpendicular to the plane of the slice. This approach has two distinct advantages. First, the model captures a complete description of the solder balls, leading to a more accurate definition of the stress and strain state within the interconnects. Second, the 3D slice is also computationally efficient. This efficiency may be evidenced in fast solution times. Alternately, the analyst may include complicating non-linear effects in the 3D slice model that could not be included in a single 3D octant model.

3D Octant. In order to avoid the boundary condition approximations of the 3D slice model, a full 3D octant model may be constructed. This model, as depicted in Fig. 3, should produce the most accurate results and is appropriate for situations where certain asymmetries, such as a rectangular die with large aspect ratios, preclude the use of a slice model. For larger packages or in the case of fine meshes, the 3D octant can have a very large number of elements and run slowly. Since the solder balls will have nonlinear material properties, computation with a 3D oc-

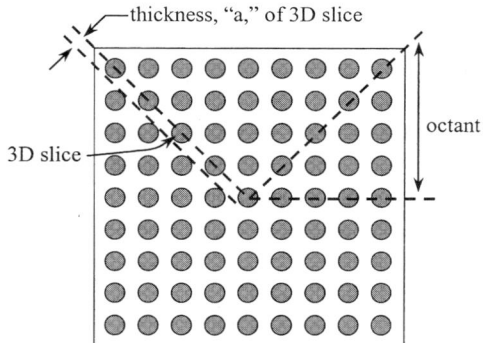

Fig. 1 Symmetry as applied to finite element models (Plan view of package with solder balls)

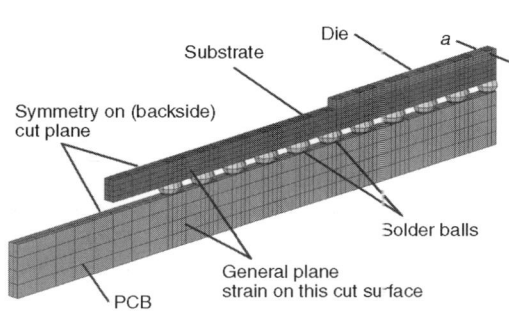

Fig. 2 3D slice model

tant may be impractical. A number of modeling strategies have been devised to make the 3D octant model more efficient. The first is to create linear superelements out of the PCB and the package (a superelement is a single linear element that offers the boundary stiffness of a region that has previously been meshed with many elements). This leaves the solder balls as discretely modeled. A further simplification is to mesh only the critical solder ball(s) of interest with a refined mesh, while the other solder balls are coarsely meshed. These simplifications may be made to other models also. Depictions of 3D octant models may also be found in Ref 5.

3D Octant with Submodel. Even with the use of superelements and selective refinement, mesh transitional issues between the refined solder joint(s) and the rest of the package and board assembly lead to extremely large 3D models. The 3D octant with submodel strategy, more formally a global-local modeling approach, makes use of two models. A global 3D octant model is relatively coarsely meshed. This model is run and the critical solder joint is determined. In a subsequent modeling step, the boundary displacements extracted from the global model for the critical solder joint are applied to a highly refined local solder joint model. The local model is then run to determine stresses or strains in the critical solder joint. The global model may possess either linear or nonlinear material properties and may or may not use superelements.

Materials Discussion

Most materials exhibit complex deformation responses when subjected to mechanical loads. In particular, solid mechanics divide these into elastic and inelastic responses. It is generally agreed that the inelastic response is the metric that defines the fatigue life of the solder. Therefore, the solder joints must be modeled with either creep or plastic or both constitutive relations. The rest of the package structure is typically modeled with exclusively elastic stress-strain relations. This assumption has negligible impact on the inelastic response of the solder interconnects while providing significant computational cost savings. The choice of solder properties used in constructing the model restricts the choices of failure theories. For example, a creep constitutive relation needs to be used in order to obtain the creep strains used by a creep-strain-based failure theory. A brief discussion of the material properties, as well as a listing of those properties used in this chapter follows.

Elastic Properties. The strain response of a material to an applied stress will depend on the magnitude of the stress, the components of the stress, the strain rate, and the temperature. Elastic deformation occurs when the strains are proportional to the stresses and when the strains appear and disappear simultaneously with the application and removal of the stresses. Elastic properties used in this study are presented in Table 1. While high lead, 90Pb-10Sn is not included in this chapter's analyses, its properties are included for reference. When applied consistently, the properties in Table 1 should be appropriate for the range of temperatures seen in accelerated thermal cycling.

Figure 4 details the elastic modulus of the Sn-Ag-Cu alloy compared with that of the common eutectic solder. The moduli of both materials demonstrate strong temperature dependence. In addition, the Sn-Ag-Cu alloy is approximately twice as stiff as the eutectic Sn-Pb solder at a given temperature. As such, for displacement-controlled loadings, stresses will be higher in the lead-free alloy than those in the eutectic solder. The modeling results presented later in this chapter will confirm this expectation.

Creep Properties. Creep is the time dependent inelastic deformation of a material. A specimen undergoing continuous deformation under fixed load is said to "creep." In general, the creep deformation will exhibit a nonlinear relationship between the rate of deformation and both temperature and stress level. Homologous temperatures, the ratio of the ambient temperature to the material's melt temperature both in K, above 0.5, coincide with the fastest creep rates in met-

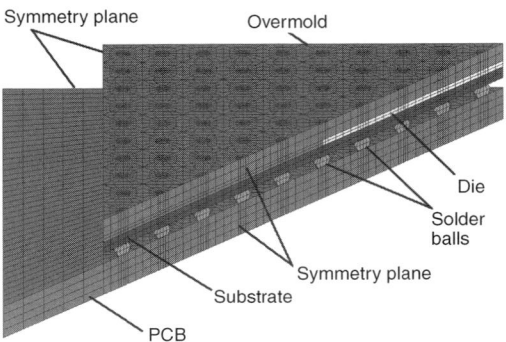

Fig. 3 3D octant model

als. Considering the standard temperature cycling range of 0 to 100 °C (32 to 212 °F), the Sn-Ag-Cu alloy has homologous temperatures spanning 0.55 to 0.76 while the eutectic Sn-Pb solder has homologous temperatures of 0.60 to 0.82. Clearly, creep can be an important component of the structural response of the solder alloys in both accelerated tests and service conditions.

Figure 5 presents an idealized creep curve. Upon loading, the material experiences a strain, ε_O, which may be composed of both elastic and

Table 1 Elastic properties of select packaging materials (Note: T is in K)

Material	Young's modulus (GPa)	Poisson's ratio	CTE (ppm/°K)	Reference No.
PC board	27	0.39	18	
Copper pad	76	0.35	17	
62Sn-36Pb-2Ag	75.94 − 0.152T	0.35	24.5	
10Sn-90Pb	23.71 − 0.047T	0.35	24.495 + 0.015T	
Sn-3.9Ag-0.6Cu	74.84 − 0.08T	0.34	16.66 + 0.017T	9, 12, used in the examples that follow
Sn-3.8Ag-0.7Cu, Sn-3.5Ag-0.75Cu Sn-3.5Ag-0.5Cu, Castin™	60.73 − 0.0574T	0.36	20	11
Sn-(3.5–3.9)Ag-(0.5–0.8)Cu	68.11 − 0.07T	0.35	16.66 + 0.017T	18
Laminate substrate		Same as PC board		
Al$_2$O$_3$ substrate	270	0.3	5.5	
Si	167	0.28	2.54	
PBGA overmold	13	0.3	15	

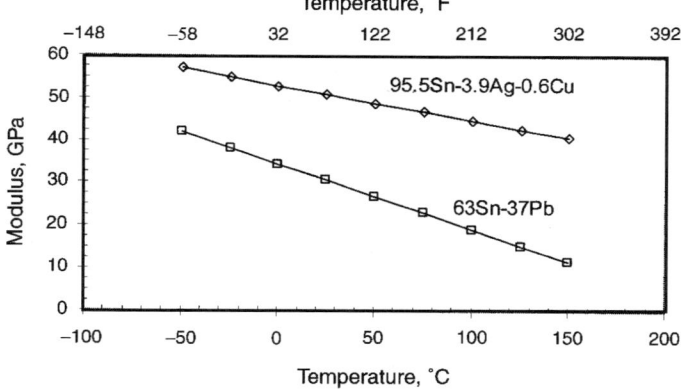

Fig. 4 Elastic moduli of the Sn-Ag-Cu and Sn-Pb solders

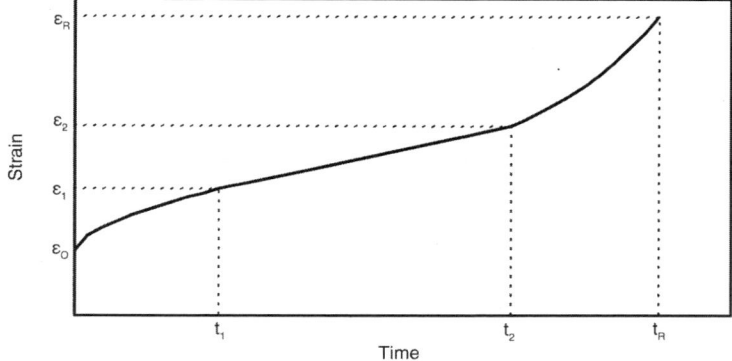

Fig. 5 Creep response of a material under uniaxial load

plastic deformations. During primary creep, between strains ε_O and ε_1, the creep rate decreases. Between strains ε_1 and ε_2, the creep rate remains nearly constant and the material is experiencing secondary or steady state creep. Finally, for times greater than t_2, the creep rate steadily increases until rupture occurs at time, t_R. This final stage is called tertiary creep. Common practice is to characterize the creep response of the solder joints solely with the steady state constitutive relations. A thorough discussion of the mechanics of creep may be found in Ref 6 and 7.

Two common equations used for the steady state creep of solders are the Garofalo-Arrhenius and the Norton, given respectively as:

$$\frac{\partial \varepsilon}{\partial t} = A[\sinh(\kappa \sigma)]^n \exp\left(\frac{-Q}{kT}\right) \quad \text{(Eq 1)}$$

$$\frac{\partial \varepsilon}{\partial t} = A\sigma^n \exp\left(\frac{-Q}{kT}\right) \quad \text{(Eq 2)}$$

where ε is the normal strain, A is an amplitude, σ is the normal stress which is raised to the power, n, and κ (Garofalo only) is a coefficient associated with power law breakdown. The activation energy, Q, is in eV, T is in degrees Kelvin and the value of the Boltzmann constant, k, is $8.617342 \times 10^{-5} \; eV/K$.

As for modeling, there is no universally accepted set of creep constitutive relations for eutectic and high lead solders, let alone the Sn-Ag-Cu alloy. Reference 8 reviewed the characterization work performed on the Sn-Ag-Cu alloy. In particular, Ref 9, 10, and 11 data were very thorough and are briefly discussed here. Reference 8 suggested that averaging the data of Ref 9, 10, and 11 would produce a good engineering creep constitutive relation for the Sn-Ag-Cu alloy. Figure 6 presents the steady state creep rates of the three constitutive relations for the temperatures of -25, 50, and 125 °C (-13, 122, and 257 °F). Reference 9 and 10 data predicts a slower creep rate at relatively lower stresses and a faster creep rate at higher stresses than does Ref 11 data. For use in the numerical studies that follow, Table 2 presents the creep constitutive properties.

Figure 7 compares the steady state creep rates of the Sn-Ag-Cu and Sn-Pb solders. For a given normal stress, the Sn-Ag-Cu solder has a creep rate that is one-half to two orders of magnitude lower than that of the Sn-Pb eutectic solder. Hence, significantly lower creep strains are expected with the Sn-Ag-Cu solder in comparison

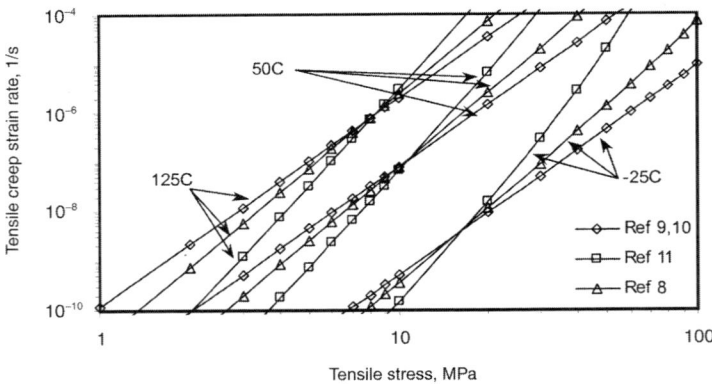

Fig. 6 Comparison of the steady state creep responses for the Sn-Ag-Cu alloy from various sources

Table 2 Creep parameters (Note: T is in K)

Alloy	A (1/s) or (1/s MPa^n)	κ (1/MPa)	n	Q (eV)	Notes
62Sn-36Pb-2Ag	926(508 − T)/T	1/(37.78 − 0.0744T)	3.3	0.55	...
90Pb-10Sn	58.08/(1.732 × 10^6)^n	N/A	8.0696−0.0122T	0.64	Norton equation
Sn-3.9Ag-0.6Cu	441000	.005	4.2	0.466	Ref 9, 12
Sn-3.8Ag-0.7Cu Sn-3.5Ag-0.75Cu Sn-3.5Ag-0.5Cu Castin™	277984	0.02447	6.41	0.56	Ref 11
Sn-(3.5–3.9)Ag-(0.5–0.8)Cu	500000	0.01	5	0.5	Ref 8, used in the examples that follow

with the leaded solder. Again, the modeling results presented later in this chapter will confirm this expectation.

Plasticity. A number of authors, Ref 12 to 18, have reported on the stress-strain response of the Sn-Ag-Cu alloys. Unfortunately, most of the reported data are limited in either temperature range or strain rate, or both. Tests performed at slower strain rates tend to capture creep effects that are reflected in lower moduli, yield, and ultimate strengths when compared with faster strain rates. In particular, Ref 12, 15, and 18 examine the effects of strain rate on the stress strain curve for the Sn-Ag-Cu alloy. The choice of stress-strain data for modeling purposes is governed by the choice of failure theory. When plasticity is combined with creep in the finite element analysis, the plasticity, i.e., the stress-strain data, should not include creep effects. Data obtained with strain rates of approximately 0.01/s or greater are recommended. In the case of failure criterion based on plastic strain alone, the stress-strain data obtained at strain rates equivalent to those to be modeled are recommended.

References 9 and 12 present stress-strain data for the temperature range -25 to $125\,°C$ (-13 to $257\,°F$) at a consistent strain rate of 4.2×10^{-5}/s. The author has demonstrated success producing fatigue life predictions on leaded solder interconnects using the Coffin-Manson relation and Cole's data (Ref 19) which was taken at a similar strain rate, 6×10^{-5}/s. Reference 9 data is shown in Fig. 8 and used in select analysis examples that follow. Note that the elastic modulus at $-25\,°C$ ($-13\,°F$) is lower than that at $25\,°C$ ($77\,°F$). Elastic modulus usually increases with decreasing temperature. Vianco notes that the material response may transition from linear elastic to elastic-plastic at $25\,°C$ ($77\,°F$) and the higher temperature data may reflect the simultaneous effects of work hardening and

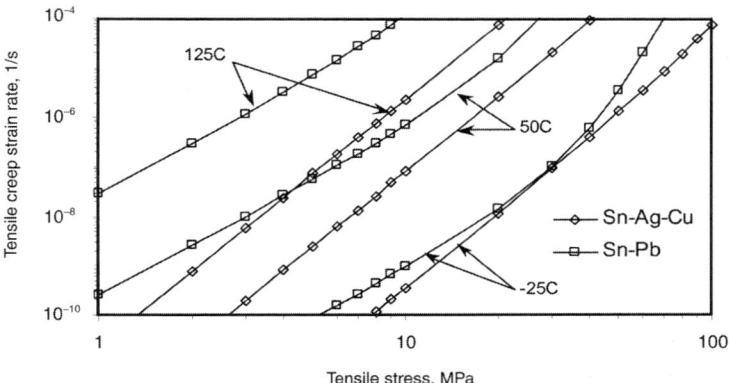

Fig. 7 Steady state creep response of the Sn-Ag-Cu and Sn-Pb solders

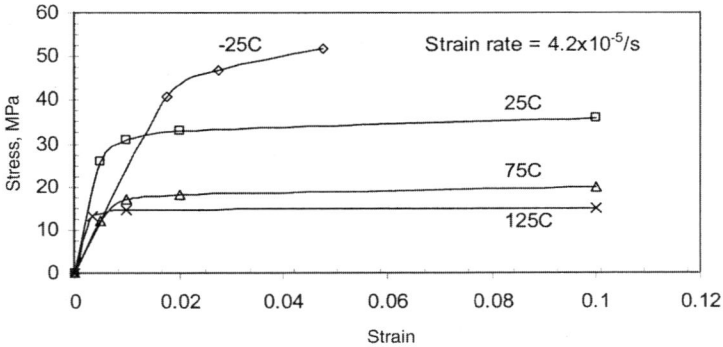

Fig. 8 Stress strain curves for the Sn-Ag-Cu solder. Source: Ref 9

dynamic recovery. Rollup effects evident in the −25 °C (−13 °F) data have been removed for analytical convenience in Fig. 8.

Failure Criteria Discussion

In obtaining the properties of materials for the construction of a stress-strain diagram, a material is typically loaded in a quasi-static manner until failure occurs. In addition to characterizing the elastic, plastic, and/or creep responses of the material, a static strength value is obtained. In reality, structural members are typically subjected to repeated loadings and it is well known that the strength of a material under repeated loadings is less than it would be under static conditions. This phenomenon of decreased strength under repeated loadings is called fatigue. Generally, the initiation and propagation of a crack through a member are the two components of fatigue failure. For large commercial structures such as pressure vessels and aircraft, fracture mechanics is the most appropriate methodology for evaluating fatigue considerations. Unfortunately, a number of issues limit the use of fracture mechanics in the practical evaluation of solder joint fatigue life. These limitations, which are discussed later in this section, are time- and knowledge-based.

In general, the calculation of solder joint fatigue has been based on a stress, strain, or energy value calculated for an intact solder joint. Most recently, these calculations have been performed through finite element analysis. The package is analyzed and the critical parameter, a stress, strain, or energy value, is incorporated into a failure theory. These failure theories are, by and large, analytic expressions that are external to the finite element code. A few of these analytic failure theories that find extensive use with finite element analysis are briefly discussed. A detailed review of current failure theories is included in Ref 20.

Originally developed for the low-cycle fatigue of traditional structural materials such as steel and nickel alloys, the Coffin-Manson equation (Ref 21–23), has found application in the evaluation of solder joint fatigue life:

$$N_f^\alpha \Delta\varepsilon_P = C \quad \text{(Eq 3)}$$

where N_f is the number of cycles to failure, $\Delta\varepsilon_P$ is the plastic strain range, and α and C are parameters. A correlation between the predictions of the Coffin-Manson equation was demonstrated (Ref 24) using analytically determined strains and test results on high-lead solder joints. Presently, finite element analysis has replaced the analytic strain calculation. The equation does not include the effects of time, but may produce very useful predictions for conditions where creep effects are not significant or are equivalent among compared packages.

A failure indicator was introduced (Ref 25, 26) that proposed that fatigue life was directly proportional to the inverse of γ_{MC}, the accumulated per-thermal-cycle matrix creep strain. The matrix creep criterion, which is consistent with the Monkman-Grant relation, takes the form:

$$N_f = \frac{C}{\gamma_{MC}} \quad \text{(Eq 4)}$$

and produced a good correlation between analysis and test. Shortly after the introduction of Eq 4, the creep of Sn-Pb solder was characterized (Ref 27) as due to the contributions of simultaneous climb and glide (these represent two separate dislocation motions within a crystal lattice). The integrated matrix creep (IMC) theory was introduced (Ref 28, 29) setting the fatigue life in proportion to the separate influences of matrix creep and glide-controlled creep. The IMC theory has been further refined (Ref 30, 31) and takes the general form:

$$N_f(0.02 D_{GBS} + 0.063 D_{MC}) = 1 \quad \text{(Eq 5)}$$

where D_{GBS} and D_{MC} are the accumulated creep strains per cycle attributable to grain boundary sliding and matrix creep, respectively. Syed has demonstrated a strong correlation between the predictions of the IMC model and plastic ball grid array (PBGA) and thin small outline package (TSOP) test data.

In 1965, Ref 32 demonstrated that fatigue life can be correlated with the mechanical energy of the hysteresis loop. As discussed in Ref 32, this concept was already relatively mature in the literature, with references as early as 1911. This approach recognizes that energy is required to create inelastic deformation and that the hysteresis loop represents the total energy dissipated during a loading cycle. While most of the imparted energy may be transformed into heat, a portion is assumed to damage the material, finally resulting in failure. Energy-based failure takes the general form:

$$N_f = \alpha(\Delta W)^\beta \quad \text{(Eq 6)}$$

where α and β are constants and ΔW is the change in strain energy per cycle. Note that for the purposes of fatigue calculation the response must be stabilized; the analysis should be run until ΔW is constant cycle-to-cycle. ΔW may be composed of the contributions of plasticity, creep, or both plasticity and creep.

The energy-based Darveaux criterion (Ref 33 and the references therein) is among the most widely used failure criteria in the electronics industry. This solder joint specific methodology combines modeling, material characterization, and failure theory into one predictive approach. A solder joint-specific methodology, combines modeling, material characterization, and failure theory into one predictive approach. Physical testing was performed on a number of solders and produced Anand constitutive relations that included the effects of both plasticity and creep. Solder joint testing of PBGA packages established a crack initiation and growth rate database. Finally, a (mesh dependent) criterion was produced that relates the volume averaged strain energy from finite element models using the Anand constitutive formulation with the empirical crack growth data. The method is one of the most comprehensive efforts focused on the finite element prediction of solder joint fatigue life. While developed for near eutectic solders, PBGA packages, and 3D strip models, analysts have applied the Darveaux method to other solder alloys and package types as well as 3D octant models. These extensions have sometimes resulted in poor test-analysis correlations.

In addition to the previously listed failure theories, other finite element-based approaches include the use of fracture mechanics and damage methods. Fracture mechanics characterize the stress intensity, strain energy release rate or J integral of discretely modeled cracks within solder joints (Ref 3). Fracture analysis is difficult to perform since the crack must be modeled, and since at least 10 to 20 elements are required along each crack face (2D case). Considering that a small solder joint crack may be on the order of tens of microns and an electronic package may be approximately tens of mm in size, the number of elements needed for the mesh becomes excessively large. The method is also deterministic in that the crack size and location must be specified a-priori and results are tied to the geometry analyzed. The definition of the crack geometry is especially difficult in the case of 3D analysis. In addition, the fracture mechanics analysis may track the crack propagation though a solder joint; the model must permit the crack to grow and a new analysis must be performed for each crack location. Consequently, a number of calculations are needed for each solder joint. Finally, the failure criteria, the critical value of K, J or G for solders have not been extensively studied. As a result of the issues associated with the required number of elements and the definition of the crack surface, 2D fracture mechanics analysis is most common.

The damage methods include those varied theories (Ref 34 to 36) that associate failure with a variable, typically designated D, whose value ranges from 0 for a pristine structure to 1, which signifies failure. These methods result from elegant, often thermodynamics-based, mathematical derivation but there is no consensus as to the damage metric. These methods have not yet obtained a foothold among practicing analysts.

While the preceding fatigue relations have been established for lead-based solders, they also constitute the starting point for an investigation of the appropriate methodologies for lead-free alloys. Different fatigue relations appear more accurate with specific package types (Ref 20). Select fatigue relations that have been applied to the Sn-Ag-Cu alloys are now briefly discussed.

Tensile tests have shown that the low-cycle fatigue life of 96.5Sn-3Ag-0.5Cu dogbone specimens follow the Coffin-Manson relation shown in Eq 7 (Ref 37):

$$\Delta\varepsilon_P N_f^{0.73} = 3.7 \quad \text{(Eq 7)}$$

where $\Delta\varepsilon_P$ is the per-cycle plastic strain range. The data used in the fit spans $\sim 0.003 \geq \Delta\varepsilon_P \geq \sim 0.02$ and has a correlation coefficient of ≥ 0.97. A fatigue life model was proposed (Ref 9):

$$N_f = \alpha(\Delta W_{CR})^\beta \quad \text{(Eq 8)}$$

based on the prevalence of similar models in use with leaded solders. ΔW_{CR} is the creep strain energy density accumulated per thermal cycle. The following life predictive equation was presented (Ref 11, 38):

$$N_f = 345(W_{CR})^{-1.2} \quad \text{(Eq 9)}$$

for Sn-Ag-Cu bumps in a flip-chip on board construction and for midsize PBGA packages mounted to a PCB and subject to thermal cycling. In the equation, W_{CR} is the minimum over the path of local maximum viscoplastic strain energy density dissipated per cycle. References 11 and 38 also produced the predictive relationship:

$$N_f = 4.5 \varepsilon_{CR}^{-1.295} \qquad \text{(Eq 10)}$$

where ε_{CR} is the critical value of the equivalent creep strain along the anticipated failure path.

Modeling Difficulties

In the order discussed, the finite element geometric modeling approaches (2D, 3D slice, and 3D octant) represent increasing levels of spatial structural detail, as well as computational commitment. Yet the ability of any of these modeling strategies to accurately predict the solder joint life for a particular package family, say PBGAs, is imprecise. In Ref 39 for example, the Darveaux failure criterion is applied to a variety of package types and accelerated thermal environments. Differences between the analysis prediction and experimental solder joint life were as small as ~15% and as large as ~±500% depending on package type and meshing parameters. Reference 40 examined the predictive capabilities of the 3D slice and octant models as applied to two package types. Again using the Darveaux criterion, no clear advantage was shown by either model type. In fact, the 3D slice was more conservative for one of the packages.

Numerical computation inevitably requires that trade-offs be made between the level and degree of detail modeled and the cost, time, or computational resources required to perform the solution. Despite the rigorous formalism associated with the development of finite element tools, the analyst must judiciously make selections regarding the element type and number, the material properties, boundary conditions, and solution parameters. A skillfully executed analysis will have compromises and approximations that are obvious to those versed in mechanics, yet produce accurate results for the quantities of interest. The ability to make these trade-offs requires an understanding of the approximations and inaccuracies in the numerical tools (Ref 41, 42) and an anticipation of the expected results based on either experimental evidence or first principles. A number of issues pertaining to the absolute accuracy of any solder joint reliability model may be easily identified, including the accuracy of the material properties, the progressive nature of fatigue damage, singularities that are present in the numerical solution, and the idealizations regarding the thermal history. Each of these issues is briefly discussed.

Accuracy of the Material Properties. Material properties are generally determined during short-time scale physical testing. As for lead-based solders, the microstructure of Sn-Ag-Cu solders will be determined by the initial processing (Ref 15) and will evolve during the course of thermal exposure (Ref 43–45). Since the mechanical properties are tied to the microstructure and since the mechanical testing does not capture long-term changes, those properties used in analysis are imprecise and appropriate only at initial times. This effect is most pronounced in the definition of the inelastic properties. In the case of the creep constitutive relations, test data may be formulated over the course of hours, days, or weeks while the analysis predicts fatigue lives over months of testing or years of service.

Progressive Damage. Solder fatigue failure is a progressive situation wherein cracks nucleate and progress through a solder joint over the period of many thermal cycles. In addition, cracking will typically occur to differing degrees in numerous solder balls. While some finite element analyses have included select fractured solder joints, the ability to include all solder joints in sufficient detail is typically not feasible. As a result, most numerical analyses base failure on a stress, strain, or energy value taken from a pristine solder joint. Those that physically incorporate a crack typically neither include crack damage in multiple solder joints nor follow the crack's growth through the critical solder joint. Consequently, the modeled structure does not necessarily represent a faithful rendition of the true geometry, especially as cracks alter that geometry.

Singularities. At multimaterial interfaces, as well as crack tips, elastic analysis will report singular stresses (Ref 46, 47). The outer surface of the solder-to-pad interface where crack initiation is most likely, is an example of such a multimaterial interface. Except in the case of a perfectly plastic material, derived stresses, strains, and strain energies will be mesh dependent. Finer elements will return higher stress values than will larger elements. Since these derived quantities are most likely used to predict solder joint fatigue life, predictions become inextricably tied to the chosen mesh.

Thermal History. The typical finite element analysis procedure makes two idealizations regarding the thermal environment and the time-temperature history within the package. The first is that the external chamber cycling environment has a trapezoidal (or other simple) temperature profile. That is, the temperature during a ramp is linearly proportional to the ramp time, and the temperature during the dwell is constant. The second simplifying assumption is that the entire package and PCB assembly have a uniform temperature distribution which is equivalent to that of the chamber environment. Transient temperature distributions and the stresses due to gradients within the assembly are ignored.

In consideration of these difficulties, it is unlikely that any model will accurately predict the solder joint fatigue life for all package types and loading scenarios. In fact, despite the level of geometric and material property complexity included in a model, approximations, simplifications and idealizations still dictate that the model produces results with some degree of imprecision. Therefore, finite element modeling should seek a calibrated methodology that provides accurate results with minimal computational effort while retaining a connection to the package geometry and materials. The goal is to obtain accurate, reliable numerical results with the greatest economy of elements and computational overhead. Well-constructed models may run in hours or less while poorly meshed models may take days or more to run.

Suggested Geometric Finite Element Model

The choice of model type is most crucial to the computational expense involved in performing the second-level reliability calculations. While some authors (Ref 31, 48) have had success with 3D octants and 3D octants combined with sub-models, these analyses are typically limited to smaller packages with depopulation and approximately 200 to 500 solder joints (or I/O). A significant amount of interest exists, however, in the reliability of larger packages, such as those with 1500 and greater I/O.

Due to their physical size, these large devices are more prone to early failures than the lower I/O packages. Constructing a 3D octant model of a 2000 I/O package with 2 planes of reflection symmetry is a costly computational proposition. The 3D slice approach easily captures the physical construction of both large and small packages in computationally efficient finite element models. Of the model types previously discussed, the 3D slice model is suggested.

In addition to the choice of general model type, the use of selective refinement can enhance computational efficiency. Selective mesh refinement is the use of very refined elements in regions where accurate definition of the stresses and strains is required and much coarser elements elsewhere within the model. Knowledge of the expected failure site(s) from previous empirical testing is indispensable. In the case of a package with a laminate substrate, failures are most likely to occur near the die shadow, that is, in balls close to the die's perimeter. More rigid substrates, such as ceramics, typically incite failures within the outer rows of solder interconnects. Since the model must produce an accurate definition of stresses and strains in the solder joint where failure is expected, a highly refined mesh is applied only to this critical solder joint. The rest of the structure simply has to provide the proper loading and boundary stiffness on the critical solder joint(s) and is meshed with much larger elements.

A 256 Input/Output PBGA package is analyzed in the next section. Consideration is paid to presenting specific characteristics of the modeling strategy, as well as discussing the physical response in the context of the material properties.

An Example Using the 256 PBGA

In this section, the structural response of a 256 PBGA package subject to part-on-board temperature cycling will be examined. A recent experimental program (Ref 49–52) studied the solder joint reliability of this package with both Sn-Pb and Sn-Ag-Cu solder balls. The elastic and creep properties of both solders will be incorporated in the analysis in order to demonstrate the influence of the new Sn-Ag-Cu material in the context of the structural responses of the familiar Sn-Pb solder.

The 256 PBGA has a 27 by 27 by 0.36 mm (1.06 by 1.06 by 0.014 in.) substrate with a 10 by 10 by 0.3 mm (0.39 by 0.39 by 0.012 in.) die and 1.17 mm (0.05 in.) thick overmold. Both the PCB and the substrate have 0.635 mm (0.025 in.) diameter pads and the assembled ball is assumed to have a height of 0.5 mm (0.02 in.) and a maximum diameter of 0.9 mm (0.03 in.). Ball pitch is 1.27 mm (0.05 in.) for the PBGA pack-

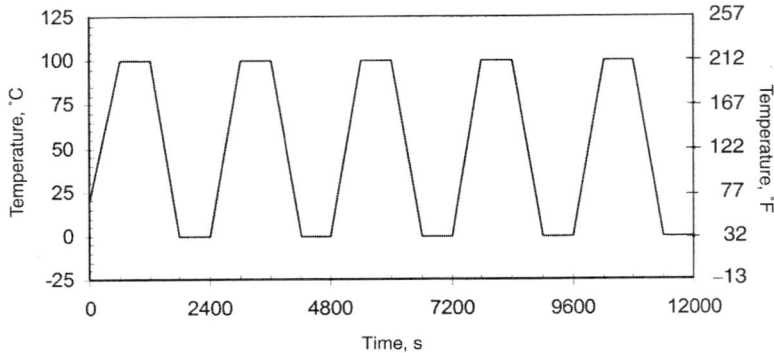

Fig. 9 Typical idealized thermal cycle environment

ages with four perimeter rows and center depopulation. Additionally, the PCB thickness is 1.6 mm (0.06 in.) for both the eutectic and lead-free PBGA models.

Chamber thermal cycling spans the temperature range of 0 to 100 °C (32 to 212 °F) with 10 min. ramps between the extremes and 10 min. dwells at temperature. Figure 9 presents the idealized thermal cycle. After starting at room temperature, each model is analyzed for 5 cycles in an attempt to obtain results that are converged cycle-to-cycle. Note that the trapezoidal temperature distribution is an idealization that differs slightly from measured part temperatures, which display damped exponential characteristics.

As shown in Fig. 10, the package is modeled as a three-dimensional strip that captures the construction along a diagonal path from the package's geometric center to a corner. Due to the symmetry at about the vertical midplane of a full strip, the models are actually half strips with the appropriate in-plane restraints placed on one symmetry plane. Coupled in-plane translations are applied to the other symmetry plane to produce a state of generalized plane strain. Using exclusively hexahedral solid elements, the models can capture the shape of the packages' solder balls and potential DNP effects while retaining significant computational efficiency over full octant models.

Experience with PBGAs has shown that the relatively high die stiffness and the flexibility of the laminate substrate generally cause first thermal cycle fatigue failures in the ball closest to the die shadow. Despite the overall economy of elements in each strip model, selective mesh refinement was used to concentrate highly refined elements in the solder joints where failure was anticipated.

There are two components to the selective refinement, the first being the refinement within a model of only the solder joint with anticipated first failure, and the second being the use of graded meshes within this critical solder joint. Figure 10 presents the location of the refined solder ball for the PBGA package as being closest to the die. In this case, the 256 PBGA model has a total of 6408 elements. Of these, 1920 are in the refined solder ball where first failure is expected and 24 elements are used in each of the other solder balls. Figures 11 and 12 present the details of the refined and coarse solder ball meshes. As depicted in Fig. 12, the elements in the refined solder ball also get smaller as they approach each of the substrate and PCB pads. This mesh grading increases the fidelity of the calculated stresses and strains at the critical solder joint-to-pad interface while saving on element count. In all refined solder interconnects discussed in this chapter, the grading produces elements, used in results reporting, with a thickness of approximately 18 μm. Since the thickness can affect the results calculation, a consistent thickness is required in order to compare

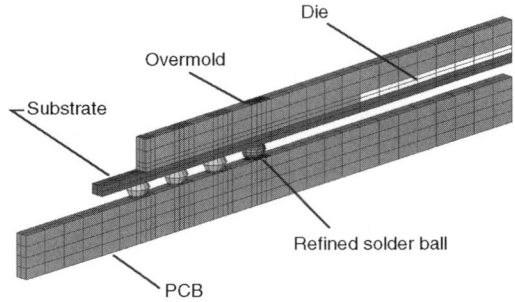

Fig. 10 256 PBGA model

results among models. The accuracy of focusing the elements to a known failure location will be discussed at the end of this section.

Figure 13 captures the out-of-plane displacement of the edge of the substrate for both solder ball compositions. From the figure, it is evident that the out-of-plane displacements are in phase with the applied thermal environment. The peak-to-peak deflections are almost identical for both solder alloys and do not differentiate the solder interconnect compositions.

The 256 PBGA models predict that failure is likely to occur in the solder ball at the pad interface. This failure site is determined through an examination of the stresses, creep strains and creep strain energy densities in the solder ball elements. These individual quantities are briefly discussed. All presented stress, strain, and energy results are for this location. These values are element averaged rather than nodal values within the critical element.

Figure 14 presents the shear stress versus time history for both packages over five thermal cycles. Similar to the modeled displacements, the shear stress is in phase with the thermal environment. Due to the relative compliance of the solder interconnects when compared with the stiffness of the PCB and the package, the solder joints are considered to be in a displacement controlled loading. As a result of the greater stiffness of the lead-free solder, the stress range within the Sn-Ag-Cu solder joint is greater than that for the Sn-Pb alloy.

The discussion of the creep constitutive properties proposed that creep strains would be lower in Sn-Ag-Cu solder balls than in Sn-Pb solder balls for like loading conditions. This expectation resulted from the faster creep rates seen in Sn-Pb solders compared with the Sn-Ag-Cu alloy. Creep shear strains for both solder joints in the 256 PBGA package are presented in Fig. 15. From the figure, the range of creep shear strains is much greater for the Sn-Pb solder than for the lead-free solder. Again, the finite element anal-

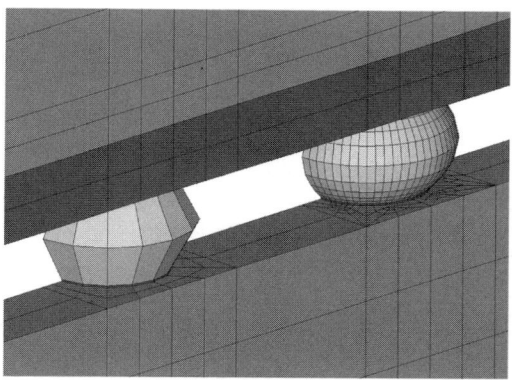

Fig. 11 Detail of assembly showing the refined and coarsely meshed solder balls

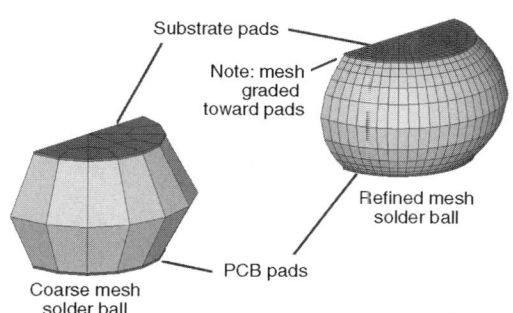

Fig. 12 Detail of refined and coarsely meshed solder balls

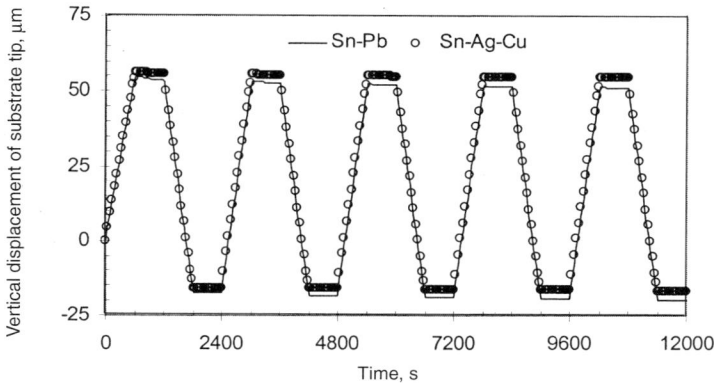

Fig. 13 Vertical displacement of the 256 PBGA substrate edge

ysis confirms the material property-based expectations for the physical responses of the two solders. For a given construction, thermal creep strains should generally span a narrower range of values with Sn-Ag-Cu solder interconnects and a comparatively broader range with Sn-Pb solder joint.

Examination of the hysteresis loops encompasses the stress and creep strain results and provides physical insight into energy-based failure theories. Figure 16 presents the shear stress-creep shear strain hysteresis loops for the 5th thermal cycle. The hysteresis loop for the Sn-Ag-Cu solder spans a broader stress but narrower strain range than does the Sn-Pb solder's hysteresis loop. The area within the hysteresis loop represents the energy that is dissipated by the creep deformations within the selected finite element and during the thermal cycle. Physically, this energy dissipated within the solder joint may be associated with damage. In one failure theory that is studied, the failure is taken to be in proportion to this amount of dissipated energy. Therefore, hysteresis loops with greater areas indicate a greater amount of damage imparted to the solder joint per cycle and, consequently, a shorter expected fatigue life. For Fig. 16, the area within the hysteresis loop is the creep shear strain energy density. Note that the area within the Sn-Pb solder joint's hysteresis loop is greater than that for the Sn-Ag-Cu solder joint.

Shear components provide one means of characterizing the hysteresis energy dissipated in the solder. Von Mises stress and creep strain components may also be used in these calculations. The choice of von Mises-based calculations is made for most of the analyses that follow. The use of von Mises stresses and creep strains has two important advantages. First, the von Mises calculations will include the effects of out-of-plane loading which would be ignored by con-

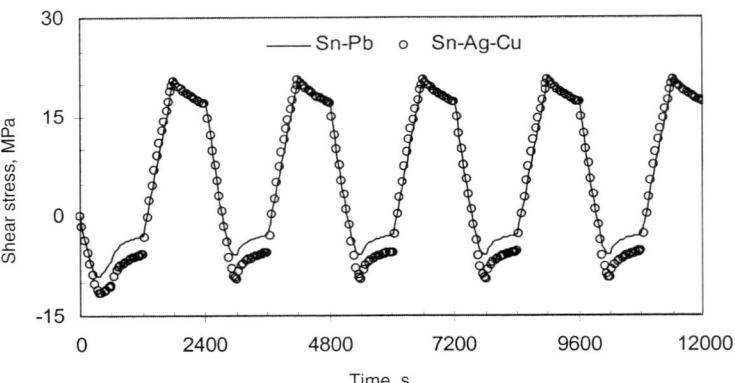

Fig. 14 Shear stress history in the 256 PBGA

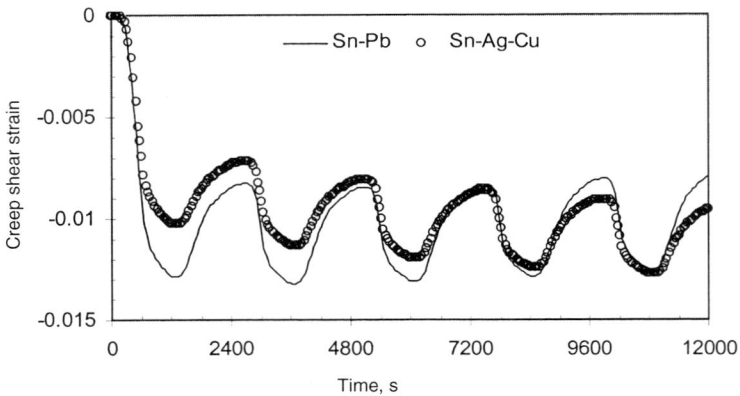

Fig. 15 Creep shear strain history in the 256 PBGA

sideration of shear alone. Second, ANSYS software automates the creep strain energy density calculation based on von Mises stresses and von Mises creep strains. The use of shear-based values requires additional calculations either within the finite element code or through the use of a spreadsheet. The von Mises-based creep strain energy density as a function of time is presented in Fig. 17 for both solder materials. From the figure, the change in creep strain energy is greater per cycle for the Sn-Pb solder than for the Sn-Ag-Cu alloy. This coincides with the observation based on the shear-based hysteresis loops. Examination of Fig. 17 reveals that the change in creep strain energy density for either package is essentially constant after a few thermal cycles. A stable change in creep strain energy density is required for the energy-based failure criteria.

It should be noted that there are likely two components associated with a solder fatigue life predictive model based on strain energy density. The first is the magnitude of the creep strain energy density dissipated per cycle For the case examined, the ΔW_{CR} per cycle is greater for the Sn-Pb solder than for the Sn-Ag-Cu solder. The second component is the material's tolerance of such irrecoverable work; a thorough understanding of this property based on isothermal fatigue testing of the lead-free solder is not yet established.

The examination of the 256 PBGA package was predicated on the expectation of first failure within the innermost solder ball As a result, the finite element model used very selective mesh refinement only for this critical solder ball. Consider the same package and board assembly utilizing the Sn-Ag-Cu solder. The assembly is subject to a simple thermal environment and the predicted failure site, as well as maximum creep strain, are determined as a function of the number of refined solder balls.

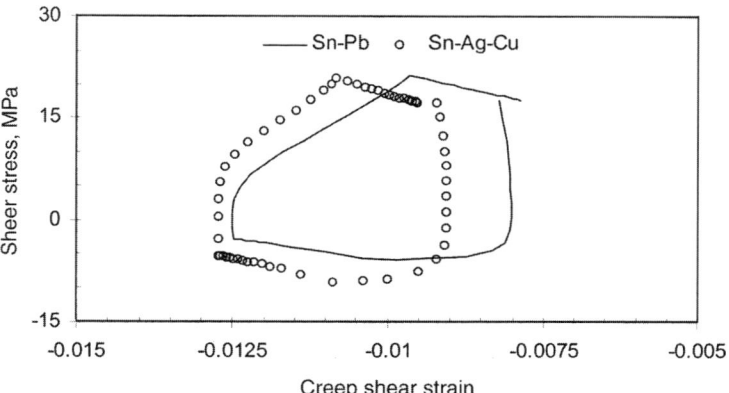

Fig. 16 Hysteresis loops in the 256 PBGA for the 5th thermal cycle

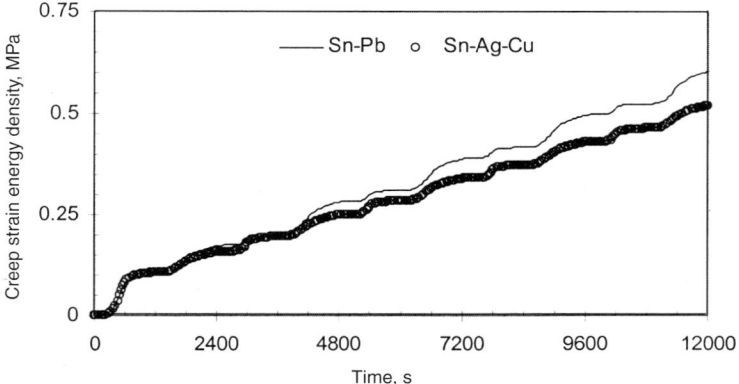

Fig. 17 Strain energy density in the 256 PBGA

The innermost solder ball is designated as number 1 and sequential numbering is applied such that the outermost ball is number 4. The analysis assumes a stress-free assembly at 20 °C (68 °F) and the model is run for one 10 min. exposure at 100 °C (212 °F). Subject to convergence criteria, the initial time step of 0.01 s may gradually increase to 0.5 s. Table 3 demonstrates that virtually no improvement in accuracy is gained over the use of select refinement (only ball 1 is refined) despite up to 3 times greater element count and run times for the models with more refined solder balls. Used judiciously, the select solder joint refinement provides the accuracy of a more complex model at a significant computational cost savings.

An Examination of Failure Theories with Literature Data

The preceding examination of the 256 PBGA demonstrated an application of the 3D slice modeling approach, as well as typical results. The power of finite element solder joint analysis is to provide accurate predictions that can be used to design new packages, to improve the performance of current designs, or to predict the performance of untested designs. A calibrated combination of modeling strategy, material properties, and failure theories is needed to make this predictive capability useful. At the very least, this calibration dictates a fit to empirical data on a similar package type through a similar temperature cycling environment. In this section, Sn-Ag-Cu solder joint reliability data from Table 4 are used to examine the predictive capability of the 3D strip model with select plasticity- and creep-based fatigue failure theories. The test vehicles are modeled and the finite element results obtained from models are incorporated into the fatigue life models through least squares fitting. The character of the correlation between each fit fatigue model and the reported fatigue lives is presented.

It should be noted that this procedure is very similar to that faced by the analyst. Typically, the analyst's company conducts a limited number of part-on-board solder joint tests. The analyst must then produce predictions for new parts or for design changes with the aid of these test results. Correlation between the model predictions and failure data lends credibility to the analyst's work.

Despite the prevalence of solders in PCB assembly, no consensus exists as to the proper failure theory for Sn-Pb solder, let alone for lead-free solders. Simple plasticity- and creep-based fatigue life failure models, are considered here:

$$\eta = A(\Delta \varepsilon_P)^E \quad \text{(Eq 11)}$$

$$\eta = B(\Delta \varepsilon_{CR})^F \quad \text{(Eq 12)}$$

$$\eta = C(\varepsilon_{CR}^{MAX})^G \quad \text{(Eq 13)}$$

$$\eta = D(\Delta W_{CR})^H \quad \text{(Eq 14)}$$

where η are the number of cycles to the characteristic life (62.3% cumulative failed), A

Table 3 Effect of select refinement

Refined solder ball(s)	Elements	Relative solution time	Max creep shear strain	Solder ball predicted to fail
1	6408	1.0	0.00555	1
1 and 4	10836	1.54	0.00555	1
1–4	20388	2.92	0.00557	1

Table 4 Select Sn-Ag-Cu solder joint reliability test parameters and results

Pkg type	I/O	Substrate size (mm)	L2 pitch (mm)	Die and overmold	PCB thickness (mm)	Thermal environment	η (cycles)	Ref No.
CBGA	937	32.5 × 32.5 × 1.5	1	None	1.83	0–100C, 2cph	1931	53, 54
CBGA	937	32.5 × 32.5 × 2.4	1	None	1.83	0–100C, 2cph	1370	53, 54
CBGA	1657	42.5 × 42.5 × 1.5	1	None	1.5	0–100C, 2cph	1670	53, 54
CBGA	1675	42.5 × 42.5 × 2.55	1	None	1.5	0–100C, 2cph	982	53, 54
CBGA	1657	42.5 × 42.5 × 3.7	1	None	1.5	0–100C, 2cph	687	53, 54
PBGA	256	27 × 27 × 0.56	1.27	10 × 10 × 0.3 die + overmold	1.57	−40–125C, 1cph	8083	44
PBGA	316	27 × 27 × 0.36	1.27	7 × 7 × 0.3 die + overmold	1.6	−40–125C, 1cph	4691	55
PBGA	313	35 × 35 × 0.56	1.27	10 × 10 × 0.3 die + overmold	1.6	−40–125C, 1cph	3926	55
PBGA	256	27 × 27 × 0.36	1.27	10 × 10 × 0.3 die + overmold	1.6	0–100C, 1.5cph	Unknown, >7830	51

through D are constants, as are the powers E through H, $\Delta\varepsilon_P$ is the change in plastic strain per cycle, $\Delta\varepsilon_{CR}$ is the accumulated creep strain per cycle, ε_{CR}^{MAX} is the maximum creep strain in a cycle, and ΔW_{CR} is the change in creep strain energy density per cycle. A number of points must be made regarding the selection of these quantities from the finite element model. The first is that the chosen values are element averaged rather than nodal. Second, values from a most critically loaded element are used, rather than averaged over a path or slice. Finally, creep analysis results are taken from the 5th thermal cycle. In general, stability is achieved by the 5th thermal cycle for energy terms, but not necessarily for the strain terms.

Table 4 summarizes select parameters of the published Sn-Ag-Cu package test results used in this study. Solder joint reliability tests were performed (Ref 53, 54) of bare ceramic ball grid array (CBGA) substrates subject to the temperature range 0 to 100 °C (32 to 212 °F) and at a frequency of two cycles per hour. The testing included 32.5 by 32.5 mm (1.28 by 1.28 in.) substrates with thicknesses of 1.5 and 2.4 mm (0.06 and 0.09 in.), and 42.5 by 42.5 mm (1.67 by 1.67 in.) substrates with thicknesses of 1.5, 2.55, and 3.7 mm (0.05, 0.10, and 0.14 in.). I/O arrays were 31 by 31 and 41 by 41 (1.22 by 1.22 and 1.61 by 1.61 in.), respectively, with 6 balls depopulated per corner. While a 1.83 mm (0.07 in.) thick PCB was used for the 32.5 mm (1.27 in.) parts, a thinner, 1.5 mm (0.05 in.) thick, board was used for the 42.5 mm (1.67 in.) substrates. In both cases, the PCB pad diameter was nominally 0.67 mm (0.03 in.) and the assembled ball height was approximately 0.38 mm (0.01 in.). Examination of the cross-sections presented in the papers suggests that the substrate pad diameter was approximately 0.76 mm (0.03 in.). The characteristic life, η, for the IBM data was deterived from the lognormal distribution parameters provided by Farooq.

The remaining literature referenced test data are for PBGA packages. A test of 256 PBGA with Sn-Ag-Cu solder balls and 10 by 10 mm (0.39 by 0.39 in.) die was performed (Ref 44). With the exception of a 0.56 mm (0.02 in.) thick substrate, the geometry of this package is the same as that for the 256 PBGA previously discussed. Testing was performed with 1.57 mm (0.06 in.) thick PCBs and a frequency of 1 cycle per hour with the temperature range of −40 to 125 °C (−40 to 257 °F). A characteristic life of 8083 cycles was determined (Ref 44).

Both 316 and 313 PBGAs with lead-free solder interconnects were tested (Ref 55). The 316 PBGA has a 27 mm (1.06 in.) body size and a 7 by 7 mm (0.27 by 0.27 in.) die. The Sn-Ag-Cu balls occupy 5 perimeter rows, as well as a 4 by 4 array of thermal balls at the package center. The 313 PBGA's body size is 35 mm (1.37 in.) and a 10 by 10 mm (0.39 by 0.39 in.) die was included in the test package. For the 313 PBGA, the ball pattern is a full grid staggered array. Both packages have 0.762 mm (0.02 in.) Sn-Ag-Cu spheres at a ball pitch of 1.27 mm (0.04 in.). These packages were subject to 1 h thermal cycles over the temperature range of −40 to 125 °C (−40 to 257 °F). FR4-based PCBs at 1.6 mm (0.06 in.) thick were used for both package types. Lee's test matrix included peak reflow temperatures of 240 and 260 °C (464 and 500 °F) for the lead-free packages. Due to the differing reflow temperatures, two slightly different characteristic lives were determined for each package. Table 4 reports the average value for each package.

In addition to the 256 PBGA model shown in Fig. 10, Fig. 18 to 20 present the other models used in the analysis. While Fig. 18 specifically represents the 937 Input/Output CBGA with a 2.4 mm (0.09 in.) substrate thickness, the other CBGA models differ in the number of solder

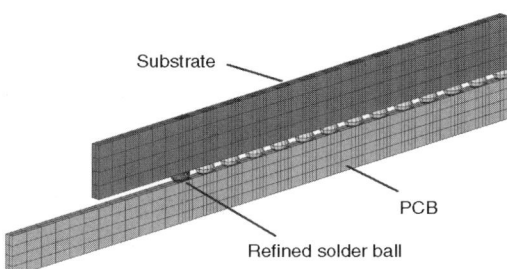

Fig. 18 CBGA model

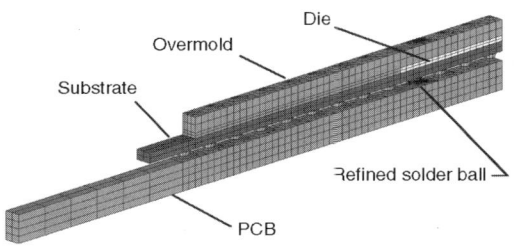

Fig. 19 313 PBGA model

balls, as well as the dimensions of the substrates and the PCB. The CBGAs are bare substrates, while the PBGA packages have die and overmold. The staggered ball placement of the 313 PBGA dictates that this model is wider than the 256 or 316 PBGA models. In all cases, the modeled PCB extends beyond the edge of the package by approximately 20% of the modeled substrate length.

While experience provides guidance for the expected first failure site, each model was run through a quick (20 s) saw-tooth temperature environment to check the creep strains at the high and low temperature extremes. During these survey runs, all solder balls had the coarse mesh, and wall clock solution times were on the order of 10 min. The critical joint was chosen to be the one with the highest creep strains. In subsequent plasticity and creep analyses, the critical solder joint used the refined mesh while others maintained the coarse mesh. This quick survey, which assures that the proper joint is chosen for refined meshing, can produce surprising results. For example, in the 313 PBGA package, the die terminates over the middle of the 5th modeled solder ball from the center. It is interesting to note that the survey run identified the 4th solder ball from the center as being most critically loaded. Figure 19 shows that this critical solder joint received the fine mesh for the analyses.

The plasticity evaluation used the stress-strain data shown in Fig. 8 with minor refinements from Ref 16 in defining the secant modulus. Following the convention recommended (Ref 56), each package is assumed stress-free at the lower temperature cycling extreme presented in Table 4. The plastic strain range, $\Delta\varepsilon_P$, is then determined at the upper extreme of the temperature range. Since the analysis is purely elastic-plastic, time does not enter into the calculation, and the analysis requires only a single load step. The highest-element-averaged von Mises plastic strain is used to characterize the critical solder joint.

Least-squares fitting is used to determine the constants in Eq 11. As suggested by Fig. 21 and 22, linear fitting of the natural logarithms of $\Delta\varepsilon_P$ and the test η produces the constants, A and E (Eq 11). Notice that the CBGA and PBGA data require distinctly different least squares fits. This may be attributed to a number of factors which include the differing package types, thermal environments, pitches, or ball sizes. As a result, separate Coffin-Manson equations characterize the two package types. Figures 23 and 24 present these equations and the correlation between

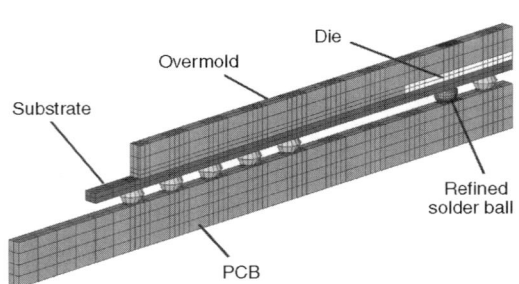

Fig. 20 316 PBGA model

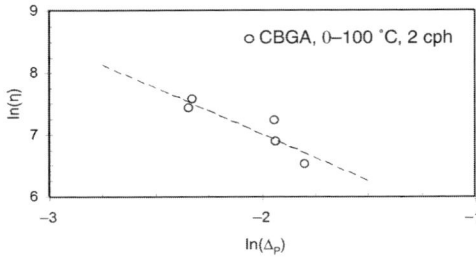

Fig. 21 Notional view of the fitting procedure—CBGA data

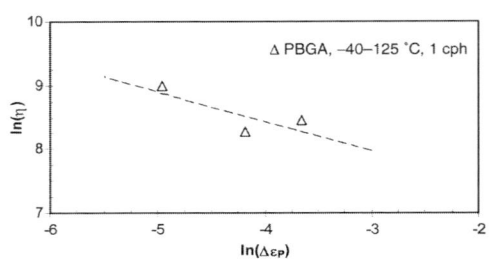

Fig. 22 Notional view of the fitting procedure—PBGA data

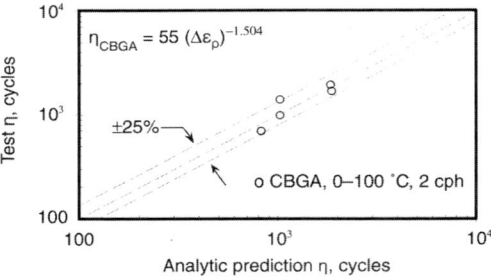

Fig. 23 Correlation between the best-fit Coffin-Manson predictions and the CBGA test data

Coffin-Manson predictions and test data. All predictions fall within 25% of the actual failure data. Subsequent creep-based analyses use the same fitting procedures and will produce separate fatigue life prediction equations for a given creep criterion and package type.

As discussed here, creep-based failure prediction (Eq 12–14) is very similar to the Coffin-Manson procedure since a single point value associated with a local maximum is used (although averaging through a volume of elements is also possible). In the Coffin-Manson approach, the plastic strain characterizes the fatigue life, and creep is not included. Consequently, the effects of cyclic frequency cannot be directly incorporated. In the creep methods, the effects of time and temperature must be expressly included in the analysis. Therefore, the comparison among tests results with different thermal environments is possible. The creep analyses are performed over five or more thermal cycles. The thermal environment follows that shown in Fig. 9, with changes to the temperatures and times as appropriate to the specific package test environment presented by Table 4.

Ideally, the creep strain or creep stain energy density extracted from the creep analysis should be stabilized, that is, consistent from cycle to cycle. Figure 25 details the evolution of the calculated von Mises stress and creep strain response of the smallest CBGA through eight thermal cycles. From the figure, the high and low strains bounding the hysteresis loop are clearly changing for the early cycles. By the 7th or 8th cycle, the maximum creep strain appears to be stabilized. Therefore, when performing an analysis based on the maximum creep strain, the analyst must recognize the quality of the convergence and the effects of non-stabilized creep strains on the failure prediction. By inference, strain convergence issues may also hold for the $\Delta\varepsilon_{CR}$ and the ΔW_{CR} criteria.

The response of the three creep strain-based criteria are presented in Fig. 26 to 31. The packages presented in the figures bound the responses of those packages studied. A number of observations are made. First, the CBGA solder joints generally have higher creep strains and creep strain energy densities than do those in the PBGA packages. This effect is due to the stiffness and low thermal expansion of the ceramic substrates when compared to the properties of the PCB. Second, the stabilization of the $\Delta\varepsilon_{CR}$ and ε_{CR}^{MAX} criteria is highly package dependent. Neither of the strain values is truly unchanged after eight thermal cycles. The critical $\Delta\varepsilon_{CR}$ stabilizes more quickly with the PBGA packages while the ε_{CR}^{MAX} appears to approach stable cycle-to-cycle values more quickly for the CBGA packages. Finally, the hysteresis energy, ΔW_{CR}, is virtually unchanged after the 2nd thermal cycle. This is a significant conclusion for creep-based analysis. Since the hysteresis energy is stable, comparative analyses that use the ΔW_{CR}

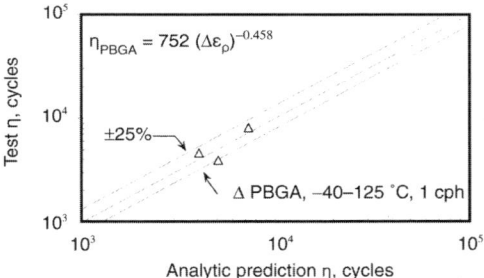

Fig. 24 Correlation between the best-fit Coffin-Manson predictions and the PBGA test data

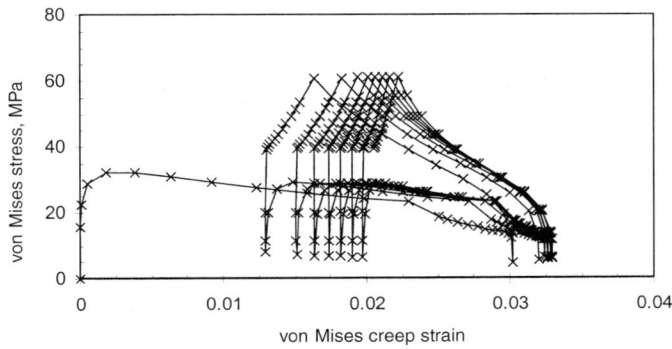

Fig. 25 Evolution of the hysteresis loops for the 32.5 × 32.5 × 1.5 mm CBGA

criterion are not cycle dependent, provided the results of the 2nd or 3rd thermal cycle are used. Equally important, since finite element solution time is proportional to the number of thermal cycles analyzed, solutions determining ΔW_{CR} over two or three thermal cycles may be solved on a moderately powerful computer over the course of a workday morning or afternoon.

Finite element creep-based results, $\Delta \varepsilon_{CR}$, ε_{CR}^{MAX} and ΔW_{CR}, are fit to Eq 12 to 14 with the published test data. Figures 32 to 37 present the least squares fit, performed separately for PBGA and CBGA packages, and the correspondence between the fit and the test data. In all cases, finite element results were taken from the 5th thermal cycle. Based on the previous discussion,

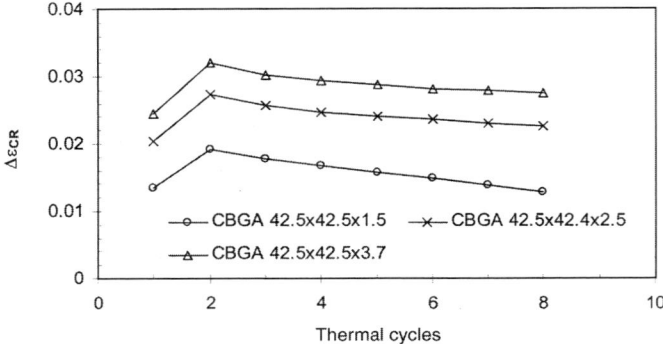

Fig. 26 Maximum calculated von Mises creep strain increment as a function of modeled thermal cycle (42.5 mm CBGA packages)

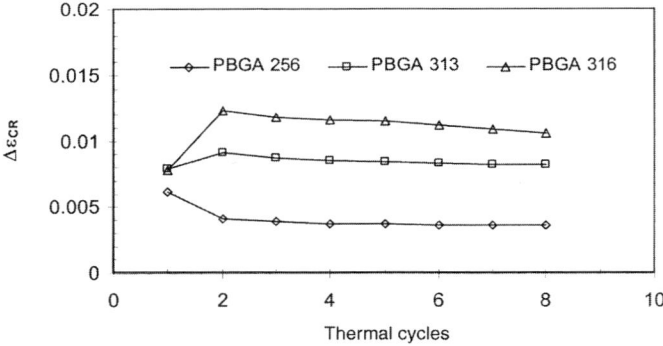

Fig. 27 Maximum calculated von Mises creep strain increment as a function of modeled thermal cycle (PBGA packages)

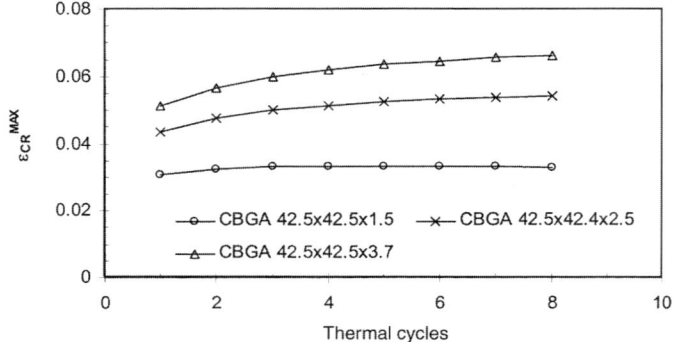

Fig. 28 Maximum calculated von Mises creep strain as a function of modeled thermal cycle (42.5 mm CBGA packages)

the choice of a different thermal cycle would affect the $\Delta\varepsilon_{CR}$ and ε_{CR}^{MAX} predictions only. From the figures, all creep-based methods produce roughly equivalent quality predictions of the test data. Therefore, the ΔW_{CR} criterion, (Eq 14), is strongly suggested, due to the stability of the hysteresis energy after just two thermal cycles for all package types considered.

In summary, the use of the 3D strip model in combination with Eq 14 has shown the desirable qualities of accuracy over a broad range of package configurations and thermal environments, as

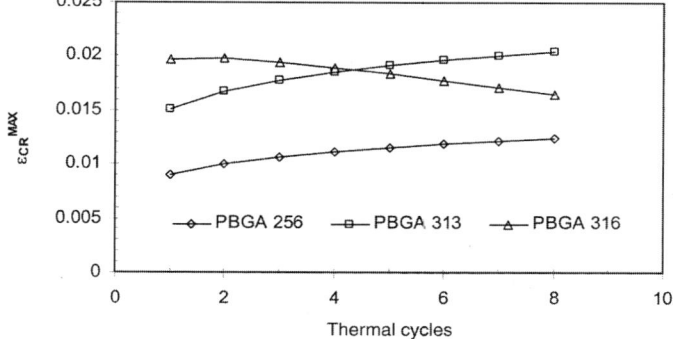

Fig. 29 Maximum calculated von Mises creep strain as a function of modeled thermal cycle (PBGA packages)

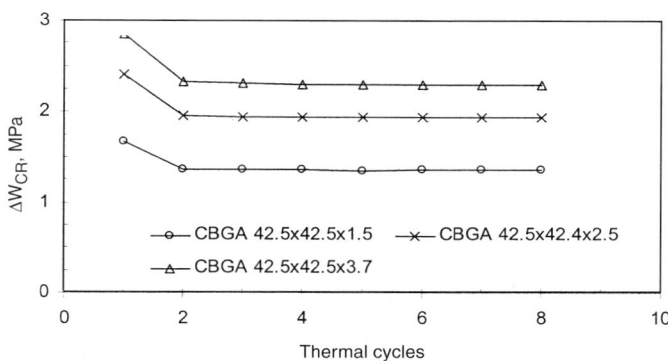

Fig. 30 Maximum calculated von Mises creep strain energy as a function of modeled thermal cycle (42.5 mm CBGA packages)

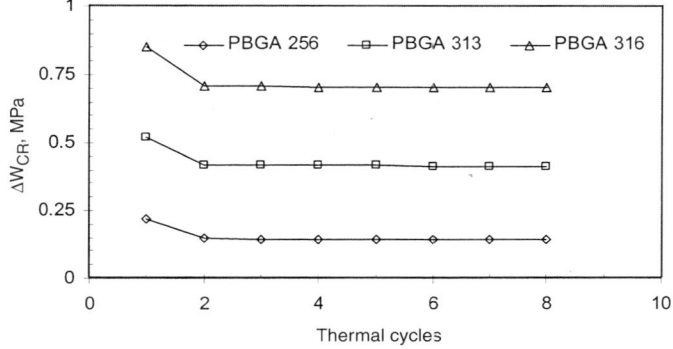

Fig. 31 Maximum calculated von Mises creep strain energy as a function of modeled thermal cycle (PBGA packages)

well as computational economy in regard to the number of thermal cycles that need to be analyzed, and to the number of elements used in the selectively refined model. Hence, based on this study, the selectively refined 3D strip model used in conjunction with the creep strain energy density fatigue life model (Eq 14) is recommended.

Design for Reliability

This chapter has focused on the numerical tools used in the design for reliability of lead-free solder joints. These tools include the finite element method, in general, and, in particular, the use of 3D strip models, specific material properties, and the use of a creep strain energy

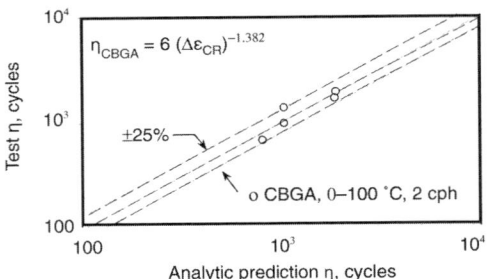

Fig. 32 Correlation between the best-fit $\Delta\varepsilon_{CR}$ based predictions and the CBGA test data

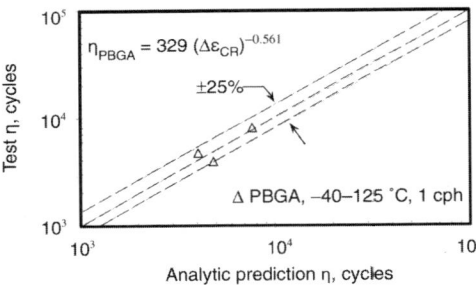

Fig. 33 Correlation between the best-fit $\Delta\varepsilon_{CR}$ based predictions and the PBGA test data

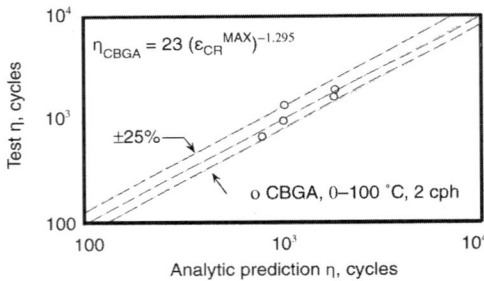

Fig. 34 Correlation between the best-fit ε_{CR}^{MAX} based predictions and the CBGA test data

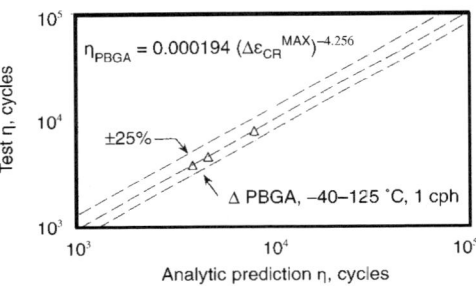

Fig. 35 Correlation between the best-fit ε_{CR}^{MAX} based predictions and the PBGA test data

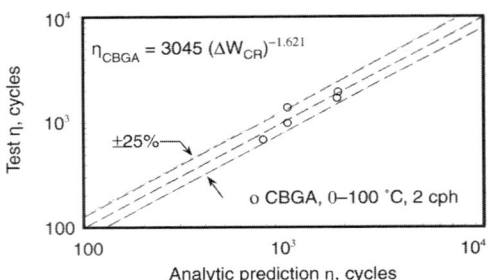

Fig. 36 Correlation between the best-fit ΔW_{CR} based predictions and the CBGA test data

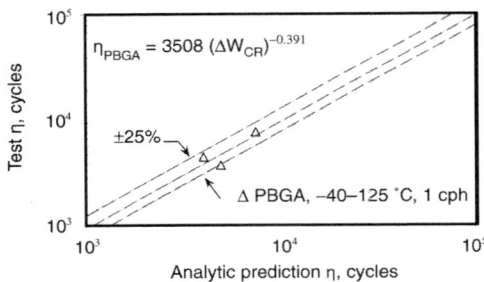

Fig. 37 Correlation between the best-fit ΔW_{CR} based predictions and the PBGA test data

density-based failure criterion. The bulk of the present study has focused on demonstrating that these suggested modeling strategies, materials, and failure criterion are both computationally efficient and accurate in the context of published reliability data.

A design for reliability exercise is considered in order to demonstrate the finite element contribution to the design for reliability process. Consider that a project team has defined a new lidless flip-chip package the geometric and material characteristics listed in Table 5.

The project manager has stated that the package must have a characteristic life of at least 1000 cycles in 0 to 100 °C (32 to 212 °F) temperature cycling performed at 2 cycles per h. The substrate thickness must be defined. A 2 mm (0.07 in.) minimum substrate thickness is required and a greater thickness would be beneficial to the electrical performance of the device. Of course, thicker ceramic substrates are associated with shorter solder joint fatigue lives. An analyst is asked whether the 2 mm (0.07 in.) thick substrate will meet the design goal and what solder joint fatigue lives may be expected with thicker substrates. The analyst's evaluation of the effect of substrate thickness on solder joint reliability will be used in the design of a test vehicle.

The package is similar to the CBGA substrates discussed in this chapter. Differences include the physical size of the substrate, the large die, and the thickness of the PCB. The solder joints are assumed to have the same characteristics (pad metallurgies, shapes, reflow environments, etc.) as those of the previously tested CBGA packages. Due to these similarities with the CBGA packages, the analyst constructs 3D slice models similar to Fig. 18. The models include the new substrate edge-length dimensions, additional solder joints, the die, and underfill. Multiple analyses are performed with substrate thicknesses of 2, 2.5, and 3 mm (0.07, 0.09, and 0.11 in.). Since creep strain energy density convergence is obtained after two cycles, only three cycles are run. The critical creep strain energy density from the 3rd cycle is used in the equation (Fig. 36):

$$\eta = 3045(\Delta W_{CR})^{-1.621} \quad \text{(Eq 15)}$$

to determine the characteristic life predictions presented in Fig. 38.

Upon consideration of both the predicted solder joint reliability and the electrical design requirements, a substrate thickness of 2.25 mm (0.08 in.) is chosen. At this point, the analyst has guided the design of a package test vehicle with the expectation of a certain level of solder joint reliability. Testing is the next and necessary step.

In addition to producing reliability performance predictions and design guidance, the finite element models interact with the subsequent reliability testing and failure analysis steps. The models guide the layout of the test vehicle's daisy chains by specifying the critical solder joints that require monitoring. Reliability testing, in conjunction with failure analysis, will produce results that either confirm or recalibrate Eq 15. During failure analysis, the models focus on the search for failure locations. Failure analysis may confirm the analytically predicted critical solder joints or may provide additional details that indicate that the model needs to consider additional failure modes. The cycle of test vehicle design, reliability test, and failure analysis is repeated until a satisfactory solder joint reliability is obtained. The design for reliability analysis should greatly shorten the number of design/build/test cycles.

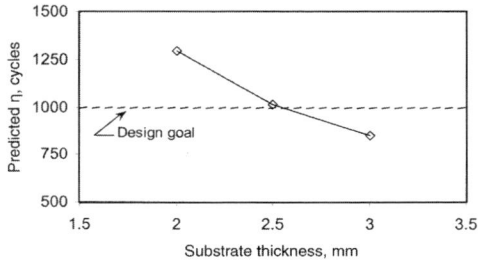

Fig. 38 Predicted characteristic lives for the "new" 37.5 mm packages

Table 5 Package description

Component	Physical dimensions (mm)	Materials	Notes
Substrate	37.5 × 37.5	Al$_2$O$_3$	Thickness needs to be specified, pad ϕ = 0.76 mm, 1272 I/O
Die	15 × 15 × 0.8	Si	...
Underfill	0.05 thick	Epoxy	...
Solder balls	Assy'd height = 0.38	Sn-Ag-Cu	...
PCB	1.7 thick	FR4	Pad ϕ = 0.67 mm

Future Research Needs

The analyses performed in this section were possible due to the published research of individuals and groups who have examined the material properties, manufacturing parameters, and the solder joint reliability of the Sn-Ag-Cu alloy. While the works cited herein reflect meticulous studies, further research is required. As an example, differing stress-strain curves can be found in Ref 9, 12, 14, 15, and 17 and differing creep responses may be found in Ref 9, 11, 13, 57, and 58. All of these data are likely correct, simply reflecting the effects of particular manufacturing parameters, minor compositional differences, test strain rates, test temperatures, etc. Consequently, continued material characterization is required to obtain a more thorough understanding of the material properties, which are necessary input to the finite element analysis.

The analyses have determined apparent correlations between finite element results, e.g., the creep strain energy density and solder joint fatigue life. These correlations were made with package test data on Sn-Ag-Cu solder in solder ball form. The characterization of the alloy's fatigue properties through isothermal fatigue testing would provide a baseline of data from which fatigue life predictions, rather than extrapolations, could be made. Continued part-on-board reliability testing is also needed.

Finally, the previous sections have discussed the methodology used to perform finite element prediction of the fatigue life of Sn-Ag-Cu solder joints in an accelerated test environment. The determination of the acceleration factor between accelerated test and product environments is an equally important component of the accelerated testing philosophy. For traditional high-lead solder interconnects, this has been performed with the Norris-Landzberg equation.

An empirically calibrated solder fatigue model that accounts for the effects of mechanical strain, temperature range, and cycling frequency on high-Pb solder fatigue life was developed (Ref 59). The strain component has its basis in the Coffin-Manson relationship, and temperature and frequency data were derived from tests on acid lead, 99Pb-1Sb and 95Pb-5Sn. A form of the Norris-Landzberg equation follows:

$$\frac{N_{product}}{N_{test}} = \left(\frac{\Delta T_{test}}{\Delta T_{product}}\right)^{1.9} \left(\frac{f_{product}}{f_{test}}\right)^{1/3} \exp\left[1414\left(\frac{1}{T^{max}_{product}} - \frac{1}{T^{max}_{test}}\right)\right] \quad \text{(Eq 16)}$$

where the subscripts "test" and "product" denote, respectively, the accelerated testing and product environments, N is the number of cycles to failure, ΔT is the temperature range, f is the cyclic frequency, and T^{max} is the maximum temperature for each temperature range. At least in the case of high-Pb solder joints, the Norris-Landzberg equation has become the de facto means to evaluate solder joint life in a product environment with accelerated test results.

The appeal of the Norris-Landzberg equation is that finite element calculations are not performed in relating accelerated test results to service lifetimes. A similar equation is needed for the Sn-Ag-Cu solder alloy. Extensive test data was published (Ref 60) that may be used in beginning to explore the development of a Norris-Landzburg type equation for the Sn-Ag-Cu alloy; further test work is suggested to cover the time and temperature spaces seen in product conditions.

Concluding Remarks

This chapter has focused on the use of finite element modeling in predicting the Sn-Ag-Cu solder joint fatigue life under accelerated thermal cycling conditions. 3D slice models combined with a creep strain energy density fatigue life criterion have shown good fit, hence predictive capability, over a range of package types, accelerated thermal environments, and solder ball pitches. In addition, only three thermal cycles need to be modeled to achieve a cycle independent accumulated creep strain energy density per cycle. This leads to rapid numerical computations. Material properties have been suggested for use in the analysis.

ACKNOWLEDGMENTS

The author gratefully acknowledges the support and encouragement provided by John Lau, without whom this work would not have happened. In addition, thanks are extended to Professor Shi-Wei Ricky Lee of Hong Kong University of Science and Technology, who provided invaluable insights regarding PBGA solder joint reliability testing, and to David Leary and Keith Newman for their thoughtful reviews of the manuscript. Finally, sincere appreciation is due the author's managers, Paul Marcoux and Wayne Richling, who have established an environment where efforts such as this research are possible.

REFERENCES

1. J. Lau, A Brief Introduction to Ball Grid Array Technologies, in *Ball Grid Array Technology,* J. Lau, Ed., McGraw-Hill, 1995
2. "Performance Test Methods and Qualification Requirements for Surface Mount Solder Attachments," IPC-9701, IPC, 2002
3. J. Lau and Y. Pao, *Solder Joint Reliability of BGA, CSP, Flip Chip, and Fine Pitch SMT Assemblies,* McGraw-Hill, 1997, Chap. 6
4. ANSYS is a product of Ansys, Inc., Canonsburg, PA. Version 7.1 was used in this study.
5. E. Madenci, I. Guven, and B. Kilic, *Fatigue Life Prediction of Solder Joints in Electronic Packages with ANSYS,* Kluwer, 2003
6. F. Garofalo, *Fundamentals of Creep and Creep Rupture in Metals,* Macmillan, 1965
7. H.J. Frost and M.F. Ashby, *Deformation-Mechanism Maps: The Plasticity and Creep of Metals and Ceramics,* Pergamon, 1982
8. J. Lau and W. Dauksher, Creep Constitutive Equations of Sn(3.5–3.9)wt%Ag(0.5–0.8)wt%Cu Lead-Free Solder Alloys, *Micromaterials and Nanomaterials,* Issue 03, 2004, p 54–62
9. P. Vianco and J. Rejent, Compression Deformation Response of 95.5Sn-3.9Ag-0.6Cu Solder, *UCLA Lead-Free Workshop,* 2002
10. J. Lau, W. Dauksher, and P. Vianco, Acceleration Models, Constitutive Equations and Reliability of Lead-Free Solders and Joints, *IEEE Electronic Components and Technology Conference Proceedings,* June 2003, p 229–236
11. A. Schubert, R. Dudek, E. Auerswald, A. Gollhardt, B. Michel, and H. Reichl, Fatigue Life Models for SnAgCu and SnPb Solder Joints Evaluated by Experiments and Simulation, *IEEE Electronic Components and Technology Conference Proceedings,* June 2003, p 603–610
12. P. Vianco, J. Rejent, and A. Kilgo, Time Independent Mechanical and Physical Properties of the Ternary 95.5Sn-3.9Ag-0.6Cu Solder, *J. Electron. Mater.,* Vol 32 (No. 3), 2003, p 142–151
13. Q. Zhang, A. Dasgupta, and P. Haswell, Viscoplastic Constitutive Properties and Energy-Partitioning Model of Lead-Free Sn3.9Ag0.6Cu Solder Alloy, *IEEE Electronic Components and Technology Conference,* 2003, p 1862–1868
14. J.H.L. Pang, B.S. Xiong, C.C. Neo, X.R. Zhang, and T.H. Low, Bulk Solder and Solder Joint Properties for Lead-Free 95.5Sn-3.8Ag-0.7Cu Solder Alloy, *IEEE Electronic Components and Technology Conference,* 2003, p 673–679
15. K.S. Kim, S.H. Huh, and K. Suganuma, Effects of Cooling Speed on Microstructure and Tensile Properties of Sn-Ag-Cu Alloys, *Mater. Sci. Eng.,* A333, 2002, p 106–114
16. T. Siewert,, S. Liu, D.R. Smith, and J.C. Madeni, *Database for Solder Properties with Emphasis on New Lead-Free Solders,* National Institute of Standards and Technology & Colorado School of Mines, 2002
17. S. Wiese, S. Rzepka, and E. Meusel, Time Independent Plastic Behavior of Solders and Its Effect on FEM Simulations for Electronic Packages, *Eighth International Symposium on Advanced Packaging Materials,* 2002, p 101–111
18. L. Xiao, J. Liu, Z. Lai, L. Ye, and A. Tholen, Characterization of Mechanical Properties of Bulk Lead-Free Solders, *Sixth International Symposium on Advanced Packaging Materials,* 2000, p 145–151
19. M. Cole and T. Caulfield, Constant Strain Rate Tensile Properties of Various Lead-Based Solder Alloys at 0, 50 and 100C, *Scr. Mater.,* Vol 27, p 903–908
20. W.W. Lee, L.T. Nguyen, and G.S. Selvadury, Solder Joint Fatigue Models: Review and Applicability to Chip Sclae Packages, *Microelectronics Reliability 40,* 2000, p 231–244
21. L.F. Coffin, A Study of the Effects of Cyclic Thermal Stresses on a Ductile Metal, *Trans. ASME,* Vol 76, 1954, p 931–950
22. L.F. Coffin, *Symposium on Internal Stress and Fatigue of Metals,* Elsevier, 1959
23. S.S. Manson, "Behavior of Materials Under Conditions of Thermal Stress," NACA Report 1170, Lewis Flight Propulsion Laboratory, 1954
24. L.S. Goldman, Geometric Optimization of Controlled Collapse Interconnections, *IBM J. Res. Dev.,* 1969, p 251–265
25. M.C. Shine and L.R. Fox, Fatigue of Solder Joints in Surface Mount Devices, *Low Cycle Fatigue—ASTM Spec. Tech. Publ.,* Vol 942, 1987, p 588–610

26. S. Knecht and L.R. Fox, Constitutive Relation and Creep-Fatigue Life Model for Eutectic Tin-Lead Solder, *IEEE Comp. Hybrid Man. Tech.,* Vol 13 (No. 2), 1990, p 424–433
27. B. Wong, D. Helling, and R. Clark, A Creep Rupture Model for Two-Phase Eutectic Solders, *IEEE Trans. Comp. Hybrids. Man. Tech.,* Vol 11 (No. 3), 1988, p 284–290
28. R. Iannuzzelli, Predicting Solder Joint Reliability, Model Validation, *IEEE Electronic Components and Technology Conference,* June 1993, p 839–851
29. R. Iannuzzelli, J. Pitarresi and V. Prakas, Application of Integrated Matrix Creep of Solder Joint Reliability Prediction, *ASME IMECE,* Nov 1995
30. A. Syed, Creep Crack Growth Prediction of Solder Joints During Temperature Cycling—An Engineering Approach, *Trans. ASME,* 1995, Vol 117, p 116–122
31. A. Syed, Thermal Fatigue Reliability Enhancement of Plastic Ball Grid Array (PBGA) Packages, *IEEE Electronic Components and Technology Conference,* 1996, p 1211–1216
32. J. Morrow, "Cyclic Plastic Strain Energy and Fatigue of Metals," ASTM STP-378, 1965, p 45–87
33. R. Darveaux, K. Banerji, A. Mawer, and G. Dody, Reliability of Plastic Ball Grid Array Assembly, *Ball Grid Array Technology,* J. Lau, Ed., McGraw-Hill, 1995, p 379–442
34. V. Stolkarts, B. Moran, and L.M. Keer, Constitutive and Damage Model for solders, *IEEE Electronic Components and Technology Conference Proceedings,* May 1998, p 379–385
35. C. Basaran and C.Y. Yan, A Thermodynamic Framework for Damage Mechanics of Solder Joints, *J. Electron. Packaging,* Vol 120, 1998, p 379–384
36. C.S. Desai, J. Chia, T. Kundu, and J.L. Prince, Thermomechanical Response of Materials and Interfaces in Electronic Packaging: Part 1—Unified Constitutive Model and Calibration and Part 2—Unified Constitutive Models, Validation and Design, *Trans. ASME,* Vol 119, 1997, p 294–300 and p 301–309
37. C. Kanchanomai, Y. Miyashita, and Y. Mutoh, Low Cycle Fatigue Behaviors of Sn-Ag, Sn-Ag-Cu and Sn-Ag-Cu-Bi Lead-Free Solders, *J. Electron. Mater.,* Vol 31 (No. 5), 2002, p 456–465
38. A. Schubert, R. Dudek, H. Walter, E. Jung, A. Gollhardt, B. Michel, and H. Reichl, Reliability Assessment of Flip-Chip Assemblies with Lead-Free Solder Alloys, *IEEE Electronic Components and Technology Conference,* 2002, p 1246–1255
39. T. Anderson, A. Barut, I. Guven, and E. Madenci, Revisit of Life Prediction Models for Solder Joints, *IEEE Electronic Components and Technology Conference Proceedings,* May 2000, p 1064–1069
40. G. Gustafsson, I.Guven, V. Kradinov, and E. Madenci, Finite Element Modeling of PBGA Packages for Life Prediction, *IEEE Electronic Components and Technology Conference Proceedings,* May 2000, p 1059–1063
41. R.D. Cook, D.S. Malkus, and M.E. Plesha, *Concepts and Applications of Finite Element Analysis,* 3rd ed., Wiley, 1989
42. R.H. MacNeal, *Finite Elements: Their Design and Performance,* Marcel Dekker, 1994
43. H.L.J. Pang, K.H. Tan, X.Q. Shi, and Z.P. Wang, Microstructure and Intermetallic Growth Effects on Shear and Fatigue Strength of Solder Joints Subjected to Thermal Cycle Aging, *Materials Science and Engineering,* A307, 2001, p 45–50
44. A. Syed, Reliability of Au Embrittlement of Lead-Free Solders for BGA Applications, *International Symposium on Advanced Packaging Materials,* 2001, p 143–147
45. S. Wiese, E. Meusel, and K.J. Wolter, Microstructural Dependence of Constitutive Properties of Eutectic SnAg and SnAgCu Solders, *IEEE Electronic Components and Technology Conference Proceedings,* June 2003, p 197–206
46. M.L. Williams, Stress Singularities Resulting from Various Boundary Conditions in Angular Corners of Plates in Extension, *J. Appl. Mech.,* 1952, p 526–528
47. M.L. Williams, The Stresses Around a Fault or Crack in Dissimilar Media, *Bull. Seis. Soc. Amer.,* 1959, p 199–204
48. J. Riebling, Finite Element Modeling as a Lifetime Prediction Tool for Ball Grid Array Packages, *HP EMAC Conference,* 1997
49. J. Smetana, R. Horsley, J. Lau, K. Snowdon, D. Shangguan, J. Gleason, I. Memis, D. Love, W. Dauksher, and B. Sullivan, HDPUG's Lead-Free Design, Materials, and Process of High Density Packages, *IPC Apex,* March 2003

50. J. Lau, W. Dauksher, R. Horsley, J. Smetana, D. Shangguan, T. Castello, D. Love, I. Memis, and B. Sullivan, HDPUG's Design for Lead-Free Solder Joint Reliability of High Density Packages, *IPC Apex,* March 2003
51. J. Lau, N. Hoo, R. Horsley, J. Smetana, D. Shangguan, W. Dauksher, D. Love, I. Memis, and B. Sullivan, Reliability Testing and Data Analysis of High-Density Packages' Lead-Free Solder Joints, *IPC Apex,* March 2003
52. J. Lau, D. Shangguan, T. Castello, R. Horsley, J. Smetana, W. Dauksher, D. Love, I. Memis, and B. Sullivan, Failure Analysis of High-Density Packages' Lead-Free Solder Joints, *IPC Apex,* March 2003
53. M. Farooq, C. Goldsmith, R. Jackson, and G. Martin, Lead-Free Ceramic Ball Grid Array: Thermomechanical Fatigue Reliability, *J. Electron. Mater.,* Vol 32 (No. 12), 2003, p 1421–1425
54. M. Farooq, L. Goldmann, G. Marin, C. Goldsmith, and C. Bergeron, Thermo-Mechanical Fatigue Reliability of Pb-Free Ball Grid Arrays: Experimental Data and Lifetime Prediction Modeling, *IEEE Electronic Components and Technology Conference Proceedings,* June 2003, p 827–833
55. S.W.R. Lee, B.H.W. Lui, Y.H. Kong, B. Baylon, T. Leung, P. Umali, and H. Agtarsp, Assembly of Board Level Solder Joint Reliability for PBGA Assemblies with Lead-Free Solders, *Solder. Surf. Mt Technol.,* Vol 14 (No. 5), 2002, p 46–50
56. H. Solomon, Life Prediction and Accelerated Testing, in *The Mechanics of Solder Alloy Interconnects,* D. Frear, H. Morgan, S. Burchett, and J. Lau, Ed., Chapman and Hall, 1994, p 199–313
57. S. Wiese, F. Feustel, and E. Meusel, Characterization of Constitutive Behavior of SnAg, SnAgCu and SnPb Solder in Flip Chip Joints, *Sensors and Actuators,* A 99, 2002, p 188–193
58. H.G. Song, J.W. Morris, and F. Hua, The Creep Properties of Lead-free Solder Joints, *J. Mater.,* 2002, p 30–32
59. K.C. Norris and A.H. Landzberg, Reliability of Controlled Collapse Interconnection, *IBM J. Res. Develop.,* 1969, p 266–271
60. J. Bartelo, S.R. Cain, D. Caletka, K. Darbha, T. Gosselin, D.W. Henderson, D. King, K. Knadle, A. Sarkhel, G. Thiel, C. Woychik, D.Y. Shih, S. Kang, K. Puttlitz, and J. Woods, Thermomechanical Fatigue Behavior of Selected Lead-Free Solders, *IPC Apex,* 2001

CHAPTER 10

Characterization and Failure Analyses of Lead-Free Solder Defects

Reza Ghaffarian, Jet Propulsion Laboratory, California Institute of Technology

Introduction

This document sets forth the technical background and specific information related to failure analyses and inspection issues that arise as a result of replacing lead (Pb) in electronic solders. Many aspects of changeover from tin-lead (Sn-Pb) to lead-free alloys were discussed in other chapters of this book. For surface mount processes, the changes are caused by differences in solderability, compatibility, material properties, higher reflow temperatures, and flux chemistries. For reliability, creep and fatigue depends on types of solders, board and package type, microstructural change, strain range, and characterization techniques used to detect defects and damage during environmental exposure. Visual, x-ray, and microstructural features that are known for Sn-Pb solders are different for lead-free and, therefore, they need to be investigated. Failure characterization, either alone or when combined with optical, scanning electron microscopy (SEM), energy dispersive x-ray spectroscopy (EDX), x-ray inspection, cross-sectioning, and microstructural exposure by chemical etching, can provide crucial information needed to make a smooth transition to lead-free solders. This chapter addresses the growing critical needs for characterization and failure analyses of Sn-Pb and lead-free solders and assemblies.

Tin-Lead and Lead-Free Alloys

Books and publications on Sn-Pb solder alloys are abundant (Ref 1–4), while those for lead-free alloys are just starting to become available (Ref 5). For continuity purpose, some introductory information is provided that may already be covered in other chapters.

Eutectic Tin-Lead Solder Alloy. Eutectic, or near eutectic, tin-lead (63Sn-37Pb) solder has been used for the past 60 years as the principal joining material for Level 2 packaging—defined as the attachment of component-level packages to a suitable substrate to produce printed wiring assemblies (PWAs). Eutectic tin-lead solder, with a unique melting point of 183 °C (361 °F) provides outstanding solderability and forms stable solder joints capable of operating in a wide variety of service environments. Billions of solder joints have been created on countless printed wiring boards using this material. In addition, the mechanical and electrical properties are well understood, and the reliability of solder joints created using eutectic tin-lead, or near-eutectic tin-lead, solders has been established and is well understood.

Lead-Free Solder Alloys. Because of the toxicity of lead and the concern that the lead in electronic products may end up in landfills, and ultimately in the water supply, the electronic industry is exploring alternative solder alloys that do not contain lead. These alternative solder alloys are typically composed of tin (Sn), with one, two, or three additives such as copper (Cu), silver (Ag), bismuth (Bi), antimony (Sb), zinc (Zn), or indium (In). Typical tin lead-free candidate solder alloys include Sn-Cu, Sn-Ag, Sn-Ag-Cu, Sn-Ag-Cu-Sb, Sn-In, and Sn-Cu-Bi-Sb. Some of these are more suitable for wave solder and others more suitable for reflow solder. Examples of the solder alloys discussed in this chapter are:

- Sn96.5Ag3.5
- Sn95.5Ag3.8Cu0.7
- Sn96.2Ag2.5Cu0.8Snb0.5 (Castin®)
- Sn77.2In20Ag2.8 (Indalloy 227)

For more detailed information about the phase diagrams, advantages and disadvantages, and other properties of these and other alloys, please refer to other chapters of this book or Ref 4 and 5.

Characterization and Analytical Techniques

Many nondestructive and destructive techniques are available for characterization of solder alloys. Characterization should start with solder paste for surface mount. During the manufacturing process, a 3D laser scanning system is used to determine solder paste characteristics as a process control tool for ball grid array and chip size packages (BGAs/CSPs). Laser scanning can inspect solder paste height and volume, thereby determining solder paste application uniformity before package placement. By inspecting these attributes, solder print process characteristics such as slumping, scooping, or peaks can be identified and controlled.

The inspection system's ability to identify, measure, and analyze defect data after assembly is also critical. Inspection of BGA solder joint integrity is important but cannot be effectively achieved by visual inspection because of hidden solder joints. For these types of packages, in addition to process control during manufacturing, nondestructive techniques such as x-ray are needed to determine the integrity of an attachment.

X-ray transmission radiography is an inspection technique in which x radiation passes through a specimen to produce a shadow image of its internal structure. Placing the specimen close to the x-ray source enables image magnification. This permits inspection of fine details. Magnifications greater than 1000× are now obtainable from commercially available equipment.

Merits of the visual and x-ray nondestructive techniques are compared in Table 1. This table summarizes various general solder joint defects and compares qualitative accuracy of the two techniques. X-ray inspection is excellent for geometric measurements and for detecting hidden features such as voids. However, for some of the unique and critical defects, such as dewetting, cracks, cold solder joints, and disturbed solder, visual inspection is superior to x-ray detection. For this reason, both optical and x-ray systems are used to characterize solder joint features of lead-free and lead-based assemblies. Ideally, a combination of various inspection techniques may be required in order to ensure quality at package and system levels.

Inspection of fine internal structures of microelectronics assemblies, the alignment of hidden interconnects, bridges, and voids in BGA assemblies, can be carried out using real time X-ray techniques. However, internal package delamination cannot be detected by x-ray and other tools such as C-mode scanning acoustic microscopy (C-SAM) are needed.

Use of nondestructive techniques is limited to research study for characterization of failure mechanisms, or when investigating root cause of a field failure. Destructive techniques commonly include cross-sectioning combined with optical inspection, and SEM and EDX. Chemical etching solutions are used to reveal features of solder alloy microstructures and interface intermetallic formation. Mechanical destructive tests such as lead pull test for leaded packages and dye-and-pry for area packages are also performed to determine quality and damage levels.

Metallography and Etching

For characterization of failure mechanism by metallography, samples are prepared by cross-sectioning, mounting in polymer, polishing to a smooth cut surface, and chemical etching to reveal microstructures. Prior to cross-sectioning,

Table 1 Key solder defect types and ability to detect visible joints

Visible features	X-ray inspection	Visual inspection
Stress marks, cracks	0	+++
Open contacts	0	++
Cold/disturbed joint	0	+++
Dull solder	0	+++
Flux residue/contamination	0	+++
Porosity and voids in solder	+++	0
Solder thickness/volume	+++	0
Heel/toe side fillets	+++	++
Solder balls	+++	++
Solder bridge	+++	++

+++ Excellent detection; ++ Good detection; 0 Poor or unacceptable

Table 2 Metallographic preparation for lead-free solder alloy samples

1. Encapsulation	a.	Place samples into clips to maintain proper orientation.
	b.	Apply mold release to sample cups, mark sample cups with proper sample ID.
	c.	Place samples into the center of sample cups.
	d.	Mix the epoxy and the hardener in accordance with manufacturers' recommendations and mix thoroughly.
	e.	Fill sample cups to two-thirds full.
	f.	Place uncured molds into vacuum chamber and degas epoxy.
	g.	Place degassed molds into tray of water to minimize epoxy exotherm.
2. Grinding (wheel speed at 500 RPM)	a.	Remove samples from cups serially, and transfer sample ID to the cured mold.
	b.	Bevel top and bottom edges of mold using 240 grit SiC grinding paper.
	c.	Rough grind samples using 240 grit SiC paper until the section plane is within approximately 4–6 mils of the targeted solder joints.
	d.	Continue grinding with 320 grit SiC paper until the section plane is within approximately 1–3 mils of the targeted solder joints.
	e.	Continue grinding with 400 grit SiC paper until the section plane is roughly one-third into the target solder joints.
	f.	Continue grinding with 600 grit SiC paper until the desired section plane (typically center of the solder joint) has been reached.
	g.	Freshen bevel with 600 grit SiC paper on the sample side of the mold.
	h.	Wash sample with abrasive-free soap and water.
		(Note that for ceramic samples [CCGA packages], grinding steps c through e are replaced with 75 µm and 15 µm metal-bonded diamond grinding disks.)
3. Rough polish (wheel speed at 500 RPM)	a.	Using a napless cloth (such as Texmet, Buehler Ltd., or equivalent), rough polish with 6 µm diamond paste and diamond extender fluid.
	b.	Wash sample with abrasive-free soap and water.
	c.	Using a napless cloth, rough polish with 3 µm diamond paste and diamond extender fluid.
	d.	Using a napless cloth, rough polish with 1 µm diamond paste and diamond extender fluid.
		(Note that this step is only needed for the ceramic packages.)
	e.	Wash sample with abrasive-free soap and water.
4. Final polish	a.	Using a soft, synthetic cloth (such as Chemomet, Buehler Ltd., or equivalent), polish with 0.05 µm alumina slurry.
	b.	Rinse sample on the polishing cloth with water.
	c.	Using a soft, synthetic cloth, polish with 0.06 µm colloidal silica.
	d.	Rinse sample on the polishing cloth with water. Blow sample dry with compressed air, being careful not to create water marks or other stains on the polished section.

the flux residue usually is removed using defluxing agent with enhanced removal by agitation. When the flux residues have been sufficiently removed, the samples are further cleaned by rinsing in isopropyl alcohol and blown dry with compressed air. Standard metallography sample preparation is done subsequent to diamond saw cutting (to reveal desired cross-section) and mounting. Sample preparation for lead-free solder alloy samples may include the steps listed in Table 2, as discussed in Ref 6.

In order to have the same microsectional quality as tin-lead, some alteration in the microsectioning process is generally required because of the differences in mechanical properties, metallurgy, and microstructure of lead-free solder alloys. Changes may include the order and grit sizes of the grinding steps, and changes in etching solutions, in order to reveal the microstructure. Generally, the harder tin-based lead-free solder alloys require a more abrasive final grinding and polishing step than their tin-lead counterparts. In some cases, samples can be prepared without cloth and diamond paste.

To reveal the microstructure of solder alloys and interfaces, a chemical etching mixture of weak-to-strong acid, diluted with isopropanol alcohol or with deionized water, is used. A list of etching solutions for a number of tin-based and lead-free solders is given in Table 3 (Ref 7).

Tin-Lead and Lead-Free Microstructural Features

Extensive histories of inspection information and photomicrographs at different environmental test intervals for tin-lead solder joints have been gathered during many years of investiga-

Table 3 List of chemical etching solutions for tin-lead and tin-based lead-free solder alloys

Solder composition	Etchant	Application time, s
Sn-Pb	25 parts 3% hydrogen peroxide, 25 parts 28% NH$_4$OH and 10 parts D.I. water	15–45
	3 parts acetic acid, 4 parts HNO$_3$, 16 parts D.I. water	5–15+
Sn-Ag	2 parts HCl, 5 parts HNO$_3$, 93 parts isopropanol	2–5+
Sn-Ag-Cu, Sn-Pb	2% nital (2 parts HNO$_3$ in 98 parts isopropanol)	3
Sn-Bi, Sn-Sb-Cu, Sn-In	2 parts HCl, 5 parts HNO$_3$, 93 parts isopropanol	2–5

D.I., deionized

tion. Data for lead-free solder joints, however, is limited, and being generated by industry for various alloys and environmental conditions.

Examples of optical and SEM inspection of tin-lead and lead-free solder joints are presented before and after exposure to environmental conditions. Assemblies include both conventional leaded and leadless packages, as well as CSP leadless packages.

Tin-Lead Solder, Conventional Leadless Package. For leaded and leadless package solder joints, the author has performed visual inspection at different magnifications to correlate damage rankings to those revealed by cross-sectioning (Ref 8, 9). Numerous leaded and leadless packages were subjected to thermal cycling, removed at intervals, inspected visually and by SEM, and the results were correlated to cross-sectioning images. An example of such correlation for a ceramic leadless package with 28 I/Os, 1.27 mm (0.05 in.) pitch, is shown in Fig. 1.

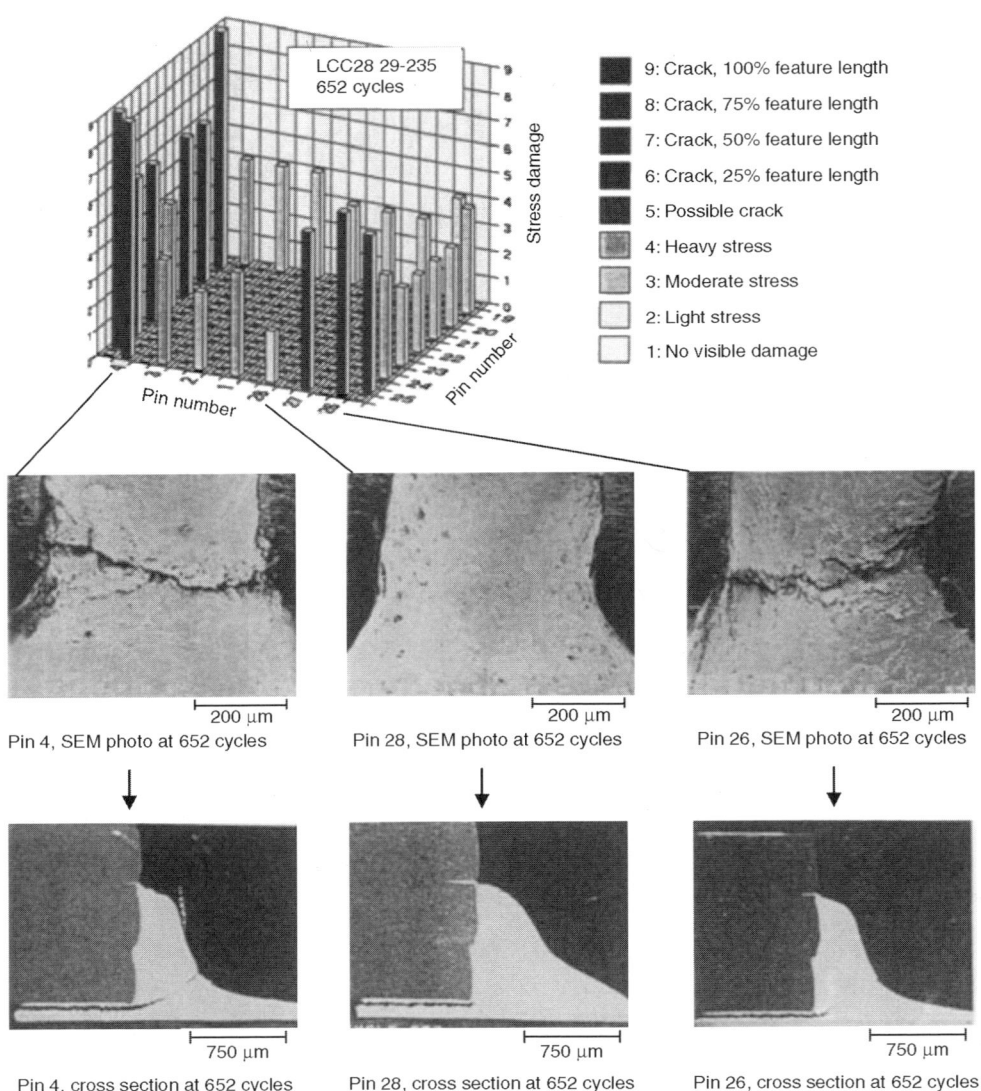

Fig. 1 Correlation between visual inspection for damage (crack) progress with thermal cycling and destructive cross-sectional microphotographs. Ceramic leadless package, 28 castellations, 1.27 mm (0.05 in.) pitch

Generally, good correlations were found between visual and cross-section rankings for cracks to 100% opening. Assemblies were subjected to a thermal cycle, ranging from −55 to 100 °C (−67 to 212 °F) with 4.2 h per cycle. This figure includes nondestructive images and visual ranks of solder joint cracking to 652 cycles and cross-section images of the same assembly at 652 cycles.

Tin-Lead Solder, CSP Leadless Package. The bottom leadless package (BLP) is a peripheral package for replacing the thin small outline package (TSOP) for dynamic random access memory (DRAM) applications. This package uses a custom-designed lead-frame with wire bond interconnection at the chip level. A package with 46 I/O, 0.5 mm (0.02 in.) pitch, was subjected to numerous thermal cycles to determine the cycles-to-failure and the failure mechanisms (Ref 10).

Figure 2 shows significant damage introduced during thermal cycling in the range of −30 to 100 °C (−22 to 212 °F). Scanning electron microscopy photos prior to and after cross-sectioning for assemblies exposed to 1500 cycles are shown. Visual inspection could reveal the outer damage and, therefore, provided an indication of internal damage. Outer and internal damage, however, were less apparent for those subjected to random vibration (Ref 11).

Tin-Lead Solder, Plastic Ball Grid Array (PBGA). Figure 3 shows various cross-sections of a plastic ball grid array (PBGA) with 313 I/Os and 1.27 mm (0.05 in.) pitch. This is a depopulated full array package having balls from center to corners. This package was first to fail among many plastic packages. The cross-sectional photos are for the balls under the die, where most damage occurs due to relatively higher local coefficient of thermal expansion (CTE) mismatch between the PWB and the encapsulated die. Photos with and without voids were also included for comparison. Voids appear to have been concentrated at the package interface under the die. Crack propagation occurred at both package and board interfaces, for sections with or without voids. The sections with voids were open as indicated by the seepage of mounting materials into voids. Except for the interface connecting cracks, there appeared to be no crack propagation among the voids.

Tin-Lead Solder, Ceramic Ball Grid Array (CBGA). Figure 4 shows microstructural feature of two ceramic packages after thermal cycling. The ceramic packages, 625 and 361 I/Os, have high-melting-point solder balls with 90Pb-10Sn composition, and a nominal diameter of 0.035 in. (0.88 mm). The high-melting-point solder balls are attached to the ceramic package with eutectic solder (63Sn-37Pb). At reflow, the eu-

Fig. 2 SEM photomicrographs of a 46 I/O bottom leadless package, 0.5 mm (0.02 in.) pitch, before and after cross-sectioning at 1500 cycles (−30 to 100 °C, or −22 to 212 °F).

tectic solder at the package side and the eutectic paste solder at the PWB side will remelt and melt, respectively, to provide the electromechanical interconnects.

Figure 4 shows typical failure sites due to two cycling conditions, one with a rapid and the other with slow ramp rate (−55 to 125 °C, or −67 to 257 °F). Failure from board and package interface cracking was observed with an increasing number of thermal cycles. The failure mechanism differences could be explained either by global or local stress conditions. Modeling in-

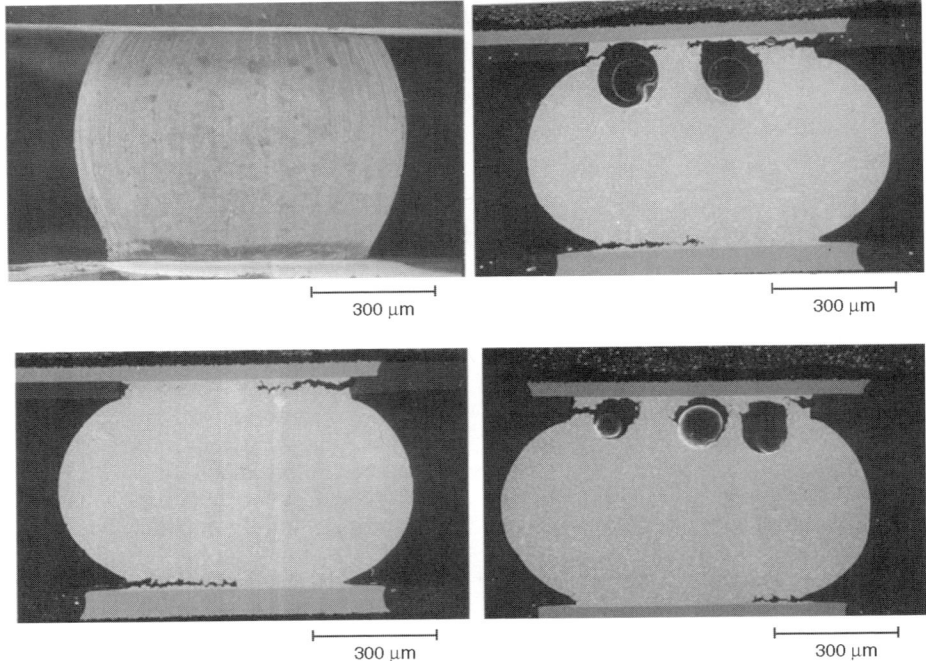

Fig. 3 SEM and cross sections of balls from PBGA 313 after 4682 cycles

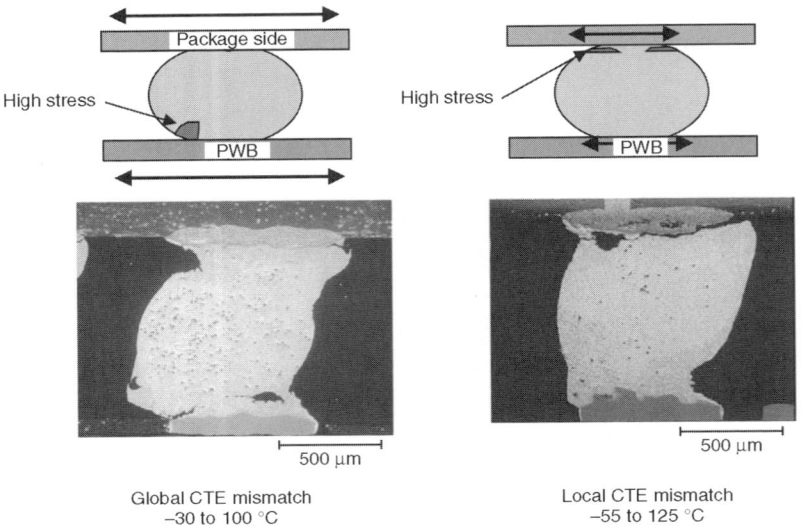

Fig. 4 Cross sections of failure sites for CBGA 625 (left) and CBGA 361 (right) after 350 cycles

dicates that the high-stress regions shifted from the board to the package themselves, when stress conditions changed from the global to local.

Figure 5 shows damage induced by vibration and subsequent thermal cycling that is different in some cases from damage induced by thermal cycling alone (Ref 12). Appearance of tensile deformation from the center of high melting solder balls was significantly different from that observed for the thermal cycling condition. Similar to the thermal cycling condition, damage is more dominant for the balls with a larger distance to the neutral point (DNP), especially for the corner balls. However, additional microcracks in the eutectic solder joints, different from the norm for thermal cycling, were induced by tensile and shear loading during random vibration.

Tin-Lead and Lead-Free Solders, Ceramic Column Grid Array (CCGA). Figure 6 shows optical photomicrographs of a CCGA package assembled with tin-lead solder, before and after thermal cycling exposures. Damage/cracking of thermal cycled joints was observed by visual inspection. Figure 7 shows SEM photomicrographs and cross-section of the assemblies after cracking due to thermal cycles. Figure 8 shows microstructure of another CCGA package assembled with lead-free (Sn-Ag-Cu) solder paste (Ref 6). As stated previously, 3D optical microscopy and visual inspection are limited to inspection of the outer rows of area array assemblies and could be performed only when enough gaps are allowed between the assembled parts. The assemblies show signs of damage/cracking after thermal cycles.

Tin-Lead Solder, Fine Pitch BGA. Figure 9 shows SEM cross-sectional photomicrographs for a fine pitch plastic ball grid array (FPBGA), 176 I/Os and 0.8 mm (0.031 in.) pitch, after thermal cycling. Similar to a full array BGA package, the first signs of cracking were observed under the die near the corner edges.

Lead-Free Solder. In an investigation of lead-free solder alloys (Ref 13, 14), four lead-free

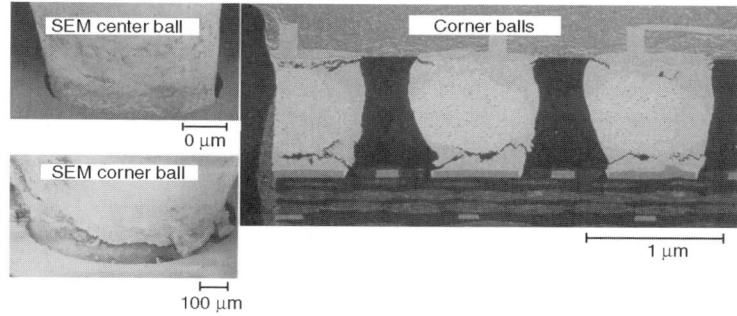

Fig. 5 Thermal-cycled samples after vibration. Note mechanical crack propagation at two locations and minimal damage at center ball.

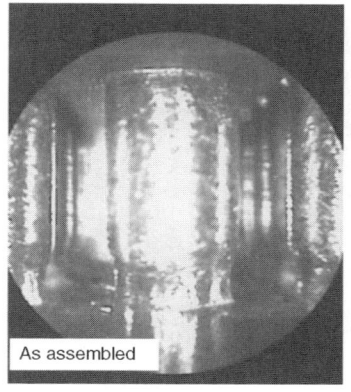 

Fig. 6 3D optical photomicrographs of CCGA with tin-lead solder as assembled and after thermal cycle

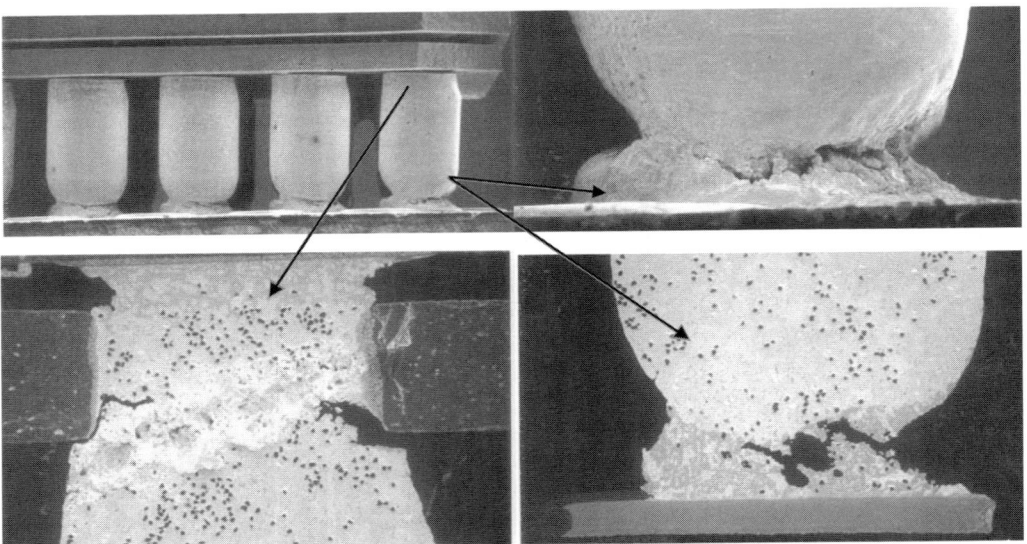

Fig. 7 SEM photomicrographs and cross section of CCGA assembled with tin-lead solder after thermal cycling with damage/cracks

solders were selected for initial screening and subsequent down selection, rebuild, and environmental testing. The four alloys were:

- Sn96.5Ag3.5 (eutectic) with 221 °C (429 °F) melting point. It has good wetting characteristics, and better strength than tin-lead, even though it may show weakness at the interface.
- Sn95.5Ag3.8Cu0.7, which was recommended by NEMI and is used by the Japanese and Europeans with a melting temperature of 217 to 218 °C (423 to 424 °F); it has no plastic range.
- Sn96.2Ag2.5Cu0.8Sb0.5 (Castin®) with a melting point of 217 to 218 °C (423 to 424 °F) improves thermal fatigue. Antimony reduces melting point, even though it may show toxicity at higher temperatures.
- Sn77.2In20Ag2.8 (Indalloy), with a large plastic melting range of 175 to 187 °C (347 to 369 °F) has a comparable melting point to tin-lead, and exhibits good ductility and creep resistance. It is costly, and its 118 °C (244 °F) eutectic melting point may deteriorate mechanical properties of the solder joint.

Figure 10 compares optical photos of solder joints for a J-lead plastic package for the four lead-free solder alloy compositions described previously, with a J-lead ceramic package attached with tin-lead solder. Figure 11 shows photos of the previously mentioned tin-based lead-free solders for a quad flat pack (QFP) plastic package. Figure 12 compares BGAs with tin-lead and lead-free solder attachment.

It may be noted that the tin-lead solder is shiny with excellent wetting. The only lead-free solder joint that comes close to this feature is Sn96.5Ag3.5, a eutectic solder. All lead-free solders show appearance of a graininess and lack of good wettability.

- Sn96.5Ag3.5 (eutectic) is grainy, but much smoother than the other lead-free compositions.

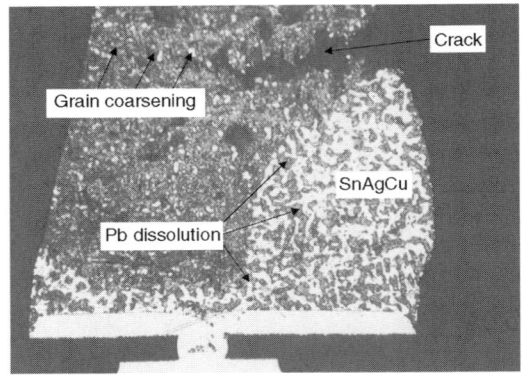

Fig. 8 1657 CCGA high-lead solder column on microvia pad PCB with Sn-Ag-Cu solder joint. Courtesy of Dr. D. Shangguan/T. Castello

Chapter 10: Characterization and Failure Analyses of Lead-Free Solder Defects / 235

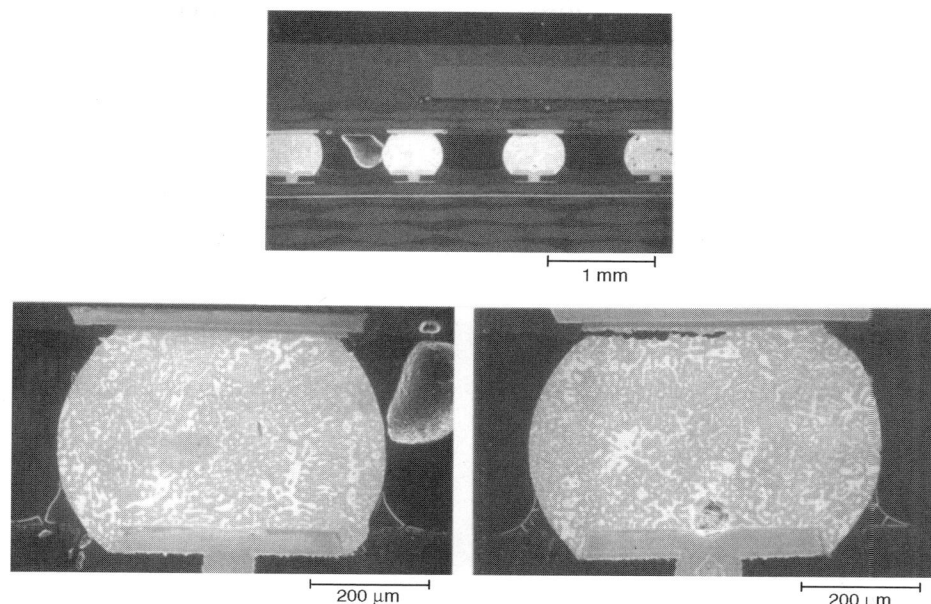

Fig. 9 Photomicrographs of 175 I/O FPBGA after thermal cycle

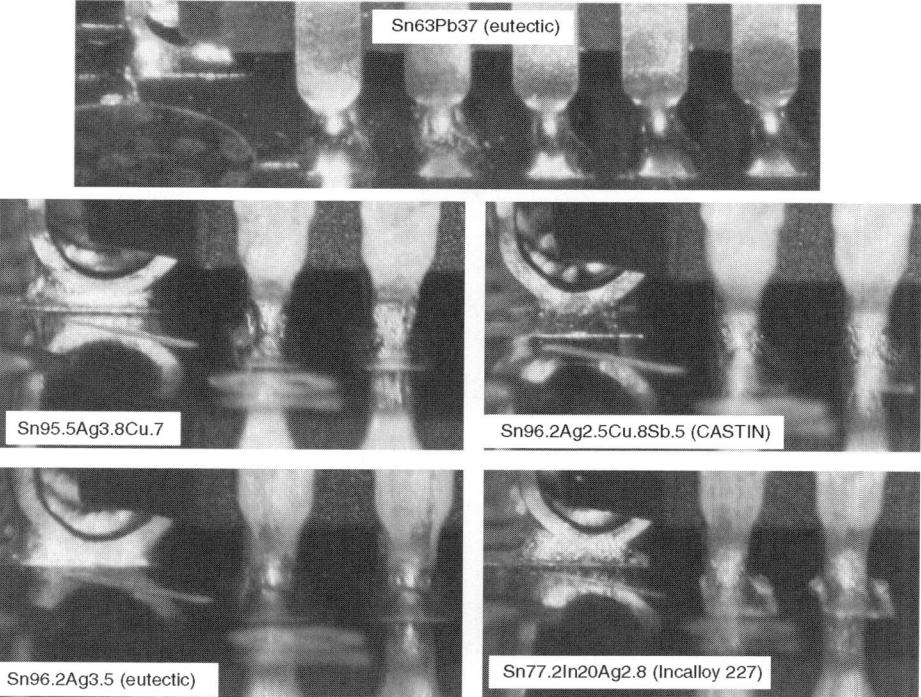

Fig. 10 Optical photographs of J-lead packages assembled with Sn-Pb solder and four lead-free solders. Note differences in solder wetting angle and appearance. The lead-free eutectic solder (Sn96.2Ag3.5) has the closest appearance to tin-lead solder.

- Sn95.5Ag3.8Cu0.7, shows graininess, rugged fillet formation, and voids.
- Sn96.2Ag2.5Cu0.8Sb0.5 (Castin®) is similar or better in appearance than Sn95.5Ag3.8Cu0.7.
- Sn77.2In20Ag2.8 (Indalloy 227) is rougher than the others (possibly due to indium) and lacks good wettability.

Reasons for Lead-Free Optical Dull Appearance

Visual inspection and automatic optical inspection (AOI) for lead-free solders is a challenge. Lead-free solder joint surfaces tend to be dull, matte, rough, and grainy, contrary to the typical shiny appearance of a tin-lead solder

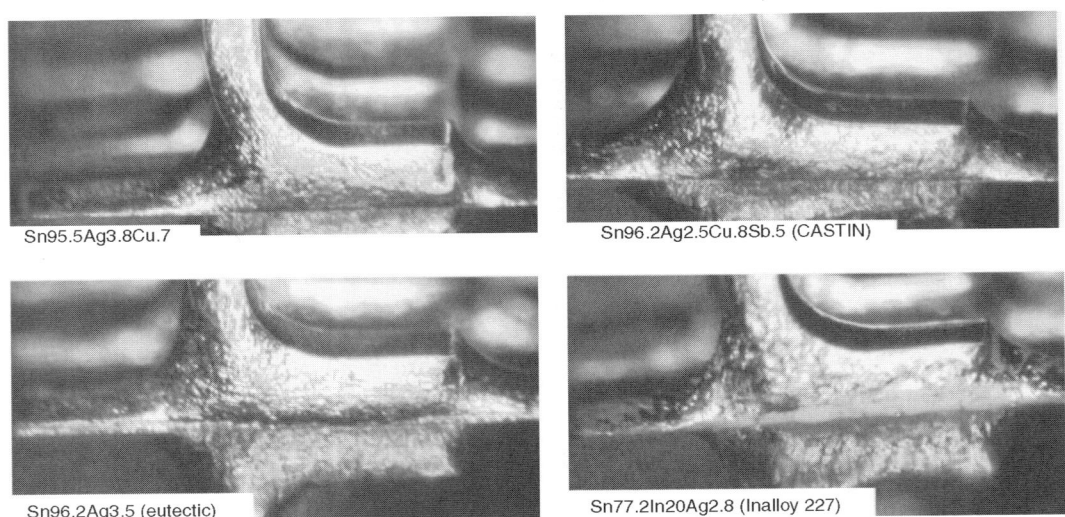

Fig. 11 Optical micrograph of QFP packages assembled with four different lead-free solders. Note differences in solder wetting angle and appearance of graininess.

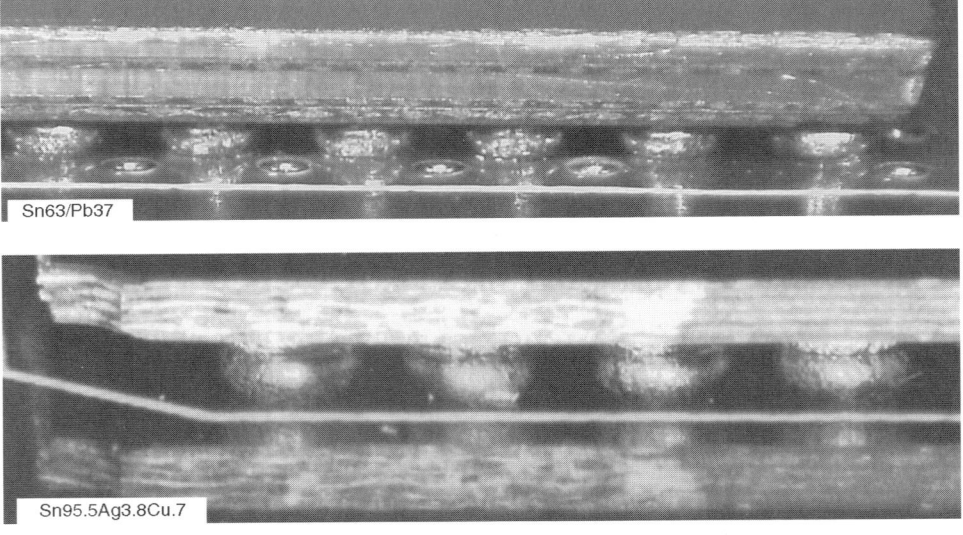

Fig. 12 Optical micrograph of BGA outer row solder balls of tin-lead eutectic and lead-free solders

joint. This is generally because certain lead-free solders are of noneutectic composition. After reflow, the solder drops in temperature slowly and nonuniformly because of the "plastic" region. For example, Indalloy 227 has a nonuniform melting temperature range of 175 to 187 °C (347 to 369 °F) (solidus-to-liquidus temperature range). Because of this, the metal composition starts to solidify nonuniformly. The liquid and solid coexist in equilibrium as defined by their phase diagram. This is the source of the dull surface that depends on the phase composition and its closeness to eutectic or pure alloy composition. In contrast, Sn63Pb37 is a eutectic alloy that melts at a constant temperature like a pure metal.

Fillet shapes of lead-free joints, which are a function of the wetting angle, are generally uneven; therefore, producing an inconsistent appearance that requires unique inspection criteria. Fillet shapes depend on many variables, including the wetting angle of solder to lead and pad surface finish, flux selection, and reflow temperature. Lead-free solder surface tension is stronger than tin-lead solder and is less prone to spreading.

Reflectivity depends on the solder surface appearance and therefore, the time solder alloy solidification characteristics. Good reflectivity will be observed when a tin-lead or eutectic lead-free alloy is used resulting in eutectic solidification.

Tin-Lead and Lead-Free Solder Microstructure after Reflow Cycles

Tin-lead and lead-free (Sn95.5Ag3.8Cu0.7) solder pastes were printed on printed wiring boards with various BGA and CSP design features. After paste printing, the boards with tin-lead and lead-free solder pastes were subjected to 1, 2, 3, 10, and 20 reflow cycles to evaluate the effect of 210 and 244 °C (410 and 471 °F) reflow temperatures, respectively, on the integrity of boards, microvias, plated through holes (PTHs), and solder microstructural changes. Marginal increase in daisy chain resistance of PTHs and microvias was observed. Figure 13 compares lead-free and tin-lead-based microstructures over a microvia after three reflows. Figure 14 shows details of intermetallic microstructure of lead-free solder after three reflows. Etching and removal of tin was performed to reveal details of intermetallic formation and growth. Plate-like intermetallic of Ag_3Sn with extensive penetration into the matrix, and extending from one side of a microvia into another, is of special interest. These plates could cause initiation of microcracks due to their lack of ductility and embrittlement at the solder joint interface, as well as within the solder joint.

Microstructure Features of Lead-free Solder after Deep Etching

Tin-lead and lead-free solder alloys exhibit different bulk solder microstructures. Tin-lead alloys exhibit distinct Sn- and Pb-rich grains; the majority of tin-based lead-free solder alloys exhibit intermetallic structures within the tin-matrix. The intermetallic morphology is not in lamellae form and may be round, plate-like, blocky, and needle-like structures. Examples of the deep etched microstructure of lead-free solder alloys follow.

Post-reflow microstructure photographs of lead-free solder joint are shown in Fig. 15. The following are key microstructural features, with the presence of porosity:

- Several of the Sn-Ag-(Cu-Sb) alloys revealed significant macro-porosity after a sin-

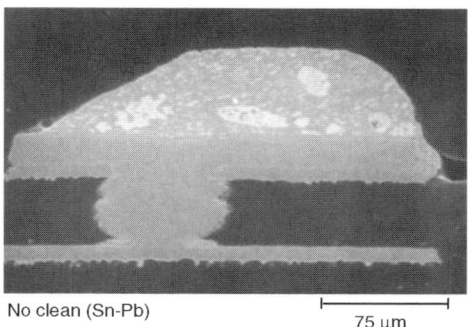
No clean (Sn-Pb)   75 μm

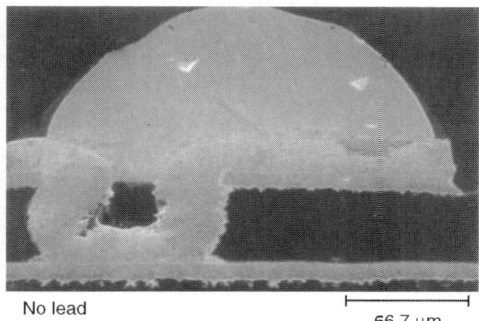
No lead   66.7 μm

Fig. 13 SEM photomicrograph of Sn-Pb and lead-free (no lead) assemblies after 3 reflows

gle reflow cycle, but not after two reflow cycles.
- Only 45 s above liquidus (minimum allowed reflow time) was used for each reflow cycle, in order to limit the amount of time the board was subjected to the elevated temperatures.

The presence of porosity was therefore attributed to insufficient time above the melting temperature.

Figure 16 provides microstructural elements determined by EDX shown over SEM photomicrographs taken near the board/solder interface. The following pertains to the interfaces:

- The light-colored region to the far left is Sn-Ag solder (high concentration of Sn and Ag).
- Intermetallic (Sn and Ni) is seen to the left of and within the dark region in the center. Elevated P from the electroless Ni bath is seen in the dark region.
- Cu (far right) is isolated behind the Ni layer.

Figure 17 shows the microstructural differences for two different surface finishes, Sn-Pb and Ni-Au. Large Cu intermetallic was formed at the interface for the Sn-Pb surface finish on Cu; however, the absence of such intermetallic is evident when Ni is used as the barrier in a Ni-Au surface finish. The latter finish is brittle and the interface mechanical strength is affected more by environmental exposures. Figure 18 shows the effect of one or two reflow cycles on the further growth of Cu intermetallic for a Sn-Pb board finish.

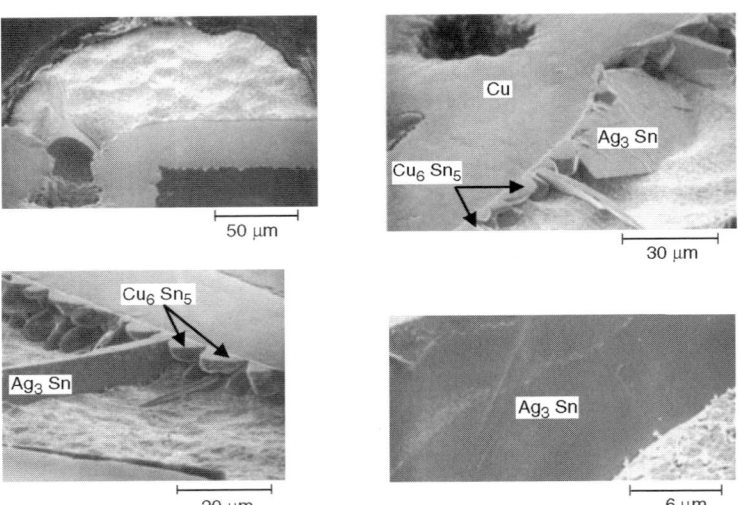

Fig. 14 SEM photomicrograph of intermetallics in lead-free solder after reflow. Note large plate-like intermetallics after 3 solder reflows

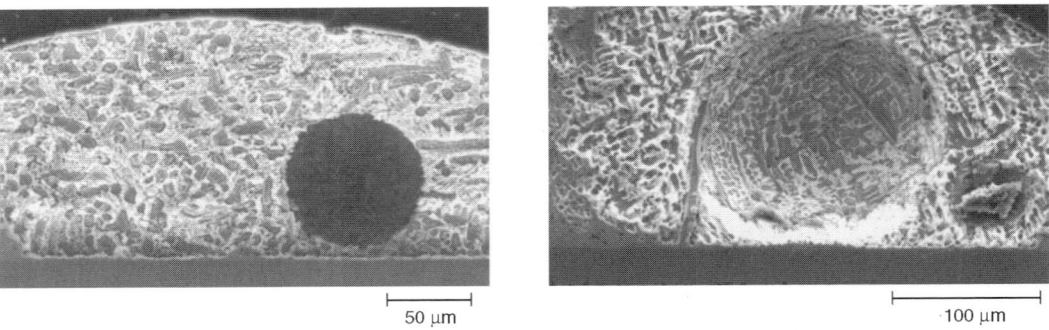

Fig. 15 Macro voids in a lead-free solder joint

Figure 19 provides a comparison of microstructural features of Sn-Ag-Cu solder on PWB pads with Ni-Au and Sn-Pb surface finishes, after one reflow cycle. Note formation of nodular Cu_6Sn_5 and Ag_3Sn needles in the Sn-matrix. Similar to previous results, coarsening occurred after two reflow cycles.

Figure 20 shows the microstructure of Sn-Ag-Cu-Sb on boards with Ni-Au and Sn-Pb surface finishes. A thinner version of Sn-Pb surface finish was also used to accelerate Cu intermetallic formation. In all cases, Cu containing intermetallics were formed within the solder, but away from the board interface.

Void Levels for Lead-Free and Mixed Assembly with Various Surface Finishes

Recent publications cite the reflow profile as the primary factor in mitigating voids for lead-free solder. The theory is that by using a high temperature soak in the reflow profile, the volatiles that create voids have an opportunity to burn off. If the volatiles do not escape before the metal becomes liquid, they will be trapped by the surface tension of the solder and become voids upon solidification. The effect of rapid ramp on void formation for the eutectic tin-lead solder was investigated a decade ago by this author and is shown in Fig. 21. This figure also includes examples of cross-sectional photos for lead-free area array assemblies with different reflow cycles (Ref 15), long soak, and rapid ramp rates (rapid-to-peak). Rapid ramp showed higher levels of voids.

Mixed assembly, i.e., tin-lead solder ball with tin-silver-copper (SAC305) solder paste, also showed a higher level of voids when compared to lead-free assemblies (Fig. 22). The tin-lead and SAC solder reach liquidus at 183 and 221 °C (361 and 429 °F), respectively. Therefore, the tin-lead solder balls become molten well before the lead-free solder pastes, possibly blocking the paste outgassing. In spite of the voids, it was shown that all lead-free assemblies and all surface finishes produced class III, 0–9% area voids, specified by IPC 7095.

Immersion-tin and OSP appeared to produce comparable voiding, with ENIG (Ni-Au) being slightly better.

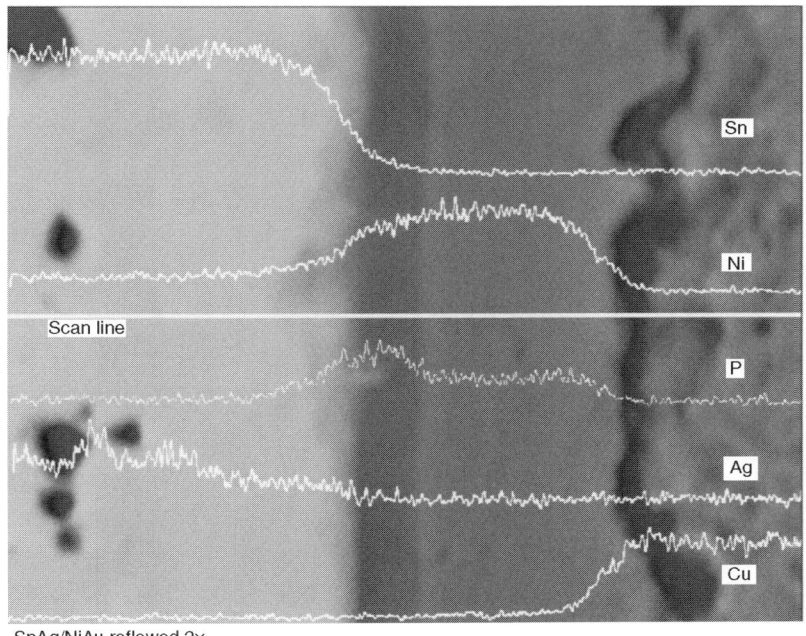

SnAg/NiAu reflowed 2×

Fig. 16 Interface between Sn-Ag and Ni-Au surface finish, elemental analysis

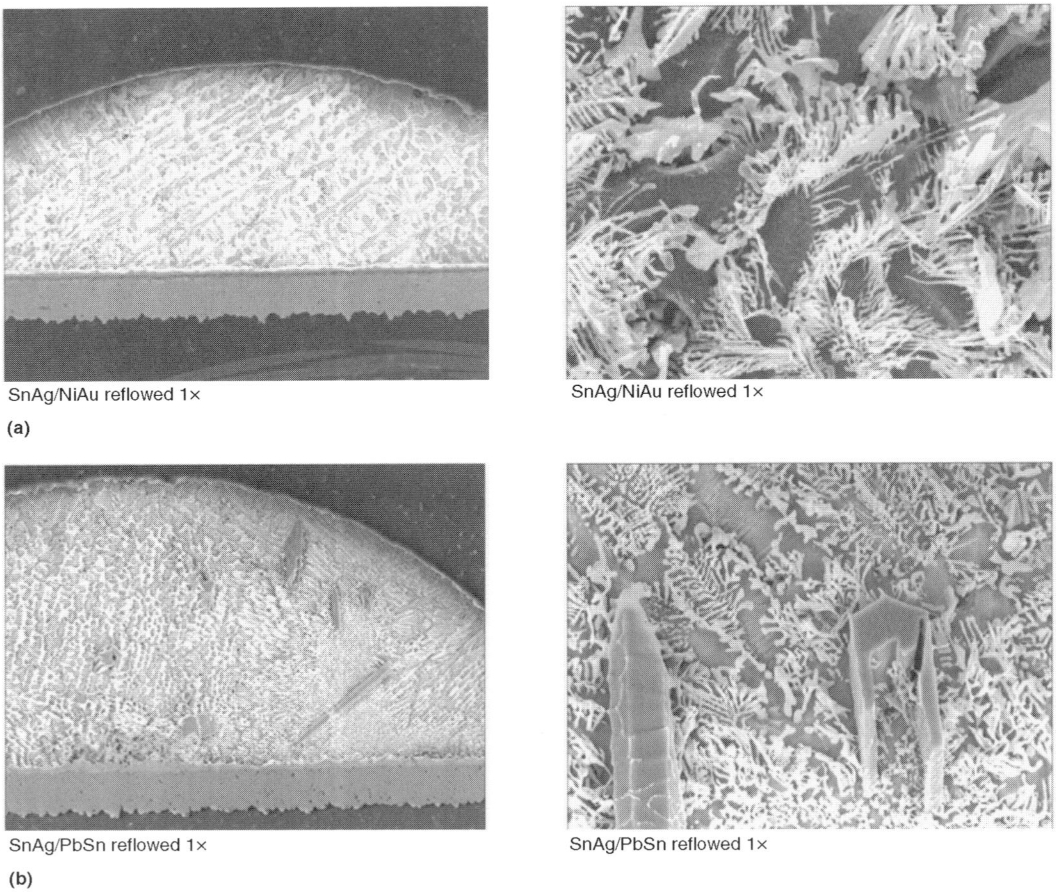

Fig. 17 Microstructural features of (a) Sn-Ag with Ni-Au surface finish, and (b) Sn-Ag with Sn-Pb surface finish

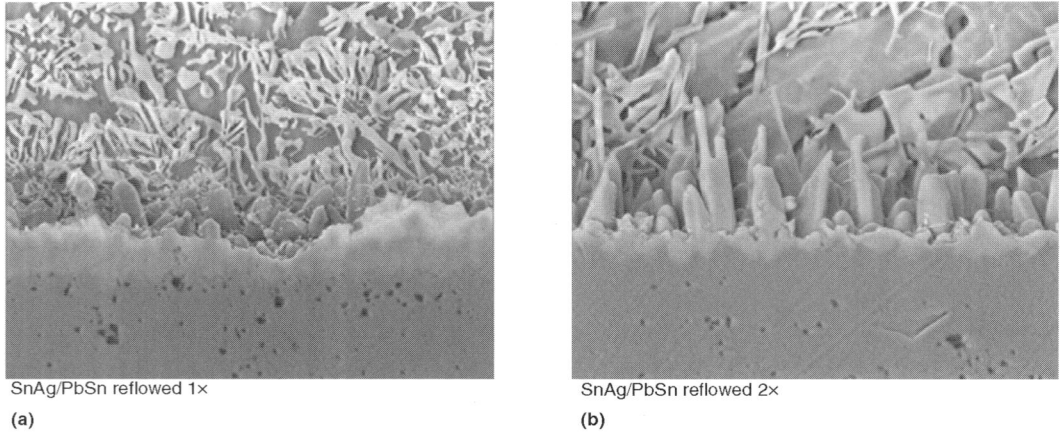

Fig. 18 Microstructural features of Sn-Ag with Sn-Pb surface finish after (a) 1, and (b) 2 reflow cycles. Note Cu intermetallic growth at interface

Microstructure Before and After Drop Test

A drop test was used to determine the effect of many process variables on the failure behavior for a CSP package with 132 I/Os and 0.5 mm (0.02 in.) pitch. Both tin-lead and lead-free (96.5Sn3.8Ag0.7Cu), type 3 solder pastes were considered in this investigation. Process variables included the effect of rapid temperature ramp-to-peak, long-soak reflow profiles, underfill, and corner adhesive bonding. Test coupons with a 31.8 g weight were dropped from a six-foot height multiple times onto a concrete surface.

Figure 23 shows photomicrographs of failed CSPs after drop tests for lead-free and tin-lead solder alloys. Without underfill, lead-free solder and tin-lead eutectic solder drop results were comparable. With capillary flow underfill, the drop test results were significantly better with the lead-free solder assemblies than with tin-lead eutectic solder assemblies.

Figure 24 shows the effect of storage of CSP assemblies with lead-free and tin-lead solders. There was degradation in the drop test results after 100 and 250 h of storage at 125 °C (257 °F) prior to the drop test.

X-ray Inspection Systems

Real time x-ray systems are categorized as 2D and 3D systems. Only 2D x-ray test results for various packages/assemblies after thermal cycling are presented here. Figure 25 shows features of 2D x-ray systems. The first example (Case 1, shown on the left side of the figure) is an x-ray inspection system with a microfocus source and an image intensifier as the detector, capable of producing offset pseudo 3D features

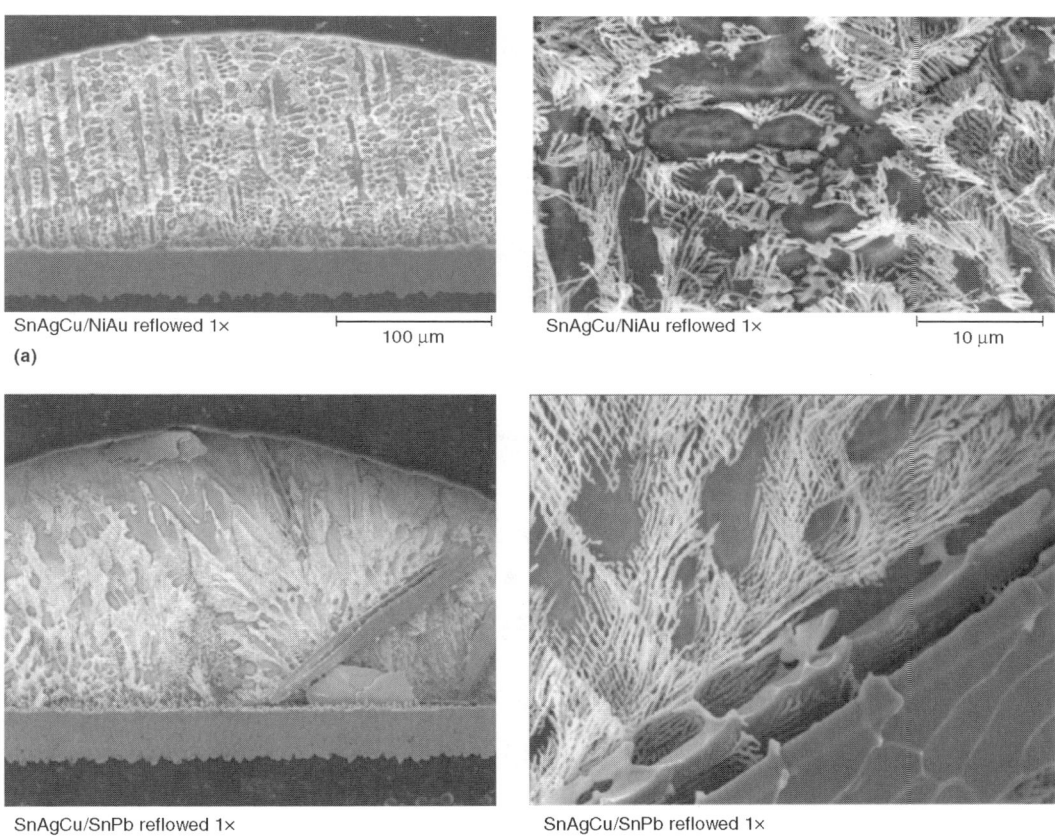

SnAgCu/NiAu reflowed 1×   100 µm
(a)

SnAgCu/NiAu reflowed 1×   10 µm

SnAgCu/SnPb reflowed 1×
(b)

SnAgCu/SnPb reflowed 1×

Fig. 19 Sn-Ag-Cu with Ni-Au and Sn-Pb surface finishes after one reflow. Note formation of Cu_6Sn_5 and Ag_3Sn needles in Sn-matrix. (a) Sn-Ag-Cu with Ni-Au surface finish. (b) Sn-Ag-Cu with Sn-Pb surface finish

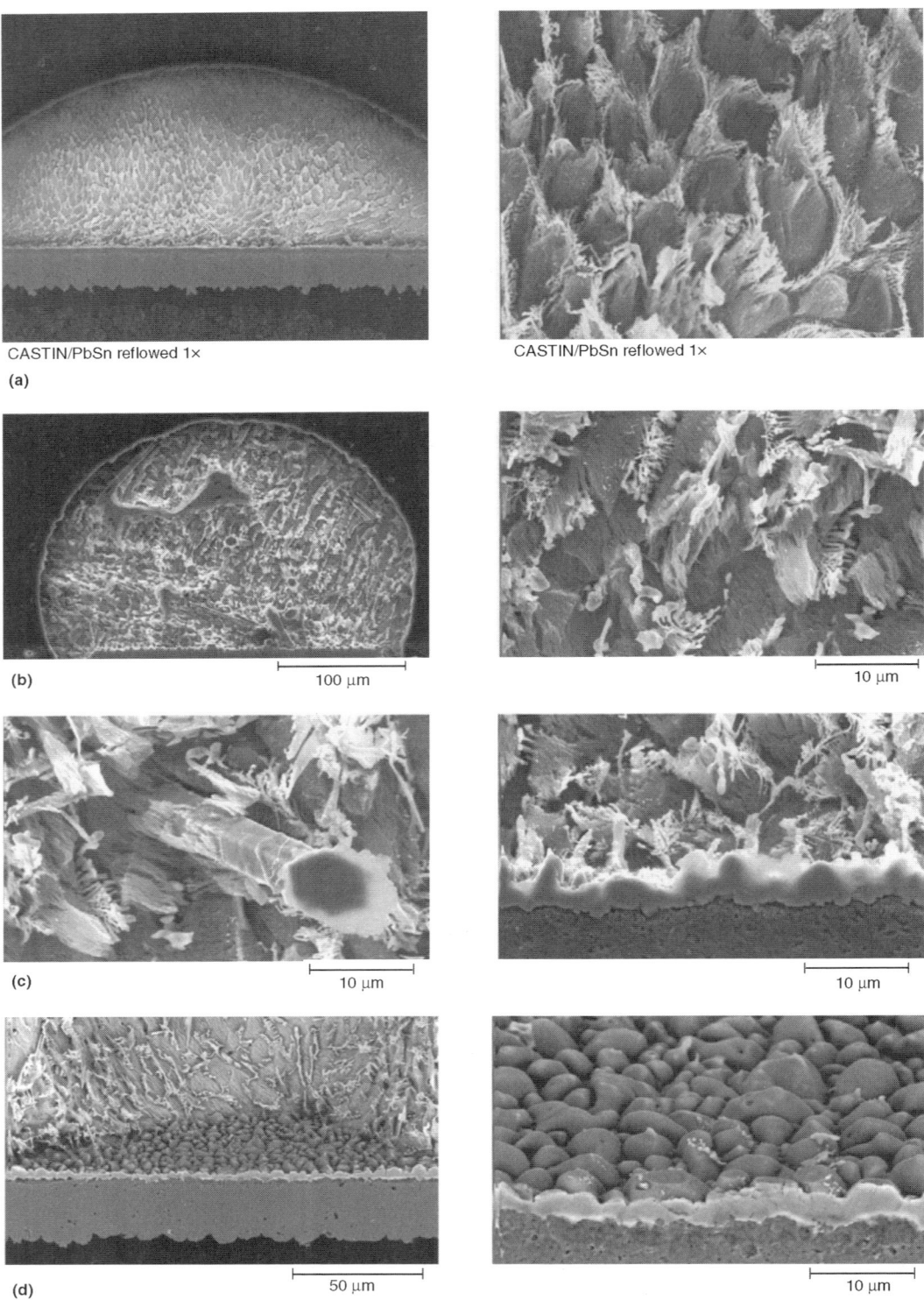

Fig. 20 Sn-Ag-Cu-Sb with Ni-Au and Sn-Pb surface finishes. Note cuboids of SbSn and Cu_6Sn_5 needles or globules. (a) Sn-Ag-Cu-Sb with Ni-Au surface finish after 1 reflow, (b) Sn-Ag-Cu-Sb with Sn-Pb surface finish after 1 reflow, (c) Sn-Ag-Cu-Sb with thin Sn-Pb surface finish after 1 reflow, and (d) Sn-Ag-Cu-Sb with thin Sn-Pb surface finish after 1 reflow

(Ref 16). The second system (Case 2, shown on the right side of the figure) that was utilized for evaluation is also a 2D x-ray tool with a similar microfocus source intensifier and stationary position, but the detector has off-axis rotational capability (Ref 17).

The x-ray system (Case 1) was limited to 2D inspections and capability of small sample rotation/tilt. The sample holder was not used since samples were larger than the capability of the sample holder. The transmission x-ray captures everything between the x-ray source and the im-

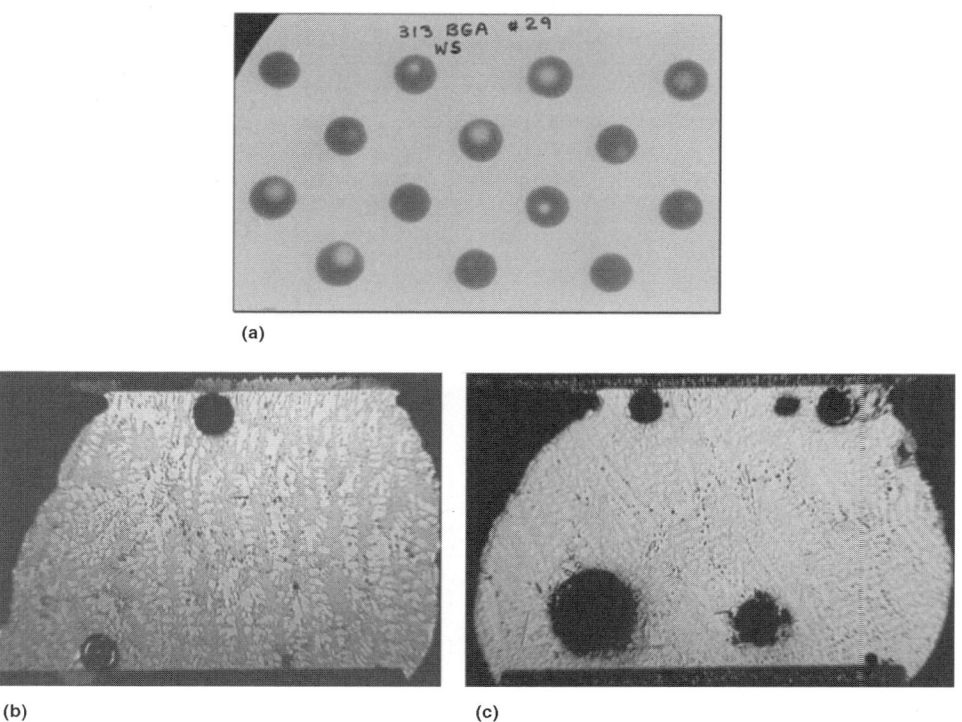

Fig. 21 Effect of reflow ramp rate on void level both in tin-lead and lead-free solders (a) tin-lead rapid ramp reflow, x-ray, (b) lead-free long-soak, and (c) ramp-to-peak. Courtesy of Rick Love

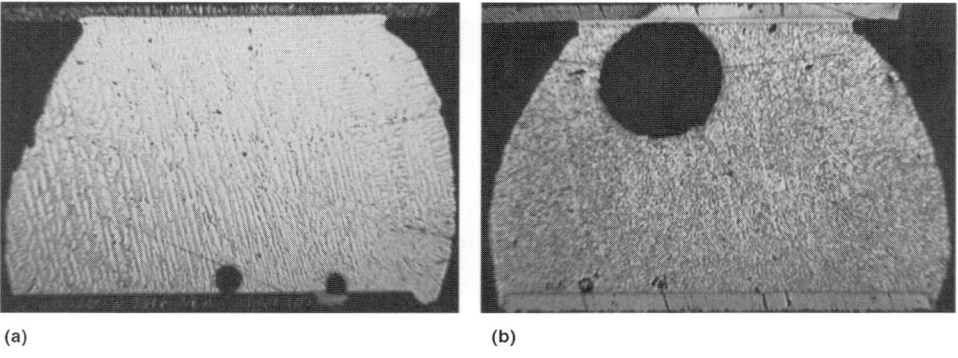

Fig. 22 Comparison of void levels for lead-free and mixed assembly, tin-lead paste and lead-free solder balls (a) lead-free, Cu OSP, long-soak and (b) mixed assembly, Cu OSP, long soak. Courtesy of Rick Love

age intensifier. X-rays are emitted from the source and travel through the sample. The higher the density of the sample, e.g., columns in CCGAs, the fewer x-rays will pass through and be captured by the image intensifier. The x-rays are displayed in a gray-scale image, with the

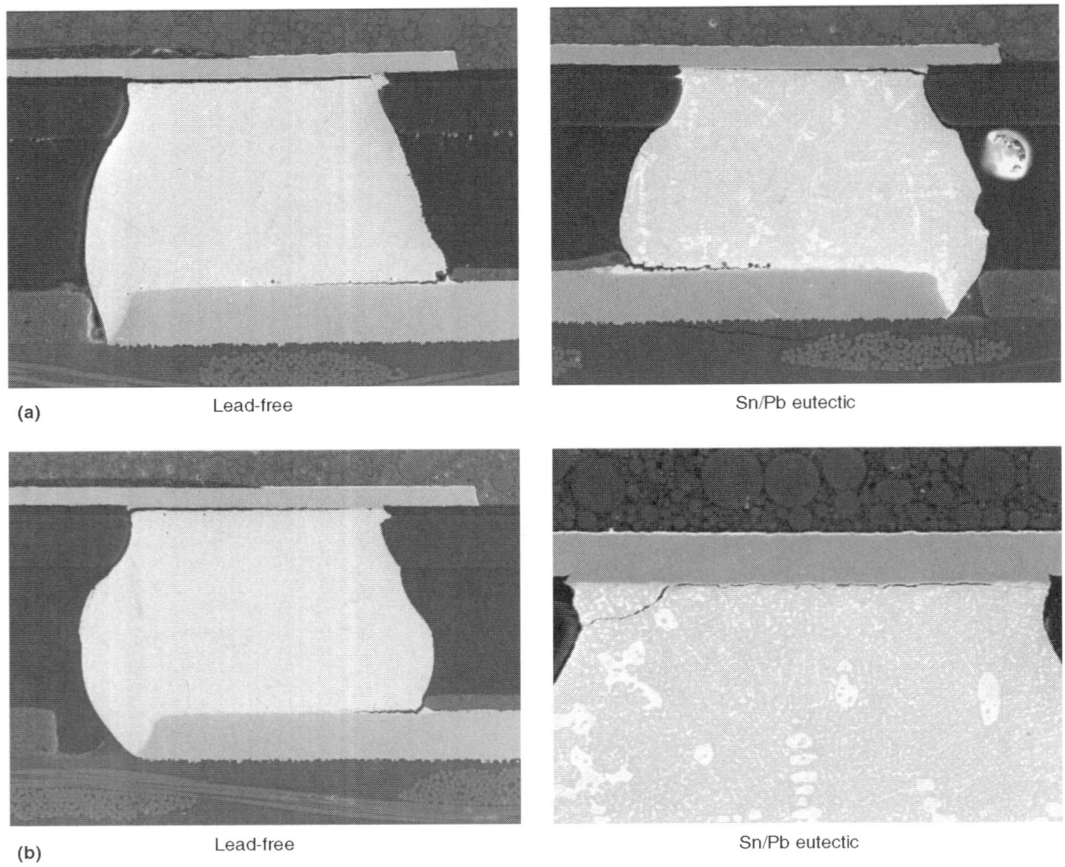

Fig. 23 Failure characteristics of lead-free and tin-lead with and without corner underfill after drop test (a) failure characteristics of lead-free and tin-lead solder assemblies of a CSP with no underfill after drop tests and (b) failure characteristics of lead-free and tin-lead assemblies of a CSP with corner stake underfill after drop tests. Courtesy of Dr. Wayne Johnson

Fig. 24 Effect of aging at 125 °C (257 °F) on failure characteristics after drop test. Courtesy of Dr. Wayne Johnson

lower density (such as voids) areas appearing brighter than the higher density areas. The voltage and current of the x-ray's intensity can be adjusted to reveal features of most sections of the sample.

The 2D x-ray systems are very effective in testing single-sided assemblies. With the use of a sample manipulator, an oblique view angle enhances inspection of both single- and double-sided assemblies with some loss of magnification due to increase in distance between source and detector. Experience is needed in discerning between bottom-side board elements and actual solder and component defects. This can be very difficult or impossible on extremely dense assemblies. As discussed previously, certain solder-related defects such as voids, misalignments, solder shorts, etc. are easily identified by transmission systems. However, even an experienced operator can miss other anomalies such as insufficient solder, apparent open connections, and cold solder joints.

Figure 26 shows an x-ray photomicrograph for an assembled CCGA, after thermal cycle exposure, using the 2D x-ray transmission system discussed previously for Case 1 where the part was kept stationary during x-ray exposure. Solder joints could not be detected because of significant x-ray intensity attenuation by the CCGA ceramic casing and solder columns.

The Case 2 x-ray features allow oblique generation of x-ray images with a higher magnification and a better intensity resolution since the focal spot remains the same and there is no loss of magnification. An isocentric manipulator keeps the field of view unchanged when the oblique view mode is used. This feature allows better characterization of defect features, including wettability and void locations in area array packages.

Figure 27 shows x-ray photomicrographs of an assembled CCGA after thermal cycles using the Case 2 x-ray system with an oblique view capability. X-ray images from two views are included. CCGA columns having high lead composition (90Pb-10Sn) are much darker than eutectic solder (37Pb-63Sn) used for attachment to the board. Within lighter solder joints, at lower sections of the columns, other lighter zigzag lines, possibly caused by cracking, are apparent. Lack of smoothness of patterns may be an indication of solder graininess. This generally occurs due to solder grain growth as thermal cycling progresses.

Figure 28 shows x-ray photomicrograph features of two lead-free solder joints for two plastic packages. They appear to have the same features as tin-lead-based solder joints. Lead-free solders have similar densities to tin-lead and therefore, like lead, impede x-rays; therefore, with some calibration, x-ray results can be interpreted for lead-free solders (Ref 18).

Conclusions

Characteristics of solder defects, and microstructural and failure features of tin-based lead-free solder alloys, are somewhat different from tin-lead-based alloys. The difference is apparent when nondestructive techniques such as optical

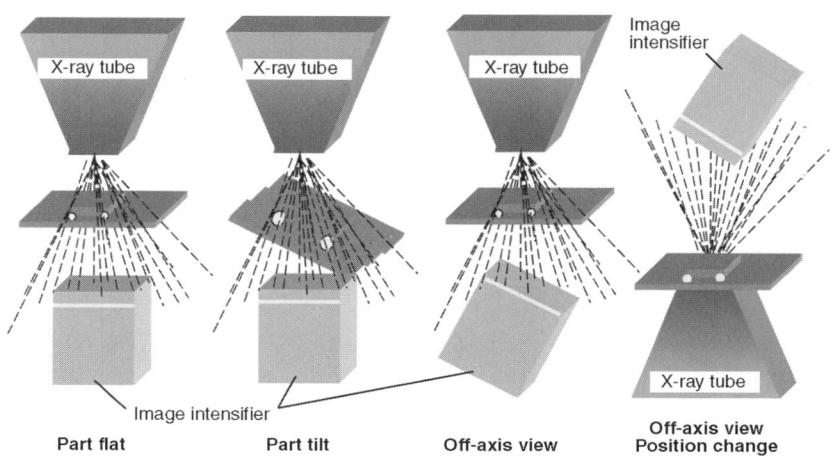

Fig. 25 Various 2D x-ray systems with stationary and rotational part and image intensifier

microscopy and x-ray are considered. Lead-free solder joints generally show graininess and a lack of wetting when compared to tin-lead solder joints. New inspection criteria for lead-free solder joints, and even solder balls in BGAs, need to be developed. Quality assurance (QA) personnel need to be trained, especially for high reliability applications, where visual/optical inspections are often used for acceptance/rejection. Automatic optical inspection (AOI) tools also need to be modified to include acceptance criteria for lead-free features.

Optical inspection, along with nondestructive evaluation systems having finer feature selection capabilities, such as x-ray and C-SAM, become critically important as electronic package/assembly become more complex, feature sizes decrease, and hidden interconnects such as BGAs are used. X-ray systems have improved significantly; however; they still have limitations. Many features of area array package assemblies, e.g., shorts and voids, can be easily detected by 2D x-ray systems. However, heavy solder joint damage/cracks in area array packages may require more advanced 2D x-ray systems with oblique view capability, or true 3D systems. In an experimental study, it was found that damage and cracks in plastic BGA and leaded package solder joints could only partially be detected when a 2D x-ray with improved defect detection with an oblique view capability was used. For ceramic column array packages, detection is limited only to the outer rows, due to the dense ceramic grid substrate that limits x-ray penetration.

X-ray features of lead-free solder for area array plastic package balls are similar to features of tin-lead-based solder. Lead-free solders have similar density to tin-lead and, like lead, impede x-rays. Therefore, with some calibration, x-ray results can be interpreted for lead-free solder. X-ray systems need to be further developed, how-

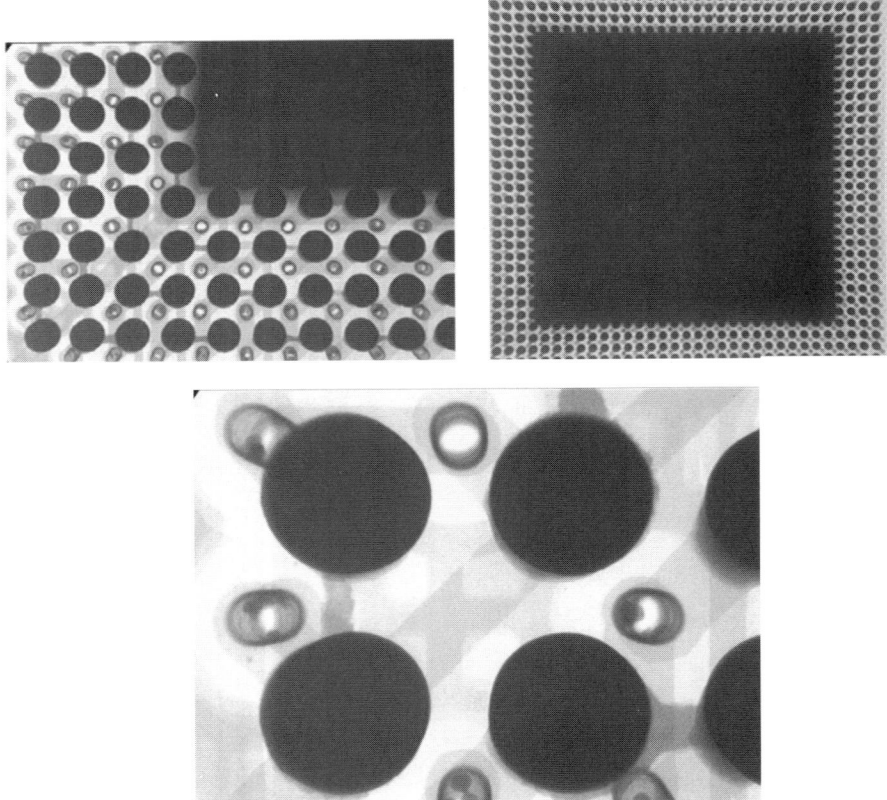

Fig. 26 2D x-ray inspection of CCGA assembly after thermal cycles. X-ray did not detect damage/cracks observed visually, by SEM, and cross-sectioning.

ever, to meet microelectronic size reduction and complexity, as well as distinguish composition of many lead-free solder interconnects. Investigations need to be carried out in order to understand the correlations between damage/cracks detected by optical/SEM/cross-sectioning and x-ray systems.

Failure analyses by destructive techniques also revealed many differences between tin-based lead-free and tin-lead solder alloys. Even the preparation methods for failure analyses are different. Alternative sample preparation is required for polishing and chemical etching in order to reveal microstructural features of lead-free solders. Microstructure of tin-based lead free solder is significantly different from a lamellae microstructure of tin-lead solder, and generally consists of a matrix of tin with dispersed intermetallic bodies in round, plate-like, blocky, or needle-like shapes. The microstructural grain coarsening, especially surrounding cracks, that is observed for tin-lead-based solders during thermal cycling, and can be interpreted as an indicator of environmental exposure, is not applicable for lead-free microstructural changes. Intermetallic formation and growth is more pronounced for lead-free solder. For tin-based lead-free solder, plate-like Ag_3Sn are typically formed in the Sn matrix and, Cu_6Sn_5, rod-like with facetted morphology, are formed at the Cu interface. Void levels are generally higher for tin-based lead-free solder than tin-lead solder joints, especially for the mixture of the two alloys and rapid ramp reflow cycle. Without underfill, mechanical drop results for tin-lead eutectic were comparable to tin-based lead-free solder. With capillary flow underfill, however, the drop test results were significantly better with lead-free solder assemblies than with tin-lead eutectic solder. Storage at 125 °C (257 °F) prior to drop test degraded the drop resistance.

ACKNOWLEDGMENTS

The author would like to acknowledge the kind assistance of Linda Del Castillo, John

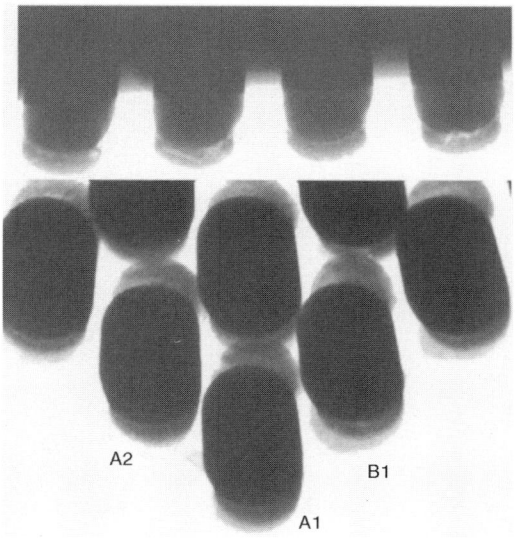

Fig. 27 X-ray photomicrographs of CCGA using a 2D x-ray system with an oblique detector at two angles. Signs of cracking/damage are somewhat apparent.

Sn95.5Ag3.8Cu.7   Sn96.2Ag2.5Cu.8Sb.5 (CASTIN)

Fig. 28 X-ray photographs of BGA assemblies with lead-free solder

(Kirk) Bonner, and Stephen Bolin at Jet Propulsion Laboratory who provided technical input into this chapter, and some of the photos used in this report and its review. Thanks to Wayne Johnson (Auburn University), Dongkai Shangguan (Flextronics), Rick Love (Cookson Electronics), and Keith Newman (Sun Microsystems) for providing photos on drop test, lead-free CGA, and voids, and review of this chapter, respectively. The author would also like to acknowledge Charles Barnes and Phillip Zulueta of the NASA Electronic Parts and Packaging (NEPP) Program for their continuous encouragement and funding provided under various NEPP programs.

The portion of research described in this publication was conducted by the Jet Propulsion Laboratory, California Institute of Technology, under a contract with the National Aeronautics and Space Administration.

Reference herein to any specific commercial product, process, or service by trade name, trademark, manufacturer, or otherwise, does not constitute or imply its endorsement by the United States Government or the Jet Propulsion Laboratory, California Institute of Technology.

REFERENCES

1. J. Fjelstad, R. Ghaffarian, and Y.G. Kim, *Chip Scale Packaging for Modern Electronics,* Electrochemical Publications, 2002
2. R. Ghaffarian, Chip Scale Package Assembly Reliability Chap. 23, *Area Array Interconnect Handbook,* Karl Puttlitz and Paul Tottar Ed., Kluwer Academic, 2002
3. R. Ghaffarian, BGA Assembly Reliability, Chap. 20, *Area Array Packaging Handbook,* K. Gilleo, Ed., McGraw-Hill
4. D. Frear, H. Morgan, S. Burchett, and J. Lau, *The Mechanics of Solder Alloy Interconnects,* Chapman and Hall, 1994
5. S. Ganesan and M. Pecht, *Lead-Free Electronics,* CALCE EPSC Press, 2003
6. J. Lau, D. Shangguan, T. Catello, R. Horsely, J. Smetana, N. Hoo, W. Dauksher, D. Love, I. Memis, and B. Sullivan, "Failure Analysis of High-Density Packages' Lead-Free Solder Joint," The Proceedings of APEX 2003, (Anaheim, CA), Mar 31–Apr 2003
7. B. Partee, "Microstructure Behavior of Lead-Free Solder Joints," The Advanced Packaging Technology Surface Mount International, July 2001
8. R. Ghaffarian, "The Interplay of Surface Mount Solder Joint Quality and Reliability of Low Volume SMAs," The Proceedings of NEPCON WEST, (Anaheim, CA), Feb 25–29, 1996
9. R. Ghaffarian, Solder-Joint Quality with Low-Volume PCB Processing *SMT Magazine,* July 1996
10. R. Ghaffarian, "Long-life Reliability of CSP Assemblies with and without Underfill," The Proceedings of Surface Mount International, (Chicago), Sept 22–26, 2002
11. R. Ghaffarian and N. Kim, "Vibration of CSP Assemblies with and without Underfill" The Proceedings of Surface Mount International, (Chicago), Sept 21–25, 2003
12. R. Ghaffarian, "Shock and Thermal Cycling Synergism Effects on Reliability of CBGA Assemblies," The Proceedings of 2000 IEEE Aerospace Conference, 2000, p 327
13. J.K. Bonner, L. Del Castillo and A. Mehta, "Hi-Rel Lead-Free Printed Wiring Assembly," The Proceedings of Surface Mount International, (Chicago), Sept 22–26, 2002, p 453
14. R. Ghaffarian, "Visual and X-ray Inspection Characteristics of Eutectic and Lead-free Assemblies," The Proceedings of Surface Mount International, (Chicago), Sept 21–25, 2003
15. G. Echeverria, D. Santos, P. Chouta, C. Shea, and R. Love, "Effect of Lead-Free Processing on Solder Joint Voiding" presented at IMAPS, Pasadena, California, May 11, 2004
16. Feinfocus GmbH, Garbsen, Germany, http://www.feinfocus.de/ (accessed Feb 2005)
17. Phoenix X-Ray Systems + Services GmbH, Wunstorf, Germany, http://www.microfocus-x-ray.com (accessed Feb 2005)
18. D. Geiger, T. Castello, and D. Shangguan, "X-ray Inspection of Area Array Packages Using Tin-Lead and Lead-Free Solders," The Proceedings of Surface Mount International, (Chicago), Sept 26–30, 2004

CHAPTER 11

Reliability of Interconnects with Conductive Adhesives

Johan Liu, Chalmers University of Technology, Gothenburg, Sweden; and Shanghai University, China
Zhimin Mo, Chalmers University of Technology, Gothenburg, Sweden

Introduction to Conductive Adhesive Technology

Recent environmental legislation has led to an increasing interest in the possibility of substituting electrically conductive adhesives for the traditional tin-lead solders in electronics manufacturing. The conductive adhesives mentioned in this chapter are not inherently conductive polymers. Instead, they are composites of insulating polymer matrix and conductive fillers. The polymer matrix and its characteristics are mostly responsible for the adhesive ability to bond and withstand mechanical stresses. The electrical conductivity of the adhesive depends particularly on the fillers. As a result, the electrical and mechanical properties can, to a large degree, be adjusted independently. Depending on the loading of fillers, conductive adhesives can be cataloged as isotropically conductive adhesive (ICA) and anisotropically conductive adhesive (ACA). Compared with traditional tin-lead solders, conductive adhesive interconnections typically possess the following merits (Ref 1):

- Low temperature processing
- Compatibility with a wide range of substrates
- No flux pretreatment or post-cleaning procedures required
- No lead or other toxic metals
- Finer pitch capability
- Solder mask not required

Isotropically Conductive Adhesives (ICA) have been successfully used for decades in the electronics industry as die-attach materials. New adhesives have been formulated to replace traditional solders in mainstream applications. The volume fraction of conductive fillers in the adhesive is between 20 and 35%, which is so high that the adhesive can conduct equally well in all directions. As a result, ICAs may be deposited only where electrical connects are required. In general, the conductivity of the adhesive improves with increasing filler loading, but at the expense of the adhesive becoming increasingly brittle. Copper, nickel, carbon, and silver are commonly used as conductive fillers. Silver is unique among these affordable fillers because of its good electrical performance, stability, and inherent conductivity of silver oxides. The matrix is mostly one- or two-component epoxies that can be cured with heat and/or IR radiation. However, polyimides, silicones, and thermoplastic adhesives can also be used as matrices.

The electrical conduction of an ICA joint is primarily established during cure (Ref 2). Instead of metallurgical connection, the joint conduction is based on mechanical contacts among conductive fillers (Fig. 1). Studies have shown that the conduction development during cure is accompanied by the decomposition of organic lubricants (Ref 3), which exposes the metallic surface of the fillers, and the cure shrinkage, which brings the fillers closer (Ref 4, 5). However, the conduction mechanism of ICA is still not fully understood, and which effect plays a dominant role is still open to question.

After the adhesive is cured, the fillers are randomly distributed and form a network within the polymer matrix. By this network, electrons can flow from one adherent to the other across the filler contact points. The overall result is to create numerous electron pathways, but with each path made up of a large number of mechanical contacts. Any factors affecting intimate contacts among fillers will surely influence the performance and reliability of ICA interconnects.

Besides die attachment, ICAs are utilized in surface mount and flip-chip packages as alternatives to traditional solders. Due to their low surface tensions, ICAs are not suitable for wave soldering (Ref 6). Despite the advantages of ICA interconnection, the wide use of this technology has not been adopted by the electronics industry. The main concern is long-term reliability.

Anisotropically Conductive Adhesives (ACA) are also prepared by dispersing conductive fillers in an adhesive matrix. Contrary to ICAs, ACAs provide unidirectional conduction, which is achieved by using a relatively low volume fraction of conductive fillers. This low filler loading is insufficient for interparticle contact and prevents conduction in the X-Y plane of the adhesive, but enough particles are present to ensure reliable conduction between bonding electrodes in the Z direction (Ref 7). Because of the anisotropy, ACAs can be deposited over the entire contact region, greatly expanding the bonding area. Also, the low filler loading improves the bonding strength. Thus, mechanically robust interconnection can be achieved with ACA assembly.

ACAs come in two distinct forms: paste and film. Pastes can be printed with screen or stencil, or dispensed with a syringe. Films are supplied by manufacturers as reel and are extremely suitable for nonplanar bonding surfaces. Both thermoplastic and thermosetting resins have been used as adhesive matrices. The principal advantage of thermoplastic ACAs is the relative ease to disassemble the interconnections for repair operation, while thermosetting adhesives possess higher strength at elevated temperature and form more robust bonds (Ref 7). The commonly used conductive fillers include silver and nickel particles, and polymer spheres coated with metal (Ni-Au). Silver particles offer moderate cost, high electrical conductivity, and low chemical reactivity. Nickel particles can break the oxide layer on the electrodes and are suitable for interconnecting easily oxidized metallizations. Metal-coated polymer spheres have fairly uniform diameter distributions. They can provide high interconnection reliability because of the large elastic deformation during bonding. Recent application of solder particles as ACA fillers has also been reported (Ref 8).

Since the conduction of ACAs is based on mechanical particle-electrode contacts, pressure is a requisite to form qualified joints. A typical ACA assembly is shown in Fig. 2. After alignment, pressure is applied on the backside of the chip. The adhesive resin is squeezed out and conductive particles are trapped and deformed between opposing electrodes. Once electrical continuity is generated, the adhesive resin is cured with heat or UV. The intimate particle-electrode contacts are maintained by the cured matrix, and the elastic deformation of particles and electrodes exerts a continuous contact pressure.

Compared with solder interconnection, ACA interconnection offers benefits in terms of pro-

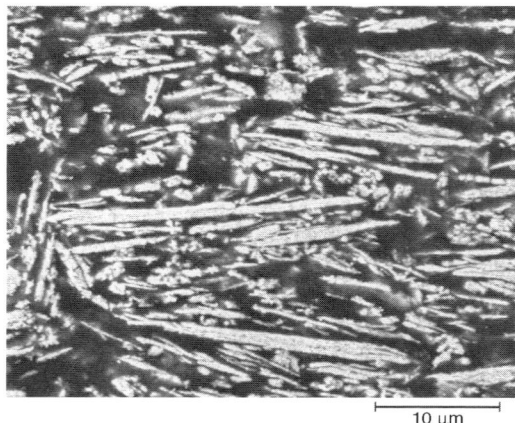

Fig. 1 Microstructure of an ICA showing silver fillers (white) embedded in the epoxy matrix (black)

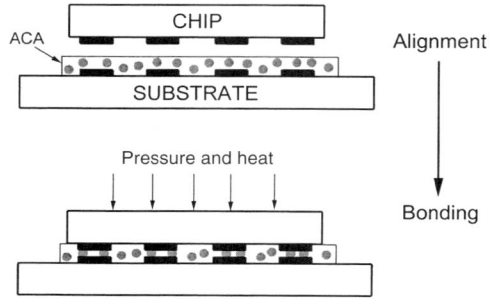

Fig. 2 Manufacturing process of an ACA assembly

cess simplicity, and resistance against vibration and shock. This technology finds particular applications with fine-pitched flip-chip techniques used to mount bare chip on various substrates such as ITO coated glass, FR4 board, and flexible films (Ref 9). ACA joining is also attractive for fine-pitched surface mount component assembly. However, the performance and reliability of ACA joints are more sensitive to the joint design, substrate/component properties, and process conditions than solder joints.

Reliability Concerns of ICA Interconnects

Effect of Metallization

In order to get a good adhesive joint, the adhesive must wet the bonding surface. A necessary condition for this is that the adhesive has lower surface tension than the bonding surface. Epoxy and polyimides are major polymers used as base matrices of ICAs. These materials have lower surface tension than Sn, Pb, Cu, Au, and Pd. Therefore, a good adhesive joint is expected when bonding on Sn, Sn-Pb, Cu, Ag-Pd; and Au surfaces.

As water molecules can easily penetrate through the adhesive and oxidize/hydrate the bonding surfaces, ICA joints with different metallizations have different reliability performances in high humidity environments. Several investigations (Ref 10–12) showed that joints with noble Au and Ag-Pd metallizations had much lower resistance increases, compared with those with non-noble Sn-Pb and Cu metallizations. With transmission electron microscopy (TEM) and electron spectroscopy for chemical analysis (ESCA), the detailed failure mechanisms were investigated by Liu et al. (Ref 10).

TEM observations on an adhesive joint with Sn-37Pb metallization show that water has penetrated to the Sn-37Pb surface after 1000 h 85 °C (185 °F) and 85% RH test. As a result, oxygen signals can be detected with EDS analysis in TEM. The corresponding electron diffraction observes a diffused ring, which indicates that Pb was converted to an amorphous structure. So the reaction product is not crystalline PbO, but $Pb(OH)_2$ or other Pb oxides like Pb_2O_3 and Pb_2O, which have amorphous structures. The ESCA analysis on the Sn-37Pb surface shows the chemical shift of the oxygen signals, confirming that the product is Pb-hydroxide. Due to the formation of amorphous $Pb(OH)_2$, which is an insulating compound and has a powdery structure, both electrical and mechanical properties of the ICA joint become deteriorative. With the same metallization, Ref 11 and 12 got similar resistance shift trends, but they primarily addressed tin oxidation according to electrochemical analysis.

TEM studies on an adhesive joint with copper metallization show the existence of an oxide layer on the Cu pad. The thickness is approximately 100 nm after 1000 h 85 °C (185 °F) and 85% RH humidity test. The rings in the diffraction pattern obtained from the oxide layer indicate that the layer consists of fine crystalline grains. The radii of the rings correspond to the spacing of the crystallographic planes of Cu_2O. It is therefore concluded that the formed oxide is Cu_2O, which is a poor conductor. This helps to explain why the joint resistance increased after the humidity test.

Contrary to non-noble metallization, the electrical resistance of ICA joints mounting a goldplated QFP80 component on an electroless Au-plated FR4 board is quite stable in the 85 °C (185 °F) and 85% RH environment. No significant increase can be observed up to 2000 h. Hence, it can be concluded that a noble metallization such as Au or Ag-Pd is preferable for normal ICAs. However, by adding corrosion inhibitors, some superior ICAs for pre-tinned metallization have been developed (Ref 13).

Effect of Curing Degree

There is no doubt that proper curing is very important for joint reliability. It was found that a minimum curing degree is required to provide a certain level of mechanical and electrical performance of adhesive joints, especially with nonnoble metallizations (Ref 14). Once this is achieved, increasing curing times does not result in significant improvement.

Figure 3(a) shows the electrical resistance shifts of epoxy-based ICA joints after 1000 h humidity test at 85 °C (185 °F) and 85% RH. These joints were on a Sn-37Pb surface and cured at 150 °C (302 °F) for various times. The corresponding curing degrees were determined by differential scanning calorimetry (DSC) measurement as between 65 and 90%. Below a critical curing degree (for this adhesive, the critical curing degree is 77%), the electrical resistance of the joint increases significantly after the humidity test. The reason is that an under-cured epoxy can absorb a significant amount of water, which in turn causes oxidation/hydration of the Sn-37Pb metallization. If a noble metallization like Ag-Pd is used, no electrical resistance shift has been observed, despite the fact that the curing degree can be very low, as can be seen in Fig. 3(b).

Similar to electrical performance, once a critical curing degree is achieved (77%), the shear strength of the joint on the Sn-37Pb bonding surface can be maintained at a constant level (Fig. 4a). However, on the noble metal bonding surface, the shear strength of the joint is almost independent of the curing degree in the range between 65 and 90%, as can be seen in Fig. 4(b). These results also indicate that for conductive adhesive joining, noble metallizations are preferable to nonnoble metallizations.

Impact Strength

Due to their high filler loading, many ICAs suffer from poor impact strength, which is one of the major drawbacks preventing their wide applications. Without adequate impact strength, ICA joints can hardly survive the significant shocks during assembly, handling, and usage. For bulk materials, impact performance is closely related to their fracture toughness and damping property. An adhesive with higher toughness and higher loss modulus normally has better impact performance. So, a simple approach is to modify base epoxy resins with elastomers to improve the impact performance of ICAs (Ref 13). However, for adhesive joints, the adhesion strength between the adhesive and adherend is also very critical. Low-impact strength can result from adhesive failure due to poor adhesion. Using conformal coating of surface mount devices is another practical way to improve the impact strength of the package (Ref 15).

To evaluate the impact performance of board-level packaging with ICAs, the National Center for Manufacturing Science (NCMS) has developed a special drop test (Ref 16). It involves dropping circuit boards onto hard ground from a height of 1.5 m (59.05 in.) and the sample surviving six drops is regarded as possessing acceptable impact strength. The test is easy to conduct, but only qualitative information could be supplied. A novel falling wedge fracture test was

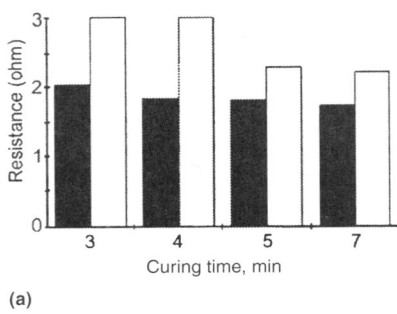
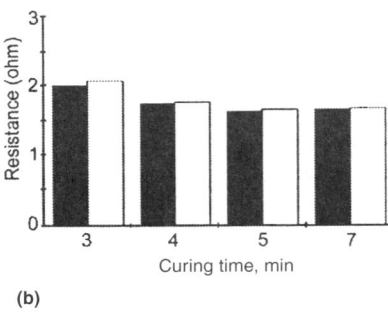

Fig. 3 Contact resistance shifts of ICA joints on (a) Sn-37Pb and (b) Ag/Pd surfaces (black: before test; white: after test). These joints were cured at 150 °C (302 °F) for various times and then aged in the 85 °C (185 °F) and 85% RH environment.

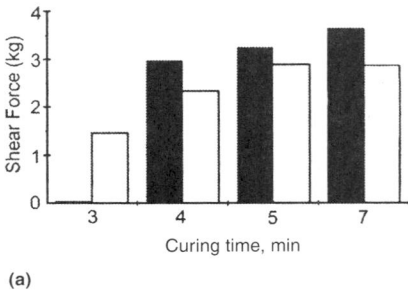
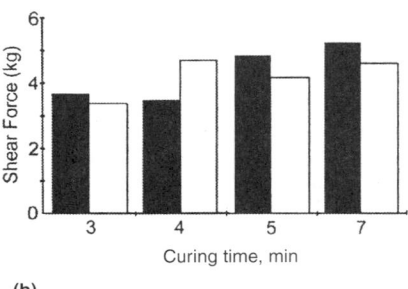

Fig. 4 Shear strength shifts of ICA joints on (a) Sn37Pb and (b) Ag-Pd surfaces (black: before test; white: after test). These joints were cured at 150 °C (302 °F) for various time and then aged in the 85 °C (185 °F) and 85%RH environment.

developed (Ref 17) that is capable of quantitatively determining the impact strength of ICA joints. They used a modified double cantilever beam (DCB) specimen with ICA and PCB boards and measured the fracture energies under different test temperatures with this new test technique. The impact fracture energy was found to keep an approximately logarithmic relationship with the loss factor that in turn can serve as a good indicator of the impact performance of ICAs.

Failure Mechanisms

Cracking. Due to the temperature fluctuation caused by the circuit power on/off cycles, ICA interconnects have to sustain cyclic stresses from thermal expansion mismatch between the substrate and component, and thermo-mechanical fatigue cracking is considered as one of the primary failure mechanisms. Based on temperature cycling tests and cross-section observations, the fatigue cracking behavior of ICA joints of leadless chip resistors was investigated (Ref 18). Early cracking was detected at the top of the vertical adhesive/termination interface (Fig. 5a), which has been reported in Ref 19. With more cycles, cracks were observed at the inner end of the horizontal interface between adhesive and ceramic resistor body (Fig. 5b). As the number of cycles increased further, bulk cracking occurred around the knee of the joint (Fig. 5c). It appears that several micro-cracks nucleated simultaneously due to the debonding of silver flakes. Then they merged together and formed the main bulk crack that propagated from the component side to the board side. After initiation, both vertical and horizontal cracks propagated toward the knee area along the adhesive/termination interface and the final merging of the three cracks (Fig. 5d) resulted in a complete failure of the entire joint. Since most crack development occurs at the interface, the adhesion of ICA is critical to the joint reliability.

In humidity aging tests, cracks have always been found associated with electrical degrada-

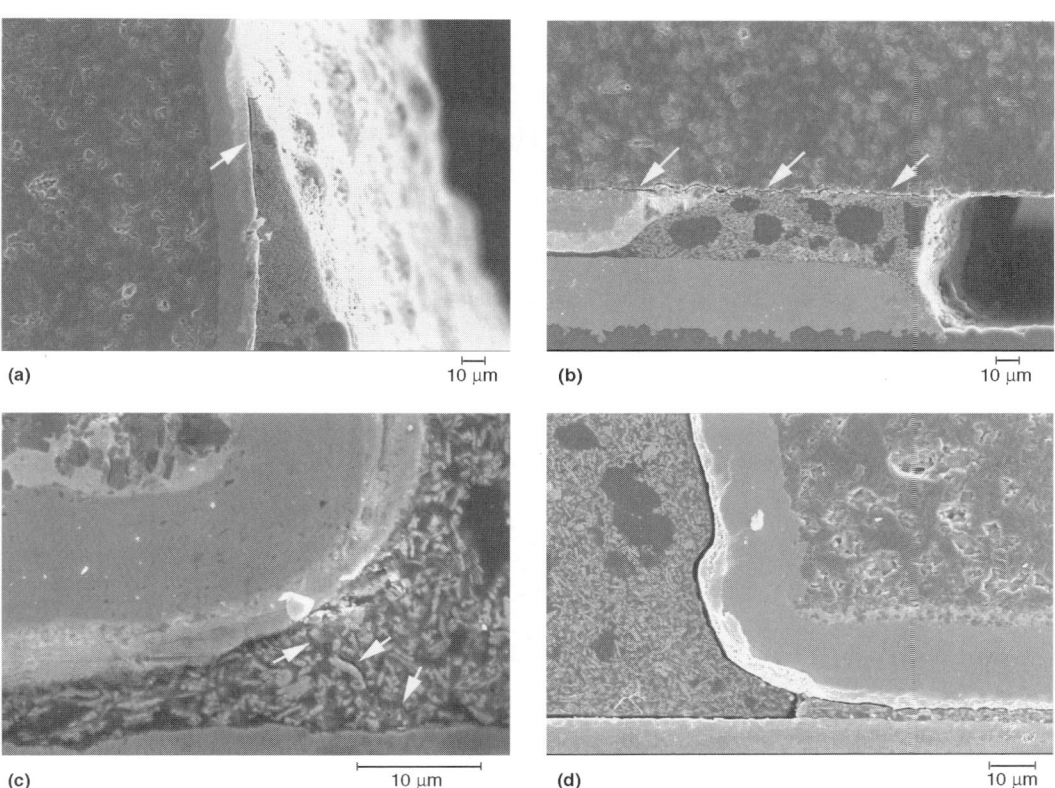

Fig. 5 Interfacial cracks initiated at (a) top and (b) inner ends of the adhesive/component interface; (c) bulk microcracks occurred around the knee of the joint. Cracks are indicated with arrows. (d) Final merging of these cracks resulted in a complete failure of the entire joint.

tion of ICA joints. Reference 20 reported that cracks existed after cure and developed due to the humidity exposure, leading to deterioration in both mechanical strength and electrical conductivity. Cracking after the humidity test was attributed (Ref 11) to the formation of oxides. In a recent investigation on the degradation of ICA joints in humid environments, Ref 21 concluded that moisture attack on the adhesive/metallization interface could be divided into three phases: displacing the adhesive due to high surface-free energy around the interface, hydrating the metal or metal oxide, and forming a weak boundary layer at the interface. If the attack occurs in the first phase, the fracture energy could recover to some extent after redrying at high temperature. However, the degradation becomes irreversible after the second phase.

Formation of Oxides. When ICAs are used together with nonnoble metallizations, the contact resistance increases significantly during the high temperature and high humidity aging. As discussed previously, various oxides have been observed forming at the interface between ICA and metallization, which leads to resistance deterioration.

In climate tests (98% RH), tin oxides occurred only in the area where adhesive was attached (Ref 11). No visible oxidation was observed at the fully air-exposed area on the top of the resistor. So they proposed that the direct contact between the noble metal (Ag filler) and the nonnoble metal (tin metallization), combined with absorbed water in the adhesive, formed a local electrochemical cell, which corroded the nonnoble metal. They also pointed out that tin oxide has no passivation effect and the contact resistance would increase progressively when tin is present in the metallization. However, metallizations of pure Cu and Pb were acceptable in high-humidity environments because oxides of Cu and Pb tend to form dense layers and hinder further corrosion.

Additional evidence supporting the electrochemical corrosion mechanism was shown (Ref 22). They found that the joint resistance could keep stable under the 85 °C (185 °F) and 85% RH condition if only one metal was involved. However, if two different metals (e.g., Ni fillers and Ag wire) were involved, the joint resistance would increase inevitably after the humidity aging. By formulating low moisture absorption resin and adding corrosion inhibitors, these authors developed new ICAs that can be used with nonnoble metallizations (Ref 13).

Formation of Intermetallic Compounds. Increase of resistance after environmental tests can also be attributed to the formation of intermetallic compounds. The heat-induced degradation of the interface between ICA and Sn-Pb-plated Cu electrode was investigated (Ref 23). Their element mapping analysis showed the apparent Sn diffusion into Ag particles. The occurrence of Ag-Sn intermetallic compounds, such as Ag_3Sn and Ag_4Sn, was identified in the x-ray diffraction pattern. They attributed this phenomenon to the Kirkendall diffusion of Sn from the plating layer into the Ag particles. At 150 °C (302 °F), the diffusion constant of Sn in Ag (2.31×10^{-17} m^2/s) is much larger than that of Ag in Sn (2.32×10^{-20} m^2/s). Therefore the preferential diffusion of Sn occurs. This results in large Kirkendall voids in the Sn-Pb plating layer, which decreases the true bonding area of the ICA joint and thus degrades both electrical and mechanical properties. These authors also pointed out that the diffusion constant of Sn in Ag_3Sn (6.37×10^{-12} m^2/s) is even higher than that of Sn in Ag and the formation of Ag_3Sn cannot hinder the Kirkendall diffusion of Sn.

Filler Motion. Several researchers (Ref 24–27) have noticed the difference in deformation behaviors of metal fillers and polymer matrix. Typically, conductive adhesive joints can sustain a shear strain of 10%, which is an order greater than solders. However, the metal fillers in ICA cannot be strained that much. Instead, they would move relatively to one another due to the compliancy of the matrix. Some possible influences on ICA reliability were proposed concerning this situation.

Compliant adhesive joints could survive more than 3000 thermal cycles without losing much mechanical strength, but the electrical resistance increased significantly (Ref 24). It was suggested that relative movement among fillers, combined with viscoplastic deformation of matrix, would pull the insulating polymer in between fillers, leading to loss of interfiller contacts.

With similar observations Ref 25 suggested that the filler motion would result in sliding along the interface between fillers. When the adhesive joint is subject to cyclic loadings, this interfacial sliding would eventually wear out the direct contact points among fillers and degrade the electrical performance of the ICA joint in the long run. Besides filler friction, the numeric simulation (Ref 26) showed that stress concentration due to filler motion would promote the initiation of micro-cracks in polymer matrix, which could weaken the constraint on fillers, loosen their intimate contacts, and therefore, increase the bulk resistance.

Mechanical low-cycle fatigue tests were performed (Ref 27) on several ICA joints and measured the resistance changes with high-sensitive micro-ohm technique. The resistance was observed to increase apparently at the initial stage of the tests, while the force required for the same deformation amplitudes decreased gradually. The authors attributed this phenomenon to the formation of wear tracks from filler frictions. However, they insisted that the influence of filler motion is limited and the dominant failure mechanism is interfacial fracture of the joint.

Ag Migration. In the presence of water and an electric field, silver is anodically dissolved at its original location and moves toward the cathode where it is deposited. This migration phenomenon can lead to the growth of dendrites between adjacent electrodes and lower the surface insulation resistance (SIR) of the board. For many years, the short circuit due to Ag migration has been a nuisance to those using silver inks and similar materials.

Because the silver fillers are encapsulated with an epoxy layer, however, Ag migration is not likely to occur in conductive adhesives under test conditions relevant in practice, e.g., 85 °C (185 °F) and 85% RH or 60 °C (140 °F) and 90% RH under 5 V bias (Ref 12). However, under more severe conditions, such as the presence of a liquid water film, higher bias, and smaller pitch spacing, Ag migration does occur. For example, short circuit between 0.20 mm (8 mil) spaced pads has been observed after 2000 h of 85 °C (185 °F) and 85% RH test with 15 V bias (Fig. 6). In Ref 28, the migration of Ag particles was also observed in ICA joints subjected to the current-induced aging (10 ~ 30A) and the consequent electrical degradation was reported.

Reliability of ACA Interconnects

Experimental Evaluation. With regard to the reliability of conductive adhesive joints, contact resistance of single joints is the most important feature. The reliability performance of ACA flip-chip joints on FR4 boards was investigated (Ref 29). In the study, nine types of ACAs (A-I) and one pure adhesive without any particles (J) were evaluated and some relevant technical data are shown in Table 1. As shown in Fig. 7, the test chip had a pitch of 100 μm, containing 18 single joints and two daisy-chains (18 joints for each). The substrate used was a 0.8 mm (0.03 in.) thick FR4 rigid board. Therefore, the work was particularly focused on the reliability of ultra-fine pitch ACA flip-chip interconnects on low-cost substrates. In total, 954 joints (53 chips) with various ACAs were subjected to temperature cycling between −40 and 125 °C (−40 and 257 °F) with a dwell time of 15 min. and a ramp rate of 110 °C/min. (230 °F/min.). The contact resistance of 36 joints (two chips) with ACA A was measured in situ during testing up to 3000 cycles. Other joints were taken out from the chamber every several hundreds of cycles and measured manually at room temperature.

Figure 8(a) shows a typical in situ resistance change of a single joint measured at both extremes of each cycle during the test. As can be seen, the resistance increased gradually with the thermal cycles. Cumulative failures of ACA A with in situ measurement are shown in Fig. 8(b). The number of failures depends heavily on the definition. If the criterion was defined as a 20% increase in joint resistance, all joints failed after 2000 cycles. This definition might be too harsh for those joints with only several mΩ where a 20% increase means only a few mΩ to vary. In some cases, the limitation is still within the error margin of the measurement. If an increase of 50 or 100 mΩ was used as the failure criterion, the mean time to failure (MTF) value became 2500 and 3500 cycles, respectively. Therefore, it is very important to define suitable criteria according to the product requirements.

As shown in Fig. 9, the reliability data of ACA A with the various failure criteria have also been analyzed with the three-parameter

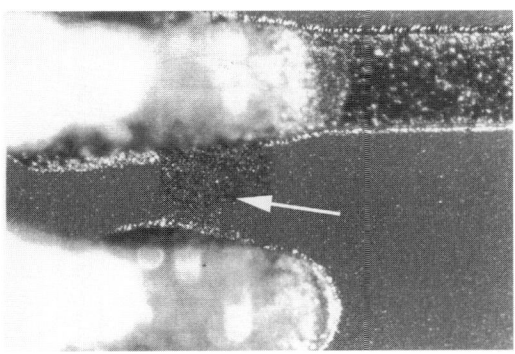

Fig. 6 Ag migration between 0.20 mm (8 mil) spaced pads after 2000 h of 85 °C (185 °F) and 85%RH test with 15 V bias

Weibull distribution. In the case of 20% increase, the Weibull analysis indicated that there were always some joints that would reach this value even before the first cycle was finished ($\gamma \cong 0$ in Fig. 9a). However, if the 50 mΩ failure criterion was used, it could be guaranteed that no failure occurred within 1757 cycles (Fig. 9b). Similarly, the first failure would occur after 2347 cycles if the 100 mΩ failure criterion was used (Fig. 9c). In both cases, it is clear that the ACA flip-chip/FR4 joints could withstand at least 1700 thermal cycles.

Cumulative failures of ACA joints measured manually at room temperature are shown in Fig. 10. The MTF (>20%) of joints with ACAs A, C, and J was 2500 cycles. Although other joints,

Table 1 Relevant technical data of ACAs subjected to reliability tests

Adhesive	Form	Conductive filler	Bonding condition
A	Film	Ni, 3 µm	180 °C (356 °F)/10s, 6 kg/mm^2
B	Film	Ni, 3 µm	180 °C (356 °F)/10s, 6 kg/mm^2
C	Film	Ni, 2–3 µm	160 °C (320 °F)/20s, 2.5 kg/mm^2
D	Film	Ni-Au plastic, 5 µm	160 °C (320 °F)/20s, 2.5 kg/mm^2
E	Paste	Ni-Au plastic, 3 µm	150 °C (302 °F)/30s, 5 kg/mm^2
F	Paste	Ni, 5 µm	150 °C (302 °F)/30s, 5 kg/mm^2
G	Paste	Ni-Au plastic, 5 µm	180 °C (356 °F)/60s, 7 ~ 15 kg/cm^2
H	Paste	Ni-Au plastic, 7 µm	180 °C (356 °F)/60s, 7 ~ 15 kg/cm^2
I	Paste	Ni-Au plastic, 11.5 µm	180 °C (356 °F)/60s, 7 ~ 15 kg/cm^2
J	Film	No filler	160 °C (320 °F)/20s, 2.5 kg/mm^2

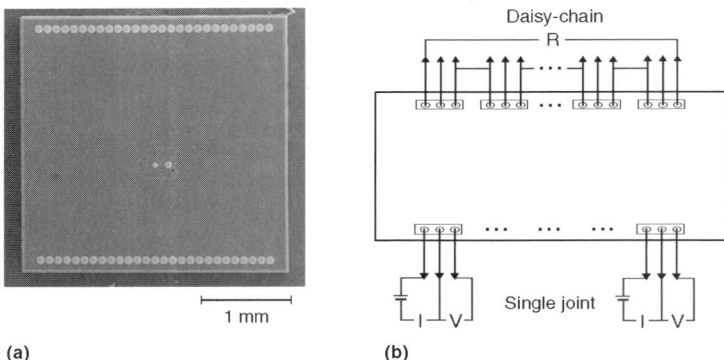

Fig. 7 Reliability testing of ACA flip-chip joints on FR4 boards. (a) Configuration and (b) measurement wiring of the test chip

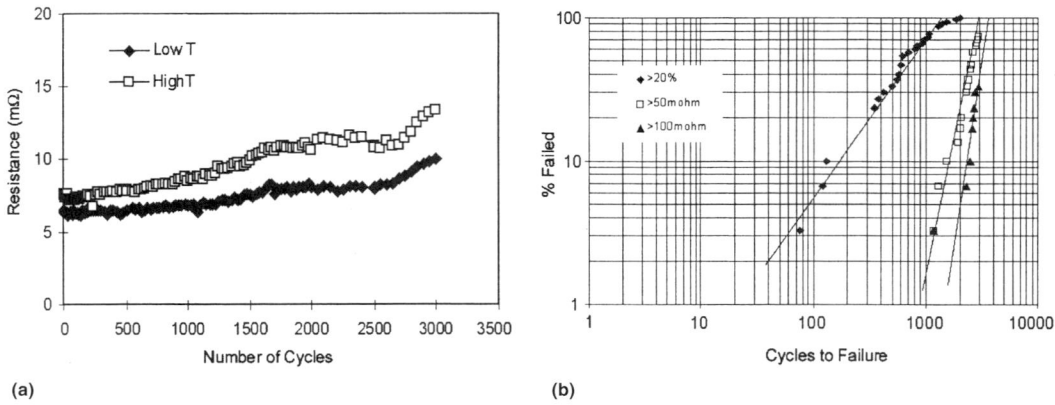

Fig. 8 Test results for ACA flip-chip joints. (a) Typical resistance evolution of a single ACA joint subjected to the temperature cycling test and (b) cumulative failures of ACA A joints during the test. Data were in situ measured up to 3000 cycles.

for example, those with ACAs G, H, and I, could also have an MTF of 2500 cycles, their infant-mortality failure rate was high, which means these joints were unreliable.

If we compare the manually measured data with the in situ data (Fig. 10a and 8b, respectively), the in situ measurement gave rather pessimistic result: 50% of joints failed before 650 cycles if a 20% increase in joint resistance was used as the failure criterion. With manual measurement, the number of cycles to 50% failure was around 2500. The exact reason is not clear. However, as they are closer to the real application situations, the in situ data are considered to be more important than the data measured manually at room temperature.

ACAs A and B are from the same vendor. The only difference is that ACA B uses the so-called double-layer technique to increase the fine pitch capability. As can be seen from Fig. 10(a) and (b), the double-layered ACA B showed similar reliability as the single-layered ACA A. If 50 and 100 mΩ failure criteria were used, the MTF of ACA B could also reach 1200 and 2200 cycles, respectively.

ACAs C and D are from another vendor. ACA C is filled with nickel particles, while ACA D is filled with nickel/gold-coated plastic balls. Both of them showed good reliability performance (Fig. 10c and d). Although the data scattering was large, especially in the case of ACA D, the MTF values could reach over 2700 and 3000 cycles for the ACAs C and D, respectively.

ACAs E and F are in a paste form, different from ACAs A to D which are in a film form. As shown in Fig. 10(e) and (f), their reliability was also very good. In the case of ACA E which is filled with nickel/gold-coated plastic balls, the MTF could reach 2000 cycles when the failure was defined as 20% resistance increase. If the 50 and 100 mΩ failure criteria were used, the MTF was about 5000 cycles. ACA F, filled with

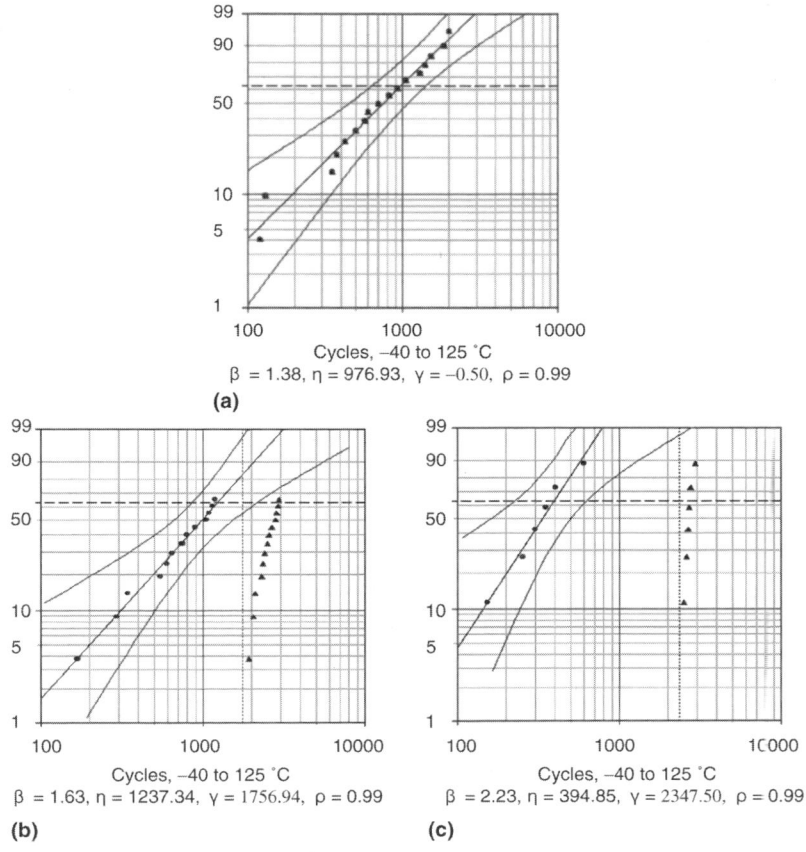

Fig. 9 Weibull analyses of ACA A joints with various failure criteria: (a) >20%, (b) 50 mΩ, and (c) 100 mΩ

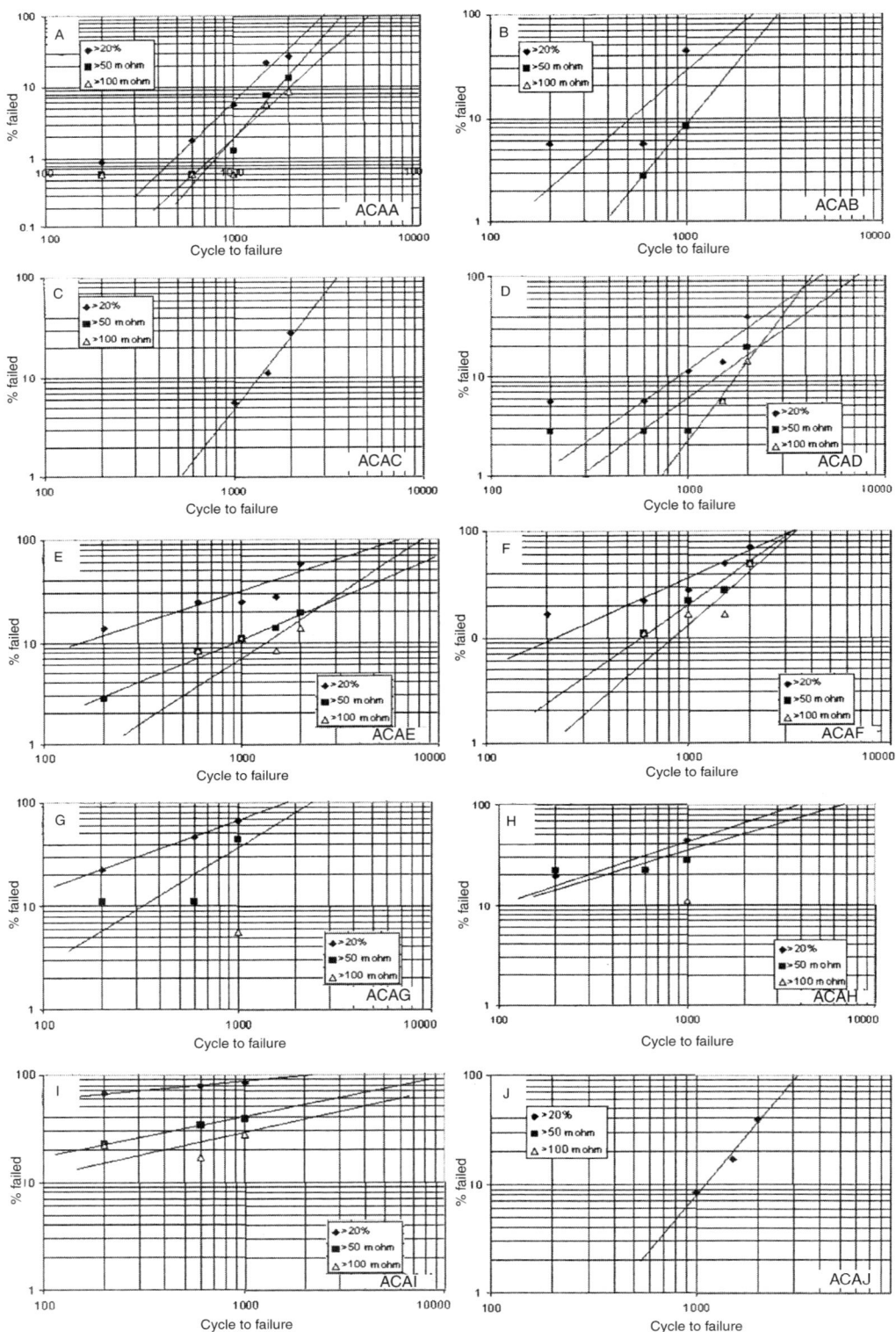

Fig. 10 Cumulative failures of various ACA joints. These joints were taken out from the chamber every several hundreds of cycles and measured manually at room temperature. See text for details.

nickel particles, also had an MTF over 1500 cycles.

The reliability curves in Fig. 10(g) to (i) show the effect of filler size on the joint reliability. These adhesives are all filled with nickel/gold coated plastic particles. It seemed that fillers larger than 5 μm had negative effect on the joint performance. With these pastes, joints could fail rather early in the test, though their MTF values were quite high.

Figure 10(j) shows the results of joints with a pure adhesive film. It is interesting to notice that this film also showed good reliability, and the MTF value was about 2000 cycles if the failure was defined as 50 mΩ increase. This indicates that the resin itself is extremely important in determining the reliability of the ACA joint.

Effects of Assembly Process. The assembly process of ACA interconnection includes alignment, bonding, and, if solder interconnects exist in the same board, reflow. Due to the low surface tension, ACA interconnection lacks the benefit of the self-alignment, which put a stringent requirement on the alignment accuracy. A normal flip-chip bonder that offers a ±5 μm accuracy is normally good enough. Nevertheless, poor alignment would result from incorrect operations. It can influence the pressure distribution and, in more serious situations, decrease the contact area for electrical interconnection (Fig. 11).

The bonding process is very critical to the ACA joint performance and reliability, since both mechanical integration and electrical interconnection are established in this process. Bonding pressure and temperature are the two most important parameters. To achieve reliable ACA joints, adequate bonding pressure should be applied uniformly, and suitable bonding temperature should be kept for sufficient time (Ref 30).

The bonding pressure is applied to force the conductive particles to contact the electrodes. The performance of the joint depends heavily on the deformation degree of particles. Ideally, the particles should be squashed enough to gain the largest contact area. However, the integration of particle body should be maintained, as cracking due to overpressure could degrade the electrical performance (Ref 31).

It is also important to keep the pressure uniform during the bonding. Non-uniform bonding pressure can cause particles being deformed unevenly, which could result in poor long-term reliability. There are several factors that can cause uneven bonding pressure, such as improper electrical routing on the substrate and incorrect operation of the bonder. This problem becomes more serious for thin and flexible substrates. Rigid and thick substrates like glass tend to distribute the bonding pressure more uniformly over the joint area, and effects of electrical routing and track design are not so critical.

The effects of particle deformation on joint electrical reliability during temperature cycling are summarized schematically in Fig. 12. Type 1 represents the best case, where the particles are deformed uniformly and atomic bonding between the particles and contacts is achieved. Type 2 joints consist of un-deformed or slightly deformed particles due to either low bonding pressure or uneven pressure distribution. The conductive character of these joints is unstable at high temperature because the epoxy matrix will expand more than the particles. Type 3 joints can result from shape or height variations of the contact areas. Some particles are not deformed enough and will shrink more than those well deformed, causing problems at low temperature. Finally, type 4 pictures a uniform height of the contact areas, but a very large variation of particle size. Due to the weak bonding between the smaller particles and contact area, electrical opens can be observed at both low and high temperatures. All of these situations have been observed experimentally (Ref 30).

The bonding temperature and time heavily influence the curing degree of the adhesive that plays an important role in the reliability of ACA joints (Ref 32). In an undercured joint, the cross-linkage of the polymer may be incomplete and neither mechanical performance nor electrical reliability can be guaranteed under high-humidity tests. To gain a certain curing degree, longer bonding time should be employed with lower bonding temperature. However, this is not pref-

Fig. 11 Bad alignment degrades the electrical performance and reliability of ACA joints.

erable due to the low productivity. On the other hand, excessive bonding temperature is not desired, either. This is because the epoxy may solidify too quickly, and the conductive particles would not have enough time to distribute themselves in between the bumps and pads. Recent work (Ref 33) also observed the chain scission due to high bonding temperature. Finding the optimum combination of bonding temperature and time is a fundamental step toward reliable ACA interconnection.

If ACA interconnection is used together with soldering technology for the final products, reflow soldering after ACA bonding is inevitable. During reflow, the package needs to be heated above 200 °C (392 °F), which is much higher than the normal bonding temperature of an ACA joint. The ability of ACA interconnection to withstand this high temperature is critical for successful packaging. The contact resistance of ACA joints increased significantly after reflow process and conduction gaps formed between the conductive particles and the electrode (Ref 34). Reference 35 reported the detrimental effects of reflow on the reliability of ACA joints. The possible reason is that, due to its much higher coefficient of thermal expansion (CTE), the adhesive matrix expands in the Z direction much more than the particles during the reflow. The induced thermal stress lifts the chip from substrate and damages the bonding structure. Therefore, the peak temperature of the reflow profile and the distance between the chip and substrate (related to bump height) are the most important factors. By optimizing process parameters and adopting ACA with lower CTE, the effects of the reflow process can be reduced to some extent (Ref 34).

Effects of Substrate and Component. Suitable substrate stiffness and bump dimensions are also important to achieve reliable ACA joints. With a soft substrate, significant deformation of the substrate may occur during the bonding, which has a direct influence on the joint quality (Ref 36). On the FR4 board, it was observed that the electrical resistance and reliability of a joint depend on the distance between the pad and glass fibers in the substrate (Fig. 13). A long distance means a thick layer of soft epoxy that may deform during bonding. Therefore, enough particle deformation cannot be obtained at that point. Figure 14(a) shows that a large force exerted on the pad causes pad sinking and almost no deformation occurring in the particles. An approach to reduce the pad sinking is to use a relatively smaller bump area compared to the pad area. Therefore, less bonding force will be transferred to the pad, as shown in Fig. 14(b).

For flip-chip solder joining, plastic strain of solder bumps is a critical parameter that governs the joint reliability. Increasing bump height can reduce the bump strain and thus increase the joint reliability, as shown in Fig. 15(a). However, a systematical study on the effect of bump height showed that the failure mechanism of ACA flip-chip joints is totally different (Ref 37). In ACA joints, the bump and pad are usually made of metals that are much stiffer than adhesives. In other words, thermal mismatch stresses can hardly deform the bump and pad, and the shear strain is localized in the adhesive between the mating bump and pad (Fig. 15b). In this case,

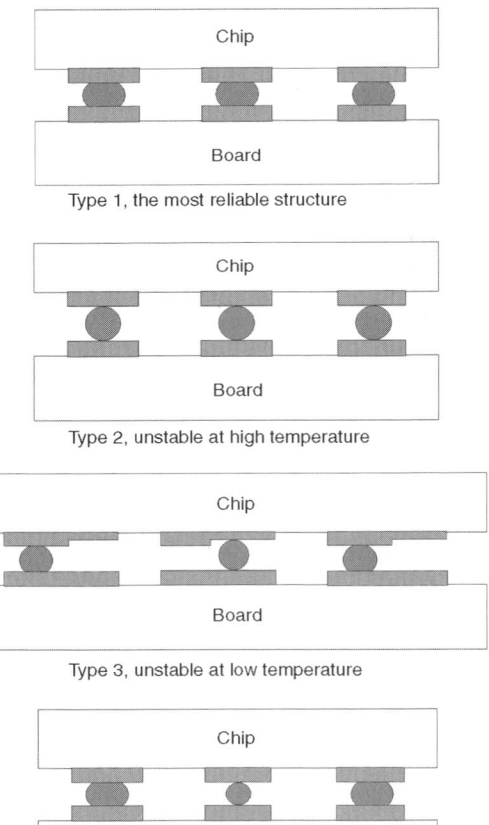

Fig. 12 Schematics of four types of ACA joints caused by variations in bonding pressure, bump geometry, and filler size

the joint reliability is governed by the shear strain in the adhesive and the influence of bump height is limited. Meanwhile, the stress in the Z-axis will be raised with bump height due to the increased adhesive volume. At elevated temperatures, this stress can lift the chip and weaken the joint. So, benefits from high bumps cannot be expected for ACA joints. Another practical problem associated with high bumps is that air bubbles are easily introduced during ACA bonding (Fig. 16).

Degradation Due to Moisture Absorption. Compared with ICAs, ACAs contain a much larger quantity of polymers. Therefore, besides the failure mechanisms observed in ICA joints, polymer degradation due to moisture absorption becomes more significant in ACA joints. Water can degrade polymers through (a) depressing of the glass transition temperature T_g and functioning as a plasticizer, (b) giving rise to swelling stresses, and (c) generating voids or promoting the catastrophic growth of voids already present.

Fig. 13 Electrical resistance and reliability of a joint depend on the distance between the pad and glass fibers in the substrate. Joint (a) has a better electrical performance than Joint (b) (5 mΩ vs. 14 mΩ) due to its location closer to glass fibers.

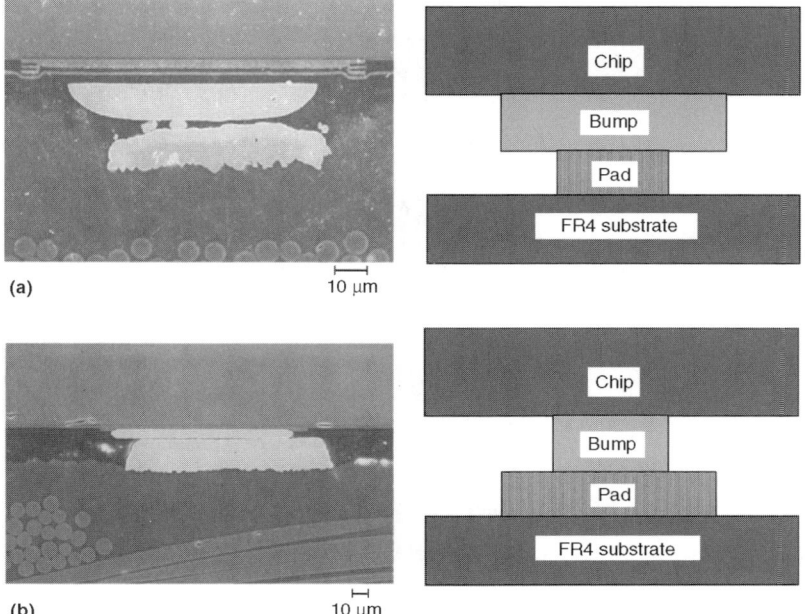

Fig. 14 Effect of bump size on pad sinking. (a) Pad sinking leads to insufficient particle deformation, and (b) using a bump smaller than the pad can decrease pad sinking.

All three occurrences have been known to lead to mechanical degradation. Moisture absorption can also contribute to the disruption of conductivity in the path between mating electrodes. This may include, for example, changes in the polymer/filler dispersion state through the expansion of the polymer matrix and formation of defects like cracks and delaminations.

The effects of moisture on an ACA film was studied with Fourier transform infrared (FTIR) spectra that provide a vast reservoir of molecular information pertaining to the chemical groups present, as well as to the structure arrangement and bonding preferences of these groups (Ref 38). The adhesive was conditioned in two environments: 85 °C (185 °F) and 85% RH and 22 °C (71 °F) and 97% RH. After a certain amount of time, samples were taken out of the chamber and FTIR spectra were collected.

Figure 17 shows, respectively, the spectra of the adhesive (a) after curing, (b) after 41 h exposure to 85 °C (185 °F) and 85% RH, and (c) the difference spectra representing the changes due to the moisture exposure. As shown, the negative bands at 868, 916, 1345, 3005, and 3058 cm^{-1} indicate the further progress of the cure reaction. Moisture degradation is believed to occur by hydrolysis of the ester linkages, which creates two end groups, a hydroxyl and a carbonyl. Though it is hard to see any new emerging carbonyl groups in this figure, the band at 3560 cm^{-1} indicates the existence of free hydroxyls. With more time exposure, curing effect is not observed, but degradation becomes more apparent. The spectra collected from samples exposed to 22 °C (71 °F) and 97% RH showed moisture absorption through hydrogen bonding, but neither further curing nor degradation is observed, implying the dominant degradation is associated with heat.

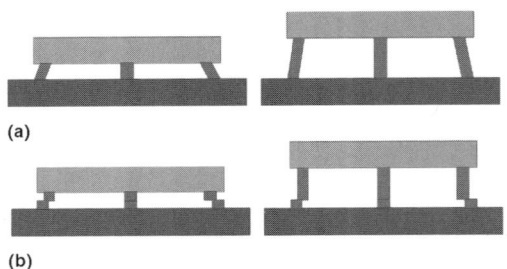

Fig. 15 Increasing bump height (a) can reduce the strain of solder joints, (b) but it has much less influence on the strain of ACA joints.

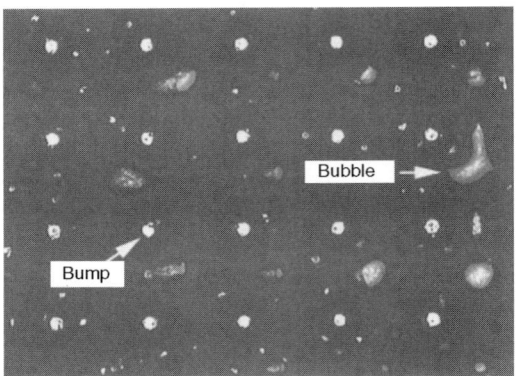

Fig. 16 Air bubbles introduced during bonding due to high bumps.

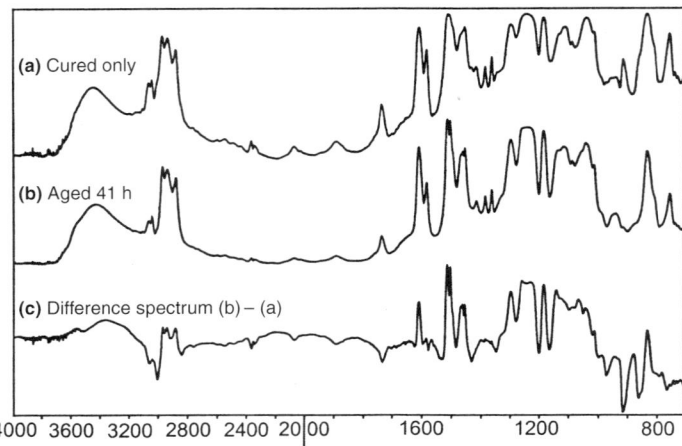

Fig. 17 FTIR spectra of an ACA film (a) after curing, (b) aged at 85 °C (185 °F) and 85% RH for 41 h, and (c) the difference spectrum (b) − (a).

Theoretical Studies and Numeric Simulations

Theoretical Treatment of Oxidation and Crack Growth

In order to correlate the electrical resistance shift as a function of humidity test time, a theoretical model has been developed based on Ref 39. It takes into account both oxidation and cracking, two primary failure mechanisms of conductive adhesive joints, and can thus explain the experimental observations quite well.

Figure 18 shows schematically the electrical conducting path through a conductive adhesive joint. Before exposure to the humid environment, the initial resistance through the joint is:

$$R_{init} = R_s + R_j + R_l \quad \text{(Eq 1)}$$

where R_s is the resistance through the substrate, R_j the resistance through the adhesive joint, and R_l the resistance through the component lead. After the humidity test, the joint resistance becomes:

$$R_{after} = R_s + R_j + R_l + R_{oxide} = R_{init} + R_{oxide} \quad \text{(Eq 2)}$$

where R_{oxide} is the resistance through the oxide layer, which can be expressed as:

$$R_{oxide} = \rho_{oxide} \frac{L}{A} \quad \text{(Eq 3)}$$

where ρ_{oxide} is the volume resistivity of the oxide layer, L the oxide layer thickness, and A the contact area.

Since polymer structures normally contain a large amount of free volume, it is reasonable to assume that the diffusion of oxygen is much faster in polymers than in metal oxides. In other words, the oxygen diffusion through the oxide layer will control the oxide growth rate and consequently the increase of the resistance in the oxide layer.

Assume the following Einstein equation holds:

$$L = \sqrt{2D_{oxide}t} \quad \text{(Eq 4)}$$

where D_{oxide} is the diffusion parameter of oxygen through the oxide layer and t is the time for the oxygen diffusion. Combining Eq 2, 3, and 4, one can obtain the relationship between the time and the resistance change:

$$\frac{R_{after}}{R_{init}} = 1 + \frac{L_e \rho_{oxide}}{A R_{init}} \sqrt{\frac{t}{t_e}} \quad \text{(Eq 5)}$$

where L_e is the oxide layer thickness at the end of the test, t the elapsed time, and t_e the total test time. Equation 5 can be used to calculate the relative electrical resistance change due to oxidation.

The crack normally occurs at the interface between the adhesive and the electrode, and decreases the real contact area gradually. Here, assume that the contact area A can be expressed as:

$$A = A_0\left(1 - \frac{t}{t_e}\right) \quad \text{(Eq 6)}$$

where A_0 is the original contact area. Therefore, taking into account the crack growth, the electrical resistance change becomes:

$$\frac{R_{after}}{R_{init}} = 1 + \frac{L_e \rho_{oxide}}{A_0\left(1 - \frac{t}{t_e}\right) R_{init}} \sqrt{\frac{t}{t_e}} \quad \text{(Eq 7)}$$

Figure 19 shows the calculated results with Eq 5 and 7, using the parameters given in Table 2. The calculations show that if no crack is formed, the electrical resistance will increase gradually with test time, but no catastrophic failure will be expected. The effect of cracking is rather small at the beginning, but then becomes more and more significant with the increase of test time. If a complete crack forms by the end of the testing, the electrical resistance will go to infinity.

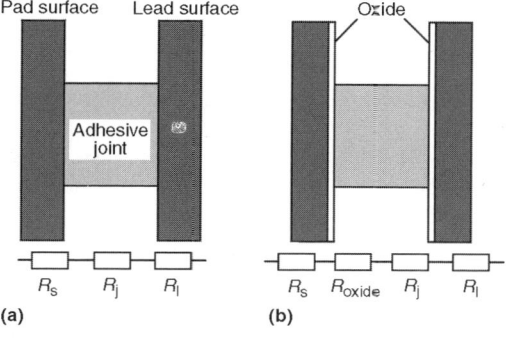

Fig. 18 Electrical conducting paths through a conductive adhesive joint (a) before and (b) after humidity exposure

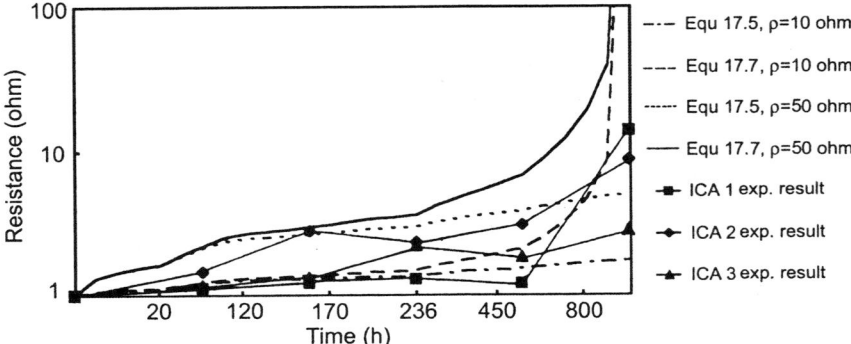

Fig. 19 Calculated and observed results of electrical resistance change as a function of humidity test time for ICA joints on copper surfaces

Table 2 Parameters used for calculation of the resistance evolution of an adhesive joint at 85 °C (185 °F) and 85% RH

Bonding surface	Oxide	ρ_{oxide}, Ωm	A_0, μm²	L_e, nm	D_{oxide}, m²/s	R_{init}, Ω
Copper	Cu$_2$O	10–50	1.1×10^{-6}	20	5×10^{-20}	0.2

For comparison, the experimental results are also given in Fig. 19. Before 500 test h, Eq 5 can predict the experimental observations quite well. However, the experimental results after 500 h cannot be explained by considering the oxidation of copper metal surface only, which means that fracture must have taken place during the humidity testing. In fact, cracks have already been observed after 158 h of exposure.

Electrical Performance Simulations

Electron Conduction through Nanoparticles in ICA. High metal loading, in the range of 20 to 35 vol%, is normally required to guarantee effective electrical conduction of ICA joints, which typically results in adhesive failure. Based on percolation theory, ICAs using a bimodal distribution of metal fillers were expected to have a decreased metal loading for better mechanical performance while the electrical property remains unchanged (Ref 40). It was, however, demonstrated experimentally that the electrical conductivity has been reduced when the volume percentage of the nano-size fillers in the system was increased (Ref 41). To explain this phenomenon, the electron conduction through nanoparticles in a normal ICA was investigated based on quantum mechanical considerations (Ref 42).

Consider a substructure in the ICA that consists of one nanoparticle sandwiched between two microparticles (Fig. 20). For the nanoparticle between the two microparticles, the quantum confinement effects must be included. Using the uniform background model, the ground sublevels across the substructure are approximately:

$$E_{nm}(z) = \frac{\pi^2 \eta^2}{2r(z)^2 m_0}(n^2 + m^2) \qquad (\text{Eq 8})$$

where $r(z)$ is the radius of the cross section at z, n and m are nonzero integers. The ground sublevel $E_{11}(z)$ is also presented in Fig. 20 where a potential barrier in the nanoparticle side of the interconnect ($z = 0$) between the micro and nanoparticles is observed.

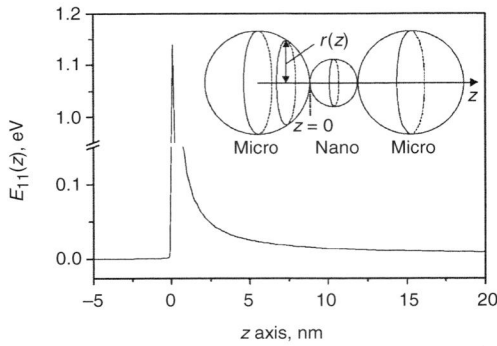

Fig. 20 Schematics of a micro-nano-micro substructure in ICA and the potential energy profile at the interconnect between the micro- and nanoparticles

When the structure is biased by V_{ex}, the local Fermi level of the left microparticle is kept unchanged (assumed to be grounded). The local Fermi level of the other microparticle becomes $E_f + eV_{ex}$, and its conduction band edge E_{11} is also lifted up by an amount of eV_{ex}. The time-dependent quantum mechanical behavior of an electron can be described by its wave packet. As the electron transports through the substructure of Fig. 20 from left to right, the electron wave packet is split into two parts after reaching the left interconnect energy barrier at $z = 0$. One is reflected back, and the other tunnels through the nanoparticle. Calculation shows that, due to the barriers, only half of the initial electron gets transmitted through the nanoparticle. This qualitatively agrees with experimental observations that the electrical conductivity has been reduced as nano-size fillers are added into the ICA.

The current-voltage characteristics of such a structure are presented in Fig. 21. The total current is obtained by subtracting the reflected current from the transmitted current. Increasing the external bias effectively lowers the energy barriers so that the transmitted current increases; however, it lifts up the conduction band edge E_{11} of the left microfiller so that the reflected current decreases very much. The final total current through the substructure, in general, increases linearly in the external bias range under investigation.

RF Performance of ACA Flip-Chip Joints. Sihlbom et al. (Ref 43) studied experimentally the high-frequency performance of a silicon flip-chip bonded on a Teflon substrate using ACA. However, due to the large variation of individual joints, the experimental study could not give a detailed picture on RF performance of an ACA joint. To elucidate the effects of conductive particles, a 3D finite difference in time domain (FDTD) modeling of the ACA flip-chip discontinuities has been performed (Ref 44).

The geometry of a single ACA joint is shown as Fig. 22, and parameters of the chip, Teflon board, and ACA joint are listed in Table 3. To avoid extremely small cells in the FDTD domain, the conductive particles are assumed to have a column shape with a height of 10 μm. A previous study (Ref 43) has shown that there is no significant difference in transmission behavior when 3 or 10 μm particles are used. In the FDTD simulation, the media under consideration were assumed to be uniform, isotropic, and homogeneous. Both the microstrip lines on the chip and on the substrate were designed as 50 Ω transmission lines. A pulse with a spectrum of 1 to 30 GHz was used as the excited source. It was applied at the feed-in microstrip that was connected with the microstrip on the chip by flip-chip transition. To avoid electromagnetic reflec-

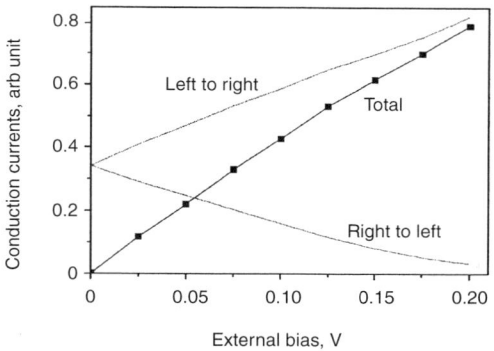

Fig. 21 Current-voltage characteristics of the micro-nano-micro substructure

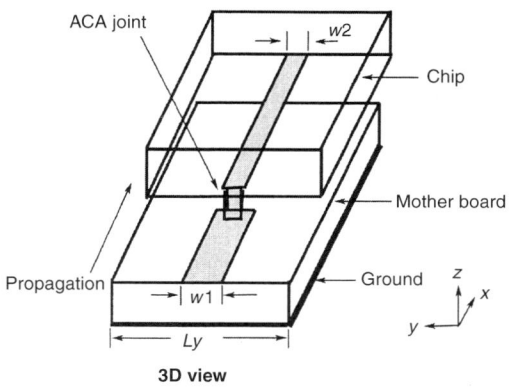

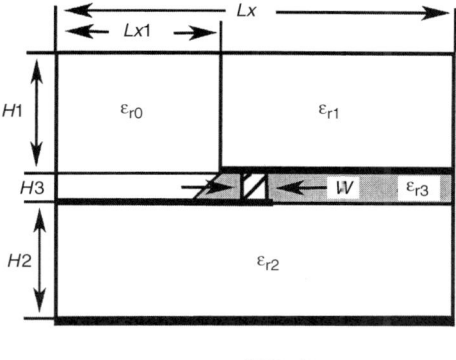

Fig. 22 Schematics of the ACA flip-chip joint for 3D FDTD modeling

Table 3 Parameters used for 3D FDTD modeling of ACA flip-chip discontinuities

	Material	Thickness, mm (in.)	ε_r	tan δ
Test chip	Silicon	0.64	11.8	0.01(a)
Test board	Teflon	0.764	2.94	0.0012
ACA resin	Epoxy	0.01	4.0	0.01
ACA filler	Nickel	0.01	...	...

(a) Silicon is assumed with high resistance to reduce the dielectric loss caused by the chip.

tions at the boundaries, the super-absorption Mur boundary condition was added at the five outer walls of the computational domain. Figure 23 shows an FDTD-simulated transmission coefficient |s21| of an ACA joint with 10 particles. The simulated result agrees quite well with the measured one from Ref 43.

Further parametric studies show that both the number and distribution of conductive particles have very little influence on the high-frequency performance of a single ACA joint, though the existence of conductive particles is necessary to decrease the signal reflection. One possible explanation is that the size of conductive particles (3 to 15 μm) is too small compared with the wavelength, even for the maximum frequency of the signal. These conductive particles can thereby be regarded as direct conductive interconnections. However, this statement is not valid for high power applications where the limitations in the current density in the conducting layer need to be considered.

Based on the FDTD simulations, an equivalent circuit model of the ACA flip-chip joint has been developed. It is shown in Fig. 24, where C denotes the capacitance between the pads at the chip and board, C1 the discontinuity capacitance at the chip, C2 the discontinuity capacitance at the Teflon board, L represents the inductance of the joint, and R, R1, and R2 are the loss in the joint, the chip, and the board, respectively. To extract the parameters of these lumped elements, their primary values were first calculated from the geometry of the ACA joint and then optimized until the s-parameters from the equivalent circuit correlate well with those from the de-embedded FDTD. Table 4 shows the typical values of these lumped elements, where the dielectric constant of the adhesive resin is chosen as 4. With this equivalent circuit model, the calculated s-parameters essentially remain unchanged when R changes from 10 to 100 mΩ, corresponding to the decreasing number of the conductive particles in a single ACA joint. When R increases further to 100 Ω, the reflection increases dramatically, especially in the low-frequency range. In this case, the calculated s-parameters are very similar to the de-embedded FDTD simulation results of the joint without particle.

Finite Element Stress Analysis of ICA Interconnection

Surface Mount Leadless ICA Joint. To gain an insight into the cracking mechanisms of the leadless ICA joint, a 2D plain strain model was constructed for linear elastic FEM analysis (Ref 18). It consisted of a 1206 chip resistor with tin terminals, an FR4 board with copper pads, and ICA joints. Table 5 summarizes all material properties used in the analysis. The standoff and fillet heights were set as 40 and 125 μm, respectively. For simplicity, the cross-sectional shape

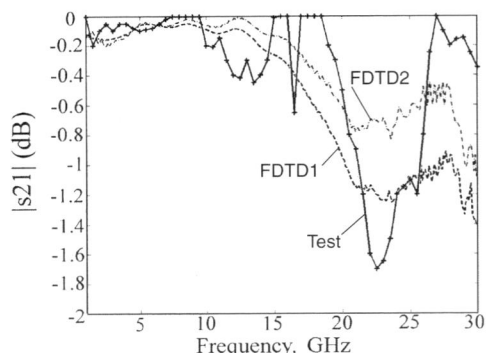

Fig. 23 FDTD simulated transmission coefficient |s21| of the ACA joint, as well as the measured one

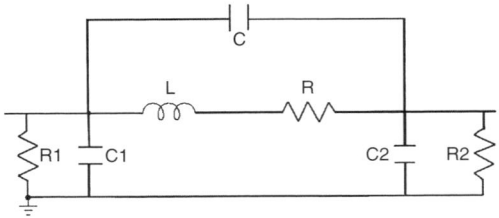

Fig. 24 Equivalent circuit model of the ACA flip-chip joint

Table 4 Typical values of the lumped elements in the equivalent circuit of ACA joint

$R1$, Ω	$C1$, pF	$R2$, Ω	$C2$, pF	R, mΩ	L, nH	C, pF
750	0.005	750	0.005	57	0.024	0.78

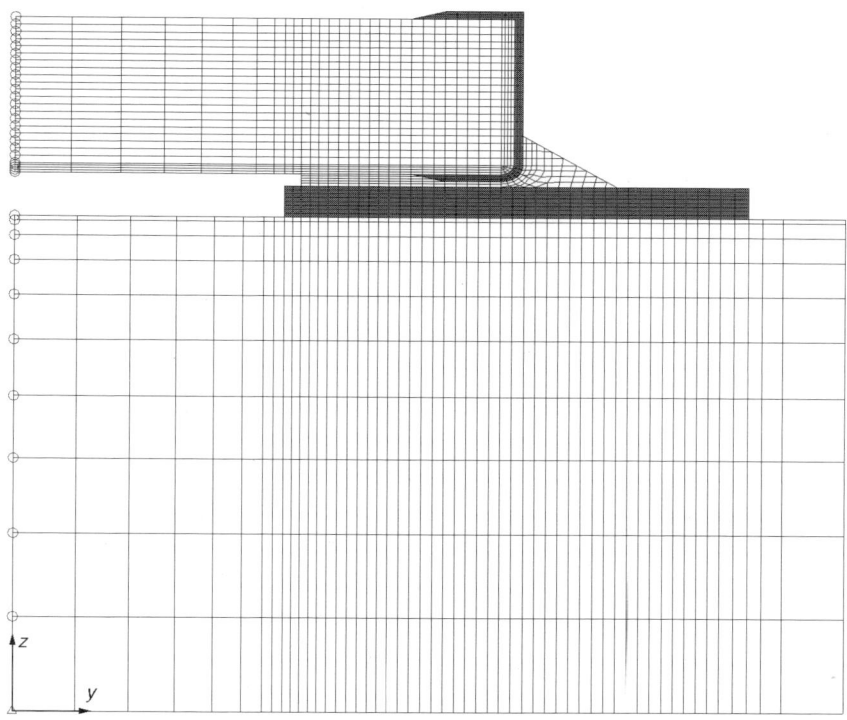

Fig. 25 2D meshed model of the surface mount ICA joint

of the fillet was assumed to be a triangle. Because of symmetry, only one-half of the package was modeled and appropriate boundary conditions were applied to the nodes on the symmetry line (Fig. 25).

Due to the choice of material properties, only a single temperature drop from 100 °C (212 °F) was considered. The stress-free temperature was set at 120 °C (248 °F), which is the curing temperature of the ICA paste. In the simulation, the temperature was gradually decreased to −55 °C (−67 °F), corresponding to the lowest temperature in the thermal cycling test. The stresses in the joint at this temperature were calculated.

The distribution of von Mises stress in the ICA joint is shown in Fig. 26(a). As shown, the maximum stress occurs around the knee of the joint, which agrees with the observation of bulk cracking. Thus, the von Mises stress can be used as an indicator to analyze the bulk cracking of this type of ICA joint.

With regard to interfacial cracking, the peel stress is considered to play a very important role. When the maximum peel stress exceeds the interfacial strength, cracking will occur. Figure 26(b) gives the variation of the peel stress, σ_{yy}, along the vertical adhesive/terminal interface. As shown, the maximum peel stress, 71 MPa (10.30 ksi), occurs at the top of the interface. Consequently, interfacial cracks would be expected to initiate at this highly stressed site. The distribution of peel stress, σ_{zz}, along the hori-

Table 5 Linear elastic material properties used in the FEM analysis of surface mount ICA joint

	Young's modulus, GPa	Poisson's ratio	CTE, ppm
FR4 board	16	0.28	16
Ceramic resistor	310	0.30	6
ICA	5.6	0.30	45
Tin terminal	40	0.33	23.8
Copper pad	117	0.33	17.6

zontal interface is shown in Fig. 26(c). The maximum value, 31 MPa (4.50 ksi), is much lower than that on the vertical interface. However, cracking could occur at this site, as long as the adhesive/ceramic interface is weakened by the commonly existing contaminants on ceramic surfaces.

With this linear elastic stress analysis, the correlation between crack initiation and stress distribution has been established. The interfacial cracking occurs at sites sustaining high peel stresses, while the bulk cracking results from the concentration of the equivalent stress. Thus, enhancing either the adhesion to the component or the bulk strength of the ICA joint will help to improve its reliability.

It should be noticed that varying the package configuration would change the magnitude and/or distribution of thermal stresses, resulting in different crack patterns and thus, different fatigue lifetimes. The effects of two important geometry parameters, standoff height and fillet height of the ICA joint, were thereby evaluated based on FEM calculations. It was found that the standoff height significantly influences the maximum von Mises stress at the knee of the joint. With the increase of standoff height, the maximum equivalent stress decreased dramatically. As a result, a complete interfacial failure mode (without bulk cracking) can be expected if a larger standoff height is adopted. The most prominent effect on the maximum peel stress at the vertical interface is from the variation of fillet height. The higher the fillet, the higher the maximum peel stress. This indicates that increasing the fillet height is not so effective to improve the lifetime of the ICA joint, since the vertical interface sustains higher peel stress and cracking occurs more easily with higher fillet. However, both geometry factors have a limited impact on the maximum peel stress at the horizontal interface.

Encapsulated ICA Flip-chip Interconnection. Stress analysis of encapsulated ICA flip-chip interconnection was also executed with 2D FEM calculations (Ref 45). Corresponding to experimental tests, four chip dimensions, 3 by 3 mm^2, 6 by 6 mm^2, 9 by 9 mm^2, and 15 by 15 mm^2, were modeled. Only one joint was considered in each model, and the distance to neutral point (DNP) is half the diagnostic length of the test chip. To consider its anisotropic properties, the PCB substrate was modeled as a sandwich structure consisting of five epoxy layers and four fiber layers. The mesh for the whole package with fine structure details is presented in Fig. 27. Since the real joints were peripherally located on the test chip, axisymmetric boundary condition was employed.

All the material properties used in the calculations are listed in Table 6. The PCB layers and chip were considered as elastic materials, while

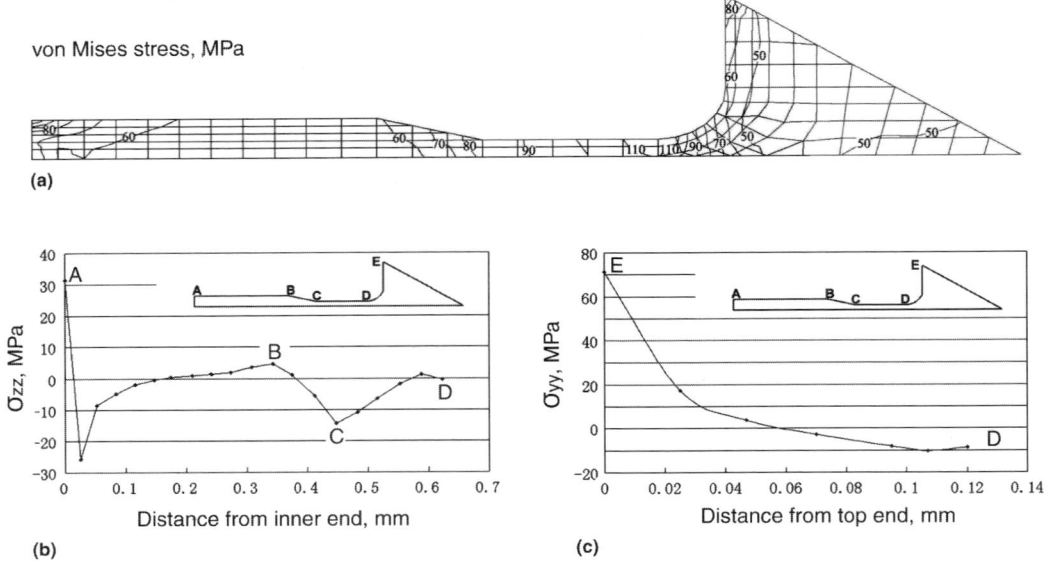

Fig. 26 Simulation of stresses. (a) von Mises stress in the ICA joint, and peel stresses along (b) vertical, and (c) horizontal adhesive/component interfaces

the bump and pad as elastic-plastic materials with a yield stress of 150 MPa (21.76 ksi). The ICA and underfill were treated as viscoelastic materials. The master curves needed for viscoelastic input in the simulation were converted from commercially available data, as shown in Ref 46. Meanwhile, to evaluate the influence of material properties on the calculated results, cases with temperature-dependent elastic underfill and ICA were also studied.

The typical distributions of von Mises stress S_{eqv} and shear stress S_{xy} in the ICA joint are shown in Fig. 28. These stress patterns were kept the same in all cases, and either chip dimension or material model only influenced the stress magnitude. The maximum stresses were found to occur in the middle part of the joint and ranged from 25 to 102 MPa (3.63 to 14.79 ksi) in different cases (Table 7). For a better comparison, the experimentally measured lifetimes are listed in Table 7 as well. It is evident that, with the same chip dimension, the stresses calculated with the viscoelastic model are approximately 30% less than those with the temperature-dependent elastic model. This difference results from ignoring stress relaxation in the elastic model, which consequentially overestimates the stresses of ICA joints. Therefore, in the case that the temperature-dependent elastic model is used for simplification (mainly to re-

Table 6 Material properties used in the FEM analysis of ICA flip-chip interconnection

	Young's modulus, GPa	Poisson's ratio	CTE, ppm/K
Silicon chip	131.0	0.3	2.7
Epoxy in FR4	3.45	0.37	69
Fiber in FR4	72.4	0.22	5.4
Chip bump	206	0.3	18
PCB pad	123.0	0.39	17.6
Underfill	0.4(a)	0.35	125(a)
	4.0(b)		35(b)
ICA	0.3(a)	0.3	110(a)
	3.0(b)		26(b)

(a) High temperature. (b) Low temperature

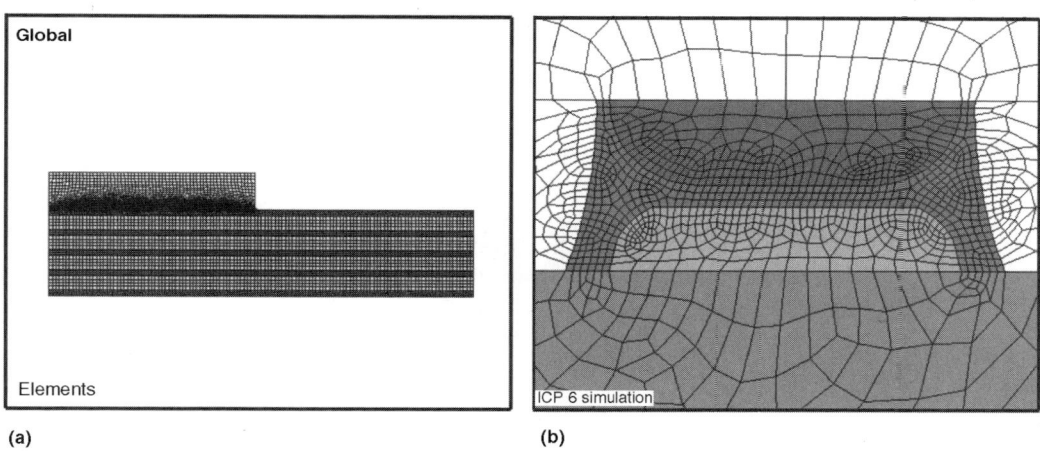

Fig. 27 2D model of an ICA flip-chip interconnection, showing the mesh for (a) global and (b) local structures

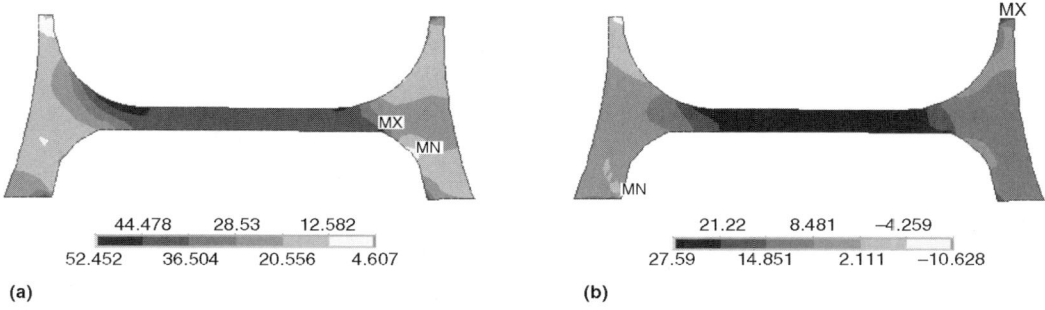

Fig. 28 Distributions of (a) von Mises stress and (b) shear stress in the ICA flip-chip joint (MPa)

Table 7 Maximum calculated stresses in the ICA flip-chip joints with various chip dimensions and material models

Chip size, mm²	Average lifetime, cycles	TempDepend model		ViscoElast model	
		S_{eqv}, MPa	S_{xy}, MPa	S_{eqv}, MPa	S_{xy}, MPa
3 × 3	13,513	69.47	32.92	45.37	25.14
6 × 6	4,716	84.94	40.64	52.45	27.59
9 × 9	2,272	93.63	46.78	59.82	31.03
15 × 15	787	101.39	51.97	67.05	34.67

duce the computation time), this approximation should be taken into consideration.

From Table 7, it can also be observed that the larger the chip, the higher the stresses and the shorter the joint lifetime. This close correlation suggests that the stress value could be used as an indicator for lifetime estimation of ICA joints. An equation similar to the Coffin-Manson formula can then be proposed:

$$N = c(S)^b \quad (Eq\ 9)$$

where N is the average joint lifetime and S is the maximum stress in the joint. The parameters c and b could be obtained by fitting the measured N with the calculated S, as illustrated in Fig. 29. By choosing equivalent stress or shear stress as S value and considering two different material models, four lifetime equations have been obtained:

$$N = 1.20 \times 10^{18} (S_{TD,eqv})^{-7.52} \quad (Eq\ 10)$$

$$N = 3.36 \times 10^{13} (S_{TD,shear})^{-6.15} \quad (Eq\ 11)$$

$$N = 9.50 \times 10^{15} (S_{Visco,eqv})^{-7.14} \quad (Eq\ 12)$$

$$N = 1.48 \times 10^{16} (S_{Visco,shear})^{-8.62} \quad (Eq\ 13)$$

Process Simulations of ACA Interconnection

Probabilities of Open and Bridging. If the ACA contains insufficient particles, there is of course a certain probability that no particle exists in the joint and an open will result. On the other hand, bridging is possible due to too many particles in a short spacing, causing a short circuit between neighboring pads. Accurate probability estimates of open and bridging are needed to define the limiting pitch of ACA interconnects. An analytical method to estimate the open probability was proposed (Ref 47). Assume that the number of particles on a pad obeys Poisson's distribution:

$$P(n) = \frac{e^{-\mu}\mu^n}{n!} \quad (Eq\ 14)$$

where $P(n)$ is the probability of finding n particles on a pad and μ is the average number of particles on a pad. If the volume fraction of particles f and the particle radius r are known, μ is given by:

$$\mu = \frac{3Af}{2\pi r^2} \quad (Eq\ 15)$$

where A is the pad area. Thus, the probability for an open ACA joint is:

$$P(0) = e^{-\mu} = e^{-(3Af/2\pi r^2)} \quad (Eq\ 16)$$

For a typical ACA with a volume fraction of particles ranging from 3 to 15 vol%, the open circuit probability on a 100 μm² pad varies from 10^{-13} to 10^{-3}, which is extremely small. However, in reality, there is always a crowding effect that must be taken into account. In this case, the particle distribution can be described using a binominal distribution model:

$$P(n) = C_n^N (1-s)^{N-n} s^n \quad (Eq\ 17)$$

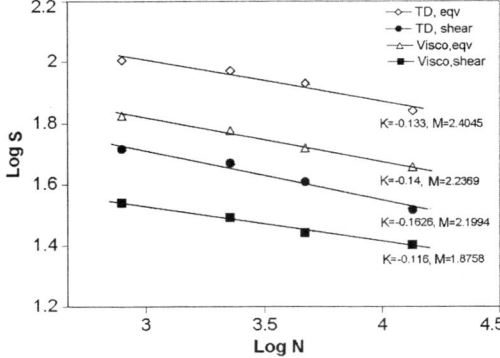

Fig. 29 Logarithmic relationships between the lifetime and calculated stresses. The curves were fitted with linear equations as $LogS = KLogN + M$.

where N is the maximum number of particles that can be contained in the pad area A. C_n^N is the binominal coefficient and s is equal to f/f_m where f_m is the volume fraction corresponding to maximum packing. In the limit that $f \ll 1$, Eq 16 and 17 give identical results for $P(0)$.

Reference 48 proposed a simplified box model for a rough estimate of bridging. As shown in Fig. 30, the volume between pads can be divided into cubic boxes with sides of the same length as the particle diameter. If k boxes are filled out of a total of N, the volume fraction of particles is:

$$f = \frac{k \frac{4}{3} \pi r^3}{N(2r)^3} \quad \text{(Eq 18)}$$

where r is the particle radius. Thus, the probability for a single box being occupied is given by:

$$\frac{k}{N} = \frac{6f}{\pi} \quad \text{(Eq 19)}$$

Determined by the number of boxes that can be fitted onto the side of a single pad and by $(6f/\pi)^q$ where q is the lowest number of particles needed to bridge the pad spacing, the bridging probability is given by:

$$p = 1 - \left(1 - \left(\frac{6f}{\pi}\right)^{d/4r^2}\right)^{hl/4r^2} \quad \text{(Eq 20)}$$

where h and d are the pad height and length, respectively, and l is the spacing between the pads.

This box model only gives an upper limit. Figure 31 shows the bridging probabilities derived from different models. It is clear that the lowest combined probability for bridging and skipping occurs in the volume fraction between 7 and 15%, depending on which model is used. This volume fraction range is also generally used for commercial ACA materials today.

ACA Flow During Bonding. As modeled (Ref 48), there are two types of adhesive flow during the ACA bonding (Fig. 32). Type I flow occurs around individual pads and bumps at the beginning of bonding, filling voids nearby. After voids are completely filled, type II flow becomes dominant, expelling the adhesive from under the chip to edges.

By solving the Navier-Stokes equations of Newtonian fluid, one can obtain the following equation, which describes the pressure distribution under the chip in the cylindrical coordinate system:

$$P(r) = \frac{2F}{\pi R^2}\left(1 - \frac{r^2}{R^2}\right) \quad \text{(Eq 21)}$$

where R is half of the side length of the chip and F is the bonding force. In reality, the ACA resin probably behaves more like power law fluids:

$$\tau_{xy} = \eta_0 \left(\frac{d\gamma}{dt}\right)^n \quad \text{(Eq 22)}$$

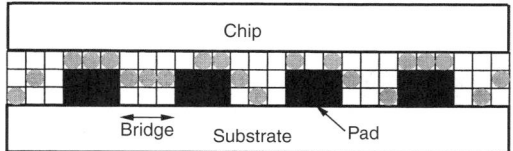

Fig. 30 Schematics of bridging in the ACA interconnection. Courtesy of S.H. Mannan, Loughborough University

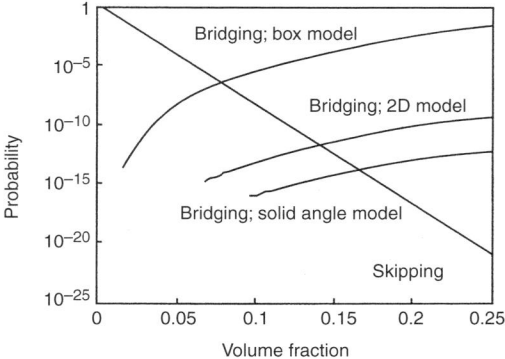

Fig. 31 Probability of particles bridging gap as a function of filler volume fraction. Courtesy of S.H. Mannan, Loughborough University

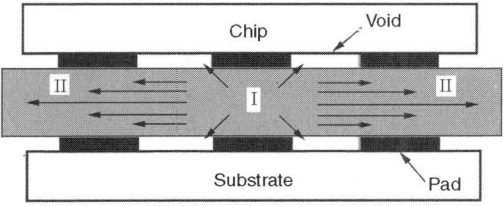

Fig. 32 Schematic drawing of the ACA flow during the bonding. Courtesy of S.H. Mannan, Loughborough University

where η_0 is termed the consistency and n the power law index. For a Newtonian fluid, n equals 1 and η_0 becomes the viscosity of the fluid. As the chip is pressed down, the ACA is squeezed out between the chip and substrate. With power law fluids, the process time t_p for reducing the gap height from h_0 to h_1 is given by:

$$t_p = \frac{2n+1}{n+1}\left(\frac{2\pi\eta_0 R^{n+3}}{F(n+3)h_0^{n+1}}\right)^{1/n}\left(\left(\frac{h_0}{h_1}\right)^{(n+1)/n} - 1\right) \quad \text{(Eq 23)}$$

This process time is important for determining the suitable heating ramp for bonding. Excess bonding temperature may cause the adhesive to solidify before particles are deformed completely, resulting in less reliable joints.

Electrical Conduction Development and Residual Stresses. ACAs contain a small volume fraction of particles; there is no conduction in any direction before bonding. The electrical resistance decreases as pressure increases due to enlarged contact areas. Several research groups have reported the deformation effect on the electrical conduction development during the ACA assembly. The first publication is from Ref 49, and the contact resistivity ρ of an ACA joint was estimated as:

$$\rho = \frac{A\rho_B\left(\sqrt{\frac{6\pi n\kappa}{\sigma A}} - \frac{1}{R_B}\right)}{4\pi nR_B} \quad \text{(Eq 24)}$$

where ρ_B is the resistivity of the conductive particle, n is the number of contacts within the contact area A, κ is the shear yield stress of a conductive particle with a radius of R_B, and σ is the pressure applied to the joint.

With a combination of analytical method and FEM, Ref 50 derived the relationship between the resistance and bonding pressure both for the rigid and deformable particle systems, as shown in Fig. 33. They also simulated the contact be-

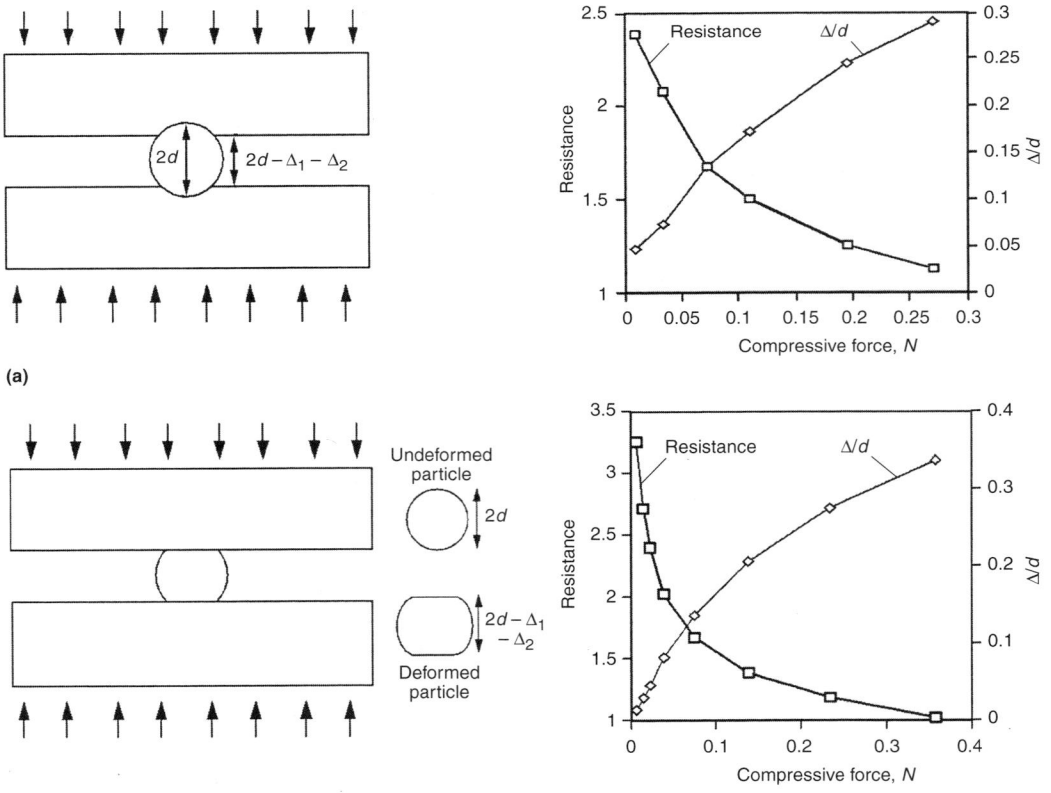

Fig. 33 Force-resistance-deformation relationships for (a) rigid particle system and (b) deformable particle system. Courtesy of C.P. Yeh, Motorola Inc.

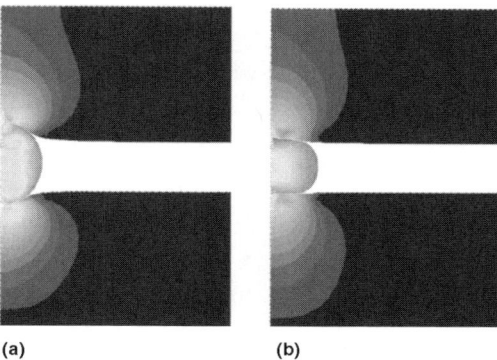
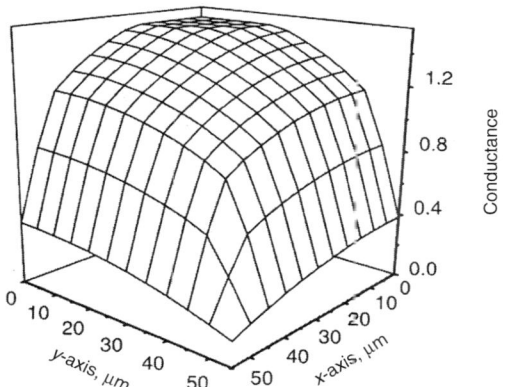

Fig. 34 Deformation distributions of (a) rigid particle system and (b) deformable particle system. Courtesy of C.P. Yeh, Motorola Inc.

Fig. 35 Electric conductance of the particle as a function of its location away from the center of the ACA joint

tween the particle and electrode with FEM. As shown in Fig. 34, significant compressive stress is found to build up in the interface between the two contacts. This stress is believed to generate peel stress in the adhesive, which is probably the reason for catastrophic failure.

Reference 51 considered multiparticle cases and found that the particle location in an ACA joint can affect its electric conductance. As shown in Fig. 35, a particle in the center of the joint contributes much more to the electrical performance than a particle close to the edge of the joint. This helps to explain why the measured resistance scatters greatly from one joint to another. Increasing the number of particles on the contact pad can improve the uniformity of the electric conduction. It, however, also increases the constriction resistances due to fellow particles. So the total conductance does not increase in the additive manner.

Conclusions

Conductive adhesive interconnection for low-cost flip-chip and surface mount applications is an emerging technology. It has been shown that conductive adhesives play a significant role in electronics packaging and are expected to do so for years to come. It is clear that the wider use of adhesives has been recognized in recent decades. An increasing amount of performance data has been generated by research efforts worldwide. More products based on adhesive joining technology are appearing, especially in the field of flip-chip assembly. However, further reliability data and real service life prediction tools are necessary for this environmentally friendly and low-cost technology to be fully accepted.

ACKNOWLEDGMENTS

The authors wish to thank their colleagues: Zonghe Lai, Hans Grönqvist, Ying Fu, Shiming Li, Liu Chen, Gang Zou, and Helge Kristiansen and Liqiang Cao who have been engaged in conductive adhesive research at Chalmers and who have generated tremendous research results during the last five years. Also, we wish to acknowledge the financial support from the SMIT Center member company: Hitachi Chemicals, Japan.

REFERENCES

1. K. Gilleo, Assembly with Conductive Adhesives, *Solder. Surf. Mt. Technol.*, No. 19, 1995, p 12–17
2. D. Klosterman, L. Li, and J.E. Morris, Materials Characterization, Conduction Development, and Curing Effects on Reliability of Isotropically Conductive Adhesives, *IEEE Transactions on Components Packaging and Manufacturing Technology Part A*, Vol 21, 1998, p 23–31
3. A.J. Lovinger, Development of Electrical-Conduction in Silver-Filled Epoxy Adhesives, *J. Adhes.*, Vol 10, 1979, p 1–15
4. E. Sancaktar and Y. Wie, The Effect of Pressure on the Initial Establishment of Conductive Paths in Electronically Conductive

Adhesives, *J. Adhes. Sci. Technol.*, Vol 10, 1996, p 1221–1235

5. D. Lu, Q.K. Tong, and C.P. Wong, Conductivity Mechanisms of Isotropic Conductive Adhesives (ICAs), *IEEE Transactions on Electronics Packaging Manufacturing*, Vol 22, 1999, p 223–227

6. P.G. Harris, Conductive Adhesives: A Critical Review of Progress to Date, Solder. *Surf. Mt. Technol.*, No. 20, 1995, p 19–21

7. D.D. Chang et al., An Overview and Evaluation of Anisotropically Conductive Adhesive Films for Fine-Pitch Electronic Assembly, *IEEE Transactions on Components, Hybrids and Manufacturing Technology*, Vol 16, 1993, p 828–835

8. P. Savolainen and J. Kivilahti, A Solder Alloy Filled Z-Axis Conductive Epoxy Adhesive, *J. Adhes.*, Vol 49, 1995, p 187–196

9. H. Kristiansen and J. Liu, Overview of Conductive Adhesive Interconnection Technologies for LCDs, *IEEE Transactions on Components Packaging and Manufacturing Technology, Part A*, Vol 21, 1998, p 208–214

10. J. Liu et al., Surface Characteristics, Reliability, and Failure Mechanisms of Tin/Lead, Copper, and Gold Metallizations, *IEEE Transactions on Components, Packaging, and Manufacturing Technology, Part A*, Vol 20, 1997, p 21–30

11. H. Botter, R.B. Van Der Plas, and A. Arunjunai, Factors that Influence the Electrical Contact Resistance of Isotropic Conductive Adhesive Joints During Climate Chamber Testing, *International Journal of Microelectronic Packaging Materials and Technologies*, Vol 1, 1998, p 177–185

12. J.C. Jagt, Reliability of Electrically Conductive Adhesive Joints for Surface Mount Applications: A Summary of the State of the Art, *IEEE Transactions on Components Packaging and Manufacturing Technology, Part A*, Vol 21, 1998, p 215–225

13. D.Q. Lu and C.P. Wong, Development of Conductive Adhesives for Solder Replacement, *IEEE Transactions on Components and Packaging Technologies*, Vol 23, 2000, p 620–626

14. C.G.L. Khoo et al., Influence of Curing on the Electrical and Mechanical Reliability of Conductive Adhesive Joints, *International Electronics Packaging Society Conference*, Sept/Oct 1996, Proceedings, Austin, TX, 1996, p 483–501

15. J. Liu and B. Weman, Modification of Process and Design Rules to Achieve High Reliable Conductive Adhesive Joints for Surface Mount Technology, *Second International Symposium on Electronics Packaging Technology (ISEPT)*, Dec 1996, Proceedings, Shanghai, China, Fudan University, 1996, p 313–319

16. S.A. Vona et al., Surface Mount Conductive Adhesives with Superior Impact Resistance, *Fourth International Symposium on Advanced Packaging Materials*, March 1998, Proceedings, Braselton, GA, Georgia Tech, 1998, p 261–267

17. S.Y. Xu and D.A. Dillard, Determining the Impact Resistance of Electrically Conductive Adhesives using a Falling Wedge Test, *IEEE Transactions on Components and Packaging Technologies*, Vol 26, 2003, p 554–562

18. Z.M. Mo et al., Thermal Fatigue Cracking of Surface Mount Conductive Adhesive Joints, *Solder. Surf. Mt. Technol.*, Vol 16 (No. 1), 2004, p 48–52

19. M.G. Perichaud et al., Reliability Evaluation of Adhesive Bonded SMT Components in Industrial Applications, *Microelectronics Reliability*, Vol 40, 2000, p 1227–1234

20. L. Li et al., Reliability and Failure Mechanism of Isotropically Conductive Adhesives Joints, *45th Electronic Components and Technology Conference (ECTC)*, May 1995, Proceedings, Las Vegas, NV, IEEE CPMT Society, 1995, p 114–120

21. S.Y. Xu, D.A. Dillard, and J.G. Dillard, Environmental Aging Effects on the Durability of Electrically Conductive Adhesive Joints, *Int. J. Adhes. and Adhes.*, Vol 23, 2003, p 235–250

22. D. Lu, Q.K. Tong, and C.P. Wong, Mechanisms Underlying the Unstable Contact Resistance of Conductive Adhesives, *IEEE Transactions on Electronics Packaging Manufacturing*, Vol 22, 1999, p 228–232

23. M. Yamashita and K. Suganuma, Degradation Mechanism of Ag-Epoxy Conductive Adhesive/Sn-Pb Plating Interface by Heat Exposure, *J. Electron. Mater.*, Vol 31, 2002, p 551–556

24. R.L. Keusseyan, J.L. Dilday, and B.S. Speck, Electric Contact Phenomena in Conductive Adhesive Interconnections, *International Journal of Microcircuits and Electronic Packaging*, Vol 17, 1994, p 236–242

25. R.S. Rörgren and J. Liu, Reliability Assessment of Isotropically Conductive Adhesive Joints in Surface-Mount Applications, *IEEE Transactions on Components Packaging and Manufacturing Technology, Part B,* Vol 18, 1995, p 305–312
26. Z.M. Mo et al., Electrical Characterization of Isotropic Conductive Adhesive under Mechanical Loading, *J. Electron. Mater.,* Vol 31, 2002, p 916–920
27. J.H. Constable et al., Continuous Electrical Resistance Monitoring, Pull Strength, and Fatigue Life of Isotropically Conductive Adhesive Joints, *IEEE Transactions on Components and Packaging Technologies,* Vol 22, 1999, p 191–199
28. H.K. Kim and F.G. Shi, Electrical Reliability of Electrically Conductive Adhesive Joints: Dependence on Curing Condition and Current Density, *Microelectronics Journal,* Vol 32, 2001, p 315–321
29. J. Liu and Z. Lai, Reliability of Anisotropically Conductive Adhesive Joints on a Flip-Chip/FR4 Substrate, *J. Electron. Packaging,* Vol 124, 2002, p 240–245
30. Z. Lai and J. Liu, Anisotropically Conductive Adhesive Flip-Chip Bonding on Rigid and Flexible Printed Circuit Substrates, *IEEE Transactions on Components Packaging and Manufacturing Technology, Part B,* Vol 19, 1996, p 644–660
31. M.J. Yim and K.W. Paik, Design and Understanding of Anisotropic Conductive Films (ACFs) for LCD Packaging, *IEEE Transactions on Components Packaging and Manufacturing Technology, Part A,* Vol 21, 1998, p 226–234
32. Y.C. Chan and D.Y. Luk, Effects of Bonding Parameters on the Reliability Performance of Anisotropic Conductive Adhesive Interconnects For Flip-Chip-On-Flex Packages Assembly I. Different Bonding Temperature, *Microelectronics Reliability,* Vol 42, 2002, p 1185–1194
33. S.C. Tan et al., Thermal Stability Performance of Anisotropic Conductive Film at Different Bonding Temperatures, *Microelectronics Reliability,* Vol 44, 2004, p 495–503
34. C.Y. Yin et al., The Effect of Reflow Process on the Contact Resistance and Reliability of Anisotropic Conductive Film Interconnection for Flip Chip on Flex Applications, *Microelectronics Reliability,* Vol 43, 2003, p 625–633
35. A. Seppälä and E. Ristolainen, Study of Adhesive Flip Chip Bonding Process and Failure Mechanisms of ACA Joints, *Microelectronics Reliability,* Vol 44, 2004, p 639–648
36. J. Liu et al., A Reliable and Environmentally Friendly Packaging Technology—Flip-Chip Joining using Anisotropically Conductive Adhesive, *IEEE Transactions on Components and Packaging Technologies,* Vol 22, 1999, p 186–190
37. Z. Lai et al., Effect of Bump Height on the Reliability of ACA Flip Chip Joining with FR4 Rigid and Polyimide Flexible Substrate, *J. Electron. Manufacturing,* Vol 8, 1998, p 217–224
38. C.G.L. Khoo and J. Liu, Moisture Sorption in Some Popular Conductive Adhesives, *Circuit World,* Vol 22 (No. 4), 1996, p 9–15
39. J. Liu, On the Failure Mechanism of Anisotropically Conductive Adhesive Joints on Copper Metallisation, *Int. J. Adhes. Adhes.,* Vol 16, 1996, p 285–287
40. Y. Fu, J. Liu and M. Willander, Conduction Modelling of a Conductive Adhesive with Bimodal Distribution of Conducting Element, *Int. J. Adhes. Adhes.,* Vol 19, 1999, p 281–286
41. L. Ye et al., Effect of Ag Particle Size on Electrical Conductivity of Isotropically Conductive Adhesives, *IEEE Transactions on Electronics Packaging Manufacturing,* Vol 22, 1999, p 299–302
42. Y. Fu and J. Liu, Electron Conduction through Nano Particles in Electrically Conductive Adhesives, *Micromaterials and Nanomaterials,* No. 4, 2004, p 104–109
43. R. Sihlbom et al., Conductive Adhesives for High-Frequency Applications, *IEEE Transactions on Components Packaging and Manufacturing Technology, Part A,* Vol 21, 1998, p 469–477
44. G. Zou, H. Grönqvist, and J. Liu, Theoretical Analysis of RF Performance of Anisotropic Conductive Adhesive Flip-Chip Joints, *IEEE Transactions on Components and Packaging Technologies,* Vol 27, 2004, p 546–550
45. L. Chen et al., Reliability Investigation for Encapsulated Isotropic Conductive Adhesives Flip Chip Interconnection, *Sixth IEEE CPMT Conference on High Density Microsystem Design and Packaging and Component Failure Analysis (HDP'04),* June/July

2004, Shanghai, China, Proceedings, IEEE CPMT Society, 2004, p 134–140
46. L. Chen et al., The Effects of Underfill and its Material Models on Thermomechanical Behaviors of a Flip Chip Package, *IEEE Transactions on Advanced Packaging,* Vol 24, 2001, p 17–24
47. S.H. Mannan, D.J. Williams, and D.C. Whalley, Some Optimum Processing Properties for Anisotropic Conductive Adhesives for Flip Chip Interconnection, *J. Mater. Sci.: Mater. Electron.,* Vol 8, 1997, p 223–231
48. S.H. Mannan et al., Models to Determine Guidelines for the Anisotropic Conducting Adhesives Joining Process, *Conductive Adhesives for Electronics Packaging,* J. Liu, Ed., Electrochemical Publications Ltd., 1999, p 78–98
49. D.J. Williams et al., Anisotropic Conducting Adhesives for Electronic Interconnection, *Solder. Surf. Mt. Technol.,* No. 14, 1993, p 4–8
50. K.X. Hu, C.P. Yeh, and K.W. Wyatt, Electro-Thermo-Mechanical Responses of Conductive Adhesive Materials, *IEEE Transactions on Components Packaging and Manufacturing Technology, Part A,* Vol 20, 1997, p 470–477
51. Y. Fu, M. Willander, and J. Liu, Statistics of Electric Conductance through Anisotropically Conductive Adhesive, *IEEE Transactions on Components and Packaging Technologies,* Vol 24, 2001, p 250–255

CHAPTER 12

Lead-Free Solder Interconnect Reliability Outlook

Dongkai Shangguan, Flextronics International

As the preceding chapters have illustrated, lead-free solder interconnect reliability is an important, yet complex, subject. A significant amount of work has been carried out over the past decade, yet more challenges still remain. Some of the most pressing issues are outlined in the following.

Solder Alloy Characteristics and Interfacial Interactions

As a wide range of compositions (3.0%Ag to 4.0%Ag) are being used for the Sn-Ag-Cu alloy across the industry worldwide, it is important to determine conclusively whether or not there is any significant difference in the reliability performance of these compositions. This is important in the effort to "standardize" the alloy composition, and if a wide range of compositions continue to be used, for determining whether or not these alloys can be used interchangeably from the perspective of the interconnect reliability, including the associated reliability models and accelerated testing methods (Ref 1).

Work is also needed to further characterize the creep and fatigue behavior of the lead-free solder alloy. The interfacial interactions (including intermetallics formation and growth, and the formation of Kirkendall voids on various PWB surface finishes) and their impact on reliability also warrant further investigation (Ref 2–3).

Further study is also needed on the ductile-to-brittle transition of the Sn-Ag-Cu alloy. Recent work (Ref 4) indicates a sharp transition in the range of -78 °C to -45 °C depending on the alloy composition, with increases in the Ag content shifting the transition towards higher temperatures. This effect is expected to be even more pronounced under complex mechanical loading conditions. Such a phenomenon can have significant implications for certain applications (such as aerospace, automotive, and marine electronics). Tin pest is another low temperature phenomenon currently under study, and its practical reliability implications are not yet clear.

Tin Whisker Growth

Many years of research points to compressive stresses as the underlying driving force for the formation and growth of tin whiskers. The compressive stress may arise from metallurgical interactions, thermomechanical factors, and/or mechanical processes. A recent nano-indentation study (Ref 5–6) suggests that the stress field, including the level of the compressive stress, as well as the stress gradient, needs to be considered.

As the mechanism for tin whisker growth becomes clearer, predictive modeling for tin whisker nucleation and growth, which can accurately account for the thermodynamic driving force and the nucleation and growth kinetics, will become a very useful tool for life prediction and optimization. Accelerated testing methods (with well established acceleration factors) and acceptance criteria for different applications

PWB Reliability

More work is needed to determine the critical materials parameters for the PWB (printed wiring board) to meet the reliability requirements after multiple exposures to the higher lead-free soldering temperatures (260 °C or 500 °F). Board-level qualification tests also need to be developed. The reliability requirements need to include laminate integrity (structural, such as laminate cracking, as well as surface defects, such as blistering) after soldering, and Cu trace reliability and electrochemical reliability (especially conductive anodic filament, or CAF) during product use. The work needs to be carried out for different PWB configurations (PWB thickness, via geometry, Cu thickness and distribution, etc.) (Ref 7–8). PWB reliability under mechanical shock (due to drop of a handheld product, for example) also warrants further investigation.

Solder Constitutive Equation and Thermal Fatigue Reliability Prediction

Data are still emerging from different studies on the parameters for the constitutive equation for lead-free solder alloys (for example, Refs 9–10). The effort is complicated by the high strain-rate sensitivity (strain hardening), and the temperature sensitivity of the Sn-Ag-Cu alloy (Ref 11). Complicating the situation further is the different stress dependency of the creep rates for Sn-Ag-Cu and Sn-Pb eutectic alloys (Ref 11), leading to different comparisons between the two alloys at low and high stress levels (Ref 12–13). Recent studies (Refs 11 and 14) also suggest that both primary creep and tertiary creep are important for the Sn-Ag-Cu alloy. Such considerations are particularly important for high reliability applications (such as aerospace, automotive, military, etc) and long life products (such as telecommunication infrastructure equipment).

For fatigue life prediction, materials parameters need to be established to correlate the number of cycles for thermal fatigue life to the amount of damage (such as the creep strain energy density) per cycle for the solder joint (Ref 15–21). It is important that the experimental work in this area is based on relevant mechanical and environmental loading conditions and realistic solder joint configurations (geometry and size, interfacial factors, etc.) and is not unduly skewed by defects in the test samples.

Once the constitutive relationship and fatigue life models for lead-free solders have been established, the "design for reliability" methodologies through multi-scale finite element analysis (FEA) can be applied to lead-free solder interconnects.

Dynamic Mechanical Loading Conditions

For handheld products in particular, dynamic mechanical loading conditions (such as when the product is dropped to the ground) are often the predominant cause for the brittle failures of the solder interconnects, due to the excessive strains, at high strain rates, caused by the mechanical shock. Due to the higher modulus of the Sn-Ag-Cu alloy, the failure location may move from the solder joint to the PWB. Neglecting reliability under dynamic mechanical loading in product development has often led to costly remedies (such as the use of underfills applied to PWBs not designed for underfill application) in the face of dramatic failures either in reliability testing or in the field. Understanding of the deformation behavior and failure, and reliability prediction under dynamic mechanical loading conditions, is an area of great importance for handheld products (Ref 22). Predictive modeling and "design for reliability" tools and methodologies, taking into consideration of the package structure, component location on the PWB, solder interconnect characteristics, and the product mechanical structure, are needed for optimized design for handheld products.

For large boards, excessive board flexure, during PWB manufacturing, testing, assembly, and use, often can cause solder interconnect failures. Establishment of the strain rate-strain limit for lead-free solder interconnection on different PWB surface finishes can greatly help safeguard the reliability of the products (Ref 23).

Reliability of lead-free solder PWB assemblies under vibration is another area for further study (including failure characterization and reliability modeling).

Accelerated Testing Profile and Acceleration Factor

Appropriate accelerated testing profiles need to be developed for lead-free solder interconnects. The accelerated tests need to be able to generate failure modes and mechanisms that can be realistically expected in actual applications; appropriate temperature extremes in the test profile are of particular relevance in this regard. Within this boundary, the accelerated test profile should be designed to generate the maximum amount of damage to the solder interconnect per unit test time, in order to maximize test efficiency; the dwell time is the primary parameter of interest in this regard, considering the creep rate of the solder alloy at the relevant temperatures and loading conditions. Preliminary studies (Ref 18) have suggested that the 10 minutes dwell time specified in IPC-9701 (Ref 24) yields the maximum testing efficiency for the Sn-Ag-Cu solder. Much work is needed to establish the acceleration factor to quantitatively correlate the accelerated tests to the field reliability for different loading conditions.

Complex Loading Conditions and Total Reliability Optimization

In actual applications, solder interconnects are often under complex loading conditions (Ref 25). For example, ball grid array (BGA) solder joints may be simultaneously under cyclic shear loading and static tensile (or compressive) loading often with vibration. The deformation behavior and reliability prediction under complex loading conditions warrant further examination. Time and path-dependent creep models are needed for the solder joints under different and often complex loading conditions (Ref 2).

The ultimate goal for reliability studies is to optimize the total reliability of the product. This involves reliability of the various elements of the system (electronic/electrical components, mechanical components, PWB, and solder joints), through different failure modes (mechanical, electrical, electrochemical, etc.), under the actual use conditions (which often involves complex loading conditions with multiple loadings simultaneously imposed, such as cyclic temperatures, humidity, atmospheric chemical exposure, electrical field, vibration, mechanical shock, etc.) (Ref 26). The interactions between the different loading conditions are not yet well characterized, and the non-linear nature of the cumulative damage to the solder interconnects by the various loading conditions makes modeling and analysis extremely challenging. Total reliability optimization in product co-design often demands trade-offs to be made. Tools for the total reliability optimization of the product (on the system level), tailored to the expected life of the product, are urgently needed.

Reliability Degradation Assessment for Component Reutilization and Repurposing

Due to the rapid introduction of new products and models, the physical reliability life of a product (particularly in consumer-oriented markets) is often different from the market life of the product. The reutilization and repurposing of components and subassemblies through multiple product life cycles is a key enabler to proactively, intelligently, and economically manage the total environmental life cycle of electrical and electronic products (Ref 27). Such an approach will require accurate reliability degradation assessment of the components and subassemblies at the end of each product life cycle, and reliability life prediction for the next product life cycle. It is believed that such capabilities towards a comprehensive and integrated solution to the total environmental life cycle of electronics products will provide competitive advantages for players in the industry worldwide.

ACKNOWLEDGEMENT

Helpful discussions with Keith Newman, Dr. Jean-Paul Clech, and Dr. Paul Vianco, are gratefully acknowledged.

REFERENCES

1. D. Shangguan, Supply China Impact of Lead-Free Soldering, *Global SMT & Packaging,* Jan 2005, p 8
2. J. Liang, N. Dariavach, and D. Shangguan, Metallurgy, Processes and Reliability of Lead-Free Solder Interconnects, to appear in *Micro- and Opto-Electronic Materials and Structures: Physics, Mechanics, Design, Reliability, Packaging,* E. Suhir, C.P.

Wong, and Y.C. Lee, Ed., Kluwer Academic Publishers, 2005

3. D. Shangguan, Global Trend in Packaging and Assembly Technology and Environmental Compliance, *Proceedings of NEPCON Shanghai,*, Shanghai, China, April 2005

4. P. Ratchev, T. Locufier, B. Vandevelde, B. Verlinden, S. Teliszewski, D. Werkhoven, and B. Allaert, A Study of Brittle to Ductile Fracture Transition Temperatures in Bulk Pb-Free Solders, *Proceedings of EMPC 2005,* Brugge, Belgium, June 12–15, 2005

5. J. Liang, C. Li, Z. Xu, and D. Shangguan, Nano-Indentation Study on Whisker Formation on Tin-Plated Component Leads, *Proceedings of the IPC/JEDEC Eighth International Conference on Lead-Free Electronic Components and Assemblies,* San Jose, CA, April 2005

6. J. Liang, X. Li, Z. Xu, and D. Shangguan, Indentation Induced Tin Whisker Formation on Tin Plated Component Leads, *Proceedings of ASME InterPack'05,* July 17–22, 2005, San Francisco, CA

7. D. Shangguan, Supply Chain Impact of Lead-Free Soldering, *Proceedings of the Tenth Annual Pan Pacific Microelectronics Symposium,* Kauai, Hawaii, Jan 2005, p 257–262

8. D. Shangguan, Supply Chain Readiness for Lead-Free Soldering and Environmental Compliance, *Proceedings of IPC/JEDEC Eighth International Conference on Lead Free Electronic Components and Assemblies,* San Jose, California, April 2005

9. P. Sharma, S. Ganti, A. Dasgupta, and J. Loman, Prediction of Rate-Independent Constitutive Behavior of Pb-Free Solders Based on First Principles, *IEEE Transactions on Components and Packaging Technologies,* Vol. 26, No. 3, September 2003, pp. 659–666

10. Q. Zhang, A. Dasgupta, and P. Haswell, Viscoplastic Constitutive Properties and Energy-Partitioning Model of Lead-Free Sn3.9Ag0.6Cu Solder Alloy, *Proceedings of IEEE 53rd Electronic Components and Technology Conference,* New Orleans, LA, May 27–30, 2003

11. J. Liang, N. Dariavach, and D. Shangguan, Deformation Behavior of Solder Alloys under Variable Strain Rate Shearing and Creep Conditions, *Proceedings of Tenth International Symposium and Exhibition on Advanced Packaging Materials,* March 16–18, 2005, Irvine, CA

12. D. Xie, M. Arra, H. Phan, D. Shangguan, D. Geiger, and S. Yi, Life Prediction of Leadfree Solder Joints for Handheld Products, *Proceedings of the Telecomm Hardware Solutions Conference & Exhibition,* SMTA/IMAPS, May 2002, Legacy Park, TX, pp. 83–88

13. J.P. Clech, Lead-Free And Mixed Assembly Solder Joint Reliability Trends, *Proceedings of IPC/SMEMA APEX 2004 Conference,* Anaheim, CA, Feb. 21–26, 2004.

14. J.P. Clech, An Extension of the Omega Method to Primary and Tertiary Creep of Lead-Free Solders, *Proceedings of ECTC '05,* Orlando, FL, May 31–June 2, 2005

15. J. Lau, R. Lee, F. Song, D. Shangguan, D.C. Lau, and W. Dauksher, Thermal-Fatigue Life Prediction Equation for Plastic Ball Grid Array (PBGA) SnAgCu Lead-Free Solder Joints. *Proceedings of ASME InterPack '05,* July 17–22, 2005, San Francisco, CA

16. J. Lau, W. Dauksher, and P. Vianco, Acceleration Models, Constitutive Equations, and Reliability of Lead-Free Solders and Joints, *IEEE Electronic Components and Technology Conference Proceedings,* New Orleans, Louisiana, June 2003, pp. 229–236

17. J.H. Lau, S.W.R. Lee, W. Dauksher, D. Shangguan, F. Song, and D.C.Y. Lau, A Systematic Approach for Determining the Thermal Fatigue-Life of Plastic Ball Grid Array (PBGA) Lead-Free Solder Joints, *Proceedings of ASME InterPack '05,* July 17–22, 2005, San Francisco, CA

18. J.P. Clech, Acceleration Factors and Thermal Cycling Test Efficiency for Lead-Free Sn-Ag-Cu Assemblies. To appear in *Proceedings of SMTA International Conference,* Chicago, IL, September 2005

19. A. Schubert, R. Dudek, E. Auerswald, A. Gollhardt, B. Michel, and H. Reichl, Fatigue Life Models for SnAgCu and SnPb Solder Joints Evaluated by Experiments and Simulation, *Proceedings of IEEE 53rd Electronic Components and Technology Conference,* New Orleans, LA, May 27–30, 2003

20. A. Syed, Accumulated Creep Strain Energy and Energy Density Based Thermal Fatigue Life Prediction Models for SnAgCu Solder Joints, *Proceedings of 54th Electronic Components and Technology Conference,* Las Vegas, NV, June 1–4, 2004, pp. 737–746

21. O. Salmela, K. Andersson, J. Sarkka, and

M. Tammenmaa, Reliability Analysis of Some Ceramic Lead-Free Solder Attachments, *Proceedings of SMTA Pan Pacific Microelectronics Symposium,* Kauai, Hawaii, January 25–27, 2005, pp. 161–169
22. D. Xie, D. Geiger, D. Shangguan, D. Rooney, and L. Gullo, Characterization of Fine Pitch CSP Solder Joints Under Board-Level Free Fall Drop. *Proceedings of ASME InterPack 2005,* May 2005, San Francisco, CA
23. M. Ahmad, R. Duggan, T. Hu, B. Ong, C. Ralph, S. Sethuraman, and D. Shangguan, Strain Gage Testing: Standardization. To appear in *SMT,* 2005
24. IPC Association Connecting Electronics Industries, Performance Test Methods and Qualification Requirements for Surface Mount Solder Attachments, January 2002
25. J. Liang, N. Dariavach, P. Callahan, D. Shangguan, and C. Li, Deformation and Fatigue Fracture of Solder Alloys under Complicated Load Conditions. *Proceedings of ASME InterPack'05,* July 17–22, 2005, San Francisco, CA
26. J. Liang, S. Downes, N. Dariavach, D. Shangguan, and S. M. Heinrich, Effects of Load and Thermal Conditions on Pb-Free Solder Joint Reliability, *J. Electron. Mater.,* Special Issue, Vol 33 (No. 12), Dec 2004, p 1507–1515
27. D. Shangguan, Managing the Total Environmental Life Cycle of Electronics Products, *SMT,* March 2005, p 49

For additional references, please refer to Chapter 1 of this book.

APPENDIX 1

Table 1. Selected acronyms and abbreviations related to surface mount and lead-free solder interconnect technology

Acronym	Definition	Acronym	Definition
ACA	anisotropic conductive adhesive	LCA	life cycle assessment
ACF	anisotropic conductive film	LCCC	leadless ceramic chip carriers
AOI	automatic optical inspection	LCD	liquid crystal diode
ASTM	American Society for Testing and Materials	LGA	land grid array
ATC	accelerated thermal cycling	MC	matrix creep
BCC	body-centered cubic	MTF	mean time to failure
BCT	body-centered tetragonal	MTO	metal turnover
BGA	ball-grid array	MTTF	mean time to failure
BLP	bottom leadless package	NCMS	National Center for Manufacturing Sciences
CAF	conductive anodic filament	NEMI	National Electronics Manufacturing Initiative
CBGA	ceramic ball grid array	NEPP	NASA Electronic Parts and Packaging
CCGA	ceramic column grid array	OEM	original equipment manufacturer
C-SAM	C mode scanning acoustic microscopy	OSP	organic solderability preservative
CSP	chip scale package	PBB	polybrominated biphenyl
CTE	coefficient of thermal expansion	PBDE	polybrominated diphenyl ether
DC	direct current	PBGA	plastic ball grid array
DC	dislocation climb	PCB	printed circuit board
DCB	double cantilever beam (test specimen)	PIP	pin-in-paste
DFR	design for reliability	PLCC	plastic leadless chip carrier
DNP	distance to neutral point	PMT	process monitoring test
DRAM	dynamic random access memory	POP	package-on-package
ECTC	Electronic Components and Technology Conference	PQFP	plastic quad flat package
EDFAS	Electronic Device Failure Analysis Society	PSD	power spectral density
EDX	energy dispersive x-ray (spectroscopy)	PTH	plated through hole
EEE	electrical and electronic equipment	PWA	printed wiring assembly
EMS	electronic manufacturing services	PWB	printed wiring board
ENIG	electroless nickel and immersion gold (Ni/Au)	QA	quality assurance
EOL	end of life	QFN	quad flat no-lead
ESCA	electron spectroscopy for chemical analysis	QFP	quad flat package
FCC	face-centered cubic	RA	reduction in area
FCOB	flip-chip on organic board	RF	radio frequency
FDTD	finite difference in time domain (modeling)	RH	relative humidity
FEA	finite element analysis	RoHS	"Directive on the Restriction of the Use of Certain Hazardous Substances in Electrical and Electronic Equipment" (European Union regulation)
FEM	finite element modeling		
FIB	focused ion beam		
FPBGA	fine pitch plastic ball grid array	SAC	Sn-Ag-Cu (alloy)
FR4	designation for a fiberglass and epoxy substrate material	SAD	selected area diffraction
		SEM	scanning electron microscopy
FTIR	Fourier transform infrared (spectroscopy)	SIR	surface insulation resistance
GBC	grain boundary creep	SMD	surface mount device
GBS	grain boundary sliding	SMT	surface mount technology
GPD	generalized plane displacement	SMTA	Surface Mount Technology Association
HASL	hot air solder leveling	S-N	stress versus cycles to failure (plot)
I-Ag	immersion silver	SOT	small-outline transistor
IC	ion chromatography	TAC	Technical Adaptation Committee (European Union)
ICA	isotropic conductive adhesive	TAL	time above liquidus
ICT	in circuit test	TBBPA	tetrabromobisphenol A
IEEE	Institute of Electrical and Electronics Engineers	T_d	decomposition temperature
IMC	intermetallic compound; integrated matrix creep	TEM	transmission electron microscopy
IPC	professional society formerly known as the Institute for Printed Circuits	T_g	glass-transition temperature
		TMF	thermal mechanical fatigue
ISEPT	International Symposium on Electronics Packaging Technology	TSOP	thin small-outline package
		UBM	under bump metallization
I-Sn	immersion tin	VOC	volatile organic compound
IST	interconnect stress test	WEEE	(European Union legislation)
ITO	indium tin oxide (coated glass)	WLCSP	wafer-level chip-scale package
JEDEC	Joint Electronic Device Engineering Council	XRD	x-ray diffraction
JEITA	Japan Electronics and Information Technology Industries Association	XRF	x-ray fluorescence

Table 2. The Chemical Elements by Symbol

Ac	Actinium	H	Hydrogen	Pr	Praseodymium
Ag	Silver	He	Helium	Pt	Platinum
Al	Aluminum	Hf	Hafnium	Pu	Plutonium
Am	Americium	Hg	Mercury		
Ar	Argon	Ho	Holmium	Ra	Radium
As	Arsenic			Rb	Rubidium
At	Astatine	I	Iodine	RE	Rare Earths
Au	Gold	In	Indium	Re	Rhenium
		Ir	Iridium	Rh	Rhodium
B	Boron			Rn	Radon
Ba	Barium	K	Potassium	Ru	Ruthenium
Be	Beryllium	Kr	Krypton		
Bi	Bismuth			S	Sulfur
Bk	Berkelium	La	Lanthanum	Sb	Antimony
Br	Bromine	Li	Lithium	Sc	Scandium
		Lr	Lawrencium	Se	Selenium
C	Carbon	Lu	Lutetium	Si	Silicon
Ca	Calcium			Sm	Samarium
Cd	Cadmium			Sn	Tin
Ce	Cerium	Md	Mendelevium	Sr	Strontium
Cf	Californium	Mg	Magnesium		
Cl	Chlorine	MM	Mischmetal	Ta	Tantalum
Cm	Curium	Mn	Manganese	Tb	Terbium
Co	Cobalt	Mo	Molybdenum	Tc	Technetium
Cr	Chromium			Te	Tellurium
Cs	Cesium	N	Nitrogen	Th	Thorium
Cu	Copper	Na	Sodium	Ti	Titanium
		Nb	Niobium	Tl	Thallium
D	Deuterium	Nd	Neodymium	Tm	Thulium
Dy	Dysprosium	Ne	Neon		
		Ni	Nickel	U	Uranium
Er	Erbium	No	Nobelium		
Es	Einsteinium	Np	Neptunium	V	Vanadium
Eu	Europium				
		O	Oxygen	W	Tungsten
F	Fluorine	Os	Osmium		
Fe	Iron			Xe	Xenon
Fm	Fermium	P	Phosphorus		
Fr	Francium	Pa	Protactinium	Y	Yttrium
		Pb	Lead	Yb	Ytterbium
Ga	Gallium	Pd	Palladium		
Gd	Gadolinium	Pm	Promethium	Zn	Zinc
Ge	Germanium	Po	Polonium	Zr	Zirconium

Index

A

ACA interconnects
  assembly process, effects of, 259–260(F)
  cumulative failures of various ACA joints, 258(F)
  degradation due to moisture absorption, 261–262(F)
  experimental evaluation, 255–259(F,T)
  manufacturing process of an ACA assembly, 250(F)
  relevant technical data of ACAs subjected to reliability tests, 256(T)
  reliability of, 255–263(F,T)
  RF performance of ACA flip-chip joints, 265–266(F,T)
  substrate and component, effects of, 260–261(F)
  Weibull analysis of ACA A joints with various failure criteria, 257(F)
Accelerated life tests (ALT), 135
Accelerated reliability tests, 16–17
Accelerated testing methodology, 165–180(F,T)
  accelerate a test, 169–170
  active temperature cycling, 173–174
  degradation of a material in creep, 167(F)
  designing a test, 176–179(F,T)
  influence of deformation mechanism on degradation, 167
  introduction, 165
  mechanical tests, 174–176(F)
  mechanisms of creep deformation, 165–166(F)
  metallurgical background, 165–168(F)
  modeling creep, 166–167(F)
  modeling degradation, 167–168
  passive temperature cycling tests, 170–173(F,T)
  test set-up for bending tests, 174(F)
  test set-up for mechanical cycling of components, 174(F)
Advanced light source (ALS), 156
American Society for Testing and Materials (ASTM), 69
Anisotropically conductive adhesives (ACA), 250
Au embrittlement
  baking, 60–61(T)
  control of, 61–62
  first reflow, 59–60(T)
  metallurgical mechanism, 59–62(T)
  overview, 56
  previous literature on, 56–57(F)
  recent observation after baking, 57–59(F)
  second reflow, 61
Automatic optical inspection (AOI), 8

B

Backward compatibility, 9, 12, 115
Ball grid array (BGA), 4
Black pads
  black pad failure, 51(F)
  brittle interfacial failure, pad not black, 52(F)
  brittle Ni-Sn intermetallic, 56
  corrosion in immersion Au bath, 55–56(F)
  defect mapping, 54
  defined, 50
  dependence on solder composition and flux, 54–55
  Electroless Ni/immersion Au (ENIG), 50–52
  failure mechanisms, 55(F)
  interface microstructure, 53(F)
  interfacial fracture, 53–54(F)
  Kirkendall voids, 56(F)
  phosphorus segregation, 55
  sensitivity to testing methods, 54
  sporadic failures, 52–53

C

CAF formation
  CAF configurations, 1404
  conductor configuration, 139–140(F)
  PWB storage and use: ambient humidity effects, 141
  solder flux/HASL fluid composition, 140
  substrate material choice, 138–139
  thermal exclusions, 140–141(T)
  voltage gradient effects, 140
CAF testing, 141–142
Ceramic ball grid array (CBGA), 108
Ceramic column grid array (CCGA), 233
Characterization and failure analysis of lead-free solder defects, 227–248(F,T)
Charpy test, 49, 122(F)

Chemical interactions and reliability testing, 129–146(F,T)
  Bi-directional coupler topology, 142(F)
  Bono and Turbini corrosion tests, differences between, 137(F)
  Bono corrosion test coupon, 136(F)
  conductive anodic filament formation, 137–142(F,T)
  corrosion test method, 135–137(F,T)
  electrochemical migration, 131–132(F)
  example of corrosion test plot, 138(F)
  flux residues and RF signal integrity, 142–143(F)
  introduction, 129
  solder flux chemistry, 129–131(F)
  summary, 144
  Surface insulation resistance (SIR), 132–135(F,T)
  T-resonator circuit, 142(F)
  Turbini corrosion test coupon, 136(F)
Chip scale package (CPS), 107, 108, 186(F)
Coble model of grain boundary creep, 152(F)
Coefficient of thermal expansion (CTE), 11, 108
Coffin-Manson
  approach to modeling degradation, 168
  classical approach for metal fatigue, 108
  fatigue life predictions, 205
  low cycle fatigue life (equations), predicting, 191, 207
  predictions/CBGA test data, 216(F)
  predictions/PBGA test data, 217(F)
  solder joint fatigue life, evaluation of, 206
Compatibility, forward/backward, 12–13(F)
Components
  additional component reliability issues, 13
  forward/backward compatibility, 12–13(F)
  internal materials, 6
  IPC/JEDEC J-STD 020, temperature qualifications, 12
  PWB reliability, 13–15
  reliability, 10–13(F)
  solder and component combinations, 13(F)
  temperature requirements, 5–6
  termination metallurgy, 5
  tin pest, 12
  tin whiskers, 10–12(F)
Conductive adhesives
  ACA flow during bonding, 271–272(F)
  ACA interconnects, reliability of, 255–263(F,T)
  adhesive technology, introduction to, 249
  anisotropically conductive adhesives, 250–251(F)
  conclusions, 273
  electrical conduction development and residual stresses, 272–273(F)
  electrical performance simulations, 264–266(F,T)
  electron conduction through nanoparticles in ICA, 264–265(F)
  encapsulated ICA flip-chip interconnection, 268–270(F,T)
  finite element stress analysis of ICA interconnection, 266–270(F,T)
  ICA interconnects, reliability concerns, 251–255(F)
  isotropically conductive adhesives, 249–250(F)
  oxidation and crack growth, theoretical treatment of, 263–264(F,T)
  probabilities of open and bridging, 270–271(F)
  process simulations of ACA interconnection, 270–273(F)
  RF performance of ACA flip-chip joints, 265–266(F,T)
  surface mount leadless ICA joint, 266–268(F,T)
  theoretical studies and numeric simulations, 263–273(F,T)
Conductive anodic filament (CAF), 13
  description, 137–138(F)
  factors that affect formation, 138–141(F,T)
  testing for, 141–142
Constitutive models for lead-free solder alloy
  creep—time-dependent inelastic behavior, 183–184(F,T)
  inelastic deformation behavior, 182(T)
  thermo-elastic behavior, 182(T)
  time-dependent plastic deformation, 182–183(T)
  unified plasticity model, 184–186(T)
Creep, 165(F)
  creep deformation, 87–103(F,T), 165–166(F)
  creep model constants for lead-free solder, 185(T)
  creep of Sn-Pb vs. lead-free solders, 116–123(F)
  creep rupture, 88(F), 90
  deformation mechanisms in creep, 166(F)
  degradation of a material in creep, 167(F)
  experiments, 96–101(F,T)
  influence of deformation mechanism on degradation, 167
  modeling creep, 166–167(F)
  properties, 202–205(F,T)
  time-dependent inelastic behavior, 183–184(F,T)
Creep deformation, 87–103(F,T)
  creep testing, ASTM method, 88
  defined, 87
  diffusion creep, 166
  dislocation climb/dislocation glide, 166(F)
  grain boundary sliding, 166
  influence of deformation mechanism on degradation, 167
  introduction, 87–91(F)
  microstructure, 91–92
  Pb-free solders, 92–101(F,T)
  primary or transient creep, 88–89(F)
  steady-state or secondary creep, 88(F), 89–90
  strain-time creep response, 87, 88(F)
  tertiary stage, 90(F)
  typical strain-time curve of creep deformation in metal alloy, 88(F)
Critical relative humidity, 134
Cu-Sn interfacial intermetallic, formation of
  interfacial intermetallic phases, identification of, 37–38
  nucleation, 37
  phase diagram, 37
Cu-Sn interfacial intermetallic, growth of, 42
Cyclic shear strain, defined, 108

D

Darveaux criterion, 207
Dendrites
  Ag migration, 255

copper dendrite grows from the cathode, 132(F)
description of, 8
examples of, 132(F)
forming metal dendrites, 131(F)
large monocrystalline dendrites, 169(F)
needle-like lead (Pb) dendrites grow from a cathode, 133(F)
Sn dendrites, 31
Design for reliability (DFR), 17
Design for reliability, finite element modeling, 220–222(F,T)
Diffusion-controlled process, 89
Double cantilever beam (DCB), 253
Drop test, 241, 252–253
Dwell time, 6, 279
Dynamic random access memory (DRAM), 231
Dynamic recrystallization, 72

E

Elastic moduli
    dynamic elastic modulus, 75–76(F,T)
    static elastic modulus, 74–75(F)
Electrical and electronic equipment (EEE), 1
Electrochemical reliability, 15
Electroless nickel and immersion gold (ENIG), 4
Electron probe microanalysis (EPMA), 47
Electron spectroscopy for chemical analysis (ESCA), 251
Electronic waste (e-waste), 1
Electronics recycling, 18
End-of-life (EOL), end-of-life management, 18
Energy dispersive x-ray (EDX), 38
Environmental compliance, 17–21(F)
    challenges, 19–21
    design for lead-free soldering, overview, 17
    electronics recycling and end-of-life management, 18
    environmental compliance requirements, 18(F)
    holistic approach, 2(F), 21
    requirements, 17–18(F)
European Union (EU), 1
Experimental data of the kinetics of IMC growth in liquid solders
    IMC growth in eutectic Sn-Pb, 42–43(F)
    IMC growth in liquid Sn, 43
    IMC growth in liquid Sn_Ag, 43–44
    theories for interfacial IMC growth, 44–45

F

Fast diffusion pathways, 71
Fatigue
    categorized as low-cycle or high-cycle fatigue, 77
    strain controlled, 76–77
    stress controlled, 76
Fatigue and creep of lead-free solder alloys
    creep deformation, 87–103(F,T)
    introduction, 67–69(F)

material deformation, 69–76(F,T)
microstructure, 70–72(F)
microstructure-properties relationship, 70
nonmaterial factors affecting solder joint performance, 67
solid-solution strengthening, 72
summary, 103
time-independent deformation, synopsis, 69–70
Fatigue deformation
    causes, 76
    grain-boundary/phase boundary sliding, 86–87(F)
    microstructure, 80
    parameters describing the fatigue cycle, 76–80(F)
    Pb-free solder microstructures, 85–87(F)
    Pb-free solders, 80–85(F)
    synopsis, 76–80(F)
Finite difference in time domain (FDTD), 265
Finite element analysis (FEA), 17
Finite-element based life prediction models, 118–119(F)
Finite element modeling, 199–226(F,T)
    3D slice model, 201(F)
    concluding remarks, 222
    creep parameters, 204(T)
    creep properties, 202–205(F,T)
    creep response of a material under uniaxial load, 203(F)
    design for reliability, 220–222(F,T)
    elastic properties, 202(F,T)
    elastic properties of select packaging materials, 203(F)
    examination of failure theories with literature data, 214–220(F,T)
    example using the 256 PBGA, 209–214(F,T)
    failure criteria discussion, 206–208
    future research needs, 222
    geometric finite element model, suggested, 209
    introduction, 199–200
    materials, 202–206(F,T)
    modeled geometry, 201–202(F)
    modeling difficulties, 208–209
    models code, ANSYS, 200
    overview, 200
    plasticity, 205–206(F)
Flip-chip on board (FCOB), 186
Flip chips
    2D model of a flip chip on organic board, 186(F)
    Au concentration, 56
    creep model constants for lead-free solder, 185(T)
    encapsulated ICA flip-chip interconnection, 268–270(F,T)
    equivalent circuit model of the ACA flip chip joint, 266(F)
    generalized plane deformation model with constraints for a flip chip on organic board, 187(F)
    reliability issues, 13
    RF performance of ACA flip-chip joints, 265–266(F,T)
    solder joining, 260–261(F)
    test results for ACA flip chip joints, 256(F)
    three dimensional model, 187–188(F)
    with underfill, 112–113(F)

Flux, 3–4
  background on solder flux chemistry, 129–131(F)
  components, 130
  corrosion test method, 135–137(F,T)
  flux residues and RF signal integrity, 142–143(F)
  low solids flux, 130–131
  rosin/resin flux, 130(F)
  water soluble flux, 130
Focused ion beam (FIB), 154
Fourier transform infared (FTIR), 262(F)
Frequency effect, defined, 78

G

Geometric modeling
  3-D model, 187–188(F)
  generalized plane deformation (GPD) or 2.5 model, 186–187(F)
  other considerations, 188
  two-dimensional model, 186(F)
Grain boundary creep (GBC), 121
Grain-boundary migration, 91
Grain boundary sliding (GBS), 121
Grain (phase) boundary sliding, 80, 91–92

H

Handbook of Chemistry and Physics, 132
Hot air solder leveling (HASL), 4
Hysteresis loops, 81, 82, 206, 212(F), 217(F)

I

ICA interconnects
  Ag migration, 255(F)
  cracking, 253–254(F)
  effect of curing degree, 251–252(F)
  effect of metallization, 251
  electron conduction through nanoparticles in ICA, 264–265(F)
  failure mechanisms, 253–255(F)
  filler motion, 254–255
  finite element stress analysis of ICA interconnection, 266–270(F,T)
  humidity aging tests, 253–254
  impact strength, 252–253
  intermetallic compounds, formation of, 254
  oxides, formation of, 254
IMC growth in liquid solders, 42–45
Immersion silver (I-AG), 5
In-circuit test (ICT), 3–4
Input/output (I/O), 108
Interconnect stress test (IST), 14
Interface reliability
  blocky Ag_3Sn, 49
  critical IMC thickness, 49
  interface reliability, 48–49
  Kirkendall voids in IMC, 49–50(F)
  Pb-rich phase band, 48–49

Interfacial IMC microstructures, 46–48
  effects of microelements on IMC between Sn-Ag-Cu solder and Cu substrate, 47–48
  intermetallic joints, 47
  morphology, 46
  spalling, 46–47
  texture, 47
Interfacial interactions, 33–62(F,T)
  Au embrittlement, 56–62(F,T)
  Black pads, 50–56(F)
  interface reliability, 48–49
  interfacial IMC microstructures, 46–48
  kinetics of Ni_3Sn_4 in liquid solder, 46
  molten solder-substrate interactions, 33–48(F,T)
  numerical analysis of interfacial IMC growth, 45–46
  solid solder-substrate reactions, 48
Intermetallic compound (IMC), 8–9
Internal state variable (ISV), 184
IPC/JEDEC Standard 020C, 6

J

JEITA (Japan Electronics and Information Technology Industries), 181
Joint reliability trends, 107–128(F,T)
  accelerated thermal cycling conditions, 120–122(F)
  alloy composition and board finish effects, 109
  area array assemblies with SAC balls and SAC or Sn-Pb paste, 114–115(F),
  area array assemblies with Sn-Pb balls and SAC or Sn-Pb paste, 116(F)
  brittle fracture under high strain rate conditions, 122–123(F)
  conclusions, 123–124(T)
  creep of Sn-Pb vs. lead-free solders, 116–117(F)
  critical component data, 111–113(F)
  cycles-to-failure, 114–115(F)
  flip-chip with underfill, 112–113(F)
  impact of Pb contaminant or Sn-Pb alloy on lead-free reliability, 113–116(F)
  introduction, 107–108
  lead-free to Sn-Pb comparison, 109–111(F)
  life prediction models, 117–120(F)
  life test data for SAC and Sn-Pb TSOP assemblies, 111–113(F)
  metrics of solder reliability studies, 124(T)
  reflowed conventional leadless SMT assemblies with Pb contaminant, 113–114(F)
  reflowed SAC vs. near-eutectic Sn-Pb, 109–110(F)
  reliability, application specific, 107
  SAC paste, 115
  SAC thermal cycling data, 108–109(F)
  Sn-Pb paste, 115
  Sn0.7Cu vs. Sn-Pb for bare chip assemblies, 110–111(F)

K

Kirkendall voids,
  black pads, 56

in IMC, 49–50(F)
in Sn-Pb plating layer, 254
propagate into cracks, 15

L

Lead-free solder
    interconnect reliability, 10–17(F)
    joint reliability trends, 107–127(F)
    microstructure, 76, 101–103(F)
    solder interconnect reliability, 10(F)
Lead-free solder alloys
    constitutive models for, 181–186(F,T)
    fatigue and creep of, 67–106(F,T)
    metallographic preparation for lead-free solder alloy samples, 229(T)
Lead-free solder defects, characterization and failure analysis of, 227–248(F,T),
    characterization and analytical techniques, 228–229(T)
    conclusions, 245–247
    eutectic tin-lead solder alloy, 227
    introduction, 227
    key solder defect types and ability to detect visible joints, 228(T)
    lead-free solder, 233–236(F)
    lead-free solder alloys, 227–228
    macro voids in a lead-free solder joint, 238(F)
    metallographic preparation for lead-free solder alloy samples, 229(T)
    metallography and etching, 228–229(T)
    microstructure before and after drop test, 241
    microstructure features after deep etching, 237–239(F)
    optical dull appearance, reasons for, 236–237
    thermal-cycled samples after vibration, 233(F)
    tin-lead and lead-free microstructural features, 229–236(F)
    tin-lead and lead-free solder microstructure after reflow cycles, 237(F)
    tin-lead and lead-free solders, ceramic column grid array, 233(F)
    tin-lead solder, ceramic ball grid array, 231–233(F)
    tin-lead solder, conventional leadless package, 230–231(F)
    tin-lead solder, CSP leadless package, 231
    tin-lead solder, fine pitch BGA, 233(F)
    tin-lead solder, plastic ball grid array (PBGA), 231
    void levels for lead-free and mixed assembly with various surface finishes, 239–240(F)
    x-ray inspection systems, 241–245(F)
    x-rays of BGA assemblies with lead-free solder, 247(F)
Lead-free solder finishes, tin whisker growth on, 147–164(F)
Lead-free solder interconnects
    chemical interactions and reliability testing for, 129–146(F,T)
    interfacial interactions, 33–62(F,T)
    thermomechanical reliability prediction, 181–198(F,T)
Lead-free solder pastes, 7
Lead-free soldering, 1–17(F)
    Components, 5–6
    equipment, 9–10
    lead-free solder alloys, 2–3
    lead-free solder pastes, 7
    other lead-free alternatives, 3
    processes, equipment, and quality, 6–10
    rework and repair, 9
    tin-copper alloy, compared to SAC, 2–3
Lead-free soldering and environmental compliance: an overview, 1–28(F)
    introduction, 1–2(F)
    summary, 21
Leadframes, finish, 147
Leadless ceramic chip carriers (LCCCs), 107
Least-squares fitting, 216(F)
Lifecycle assessment (LCA), 1
Loading conditions and thermomechanical stresses
    accelerated thermal cycling, 189(F)
    deformation mechanisms, 190–191
    field use simulation, 189–190(T)
    process profile, 188–189

M

Macro-voids, effects of on solder joints, 15
Material deformation
    elastic moduli, 74–76(F,T)
    fatigue deformation, 76–87(F,T)
    microstructure-properties relationship, 70–72(F)
    Pb-free solders, 72–76(F,T)
Matrix creep (MC), 121
Mean time to failure (MTF), 255
Mechanical tests
    mechanical cycling, 174(F)
    vibration, 174–176(F)
Metal turn over (MTO), 4
Microstructural evolution in lead-free solder interconnects, 29–33(F)
    growth, 30–31
    introduction, 29
    nucleation, 30
    phase diagram and equilibrium solidification, 29–30
    solid state aging, 32–33
    solidification microstructure, 31–32(F)
Microstructure-properties relationship, 70
Military devices, use of Sn-based banned, 148
Modeled geometry
    2D slice, 201
    3D octant, 201–202(F)
    3D octant with submodel, 201–202(F)
    3D slice, 201
    3D slice model, 201(F)
    3D octant model, 202(F)
Modeling difficulties, finite element modeling
    accuracy of the material properties, 208
    progressive damage, 208
    singularities, 208
    thermal history, 209
Moiré interferometry, description, 194(F)
Moisture sensitivity level (MSL), 12

Molten solder-substrate interactions,
  the characteristic time, 36–37
  Cu dissolution in solder joints, 35–36
  dissolution in solder bath, 34–35(F,T)
  experimental data of the kinetics of IMC growth in liquid solders, 42–44(F)
  formation of Cu-Sn interfacial intermetallic, 37–38
  formation of Ni-Sn interfacial intermetallic, 38–42(F)
  Ni dissolution in solder joints, 35–36
  Reaction between Ni and Sn when Cu present, 39–42(F)
  solubility of Cu in molten solders, 33–34(T)
  solubility of Ni in molten solders, 34

N

Nabarro-Herring model of lattice creep, 152(F)
National Center for Manufacturing Science (NCMS), 252
National Electronics Manufacturing Initiative (NEMI), 181
National Electronics Manufacturing Initiative website, 148
Ni-Sn interfacial intermetallics, formation of
  metastable $NiSn_3$ phase, 39
  reaction between Ni and Sn, eutectic Sn-Pb, and Sn-Ag, 38–39
  reaction between Ni and Sn when Cu is present, 39–42(F)

O

Octant symmetry, 201
Ohm's law, 131
Organic solderability preservative (OSP), 4
Original equipment manufacturers (OEM), 20

P

Package-on-package (POP), 7
Passive temperature cycling tests
  dwell temperatures, 172–173(F)
  dwell time, 170–171(F,T)
  temperature gradient, 171–173(F)
Pb-free solder, microstructure, 101–103(F)
Pb-free solder microstructures, 76
Pb-free solders, 72–76(F)
  creep deformation, 92–101(F,T)
  creep experiments, 96–101(F,T)
  fatigue deformation, 80–85(F)
  lead-free to Sn-Pb comparison, 109–111(F)
  microstructures, 85–87(F)
  negative creep, 93–94(F)
  shear creep performance of 96.5Sn-3.5Ag solder, 95–96
  steady-state or secondary creep, 94–95
  test methodologies, 96
Peak temperature, proposed, 6
Pin-in-paste (PIP), 7–8
Plastic ball grid array (PBGA), 231
Plated through hole (PTH), 14
Poisson's distribution, 270
Polybrominated biphenyls (PBB), 1
Polybrominated diphenyl ethers (PBDE), 1, 2
Power spectral density (PSD), 175
Precipitation strengthening or hardening, 71
Printed circuit board, 2
Printed wiring assemblies (PWAs), 67
Printed wiring board (PWB). see PWB
Process monitoring test (PMT), 54
Pull test, 54
PWB
  reliability, 13–15, 278
  storage and use: ambient humidity effects, 140–141(T)
  surface finishes, 4–5

R

Radio frequency (RF), 4
Random vibration, 175–176(F), 179(F)
Reliability of interconnects with conductive adhesives, 249–276(F,T)
  relevant technical data of ACAs subjected to reliability tests, 256(T)
  Weibull analysis of ACA A joints with various failure criteria, 257(F)
Reliability outlook
  accelerated testing profile and acceleration factor, 279
  complex loading conditions and total reliability optimization, 279
  dynamic mechanical loading conditions, 278
  PWB reliability, 13–15, 278
  reliability degradation assessment for component reutilization and repurposing, 279
  solder alloy characteristics and interfacial interactions, 277
  solder constitutive equation and thermal fatigue reliability prediction, 278
  tin whisker growth, 277–278
Restriction of the use of Certain Hazardous Substances (RoHS), 1

S

SAC paste, 115
Scanning electron microscope (SEM), 38
Selected area diffraction (SAD), 38
Short-circuit diffusion pathways, 71
Sinh expression, 96
Sinusoidal vibration, 174–175, 178–179
SIR (surface insulation resistance), defined, 133
SIR (surface insulation resistance) test procedures, 134–135(T)
  contamination, 134
  geometry, 134
  relative humidity, 134
  voltage, 134–135
SIR test parameters, 134(T)
SMT reflow, 6–8

case studies, 7–8
dwell time or time above liquidus, (TAL), 7
IPC 610 standards, 8
quality, 8
reflow process, 6–7
second reflow, 8
Sn-Ag-Cu (SAC), 2
Sn-Pb, creep behavior, 168(T)
Sn-Pb paste, 115
Solder interconnects, function of, 1
Soldering, defined, 129
Solid solder-substrate reactions
experimental data, 48
solders/Cu interface, 48
solders/Ni interface, 48–49
Solid state aging
grain growth rate, 32
microstructural evolution in, 32–33
Solidification microstructure
Cu_6Sn_5, 32
large Ag_3Sn plates, 31–32
Sn dendrites, 31
Spalling, 46–47
Substrates and solders, interactions between
introduction, 33
molten solder-substrate interactions, 33–37(F,T)
Surface hillocks, 149 see also Tin whisker
defined, 152
Surface insulation resistance (SIR), 9
Surface mount devices (SMD), 7
Surface mount technology (SMT), 107

T

Test Methods Manual (IPC_TM_650), 134
Tests, designing
active thermal cycling, 177(F)
dwell time, determined by, 177(F,T)
mechanical cycling, 177–178(F)
passive thermal cycling, 177(T)
preparation, 176–177
random vibration, 179(F)
sinusoidal vibration, 178–179
Tetrabromobisphenol A (TBBPA), 4
Theories for interfacial IMC growth
interdiffusion between Sn and Cu in interfacial IMC growth, 45
overview, 44
Schafer's model, 44–45
Thermal cycling test, 170–173(F,T)
Thermal mechanical fatigue (TMF), 69, 78–79
Thermal stability, 12
Thermomechanical reliability in packages
damage mechanics-based approach, 194
fatigue behavior of solder, 191(F)
life-prediction models for lead-free solder, 191–194(T)
Thermomechanical and dynamic mechanical reliability, 15–16
Thermomechanical models, development of, 182
Thermomechanical reliability prediction, 181–198(F,T)

conclusions, 194–196
constitutive models for lead-free alloy, 181–186(F,T)
creep model constants for lead-free solder, 185(T)
elastic constants for lead-free solder, 182(T)
geometric modeling, 186–188(F)
introduction, 181
loading conditions and thermomechanical stresses, 188–191(F,T)
model validation, 194(F)
thermomechanical reliability in packages, 191–194(T)
typical accelerated thermal cycle profile, 189(F)
typical creep strain rate curve, 183(F)
Thieving pads, 8
Thin small outline package (TSOP), 206
Time above liquidus (TAL), 7
Tin-based solder alloys
creep of lead-free alloys, 168–169(F,T)
creep of Sn-Pb, 168(T)
deformation of, 168–169(F,T)
Tin pest, 12, 277
Tin whisker growth on lead-free solder finishes, 147–164(F)
analysis of stress around whisker growth, 159(F)
conclusions, 162
cross-sectional scanning and transmission electron microscopy study of Sn whiskers, 153–155(F)
deviatoric stress around whisker, 160(F)
effect of surface oxide on whisker growth, 155(F)
introduction, 147–148(F)
measurement of parameters affecting growth, 153–161(F)
micro-beam diffraction scan of a whisker and its matrix, 157(F)
morphology, 148–149(F)
Nabarro-Herring model of stress relaxation, 152(F)
orientation distribution on solder finish around root of whisker, 158(F)
room temperature reaction between Sn and Cu to form Cu_6Sn_5, 151–152(F)
SEM image of short whisker on pure Sn finish, 148(F)
spontaneous whisker growth, 148
stress generation (driving force), 149–151(F)
stress relaxation (kinetic process), 151–152(F)
volume fraction of Cu_6Sn_5 in the Pb-free finish layer, 160–161(F)
whiskers formed on SnCu and Cu leadframes, 149(F)
Tin whiskers
cross section of a tin whisker, 153(F)
cross-sectional scanning and transmission electron microscopy study of Sn whiskers, 153–155(F)
defined, 152
effect of surface oxide on growth, 155–156(F)
growth, 10–12
growth, accelerated testing of, 161–162(F)
kinetic model of whisker growth, 159–160(F)
mitigation techniques, 11–12
suppression of Sn whisker growth, 161
synchrotron radiation micro-diffraction study, 156–159(F)
tin whisker morphologies, 148(F)
volume fraction of Cu_6Sn_5 in the Pb-free finish layer, 160–161(F)

Toothpick test, 50
Transmission electron microscope (TEM), 38

U

Under bump metallization (UBM), 34

V

Vibration
  random, 174, 175–176(F)
  sinusoidal, 174–175(F)
Volatile organic compound (VOC), 4

Von Mises (stresses and creep strains), 213(F), 217(F), 267(F), 269(F)

W

Waste electrical and electronic equipment (WEEE), 18
Wave soldering, 3, 8–9
  equipment pre-heating capacity, 9–10
  pot temperature, 6, 8
Wavelength dispersive spectrometry (WDS), 38

X

X-ray diffraction (XRD), 39
X-ray fluorescence (XRF), 5
X-ray inspection systems, 241–245(F)